With best wishes,

2 vols

CENSUS OF THE EXACT SCIENCES IN SANSKRIT

MEMOIRS OF THE
AMERICAN PHILOSOPHICAL SOCIETY
Held at Philadelphia
For Promoting Useful Knowledge
Volume 111

CENSUS OF THE EXACT SCIENCES IN SANSKRIT

SERIES A, VOLUME 3

DAVID PINGREE

Professor of the History of Mathematics
Brown University

AMERICAN PHILOSOPHICAL SOCIETY
INDEPENDENCE SQUARE · PHILADELPHIA
1976

Library of Congress Catalog Card Number 75–46233
International Standard Book Number 0–87169–111–6
US ISSN 0065–9738

CONTENTS

INTRODUCTION

This, the third volume of Series A of the *Census of the Exact Sciences in Sanskrit*, is devoted to those authors whose names begin with a cerebral (*c*, *ch*, *j*, and *jh*), a reflexive (*ṭ*, *ṭh*, *ḍ*, and *ḍh*), or a dental (*t*, *th*, *d*, *dh*, and *n*). Preceding the material relating to these authors is a section supplemental to volume one (Memoirs of the American Philosophical Society, vol. 81, Philadelphia, 1970) and to volume two (Memoirs of the American Philosophical Society, vol. 86, Philadelphia, 1971). This section contains abbreviations of new periodicals and series that have been consulted (p. 2), a bibliography of books and articles that have appeared or have been belatedly noticed since volume two went to press (pp. 3–6), and a list of additional catalogs that it has been possible to utilize (p. 7). In the rest of the volume will be found supplementary information concerning about 100 authors already noted in the two previous volumes (marked by asterisks) and all the data currently available concerning almost 800 new authors. The total number of authors discussed in *CESS* as the first half of Series A is concluded, therefore, is about 1450—a number which fully justifies the traditional Indian concept of an ocean of knowledge. This particular raft to rescue those in danger of drowning in it will continue in volume four with authors whose names begin with labials (*p*, *ph*, *b*, *bh*, and *m*).

Providence, R. I., Jan. 1974

ABBREVIATIONS OF JOURNALS AND SERIALS

AG—Acyutagranthamālā
AN—Ancient Nepal
BMI—Bulletin of the Mithila Institute of Post-Graduate Studies and Research in Sanskrit Learning, Darbhanga
Bombay SS—Bombay Sanskrit Series
JAIH—Journal of Ancient Indian History
JBIT—Journal of the Birla Institute of Technology
JCOI—Journal of the K. R. Cama Oriental Institute
JKUORIML—Journal of the Kerala University Oriental Research Institute and Manuscripts Library
JMJSG—Jñānapīṭha Mūrtidevī Jaina Saṃskṛta Granthamālā
JNSI—Journal of the Numismatic Society of India
JRU—Journal of Ranchi University
LDS—Lalbhai Dalpatbhai Series
MSVG—Mithilā Saṃskṛta Vidyāpīṭha Granthamālā
PEFEO—Publications de l'École Française d'Extrême-orient
RSO—Rivista degli Studi Orientali
SBJ—Sacred Books of the Jainas

BIBLIOGRAPHY

Abhedananda, Swami. [A3. 1968]. *A Study of Heliocentric Science*, Calcutta 1968.

Adhikari, Rabindra. [A3. 1967]. *Kī bhābe koṣṭhī dekhabo*, Kalikātā 1967.

Agarwal, R. C. [A3. 1969]. "Sūrya with Serpent Hood Canopy: A Rare Device," *Bhāratīya Vidyā* 29, 1969, 79–81.

Anjaneyulu, M. S. R. [A3. 1968]. "Hemu—The Indian Meteor," *VIJ* 6, 1968, 112–116.

Apte, B. D. [A3. 1943]. "Śrīpatikṛta Dhīkoṭidakaraṇa," *Jyotiṣatattvadarśana*, Poona 1943, pp. 224–228.

Apte, D. V. [A3. 1943]. "Bhāratīya jyotirgaṇitācā abhyāsa," *Jyotiṣatattvadarśana*, Poona 1943, pp. 94–97.

Apte, Govind Sadashiv. [A3. 1941]. "Āpaleṃ jyotiṣa," *Vedaśāstradīpikā*, Poona 1941, pp. 194–210.

Arsha, P. [A3. 1946]. *Vedic Jyautisa Sastra*, Jwalapur 1946.

Awasthi, A. B. L. [A3. 1969]. "Ancient Indian Cartography," *Dr. Satkari Mookerji Felicitation Volume*, Varanasi 1969, pp. 275–278.

Ayer, V. A. K. [1946]. Eighteenth edition, Bombay 1958.

Bag, A. K. [A3. 1969a]. "Source Materials concerning Astronomy and Mathematics," *IJHS* 4, 1969, 1–4.

——. [A3. 1969b]. "Sine Table in Ancient India," *IJHS* 4, 1969, 79–85.

Bambawale, T. A. [A3. 1962]. *Veda Rahasya or The Secret of the Vedas*, Poona [1962].

Bapat, Dhundhiraj. [A3. 1943]. "Yajñapaddhatīṃta nakṣatrāṃceṃ prādhānya āṇi yajñadevatā va nakṣatradevatā yāṃcā saṃbaṃdha," *Jyotiṣatattvadarśana*, Poona 1943, pp. 89–92.

Barker, Robert. [1777]. Reprinted in Dharampal [A3. 1971] 1–8.

Behari, R. [A3. 1955]. *Ancient India's Contribution to Mathematics*, Delhi 1955.

Belvalkar, S. K. [A3. 1939]. "The Cosmographical Episode in Mahābhārata and Padmapurāṇa, "*A Volume of Eastern and Indian Studies Presented to Professor F. W. Thomas*, Bombay 1939, pp. 19–28.

Bender, Ernest. [A3. 1959]. "An Eighteenth-Century Indian Painting," *JAOS* 79, 1959, 26–29.

——. [A3. 1968]. "A Lunar Illustration Occurring in Several Manuscripts of the Dhanya-Śālibhadracarita, an Old Gujarātī Work of the XVIIth–XVIIIth Cent., A.D.," *JAOS* 88, 1968, 709–711.

Bhasin, J. N. [A3. 1970]. *Medical Astrology: A Rational Approach*, New Delhi 1970.

Bhat, Mariappa Manappa. [A3. 1942]. "A Mathematical Work in Kannada," *Gopalakrishnamacharya Book of Commemoration*, Madras 1942, pt. 4, pp. 75–77.

Bhat, M. Ramakrishna. [A3. 1967]. *Fundamentals of Astrology*, Delhi-Varanasi-Patna 1967.

Bhatnagar, Virendra Swaroop. [A3. 1960]. "The Date of Aśvamedha Performed by Sawāī Jai Singh of Jaipur," *JBRS* 46, 1960, 151–154.

Bhatt, Harihar, and Chhotubhai Suthar. [A3. 1969/70]. "Triśaṅku (A Surmise)," *JOI Baroda* 19, 1969–70, 357–360.

Bhattacharjee, U. C. [A3. 1937]. "Space, Time and Brahma," *Jha Commemoration Volume*, Poona 1937, pt. 2, pp. 69–83.

Bhattacharya, Bhabatosh. [A3. 1939]. "The Place of the Kṛtyakalpataru in Dharmaśāstra Literature," *A Volume of Indian and Iranian Studies Presented to Sir E. Denisson Ross*, Bombay 1939, pp. 59–61.

——. [A3. 1950]. "The Devotional Element in Raghunandana's Works," *Siddha-Bhāratī*, Hoshiarpur 1950, vol. 1, pp. 225–229.

——. [A2. 1967a]. See *PAIOC* 23, 1966, 287–289.

Bhattacharyya, Jagadbandhu. [A3. 1965]. "The Basic Concept of Nakṣatra in Ancient and Medieval India," *PAIOC* 22, 2, 1965, 253–256.

Bhattacharyya, Narendra Nath. [A3. 1971]. *History of Indian Cosmogonical Ideas*, New Delhi 1971.

Bhattacharyya, S. P., and S. N. Sen. [A3. 1969]. "*Ahargaṇa* in Hindu Astronomy," *IJHS* 4, 1969, 144–155.

Billard, Roger, [A3. 1971]. *L'astronomie indienne*, *PEFEO* 83, Paris 1971.

Burrow, Reuben. [A3. 1783?]. "Hints Concerning the Observatory at Benares," in Dharampal [A3. 1971] 70–86.

——. [1790]. Reprinted in Dharampal [A3. 1971] 94–103.

Canteenwala, Phyruz. [A3. 1970]. *The Basic Truths of Astrology*, Bombay 1970.

Chakravarty, Chunilal. [A3. 1969/70]. "The Meru," *JAIH* 3, 1969–70, 123–129.

Chakravarthy, G. N. [A3. 1966]. *The Concept of Cosmic Harmony in the Ṛg Veda*, Mysore 1966.

Chanana, Dev Raj. [A3. 1969]. "Kavīndrācārya Sarasvatī. A Problem of Scholarship and Personal Integrity," *Dr. Satkari Mookerji Felicitation Volume*, Varanasi 1969, pp. 242–254.

Chatterji, S. C. [A3. 1971]. "Evolution of the Science of Motion in India. Historical Retrospect," *XII Congrès International d'Histoire des Sciences. Actes*, vol. 4, Paris 1971, pp. 39–43.

Chaudhuri, Vidya Bhushan. See N. N. K. Rau and V. B. Chaudhuri [A3. 1962].

Chauhan, D. V. [A3. 1971]. "Al-Djummal and Decimal Notation in Indo-Muslim Epigraphy," *ABORI* 52, 1971, 87–96.

Colebrooke, Henry Thomas. [1817]. Preface partially reprinted as "Hindu Algebra" in Dharampal [A3. 1971] 104–137.

Crookall, Robert. [A3. 1968]. *The Mechanisms of Astral Projection*, Moradabad 1968.

Ḍabarāla, Mohana. [A3. 1969/70]. "Aṃgrejī ke bhaugolika śabdakośa—eka sarvekṣaṇa," *Vidyā* 1, 1969–70, 56–60.

Dandekar, R. N. [1940]. See *ABORI* 20, 1938–39, 293–316.

Dash, M. P. [A3. 1967]. "Notices of Palm-leaf Manuscripts Found at Bhingarpur and a Note on Some More Works of Raghunatha Dasa," *OHRJ* 15, 1967, 45–52.

Datta, B. B. [A3. 1935]. "Mathematics of Nemicandra," *Jaina Ant* 1, 2, 1935, 25–44.

——. [A3. 1936]. "A Lost Jaina Treatise on Arithmetic," *Jaina Ant* 2, 2, 1936, 38–41.

Datta, Kalidas. [A3. 1933]. "Two Saura Images from the District of 24 Parganas," *IHQ* 9, 1933, 202–207.

de Luce, R. [A3. 1963]. *Constellational Astrology According to the Hindu System*, Los Angeles 1963.

Devasthali, G. V. [A3. 1943]. "Rāghava-bhaṭṭa and his Tithinirṇayasāroddhāra," *ABORI* 24, 1943, 233–236.

Dhaky, M. A. [A3. 1971]. "Prāsāda as Cosmos," *Brahmavidyā* 35, 1971, 211–226.

Dharampal. [A3. 1971]. *Indian Science and Technology in the Eighteenth Century: Some Contemporary European Accounts*, Delhi 1971.

Dikshit, G. S. [A3. 1969]. "The *Śivatattvaratnākara* as a Source for Sciences in Ancient and Medieval India," *IJHS* 4, 1969, 11–14.

Diskalkar, D. B. [A3. 1937]. "Foundation of an Observatory at Lucknow," *JUPHS* 10, 1937, 7–32.

Dube, P. [A3. 1928]. "Astrolabes in the State Library, Rampur," *JUPHS* 4, 1928, 1–11.

Dutt, Fakir Chandra. [A3. 1935]. *Prenatal Astrology*, Calcutta 1935.

Dwiwedi, G. [A3. 1969]. "Geographical Data in the *Kauṭilīya Arthaśāstra*," *Proceedings of the Twenty-sixth International Congress of Orientalists*, vol. 3, Poona 1969, pp. 222–226.

Esnoul, Anne-Marie. [A3. 1968]. "La divination dans l'Inde," *La Divination*, vol. 1, Paris 1968, pp. 115–139.

Filliozat, Jean. [1962]. English translation in *JCOI* 42, 1969, 100–132.

——. [A3. 1969]. "Le temps et l'espace dans les conceptions du monde indien," *Revue de Synthèse* 90, 1969, 281–295.
——. [A3. 1970]. "Influence of Mediterranean Culture Areas on Indian Science," *IJHS* 5, 1970, 326–331.
Fischer, Klaus. [A3. 1969/70]. "Celestial Symbolism in a Late Mediaeval Jaina Temple," *JJ* 4, 1969–70, 169–173.
Ganguly, K. K. [A3. 1965]. *Some Aspects of Sun Worship in Ancient India*, Calcutta 1965.
Ghosh, Batakrishna. [A3. 1945]. "Endingless Numerals in Ṛgveda," *Bhārata Kaumudī*, 2 vols., Allahabad 1945–47, vol. 1, pp. 253–258.
Gode, P. K. [A3. 1937]. "Some Contemporary Evidence Regarding the Aśvamedha Sacrifice Performed by Sawai Jaya Singh of Amber (A.D. 1699–1744)," *JIH* 15, 1937, 364–367. Reprinted in P. K. Gode [1953/56], vol. 2, pp. 288–291.
——. [A3. 1943]. "Rāghava Āpā Khāṇḍekar of Puṇyastambha—his Works and Descendents (From A.D. 1750 to 1942)," *ABORI* 24, 1943, 27–44.
——. [A2. 1945]. Reprinted in P. K. Gode [A3. 1960/69], vol. 3, pp. 71–76.
——. [A3. 1960/69]. *Studies in Indian Cultural History*, vol. 1, *VIS* 9, Hoshiarpur 1961; vol. 2, Poona 1960; and vol. 3, Poona 1969.
Gonda, J. [A3. 1951]. "Remarks on al-Biruni's Quotations from Sanskrit Texts," *Al-Bīrūnī Commemoration Volume*, Calcutta 1951, pp. 111–118.
Gupta, R. C. [A3. 1966/67]. "The Hindu Method of Solving Quadratic Equations," *JBIT*, 1966–67, 26–28.
——. [A3. 1971]. "Fractional Parts of Āryabhaṭa's Sines and Certain Rules Found in Govindasvāmi's Bhāṣya on the Mahābhāskarīya," *IJHS* 6, 1971, 51–59.
Gupta, T. R. [A3. 1927/28]. "Life and Work of Bhaskaracharya," *BMAUA* 1, 1927–28, 25–46.
Hillebrandt, Alfred. [A3. 1880]. *Das altindische Neu-und Vollmondsopfer in seiner einfachsten Form*, Jena 1880.
Irāmacāmippiḷḷai, K. [A3. 1967]. *Cukarnāṭi*, Cennai 1967.
Iyer, P. R. Chidambara. [A3. 1969]. "The Navagraha in Thailand," *BITCM*, 1969, 186–188.
Iyer, S. Venkitasubramonia. [A3. 1971]. "The Sāstrakāvyas of Kerala," *IA*, 3rd ser., 5, 1971, 23–34.
Jacobi, Hermann. [A2. 1873]. Reprinted in H. Jacobi [A3. 1970], vol. 1, pp. 405–408.
——. [A2. 1876]. Reprinted in H. Jacobi [A3. 1970], vol. 2, pp. 882–887.
——. [1888]. Reprinted Kiel, 1891, and H. Jacobi [A3. 1970], vol. 2, pp. 911–947.
——. [1888/92]. Reprinted in H. Jacobi [A3. 1970], vol. 2, pp. 949–1005.
——. [1892/94]. Reprinted in H. Jacobi [A3. 1970], vol. 2, pp. 1006–1017.
——. [1893]. Reprinted in H. Jacobi [A3. 1970], vol. 1, pp. 258–264.
——. [A3. 1894]. "On the Date of the Rig-Veda," *IA* 23, 1894, 154–159.
——. [1895a]. Reprinted in H. Jacobi [A3. 1970], vol. 2, pp. 905–910.
——. [1895b]. Reprinted in H. Jacobi [A3. 1970], vol. 1, pp. 265–277.
——. [1896]. Reprinted in H. Jacobi [A3. 1970], vol. 1, pp. 278–292.
——. [1900]. Reprinted in H. Jacobi [A3. 1970], vol. 2, pp. 1075–1076.
——. [1911/12]. Reprinted in H. Jacobi [A3. 1970], vol. 2, pp. 1017–1032.
——. [A3. 1911/12a]. "Dates of Chola Kings," *EI* 11, 1911–12, 120–132. Reprinted in H. Jacobi [A3. 1970], vol. 2, pp. 1077–1089.
——. [A3. 1911/12b]. "Dates of Pandya Kings," *EI* 11, 1911–12, 132–139. Reprinted in H. Jacobi [A3. 1970], vol. 2, pp. 1089–1096.
——. [1913/14]. Reprinted in H. Jacobi [A3. 1970], vol. 2, pp. 1033–1074.
——. [1920]. Reprinted in H. Jacobi [A3. 1970], vol. 2, pp. 888–904.
——. [A3. 1970]. *Kleine Schriften*, ed. Bernhard Kölver, *Glasenapp-Stiftung* 4, 2 vols., Wiesbaden 1970.
Jain, G. R. [A3. 1942]. *Cosmology Old and New*, Lucknow 1942.
Jain, Laxmi Chandra. [A3. 1967]. "On the Jain School of Mathematics," *Chotelal Smṛti Grantha*, Calcutta 1967, pp. 265–292.
——. [A3. 1969]. "Research on Jaina Mathematics," *Jñānapīṭha Patrikā*, Oct.–Nov. 1969., 33–41.
Jain, Nemichandraji. [A3. 1950]. "Jaina jyotiṣakī vyāvahārikatā," *Shrī Mahāvīra Commemoration Volume* pt. 1, Agra 1950, pp. 196–202.
Jaina, Udayacandra. [A3. 1967/68]. "Mālavaśiromaṇiḥ rājā Bhojaḥ," *Prajñā* 13, 1967–68, 1, 116–118.
Jaini, J. L. [A3. 1948]. *The Jaina Universe*, *SBJ* 13, Lucknow 1948.
Jha, Parameshwar. [A3. 1969]. "Āryabhaṭa I: His School," *JBRS* 55, 1969, 102–114.
Jha, Sitaram. [A3. 1946]. "Jyotiḥśāstraprayojanam," *Kashi Vidyapith* (*Silver Jubilee*), Banaras Saṃ. 2003 = A.D. 1946, Sanskrit section, pp. 24–33.
Jhaveri, Indukala H. [A3. 1956]. "The Concept of Ākāśa in Indian Philosophy," *ABORI* 37, 1956, 300–307.
Johansson, K. F. [A3. 1926]. "Die mit 'ni'—gebildeten hohen Zahlen im Altindischen," *Festgabe Jacobi*, Bonn 1926, pp. 429–439.
Joshi, M. C. [A3. 1970]. "Two Interesting Sun Images from Nachna," *JIH* 48, 1970, 81–87.
Joyis, M. N. Viśveśvara. [A3. 1969]. "Bṛhajjātakasubodhinīṭīkā," *MO* 2, 1969, 53–54.
Kane, P. V. [1930/62]. Revised and Enlarged Edition of vol. 1, pt. 1, Poona 1968.
——. [A3. 1952]. "Muhūrta," *Sri Swami Kevalananda Abhinandana Grantha*, Wai 1952, pp. 12–23.
Kanhaiyālāla, Muni. [A3. 1968]. *Gaṇitānuyoga*, with a Hindī translation by Mohanalāla Mehatā, edited by Śobhācandra Bhārilla, Sāṇḍerāva VE 2495 = A.D. 1968.
Kapadia, H. R. [A3. 1936/37a]. "Foliation of Jaina Manuscripts and Letter-numerals," *ABORI* 18, 1936–37, 171–186.
——. [A3. 1936/37b]. "A Note on Four Problems Given by Śrī Ratnaśekhara Sūri in his Work Ācārapradīpa," *ABORI* 18, 1936–37, 399–401.
Karambelkar, V. V. [1952]. See *NPP* 53, 1952, 286–299.
Karmarkar, A. P. [A3. 1945]. "Purāṇic Cosmogony," *Bhārata Kaumudī*, 2 vols., Allahabad 1945–47, vol. 1, pp. 323–332.
Kashikar, Sadashivsastri. [A3. 1943]. "Jyotiṣaśāstra heṃ mūla bhāratīyāṃceca," *Jyotiṣatattvadarśana*, Poona 1943, pp. 67–72.
Khousikan. [A3. 1971]. *Āyirattut toḷḷāyirattu elupattu onru elupattu mūnril Cani Riṣapa sañcāra palan*, Cennai 1971.
Krishnamurti, Kuthur Subbaraya Iyer. [A3. 1971a]. *Fundamental Principles of Astrology: Hindu, Western and Stellar*, Madras 1971.
——. [A3. 1971b]. *Krishnamurti Paddhati* (*Predictive Stellar Astrology*), Madras 1971.
——. [A3. 1971c]. *Marriage, Married Life & Children* (*Stellar Astrology*), Madras 1971.
——. [A3. 1971d]. *Transit* (*Gocaraphala nirnayam*), Madras 1971.
Krishnamurti, Y. G., and Chandrakanta Sharma. [A3. 1971]. *Samudrika: the Hindu Art of Sex and Body-signs Predications* (sic!), Delhi 1971.
Kulkarni, B. R. [A3. 1943]. "Antiquity of Karkācārya," *ABORI* 24, 1943, xxxvi–xxxvii.
Lahiri, N. C. [A3. 1969]. "Seminar on Panchang," *SC* 35, 1969, 194–196.
Law, Bimala Churn. [A3. 1933/34] "Geographical Data from Sanskrit Buddhist Literature," *ABORI* 15, 1933–34, 1–38.
——, [A3. 1937]. *Geographical Essays*, vol. 1, London 1937.

Luṃsala, Devīprasāda. [A3. 1969a]. "Jalavijñāna," *AN* 7, April 1969, 38–41.

——. [A3. 1969b]. "Vivāhapaṭala," *AN* 8, July 1969, 43–45.

——. [A3. 1969c]. "Hindū Vivāhapaddhatiko Vaijñānikatā," *AN* 9, October 1969, 41–46.

Mahadevan, T. M. P. [A3. 1969]. "The Advaita View of Time," *Dr. Satkari Mookerji Felicitation Volume*, Varanasi 1969, pp. 500–503.

Mankad, Harilal Rangildas. [A3. 1936/37]. "Saptadvīpa Pṛthivī," *ABORI* 18, 1936–37, 225–240.

Mirashi, V. V. [A3. 1968/69]. "Is Vijaya Mentioned in Nagarjunakonda Inscription the Name of Cyclic Year," *JOI Baroda* 18, 1968–69, 318–322.

Mishra, Umesha. [A3. 1930]. "Caṇḍeçvara Ṭhakkura and Maithili," *AUS* 4, 1, 1930, 349–357.

Misra, A. [A3. 1931/32]. "On Hindu Values of π," *BMAUA* 5, 1931–32, 12–18.

Mohan, Brij. [A3. 1967]. "History of Plus and Minus Signs," *IJHS* 2, 1967, 47–51.

Mukherjee, B. N. [A3. 1969]. "A Note on the Date of Kanishka I," *OH* 17, 1969, 33–38.

Mukherji, Kalinath. [1905]. Reprinted Calcutta 1969.

Murthy, K. R. Srikanta. [A3. 1970]. "Ancient Indian Sciences," *MO* 3, 1970, 131–137.

Murty, Jyothula Suryanarayana. [A3. 1969]. *Sūryasāmudrikamu*, Dakālayamu 1969.

Ojhā, Mīṭhālāla Himmatarāma. [A3. 1969]. "Jyautiṣe Phalānubhavakālavicāra," *Saṃskṛti*, 3 vols., Dillī 1969, vol. 1, pp. 374–381.

Pade, J. S. [A3. 1971/72]. "Praśnavidyā of Bādarāyaṇa," *JOI Baroda* 21, 1971–72, suppl. Reprinted as *M. S. University Oriental Series* 10, Baroda 1972.

Palaye-Joshi, Padmanabhashastri. [A3. 1948]. "Varāhamihira," *Vikrama Volume*, Ujjain 1948, pp. 361–376.

Panchamukhi, R. S. [A3. 1929/30]. "Kotavumachgi Inscription of Vikramaditya V," *EI* 20, 1929–30, 64–70.

Pande, Shyam Narain. [A3. 1970]. "Identification of the Ancient Land of Uttarakuru," *JGJRI* 26, 1970, 725–735.

Pandey, Lalta Prasad. [A3. 1971]. *Sun-worship in Ancient India*, Delhi-Patna-Varanasi 1971.

Pandey, Ramji. [A3. 1970]. "The Concept of the Earth in Purāṇas," *Purāṇa* 12, 1970, 252–266.

Paradkar, M. D. [A3. 1969]. "Kavīndrācārya Saraswatī—A Native of Mahārāṣṭra," *JGJRI* 25, 1969, 377–380.

Patvardhan, Madhav T. [A3. 1933/34]. "Sāra and Varāhamihira," *ABORI* 15, 1933–34, 249.

Patvardhan, Ramchandra Vinayak. [A3. 1943a]. "Pañcagraha āṇi cāndranakṣatreṃ yāṃce saṃbandhīṃ ṛgvedāntīla ullekha," *Jyotiṣatattvadarśana*, Poona 1943, pp. 73–79.

——. [A3. 1943b]. "Madhumādhavādi prācīna vaidika māsa," *Jyotiṣatattvadarśana*, Poona 1943, pp. 81–86.

Pearse, Thomas Deane. [A3. 1783]. "On the Sixth Satellite of Saturn," in Dharampal [A3. 1971] 87–93.

Pillai, K. Raghavan. [A3. 1970]. "Aṃśakaphala," *JKUORIML* 18, 1970, 3–19.

Pingree, David. [A3. 1971]. "On the Greek Origin of the Indian Planetary Model Employing a Double Epicycle," *JHA* 2, 1971, 80–85.

——. [A3. 1972a]. "Varāhamihiraviracitā Bṛhadyātrā," *BGOML Madras* 20, 1972, 1, app., pp. 1–92, and 2, app., pp. i–xiv and 93–130; reprinted Madras, 1972.

——. [A3. 1972b]. "Precession and Trepidation in Indian Astronomy before A.D. 1200," *JHA* 3, 1972, 27–35.

Pisani, Vittore. [A3. 1933/34]. "Svarbhānu-Rāhu," *RSO* 14, 1933–34, 310–311.

Playfair, John. [1790]. Reprinted in Dharampal [A3. 1971] 9–69.

Plunket, E. M. [A3. 1900]. "Ancient Indian Astronomy," *Proceedings of the Society of Biblical Archaeology* 22, 1900, 47–58. Reprinted in E. M. Plunket [1903] 162–184.

Prakash, Buddha. [A3. 1969]. "Science and Technology in Ancient India: Social and Political Influences," *VIJ* 7, 1969, 143–156.

——. [A3. 1970]. "India as Described by a Tenth Century Persian Geographer," *Proceedings of the Twenty-sixth International Congress of Orientalists*, vol. 3, pt. 2, Poona 1970, pp. 578–587.

Prasad, B. N., and R. Shukla. [A3. 1951]. "Aryabhata of Kusumapura," *BMAUA* 15, 1951, 24–32.

Raghavan, V. [A3. 1970]. "Worship of the Sun," *Purāṇa* 12, 1970, 205–230.

Rahurkar, V. G. [A3. 1969/70]. "The Saptarṣis in the Vedic and the Post-Vedic Literature," *Ṛtam* 1, 1969–70, 2, 15–21.

Raman, Bangalore Venkata. [A3. 1950a]. *Studies in Jaimini Astrology*, Bangalore 1950; 2nd ed., Bangalore 1958.

——. [A3. 1950b]. "Is Astrology a Science?" *Maha-Raval*, Dungarpur 1950, pp. 477–488.

Ranade, Purushottama. [A3. 1943]. "Jyotiṣaśāstradharmaśāstrayor mithaḥ sambandhaḥ," *Jyotiṣatattvadarśana*, Poona 1943, pp. 206–208.

Rao, B. Vidyadhara. [A3. 1968/69]. "Occurrence of Sexagenary Cycles in Two Inscriptions of Nagarjunakonda," *JOI Baroda* 18, 1968–69, 323–325.

Rau, Nemmara N. Krishna, and Vidya Bhushan Chaudhuri. [A3. 1962]. *Shodasa Varga & Dwadasa Varga Tables*, Bombay 1962.

Rele, V. G. [A3. 1924]. *An Exposition of the Directional Astrology of the Hindus as Propounded in Vimshottari Daśā*, Bombay 1924. See V. G. Rele [1935].

Rocher, Ludo. [A3. 1969]. "The Cyclical Concept of Time in Hinduism: A New Interpretation," *Proceedings of the Twenty-sixth International Congress of Orientalists*, vol. 3, pt. 1, Poona 1969, pp. 518–519.

Saha, A. K. [A3. 1969]. "The National Calendar," *SC* 35, 1969, 122–131.

Saraswati, T. A. [A3. 1962]. "Mahavira's Treatment of Series," *JRU* 1, 1962, 39–50.

——. [A3. 1969]. "The Development of Mathematical Ideas in India," *IJHS* 4, 1969, 59–78.

Sardesai, Narhar Gopal. [A3. 1917]. "The land of the seven rivers," *R. G. Bhandarkar Commemoration Volume*, Poona 1917, pp. 93–96.

Sarvari. [A3. 1970]. *Secrets of Palmistry*, Madras [1970].

Sastri, S. Srikantha. [A3. 1948]. "The Date of Śrīdharācārya," *Jaina Ant* 13, 1948, 12–17.

Sastri, T. S. Kuppanna. [A3. 1969a]. "A Historical Development of Certain Hindu Astronomical Processes," *IJHS* 4, 1969, 107–125.

——. [A3. 1969b]. "The School of Āryabhaṭa and the Peculiarities Thereof," *IJHS* 4, 1969, 126–134.

——. [A3. 1969c]. "The System of the *Vaṭeśvara Siddhānta*," *IJHS* 4, 1969, 135–143.

Schlerath, B. [A3. 1962/63]. "Die 'Welt' in der vedischen Dichtersprache," *IIJ* 6, 1962–63, 103–109.

Schroeder, L. [A3. 1912/13]. "Der siebente Āditya," *Festgabe Berthold Delbrück*, Strassburg 1912–13, pp. 178–193.

Sen, S. N. See S. P. Bhattacharyya and S. N. Sen [A3. 1969].

——. [A3. 1968]. "Praśastapāda's Impetus Theory of Motion," *Actes du XIe Congrès Internationale d'Histoire des Sciences*, 1968, vol. 3, pp. 327–331.

——. [A3. 1970a]. "Influence of Indian Science on Other Culture Areas," *IJHS* 5, 1970, 332–346.

——. [A3. 1970b]. "The Introduction of Western Science in India during the 18th and 19th Century," *Science, Technology and Culture*, ed. Surajit Sinha, New Delhi 1970, pp. 14–43.

——. [A3. 1971]. "A Survey of Source Materials"; "Astronomy"; and "Mathematics," *A Concise History of Science in India*, ed. D. M. Bose, S. N. Sen, and B. V. Subbarayappa, New Delhi 1971, pp. 1–212. Reprinted Calcutta [1972].

Sengupta, B. K. [A3. 1970]. "A Coherent Study of the *Lakṣmaṇasaṃvat*," *Proceedings of the Twenty-sixth International Congress of Orientalists*, vol. 3, pt. 2, Poona 1970, pp. 751–753.

Shah, Harilal Amritlal. [A3. 1939/40]. "Vedic Lores," *ABORI* 21, 1939–40, 262–263.

Shah, U. P. [A2. 1956]. See *Āchārya Vijayavallabhasūri Commemoration Volume*, Bombay 1956, pp. 91–140.

Shamasastry, R. [A3. 1915/16]. "Orientation of Sacrificial Halls," *Sanskrit Research* 1, 1915–16, 71–76.

——. [A3. 1940]. "The eclipse cult and Indian philosophies," *Ramalinga Reddy Ṣaṣṭyabdapūrti Commemoration Volume*, pt. II, Waltair 1940, pp. 310–314.

——. [A3. 1947]. "Vedic chronology," *Bhārata Kaumudī*, 2 vols., Allahabad 1945–47, vol. 2, pp. 855–863.

Sharma, Chandrakanta. See Y. G. Krishnamurti and C. Sharma [A3. 1971].

Sharma, M. [A3. 1967]. *Jyautiṣa śabda kośa*, Garhwal 1967.

Shastri, Ajay Mitra. [A3. 1967]. "Coins in Bṛhatsaṃhitā of Varāhamihira," *JNSI* 29, 1967, 2, 41–45.

Shastri, Biswanarayan. [A3. 1969]. "Kāmarūpa School of *Dharma Śāstra*," *Proceedings of the Twenty-sixth International Congress of Orientalists*, vol. 3, pt. 1, Poona 1969, pp. 253–255.

Shastri, Manoranjan. [A3. 1960]. "Vedacarya and Samvatsara," *JARS* 14, 1960, 63–77.

Shastri, Satyavrat. [A2. 1967/68]. See *PAIOC* 23, 1966, 361–367.

Shembavnekar, K. M. [A3. 1935/36]. "The Metamorphosis of Uṣas," *ABORI* 17, 1935–36, 351–357.

Shukla, K. S. [A3. 1966]. "Hindu Methods of Finding Factors or Divisors," *Gaṇita* 17, 1966, 109–117.

——. [A3. 1969]. "Astronomy in Ancient and Medieval India," *IJHS* 4, 1969, 99–106.

——. [A3. 1969/70]. "Nārāyaṇa Paṇḍita's Bījagaṇitāvataṃsa. Part I," *Ṛtam* 1, 1969–70, 2, suppl.

——. [A3. 1971/72]. "Hindu Mathematics in the Seventh Century as Found in Bhāskara I's Commentary on the *Āryabhaṭīya*," *Gaṇita* 22, 1, 1971, 115–130; 22, 2, 1971, 61–78; 23, 1, 1972, 57–79; and 23, 2, 1972, 41–50.

Shukla, R. See B. N. Prasad and R. Shukla [A3. 1951].

Singh, A. N. [1933a]. Read: *JUPHS* 7, 1934, 42–53.

——. [1933b]. Read: *JUPHS* 7, 1934, 54–64.

——. [1949]. Read: *Jaina Ant* 15-16, 1949–50, 46–69.

Sinha, S. R. [A3. 1951]. "Bhāskara's Līlāvatī." *BMAUA* 15, 1951, 9–16.

Sircar, D. C. [A3. 1936/37]. "Sewai Jaysingh of Amber, A.D. 1699–1743," *IC* 3, 1936–37, 376–379.

——. [A3. 1946/47]. "The Ponduru Grant and the Gaṅga Era," *JKHRS* 1, 1946–47, 219–221.

Somayaji, D. A. [A3. 1971]. *A Critical Study of the Ancient Hindu Astronomy in the Light and Language of the Modern*, Dharwar 1971.

Śrīnivāsācāryulu, Kīḷāttūru. [A3. 1970]. *Mī puṭṭina tedī, mī jīvita rahasyālu*, Karaveni [1970].

Srivastava, V. C. [A3. 1969]. "Solar Symbols in Sūryamitra-Bhānumitra Coins," *JNSI* 31, 1969, 1, 9–14.

Subbarayappa, B. V. [A3. 1971]. "The Physical World: Views and Concepts," *A Concise History of Science in India*, ed. D. M. Bose, S. N. Sen, and B. V. Subbarayappa, New Delhi 1971, pp. 445–483.

Suthar, Chhotubhai. See H. Bhatt and C. Suthar [A3. 1969/70].

Thakur, U. [A3. 1969]. "Caṇḍeśvara and his Rājanītiratnākara," *VIJ* 7, 1969, 56–68.

Thomas, Edward. [1874]. Reprinted Varanasi 1970.

Thurston, E. [A3. 1913]. "The number seven in southern India," *Essays and Studies Presented to William Ridgeway*, Cambridge 1913, pp. 353–364.

Tikkimal, H. C. [A3. 1969]. "Sawai Jai Singh and the Marwar Affairs in the Reign of Emperor Muhammad Shah (1723–1724 A.D.)," *PIHC* 31, 1969, 204–207.

Tripāṭhī, Avadhavihārī. [A3. 1969]. "Bhāratīyajyotiṣasya Vikāsakramaḥ," *Saṃskṛti*, 3 vols., Dillī 1969, vol. 1, pp. 367–373.

Tripāṭhī, Māyā Prasāda. [A3. 1958/59]. "Science of Geography in the Ṛgveda," *JGJRI* 16, 1958–59, 185–200.

——. [A3. 1969]. *Development of Geographic Knowledge in Ancient India*, Varanasi 1969.

——. [A3. 1970]. "Identification of an Avestan Daēva Tauru," *Ṛtam* 1, 2, 1970, 99–102.

Uṇṇi, K. P. K. [A3. 1971]. *Lagnaphaladīpika*, Ālappul 1971.

Upādhyāya, Baladeva. [A3. 1970]. "Bṛhatsaṃhitāyā vimarśaḥ," *JGJRI* 26, 1970, 781–790.

Upādhyāya, Bāñke Lāla. [A3. 1971]. *Prācīna bhāratīya gaṇita*, Dillī 1971.

Upādhyāya, Rājamohana. [A3. 1966/67]. "Bhāratīya pañcāṅgaḥ eka adhyayana," *Prajñā* 12, 1966–67, 1, 166–169.

Upadhye, A. N. [A3. 1938]. "Jambudvīpa-prajñapti-saṃgraha of Padmanandī," *Winternitz Memorial Number*, Calcutta 1938, 188–191.

Upadhye, P. M. [A3. 1969]. "Geography Known to the *Paumacariya*," *Proceedings of the Twenty-sixth International Congress of Orientalists*, vol. 3, pt. 1, Poona 1969, pp. 324–326.

Velankar, H. D. [A3. 1946]. "Varāhamihira and Utpala (in relation to Sanskrit metres)," *Dr. C. Kunhan Raja Presentation Volume*, Madras 1946, pp. 141–152.

Vīrakkoḍi, D. T. [A3. 1969]. *Nava pubuduva*, ? [1969].

Vogel, Claus. [A3. 1971]. "Die Jahreszeiten im Spiegel der altindischen Literatur," *ZDMG* 121, 1971, 284–326.

Volodarskii, A. A. [A3. 1972]. "Drevneindiiskie sistemy numeralii," *Indiiskaya Kultura i Buddizm*, Moskva 1972, pp. 82–89.

Vyas, Suryanarayan [A3. 1970/71]. "The Basis of Kṛta-kāla Gaṇana—An Analysis," *JOI Baroda* 20, 1970–71, 12–26.

Whitehead, R. B. [A3. 1947]. "The so-called Sun-god of Multan," *India Antiqua*, Leyden 1947, pp. 326–329.

Yabuuti, Kiyosi. [A3. 1954]. "Indian and Arabian Astronomy in China," *Silver Jubilee Volume of the Zinbun Kagaku Kenkyusyo, Kyoto University*, Kyoto 1954, pp. 585–603.

Yano, Michio. [A3. 1972]. "On *Saptarṣi* or the Great Bear," *JIBS* 20, 1972, 967–975.

LIST OF CATALOGS OF SANSKRIT MANUSCRIPTS AND BOOKS

Alwar (1884): in BORI A 1883/84, pp. 91–111.

*AS Bengal: H. Shastri, *A Descriptive Catalogue of Sanskrit Manuscripts in the Government Collection*, vols. 1–13, Calcutta 1917–1966.

*Benares (1956): *A descriptive Catalogue of the Sanskrit Manuscripts Acquired for and Deposited in the Government Sanskrit College Library, Sarasvati Bhavana, Benares, during the Years 1791–1950*, ed. Kuberanath Shukla, vol. 3 (dharmaśāstra), Benares 1956.

*BM (Gujarātī): J. F. Blumhardt, *Catalogue of the Marathi, Gujarati, Bengali, Assamese, Oriya, Pushtu, and Sindhi Manuscripts in the Library of the British Museum*, London 1905.

*Calcutta Sanskrit College (Smṛti): Hṛishīkeśa Śāstrī and Śiva Chandra Gui, *Descriptive Catalogue of Sanskrit Manuscripts in the Library of the Calcutta Sanskrit College*, vol. 2, Calcutta 1898.

Germany (Singhalese): M. Bidoli and H. Bechert, *Singhalesische Handschriften*, Teil 1, Wiesbaden 1969.

GJRI: Umesha Mishra, *Descriptive Catalogue of Sanskrit Manuscripts in Ganganatha Jha Research Institute Allahabad*, vol. 1, Allahabad 1967.

Kunte: Kashi Nath Kunte, *Report on the Compilation of the Catalogue of Sanskrit Manuscripts, for the Official Year 1881–82*, Lahore 1882 (A—Gujrānwāla and Delhi Districts; B—Lahore Division).

*LDI: Puṇyavijayaji, *Catalogue of Sanskrit and Prakrit Manuscripts*, pts. 1–4, *LDS* 2, 5, 15, and 20, Ahmadābād 1963–1968. Pt. 4 includes Vijayadevasūri's Collection (VDS) and Kṣāntisūri's Collection (KS).

Leningrad (1914): N. D. Mironov, *Catalogus codicum manu scriptorum Indicorum qui in Academiae Imperialis Scientiarum Petropolitanae Museo Asiatico asservantur*, Fasc. I, Petropoli 1914.

Maheshanagar: Badrinath Jha, "A Descriptive Catalogue of the Sanskrit Manuscripts in the Manuscript Library of the Mithila Institute of Post-Graduate Studies and Research in Sanskrit Learning, Maheshanagar, Darbhanga," *BMI* 4, 1968, 29–141.

Mithila I: Kashiprasad Jayaswal and Ananta Prasad Śāstrī, *A Descriptive Catalogue of Manuscripts in Mithila*, vol. 1 (Smṛti), Patna 1927.

*Paris BN: Jean Filliozat, *Catalogue du fonds sanscrit*, fasc. II, Paris 1970.

*NCC: vol. 6, Madras 1971.

NPS: *Hastalikhita hindī pustakoṃ kā saṃkṣipta vivaraṇa*, 2 vols., Kāśī Saṃ. 2021 = A.D. 1964.

*PrSB: K. L. Janert and N. N. Poti, *Indische und Nepalische Handschriften*, Teil 2, Wiesbaden 1970.

*Śāstrī, Not. 1911: Haraprasāda Śāstrī, *Notices of Sanskrit MSS., Second Series*, vol. 4, Calcutta 1911.

*Tanjore: P. P. S. Sastri, *A Descriptive Catalogue*, vol. 18, Srirangam 1934.

CENSUS OF THE EXACT SCIENCES IN SANSKRIT

AKHAIRĀMA (*fl.* 1755)

A Brāhmaṇa of the Gargagotra and a resident of Cūnanagara (?) in Mathurā, Akhairāma was a protégé of Sujānasiṃha (Sūrajasiṃha), the ruler of Bharatapura. He wrote the following works in Hindī on jyotiṣa.

1. *Muhūrtacintāmaṇi*. Manuscript:

NPS 1 A of 1938–40. Copied in Saṃ. 1938 = A.D. 1881. Property of Paṇḍita Revatīnandana (Revatīramaṇa Miśra) of Berī, Barārī, Mathurā.

2. *Laghujātaka*, in Saṃ. 1812 = A.D. 1755. Manuscript:

NPS 1 B of 1938–40. Copied in Saṃ. 1926 = A.D. 1869. Property of Paṇḍita Nandalāla of Bājanā, Mathurā.

3. *Svarodaya*. Manuscript:

NPS 4 A of 1932–34. Copied in Saṃ. 1901 = A.D. 1844. Property of Paṇḍita Giradhara Miśra of Candramanagaḍhī, Achanerā, Āgarā.

**AGASTYA*

An excerpt from the *Agastyasaṃhitā* (see *CESS* A 1, 35a, and A 2, 11a) is the *Prasūtigaṇḍadoṣaśānti*. Manuscript:

GOML Madras D 3384. 5pp. Telugu.

**ACYUTĀNANDA JHĀ* (*fl.* 1939/1958)

Devana Jhā of the Kāśyapagotra, a resident of Videha honored by a narapati Mukuṭa, had three sons: Bhavi, Rudi, and Jayadatta. Jayadatta was the father of Bhrātṛnātha, the father of Gosvāmin, who was raised by Gūna Jhā in Caugama and educated in Jariso in Darabhaṅgā in Mithilā, where he married the daughter of Vedamaṇi Jhā. Among their five sons was Baladeva (d. 1936), the father of seven sons: Raghuvaṃśa, Acyutānanda, Bhuvaneśvara, one who died in childhood, Harihara, Śivānanda, and Kīrtyānanda. Acyutānanda studied under Genādilāla, and taught at the Rāmasādhu Saṃskṛta Vidyālaya at Kāśī and then at the Rādhākṛṣṇa College at Khurjā in Bulandaśahara. He wrote the following works.

1. *Calanakalanapraśnottaravivaraṇa*, written in 1939; see *CESS* A 1, 39a.

2. *Subodhinī*, a ṭīkā with a Hindī version on the *Ududāyapradīpa* of Parāśara, and *Pārāśaryyartha*, a Hindī ṭīkā on the *Madhyapārāśarī*; these works, written in 1941, were published with the mūlas as *HSS* 135, Banārasa 1941; 2nd ed., Banārasa 1948.

3. *Subodhinī*, a ṭīkā with a Hindī rendering on the *Vāsturatnāvalī* of Jīvanātha (*fl.* 1744), and *Vidhivivekādhyāya*, a pariśiṣṭa to the same; these works, written at Kāśī in Saṃ. 1998 = A.D. 1941, were published with the mūla as *HSS* 152, Banārasa 1941; 2nd ed., Banārasa 1949.

4. *Paddhatiprakāśa*.

5. *Vimalā*, a ṭīkā with a Hindī rendering on the *Jaiminisūtra* of Jaimini; this was published with the mūla as *HSS* 159, Banārasa 1943; 2nd ed., Banārasa 1952.

6. *Vimalā*, a Hindī ṭīkā on the bhāvaphalādhyāya from the *Lomaśasaṃhitā*; this was published with the mūla as the second part of *HSS* 163, Banārasa 1944; 2nd ed., Vārāṇasī 1963.

7. *Vividhavāsanā*, a ṭīkā on the *Cāpīyatrikoṇagaṇita* of Nīlāmbara Jhā (b. 18 July 1823); this was published with the mūla as *KSS* 139, Banāras 1944.

8. *Vimalā*, a Hindī ṭīkā on the *Bṛhajjātaka* of Varāhamihira (*fl. ca.* 550); this work, written in Saṃ. 2002 = A.D. 1945, was published with the mūla as *HSS* 151, Banārasa 1945; 2nd ed., Banārasa 1957.

9. *Vimalā*, a ṭīkā with Hindī notes on the *Bījagaṇita* of Bhāskara (b. 1114); this work, written at Jariso in Saṃ. 2006 = A.D. 1949, was published with the mūla and the ṭīkā, *Subodhinī*, of Jīvanātha Jhā (*fl. ca.* 1846/1900) as *KSS* 148, Banārasa 1949.

10. *Vimalā*, a Hindī ṭīkā on the *Jātakābharaṇa* of Ḍhuṇḍhirāja (*fl. ca.* 1525), and a pariśiṣṭa to the same; these works, written at Jariso in Saṃ. 2008 = A.D. 1951, were published with the mūla as *HSS* 212, Banārasa 1951.

11. *Vimalā*, a Hindī ṭīkā on the *Ramalanavaratna* of Paramasukha (*fl.* 1810); this work, written in Saṃ. 2010 = A.D. 1954, was published with the *Ramalapraśnasaṅgraha* as *HSS* 245, Banārasa 1954.

12. *Saralatrikoṇa*.

13. *Vimalā*, a ṭīkā on the *Golīyarekhāgaṇita*.

14. *Vimalā*, a Hindī ṭīkā on the *Bṛhatsaṃhitā* of Varāhamihira (*fl. ca.* 550); this work, written in Saṃ. 2015 = A.D. 1958, was published with the mūla as *VSG* 41, Vārāṇasī 1959.

**AJAYARĀJA = AJERĀJA*

Author of a *Bhāṣāsāmudrika* in Hindī. Additional manuscripts (see *CESS* A 1, 39a):

NPS 4 A of 1929–31. Copied in Saṃ. 1924 = A.D. 1867. Property of Paṇḍita Rāmalāla of Turakaiyā, Achanerā, Āgarā.
NPS. Property of Paṇḍita Sohanalāla Śarmā of Nagalā Aniyā, Karahala, Mainapurī.

*ANANTA (fl. 1534)

The manuscript of his *Kālanirṇayāvabodha* is Anup 1689, not 1698 as mistakenly recorded in *CESS* A 1, 40b.

*ANANTA (fl. ca. 1600)

Additional manuscripts of his *Nakṣatrasattraprayoga* (see *CESS* A 1, 40a and 41a, and A 2, 11b):

Benares (1953) 3086. 10ff. Copied in Saṃ. 1744=A.D. 1687.
Benares (1953) 3693. 41ff. Copied in Saṃ. 1784=A.D. 1727.
Calcutta Sanskrit College (I) 273. 16ff. Copied by Rāma, the son of Bālasarasvatibhaṭṭa Gahvara, at Kāśī in Saṃ. 1801 = A.D. 1744.
AS Bengal 722 (G 2410) = Mitra, Not. 4181. Copied by Yajñeśvara Bhāgavata on 3 kṛṣṇapakṣa of Śrāvaṇa in Śaka 1674 = *ca.* 15 August 1752.
Calcutta Sanskrit College (I) 274. 8ff. Copied on Monday 9 kṛṣṇapakṣa of Māgha in Saṃ. 1811 = 6 February 1755.
Baroda 7586. 39ff. Copied in Śaka 1684 = A.D. 1762.
Benares (1953) 3525. 18ff. Copied in Śaka 1693 = A.D. 1771. (*Nakṣatreṣṭipaddhati* of Anantadeva).
Baroda 7568. 36ff. Copied in Śaka 1717 = A.D. 1795.
Calcutta Sanskrit College (I) 272. 35ff. Copied on 1 śuklapakṣa of Jyeṣṭha in Saṃ. 1875 = *ca.* 3 June 1818.
AS Bengal 721 (G 765) = Mitra, Not. 1570. 17ff. (ff. 14–16 missing). Ascribed to Anantadeva.
Baroda 467. 29ff.
Baroda 1478. 51ff.
Baroda 6789(f). Ff. 67b–68. Grantha.
Baroda 10148. 22ff.
Benares (1953) 3600. 6ff.
IO 4713 (Aufrecht 32b). Pp. 18–40. Copied from Munich 196. From T. Aufrecht.

*ANANTA (fl. ca. 1625/1650)

Additional manuscripts of his *Samayakāṇḍa* = *Kālakāṇḍa* (see *CESS* A 2, 11b):

Anup 2563. Ff. 1–101 and 101b–111. Copied in Saṃ. 1725 = A.D. 1668.
AS Bengal 2192 (G 6484). 120ff. Copied on Sunday 2 kṛṣṇapakṣa of Śrāvaṇa in Saṃ. 193*X*. Formerly the property of Bālamukunda.
Anup 2564. 117ff. Property of Vaidyanātha, the son of Anantabhaṭṭa.
Tanjore D 18361 = Tanjore BL 47. 171ff.
Tanjore D 18362 = Tanjore BL 48. 134ff.
Tanjore D 18363 = Tanjore BL 49. 109ff.
Tanjore D 18364 = Tanjore BL 50. 108ff.

Manuscripts of the *Tithikāṇḍa* of his *Rāmakalpadruma*:

Benares (1956) 13543. 163ff. Copied in Saṃ. 1860 = A.D. 1803.
Benares (1956) 13542. 145ff.

*ANANTADEVA (fl. ca. 1675)

Additional manuscripts of his *Tithidīdhiti* (see *CESS* A 1, 41b–42a, and A 2, 11b–12a):

*AS Bengal 2087 (G 2033) = Mitra, Not. 3325. 66ff. Copied in Śaka 1731 = A.D. 1809.
Baroda 1951. 86ff. Copied in Śaka 1738 = A.D. 1816.
Anup 2639. 53ff.
AS Bengal 2088 (G 5955). 40ff.
PL, Buhler III E 111. 168ff. No author mentioned. Property of Kṛṣṇarāva Bhīmāśaṅkara of Vaḍodarā.
Tanjore D 18429 = Tanjore BL 407. 71ff.
Tanjore D 18430 = Tanjore BL 408. 72ff.
Tanjore D 18431 = Tanjore BL 409. 60ff.
Tanjore D 18432 = Tanjore BL 410. 58ff.

Additional manuscripts of his *Saṃvatsaradīdhiti* (see *CESS* A 2, 12b):

Anup 2635. 409ff. Copied at Janasthāna in Saṃ. (read Śaka) 1601 = A.D. 1679.
Anup 2637. Ff. 1–52, 52b–421, and 442–462. Copied in Kauṃkaṇa in Saṃ. 1746 = A.D. 1689.
Kunte B 94. 337ff. Copied in A.D. 1699. Property of Pandit Jwālā Datta Prasāda of Lahore.
Baroda 1499. 390ff. Copied in Śaka 1751 = A.D. 1829.
Anup 2636. Ff. 1–11, 11b, 11c, and 11d–366.
Anup 2638. Ff. 1–61, 65–208, 208b–237, and 237b–294.
AS Bengal 2089. (G 6476). Ff. 1–103 and 108–199, 78ff., and ff. 236–374.
Benares (1956) 12212. Ff. 1–71, 71b–124, 1–90, 1–81, and 91–127. (*Varṣadīdhiti* from the *Smṛtikaustubha*). Incomplete. No author mentioned.
Jammu and Kashmir 2529. 225ff. Incomplete.
Tanjore D 18433 = Tanjore BL 411. 320ff.
Tanjore D 18434 = Tanjore BL 9192. 539ff. Grantha.
Tanjore D 18435 = Tanjore JL 1366. 51ff.
Tanjore D 18436 = Tanjore TS 532. 186ff.

Additional manuscripts of his *Nirṇayabindu* = *Tithinirṇayabindu* (see *CESS* A 2, 12b):

Benares (1956) 14071. 8ff. Copied in Saṃ. 1848 = A.D. 1791.

Some other sections of the *Smṛtikaustubha* are found in manuscripts:

Jammu and Kashmir 2531. 11ff. Incomplete (adhimāsakṛtya).
Jammu and Kashmir 2533. 20ff. Incomplete (nakṣatranirṇaya).

**ANANTABHAṬṬOPĀDHYĀYA* (*fl.* before 1385)

Additional information on the manuscripts of his *Tithinirṇaya* (see *CESS* A 1, 42a, and A 2, 12b):

*AS Bengal 2656 (G 5846). Ff. 4–41 and 70–91. Copied by Nārāyaṇa, the son of Paṇḍita Saravaṇa of the Bhaṭṭanāgarajñāti, at Tilakavāḍāgrāma in Śrīnandapaṭṭamaṇḍala on Wednesday 15 kṛṣṇapakṣa of Pauṣa in Saṃ. 1441 = 11 January 1385.
Benares (1956) 12328. 63ff. Copied in Saṃ. 1502 = A.D. 1445.
Benares (1956) 13907. 98ff. Copied in Saṃ. 1603 = A.D. 1546.
*AS Bengal 2657 (G 5998). Ff. 1 and 3–10. Copied on Sunday 2 kṛṣṇapakṣa of Māgha in Saṃ. 16 (1816? handwriting of early nineteenth century). Incomplete (saṅkrāntinirṇaya only).
Benares (1956) 13501. Ff. 7–38. Incomplete.
Paris BN 212 F (Sans. dév. 311). F. 1. Incomplete. Acquired May 1842.

The second verse is:

nirṇayo 'yaṃ kṛtaḥ sarvo bhaṭṭānantena dhīmatā/
ajñānānāṃ prabodhārtham adṛṣṭārthaṃ tathaiva ca//

**ANAVAMADARŚIN SAṄGHARĀJA* (*fl.* 1241)

Manuscripts of his *Daivajñakāmadhenu* (see *CESS* A1, 42b–43a):

BM 557 (Or. 5419). 83 ff. Siṃhalese script.
BM Or. 6613 (29). From the Nevill Collection.

ANDHUKA (*fl.* 1030/1033)

An authority on kāla in dharmaśāstra cited by Jīmūtavāhana (*fl.* 1092); see P. V. Kane [1930/62], vol. 1, p. 325.

APPĀDHVARIN (*fl. ca.* 1700)

The son of Cidambara and a resident of Māyūra, Appādhvarin wrote for Shāhijī, the Mahārāja of Tanjore from 1684 to 1711, an *Ācāranavanīta*, of which part 4 is a *Kālanirṇaya*. Manuscript:

Tanjore D 18048 = Tanjore BL 9320. 377ff. Grantha.

ABHINAVAKĀLIDĀSA (= *UMĀMAHEŚVARA*)

A member of the Vellāla family and a pupil of Akkayasūri, the son of Veṅkaṭārya of the Mokṣaguṇḍa family, Abhinavakālidāsa wrote a *Santānadīpikā*. Manuscript:

Mysore (1922), p. 355. See NCC, vol. 1, rev. ed., p. 299.

**ABHIMANYU RĀJAN*

The Lahore manuscript (see *CESS* A 1, 45a) of his *Praśnaprakāśa* is:

Kunte B 78. 15ff. Copied in A.D. 1524. Property of Pandit Dilarām of Gujrānwāla.

AMARASIṂHA (*fl.* 1842)

Author of a *Svapnabheda* in Hindī in Saṃ. 1899 = A.D. 1842. Manuscript:

NPS 5 of Saṃ. 2004–2006. Property of the Nāgarīpracāriṇī Sabhā in Vārāṇasī.

AMṚTANĀTHA JHĀ ŚARMAN (b. 1755)

The son of Māṇika Śarman and a resident of Cayanapura in Bhāgalapura, Mithilā, Amṛtanātha wrote the *Kṛtyasārasamuccaya* according to tradition at the age of seventy in Śaka 1747 = A.D. 1825. This was published at Benares in 1877 (see NCC, vol. 1, rev. ed., p. 348), and edited with the notes and pariśiṣṭa of Gaṅgādhara Miśra Śarman (*fl.* 1929/41) by Kṛṣṇamohana Śāstrin as *KSS* 129, Banārasa 1953. Verse 2 is

tārkikāmṛtanāthaśarmabudho hi kṛtyasamuccayaṃ
vyātanoti vilokya pūrvanibandhagranthacayān amum/
maithilavyavahārasiddhasukarmakāṇḍavirājitaṃ
dhīdhanādimudapradaṃ saraloktito bahuyatnataḥ//

The colophon begins: iti mithilādeśāvayavabhāgalapuramaṇḍalāntargatacayanapuragrāmanivāsipugalavāḍamūlotpannasacchābdikaśrīmāṇikaśarmātmajasattārkikamahāmahopādhyāyaśrīmadamṛtanāthaśarmakṛta.

Manuscripts:

CP, Kielhorn XIX 67. 87ff. Copied in Saṃ. 1888 = A.D. 1831. Property of Bābūjī Ojhā of Maṇḍalā.
Mithila I 77 C. 82ff. Maithilī. Copied in Śaka 1765 = A.D. 1843. Property of Jagadeva Ṭhākur, previously of Pandit Gokulanāth Jhā of Nanaur, Tamuriā, Darbhanga.
Mithila I 77 N. 80ff. Maithilī. Copied in Śaka 1776 = A.D. 1854. Property of Pandit Balabhadra Jhā of Jogiara, Darbhanga.
Mithila I 77. 65ff. Maithilī. Copied in Sāl. San. 1274 = *ca.* A.D. 1866. Property of Pandit Jayakṛṣṇa Jhā of Champā, Benipati, Darbhanga.
Mithila I 77 O. 72ff. Maithilī. Copied in Sāl. San. 1282 = *ca.* A.D. 1874. Incomplete. Property of Pandit Dīnakānta Miśra of Śalampur, Ghatāho, Darbhanga.

Mithila I 77 A. 44ff. Maithilī. Copied in Sāl. San. 1283 = *ca.* A.D. 1875. Property of Pandit Janārdana Miśra of Chanaur, Manīgachī, Darbhanga.
Mithila I 77 L. 60ff. Maithilī. Copied in Sāl. San. 1283 = *ca.* A.D. 1875. Property of Pandit Dāmodar Jhā of Tharheṭ, Andhraṭhārhī, Darbhanga.
CP, Hiralal 998. Property of Janaknandan of Phulchur, Bilāspur.
CP, Hiralal 999. Ascribed to Vācaspati. Property of Viśvambharnāth of Ratanpur, Bilāspur.
GJRI 3481/119. 8ff. Incomplete.
GJRI 3482/120. 25ff. Maithilī. Incomplete.
Mithila I 77 B. 54ff. Maithilī. Incomplete. Property of Pandit Mukunda Singh Jhā of Chanaur, Manīgachī, Darbhanga.
Mithila I 77 D. 26ff. Maithilī. Incomplete. Property of Pandit Bālagopāl Jhā of Taraun, Sakri, Darbhanga.
Mithila I 77 E. 38ff. Maithilī. Property of Pandit Balabhadra Jhā of Pachadhi, Pandaul, Darbhanga.
Mithila I 77 F. 58ff. Maithilī. Property of Pandit Kapileśvar Jhā of Sakhabad, Manīgachī, Darbhanga.
Mithila I 77 G. 46ff. Maithilī. Incomplete. Property of Pandit Viśvanātha Jhā of Mahinathpur, Deodhā, Darbhanga.
Mithila I 77 H. 50ff. Maithilī. Incomplete. Property of Babu Govardhana Jhā of Naduar, Jhanjhārpur, Darbhanga.
Mithila I 77 I. 6ff. Maithilī. Incomplete. Property of Pandit Gonū Miśra of Lālganj, Jhanjhārpur, Darbhanga.
Mithila I 77 J. 18ff. Maithilī. Property of Pandit Gopīnāth Jhā of Naduār, Jhanjhārpur, Darbhanga.
Mithila I 77 K. 43ff. Maithilī. Incomplete. Property of Pandit Gaṇānand Jhā of Lohnā, Jhanjhārpur, Darbhanga.
Mithila I 77 M. 14ff. Maithilī. Incomplete. Property of Pandit Śrīnandan Miśra of Kanhauli, Sakri, Darbhanga.
Mithila I 77 P. 102ff. Maithilī. Incomplete. Property of Pandit Ravināth Jhā, Professor at M. R. Vidyālaya and resident of Andhraṭharhī, Darbhanga.
Mithila I 77 Q. 156ff. Maithilī. Incomplete. Property of Pandit Gopāla Miśra of Tabhaka, Dalsingh Sarai, Darbhanga.
Mithila I 78. 36ff. Maithilī. Property of Pandit Manohar Ṭhākur of Tabhaka, Dalsingh Sarai, Darbhanga.

*AMṚTĀNANDA

Additional manuscripts of his *Amṛtacaṣaka* (see *CESS* A 1, 46a–46b):

IM Calcutta 4429. See NCC, vol. 1, rev. ed., pp. 347, 355, and 460.
Mithilā. See NCC, pp. 355 and 460.
Mysore (1942), p. 21. See NCC, pp. 355 and 460.

ARAKṢITA DĀSA (*fl.* 1970)

Author of a *Bṛhat o sacitra kākacarita* in Saṃskṛta and Uḍiyā, published at Kaṭaka in 1970.

ARUBHADRA (*fl.* 1621)

Author of a *Koka sāmudrika* in Hindī in Saṃ. 1678 = A.D. 1621. Manuscript:

NPS 17 of 1929–31. Copied in Saṃ. 1850 = A.D. 1793. Property of Paṇḍita Lakṣmīnārāyaṇa Vaidya of Bāha, Āgarā.

ARKASOMAYĀJIN DHŪLĪPĀLA (*fl.* 1964)

The son of Mañgamā and Bāpaya of the Dhūlīpālakula and the younger brother of Veñkaṭarāma and Subrahmaṇya, Arkasomayājin was born at Valiceru, Naikaṭya, Madras. He wrote a *Jyotirvijñānam* published as *SG* 5, Varanasi 1964.

*ALLĀḌANĀTHA (*fl.* 1410?)

Additional manuscripts of his *Nirṇayāmṛta* (see *CESS* A 1, 47a, and A 2, 13b–14b):

BORI 227 of 1884/87. Ff. 2–242. Copied in Saṃ. 1593 = A.D. 1536. From Gujarāt.
AS Bengal 2152 (G 866). 155ff. Copied by Anantadāsa Dīttū on Wednesday 10 śuklapakṣa of Pauṣa in Saṃ. 1641 = 30 December 1584. Formerly the property of Paṇḍita Bhagavāna Dāsa Malavalīyā.
Benares (1956) 12776. Ff. 132–162 and 162b–273. Copied in Saṃ. 1645 = A.D. 1588. Ascribed to Gopīnārāyaṇa, but said to have been composed by Allāḍanātha in Saṃ. 1467 = A.D. 1410.
Benares (1956) 12117. Ff. 1–174 and 176–215. Copied in Saṃ. 1845 = A.D. 1788. Ascribed to Sūryasena.
BORI 122 of 1892/95. 338ff. Copied in Saṃ. 1848 = A.D. 1791.
Benares (1956) 12397. Ff. 1–80, 80b–174, and 176–202. Copied in Saṃ. 1850 = A.D. 1793. Incomplete.
Benares (1956) 13551. 173ff. Copied in Saṃ. 1850 = A.D. 1793. Ascribed to Sūryasena.
BORI 77 of 1899/1915. 242ff. Copied in Saṃ. 1907 = A.D. 1850. No author mentioned.
PrSB 796 (Göttingen Mu II 30). Ff. 214v–345. Śāradā. Copied on Monday 2 kṛṣṇapakṣa of Jyeṣṭha in (Saptarsi) Saṃ. ⟨49⟩64 = 25 June 1888.
Anup 2448. 46ff.
Anup 2449. Ff. 4–10 and 12–133.
Anup 2450. 133ff. Incomplete.
Anup 2451. 182ff. Property of Anūpasiṃha (1674/1698).
Anup 2452. 216ff. Incomplete.
AS Bengal 2153 (G 8680). 86ff. Incomplete.

AS Bengal 2154 (G 2995). 2ff. Incomplete (parvanirṇaya).
Benares (1956) 13038. Ff. 2–42. Incomplete.
Benares (1956) 13255. Ff. 1–15 and 27–31. Incomplete.
Benares (1956) 13348. 142ff. Ascribed to Sūryasena.
Benares (1956) 13937. 213ff. Ascribed to Sūryasena.
Berlin 2251 (or. fol. 1424). Ff. 74–75, 81–86, and 91–162. Incomplete.
BORI 335 of 1880/81. 189ff.
BORI 130 of 1895/1902. 128ff. No author mentioned.
Calcutta Sanskrit College (Smṛti) 86. 72ff. Ascribed to Gopīnārāyaṇa.
Florence 431 (Istituto di Studi Superiori 15). 420ff. Ascribed to Sūryasena.
GJRI 3514/152. 37ff. Incomplete.
GJRI 3515/153. 158ff. Maithilī.
GJRI 3516/154. 14ff. Maithilī. Incomplete.
GJRI 3517/155. 84ff. Incomplete.
IIL Oxford Stein 80. 337ff. Śāradā. Bought from Paṇḍita Dāmodara in 1889.
Kurukṣetra 511 (50683).
Mithila I 247. 210ff. Incomplete. Property of Babu Jagadīśa Jhā of Thārhī, Andhraṭharhī, Darbhanga.
Mithila I 248. 72ff. Maithilī. Incomplete. Property of Pandit Maṇīśvara Jhā of Lālaganj, Jhanjhārpur, Darbhanga.
Oudh XIII (1881) IX 11. 334 pp. Property of Mahanta Nanda Gopāla of Lucknow Zila.
Oudh XVIII (1885) IX 22. 472 pp. Property of Paṇḍita Gopīnātha of Lucknow Zila.
PrSB 795 (Göttingen Mu I 26). Ff. 27–222.
Tanjore D 18214 = Tanjore BL 129. 235ff.
Tanjore D 18215 = Tanjore BL 130. 205ff.
Tanjore D 18216 = Tanjore BL 131. 128ff.
Tanjore D 18217 = Tanjore JL 1349. 296ff. Incomplete.
Tanjore D 18218 = Tanjore JL 1350. 178ff.
Tanjore D 18219 = Tanjore TS 259. 76ff.

AŚVADHARA TRIPĀṬHIN

Author of a ṭīkā on a *Kṣaṇikagrahānāyanaśloka*. Manuscript:

IM Calcutta 1291. See NCC, vol. 1, rev. ed., p. 440, and vol. 5, p. 145.

**ĀTREYA*

Author of a *Nakṣatraparidyūna*. Manuscript:

Kurukṣetra 477 (19624).

**ĀDITYADEVA*

Additional manuscripts of his *Narapatijayacaryā* (see *CESS* A 1, 48a):

GOML Madras D 13939. Ff. 1–3. Incomplete (2, 1–50).
IO 6425 (Mackenzie II 43). 186 and 87ff. With the ṭīkā of Narahari (!). From Colin Mackenzie.
IO 6426 (Mackenzie III 236a). 36ff. Telugu. With an Āndhraṭīkā. Incomplete. From Colin Mackenzie.
IO 6427 (Mackenzie III 97). Ff. 54–159. Telugu. With the Āndhraṭīkā of Daivajñadāsa. From Colin Mackenzie.
IO 6428 (Mackenzie III 236b). 9ff. Telugu. With the Āndhraṭīkā of Daivajñadāsa. Incomplete. From Colin Mackenzie.

**ĀDITYABHAṬṬA* (*fl.* between 1200 and 1325)

Additional information on the manuscripts of his *Kālādarśa* (see *CESS* A 1, 48a–48b, and A 2, 14b–15a):

Benares (1956) 11956. Ff. 1–103, 105–113, and 115–218. Copied in Saṃ. 1544, Śaka 1409 = A.D. 1487. Incomplete. No author mentioned.
Benares (1956) 13534. 197ff. Copied in Saṃ. 1559 = A.D. 1502.
*AS Bengal 2655 (G 10442). 25ff. Copied at Śivapurī by the son of Viṣṇu Agnihotrin in the Manmathasaṃvatsara, Śaka 10057 (read 1457 = A.D. 1535).
Oudh (1879) IX 10. 60 pp. Copied in A.D. 1619. Property of Paṇḍit Śyām Lāl of Lucknow Zila.
*Tanjore D 18577 = Tanjore JL 1879. 219ff. Copied by Mallāribhaṭṭa, the son of Mahābaleśvara Śivabhaṭṭa, in Śaka 1590 = A.D. 1668.
Benares (1956) 14057. 8ff. Incomplete (parvadvayavinirṇaya). No author mentioned.
*GOML Madras D 3114. Ff. 1–20. Grantha. Incomplete.
*GOML Madras D 3115. 145ff. Telugu. Copied by Purāṇam Padmanābhuḍu. With a vyākhyā.
Tanjore D 18575 = Tanjore BL 663. 111ff.

**ĀDIŚARMAN* (*fl.* 1456)

Additional manuscript of his *Jātakāmṛta* (see *CESS* A 1, 49a, and A 2, 15a):

SOI 9515. (*Ādiśarmoktāyuḥ*).

**ĀPADEVA* (*fl.* before 1746)

Additional manuscript of his *Kheṭapīṭhamālā* (see *CESS* A 1, 49b–50a):

Kerala 4525 (9707). 40 granthas.

**ĀPASTAMBA*

His *Śulbasūtra* (see *CESS* A 1, 50a) with the ṭīkās of Kapardisvāmin, Karavinda, and Sundararāja was edited by Satya Prakash and Ram Swarup Sharma with an English translation by Satya Prakash, New Delhi 1968.

ĀRYABHAṬA (b. 476)

See also B. N. Prasad and R. Shukla [A 3. 1951]; P. Jhā [A 3. 1969]; and T. S. Kuppanna Sastri [A 3. 1969b].

Additional manuscript of his *Āryabhaṭīya* (see *CESS* A 1, 50b–54a, and A 2, 15b):

Jaipur (II). With the *Bhaṭadīpikā* of Parameśvara.

ĀŚĀDHARA (fl. 1132)

Additional manuscript of his *Grahajñāna* (see *CESS* A 1, 54b, and A 2, 16):

RORI Cat. III 15486. 38ff. (f. 4 missing). (*Āśādhar-īsāriṇī*).

INDRADATTOPĀDHYĀYA

Additional information on the manuscript of his *Jyotiṣaratnamālādīdhiti* (see *CESS* A 1, 55a).

*AS Bengal 2680 (G 6400). Ff. 15–28. Copied by Gaṅgādhara in Saṃ. 1843 = A.D. 1786. Incomplete (adhyāyas 12–18).

The colophon begins: iti śrīmadgargakulasukula-padavīkaśrīmuralīdharātmajamohanalālatanaya-śrīmadupādhyāyalālamaṇiśarmasūnunā kṣemāva-tīdevīgarbhasaṃbhavaśrīmadindradattopādhyā-yakṛtā.

INDRAVĀMADEVA

Additional manuscripts of his *Trailokyadīpaka* (see *CESS* A 1, 55a–55b, and A 2, 16a–16b):

LDI 2989 (169). 37ff. Copied in Saṃ. 1684 = A.D. 1627.
BORI 1084 of 1891/95. 133ff.

ĪŚA (fl. 1955)

An astrologer resident in Jālandhara, "Professor" Īśa wrote in Hindī a *Navatārikā* published at Jālandhara in 1955.

ĪŚVARACANDRA VIDYĀSĀGARA

Author of a *Bhūgolakhagolavarṇana*, edited by Nārāyaṇacandra Vidyāratna, 2nd ed., Calcutta 1893 (BM 14053. b. 30).

ĪŚVARADĀSA (fl. 1663)

Additional information on the manuscripts of his *Muhūrtaratna* (see *CESS* A 1, 55b):

*AS Bengal 2724 (G 864) = Mitra, Not. 1694. 84ff.

ĪŚVARADĀSA (fl. 1699).

Author of a *Grahaphalavicāra* in Hindī in Saṃ. 1756 = A.D. 1699. Manuscript:

NPS 159 of 1929–31. Copied in Saṃ. 1902 = A.D. 1845. Property of Bābū Kedāranātha Agravāla of Bāha, Āgarā.

ĪŚVARANĀTHA GARGA (fl. 1771)

A resident of Sareṭhī, Īśvaranātha wrote a *Raṇabhūṣaṇa* in Hindī in Saṃ. 1828 = A.D. 1771. Manuscript:

NPS 174 of 1923–25. Copied in Saṃ. 1878 = A.D. 1821. Property of Paṇḍita Śatrughna of Sikandarapura, Sisaiyā, Baharāīca.

ĪŚVARĪPRASĀDA

Author of a *Yogasāgara*. Manuscript:

Kurukṣetra 1284 (50121). Incomplete (sūtikādhyāya).

ĪŚVARĪPRASĀDA PĀṆḌEYA (fl. 1958)

Author of a Hindī ṭīkā on the *Gaurījātaka*; this was published with the mūla at Bambaī in 1958.

UTTAMADĀSA

Author of a *Sāmudrika* in Hindī. Manuscript:

NPS 200 of 1920–22. Copied in Saṃ. 1896 = A.D. 1839. Property of the Mahārāja Jagadambāpratāpasiṃha kā Pustakālaya in Ayodhyā.

UTTAMADĀSA MIŚRA

The son of Hīrāmaṇi Miśra, Uttamadāsa wrote a *Svarodaya* in Hindī. Manuscript:

NPS 340 A of 1906–08. Copied in Saṃ. 1940 = A.D. 1883. Property of Vihārī Sunāra of Ajayagaḍha. NPS notes another manuscript belonging to Lālā Jagatarāja of Ṭīkamagaḍha.

UDAYACANDA CAUBE (fl. 1773)

A resident of Āgarā, Udayacanda wrote a *Svarodaya* in Hindī in Saṃ 1830 = A.D. 1773. Manuscript:

NPS 434 of 1923–25. Copied in Saṃ. 1834 = A.D. 1777. Property of Paṇḍita Badrīnārāyaṇa Bhaṭṭa of the Lakhanaū Viśvavidyālaya in Lakhanaū.

UDAYASĀGARA (fl. 1599)

Additional manuscripts of his *Bālāvabodha* on the *Kṣetrasamāsa* (see *CESS* A 1, 58a, and A 2, 16b):

LDI 3040 (4913). 53ff. Copied in Saṃ. 1688 = A.D. 1631.

LDI 3041 (2643). 39ff. Copied in Saṃ. 1706 = A.D. 1649.
LDI 3044 (901). 56ff. Copied in Saṃ. 1826 = A.D. 1769.
LDI 3043 (1813). 36ff. Copied by Vīracandra at Daityāridurga under Vāmāñgajina in Saṃ. 1883 = A.D. 1826.
LDI 3042 (3529). 57ff.

UMĀ (*fl. ca.* 1400/1450)

The daughter of Ramārūpā and Mahādeva, the son of Kṛṣṇa, the son of Gaṇeśa of the Mudgalagotra, and a resident of Kheraḍa, Umā wrote a vyākhyā on the *Kālamādhava* of Mādhava (*fl. ca.* 1375). Her grandfather, apparently, was a pupil of Mādhava. Manuscript:

AS Bengal 2667 (G 8852). 19ff.

The first three verses are:

śrīmādhavaṃ guruṃ natvā lakṣmīr lakṣmīśiśuprabhuḥ/
kheraḍe mudgalāpatyagaṇeśāpatyakṛṣṇakaḥ//
mahādevaḥ sutas tasya vedamūrtir jaṭāntavit/
śrautasmārtārthanipuṇo dīkṣito rājapūjitaḥ//
patnī yasya ⟨ra⟩mārūpā sādhvy umā tasya kanyakā/
kālamādhavasadvyākhyāṃ tanute sarvasaṃvide//

UMĀDATTA JOŚĪ

Author of a ṭīkā, *Sudhādhavalā*, on the madhyamādhikāra of the *Sūryasiddhānta*; this was published at Hadiyāvāda in Kapūrthalā [ND].

URVĪDATTA (*fl.* 1923)

Brāhmaṇa author of a Hindī ṭīkā and udāharaṇa on the *Mukundapaddhati* of Mukunda (*fl.* 1922), which he completed on Monday 10 śuklapakṣa of Phālguna in Śaka 1844 = 26 February 1923. This was published with the mūla at Mumbayī in 1928.

ṚṢABHADEVA

Author of a *Ramalapraśnāvalī* in Hindī. Manuscript:

NPS 408 of 1926–28. Copied in Saṃ. 1912 = A.D. 1855. Property of Rāmaprasāda Murāū of Puravā Viśrāmadāsa, Pariyāvāṃ, Pratāpagaḍha.

ṚṢIKEŚA

A resident of Vṛndāvana, Ṛṣikeśa wrote a *Śanikathā* in Hindī. Manuscript:

NPS 190 B of 1932–34. Copied in Saṃ. 1916 = A.D. 1859. Property of Paṇḍita Dīpacandra, adhyāpaka at Bhāratagalī, Phatehapurasīkarī, Āgarā.

ṚṢIKEŚA (*fl.* 1761)

A resident of Āgarā, Ṛṣikeśa wrote a *Svarodaya* or *Ṣaṭprakāśa* in Hindī in Saṃ. 1808 = A.D. 1761. Manuscripts:

NPS 221 of 1906–08. Copied in Saṃ. 1920 = A.D. 1863. Property of Lālā Paramānanda of Purānī Teharī, Ṭīkamagaḍha.
NPS 165 of 1917–19. Property of Paṇḍita Candrasena Pujārī of Gañgājī kā Mandira, Khurajā, Bulandaśahara.
NPS 28 of Saṃ. 2001–2003. Property of Ambikādatta Śukla of Śeragaḍha, Mūratagañja, Ilāhābāda.

Ṛṣikeśa also wrote a *Kālajñāna* in Hindī. Manuscript:

NPS 127 of 1938–40. Property of Kṛṣṇaprasāda of Māṭa, Mathurā.

**OṂKĀRA BHAṬṬA* (*fl.* 1840/41)

Additional manuscript of the *Bhūgolasāra* of Oṃkāra, a resident of Astha, Mālavā (see *CESS* A 1, 60b, and A 2, 18a):

NPS 219 of 1909–11. Property of Lālā Mahādevaprasāda, ḥakīm and jyotiṣī of Managarī, Lakhanaū.

ORĪLĀLA ŚARMAN

Author of a *Ramalajātaka* = *Ramalasaṃhitā* = *Ramalārṇava* in Hindī. Manuscripts:

NPS 218 of 1909–11. Copied in Saṃ. 1957 = A.D. 1900. Property of Paṇḍita Ayodhyāprasāda Jyotiṣī of Sāgara Geṭa, Jhāṃsī.
NPS 79 of the Pañjāba Khoja Vivaraṇa, 1922–24.

**KAPARDISVĀMIN* (*fl.* before 1250)

His *Kapardibhāṣya* (see *CESS* A 2, 19b) was edited by Satya Prakash and Ram Swarup Sharma, New Delhi 1968.

KAPILEŚVARA ŚĀSTRIN CAUDHARĪ (*fl.* 1940/1948)

Gopīnātha Khauāla of Candrapura near Videhanagara in Mithilā was the ancestor of Rañjana, who went to the court of Nabāba Vādaśāha at Vyāghravāsa and took the surname Caudharī. Rañjana was the father of Veṇīdatta, whose sons were Bhagavaddatta, Kāśīdatta, and Gañgādatta. This last was the father of Navati and Girinātha. Girinātha married Jagadambā, the daughter of Sādhuśarman Budhavāra; their first son was Kapileśvara. When the son was eight years old the family moved to Vāsukīvihārī. Kapileśvara was patronized by Nārāyaṇadāsa, the lord of Corauta, and studied there under Śrīkānta; he became a professor at the Viśveśvara Catuṣpāṭhi Mahāvidyālaya in Kāśī and later at the

Jñānodaya Mahāvidyālaya in Patna. He wrote the following works:

1. *Amṛtadhārā*, a Hindī tīkā on the *Vanamālā* of Jīvanātha Jhā (*fl. ca.* 1850/1900); this work, written at Kāśī in Saṃ. 1997 = A.D. 1940, was published with the mūla as *HSS* 147, Banārasa 1941.

2. *Sudhāśālinī*, a ṭīkā on the *Jātakapārijāta* of Vaidyanātha (*fl. ca.* 1450); this work, written in Saṃ. 1999 = A.D. 1942, was published with the mūla and a Hindī ṭīkā by Mātṛprasāda Śāstrin as *KSS* 10, Banārasa 1942; 3rd ed., Banārasa 1953.

3. *Tattvāmṛta*, a ṭīkā on the *Sūryasiddhānta*; this work, written at Paṭanā in Saṃ. 2003 = A.D. 1946, was published as *KSS* 144, Banārasa 1946.

4. *Pañcāmṛta*, a ṭīkā with a Hindī rendering on the *Muhūrtamārtaṇḍa* of Nārāyaṇa (*fl.* 1571/1572); this work, written at Pāṭaliputra in Saṃ. 2004 = A.D. 1947, was published as *KSS* 145, Banārasa 1947.

5. *Maṇiprabhā*, a Hindī ṭīkā on the *Muhūrtacintāmaṇi* of Rāma (*fl.* 1600); this work, written at Kāśī in Saṃ. 2005 = A.D. 1948, was published as *HSS* 135, Banārasa 1948.

**KABĪRADĀSA = KABĪRA* (1398/1448)

A resident of Kāśī, and the pupil of Rāmānanda, and the teacher of Dharmadāsa, Kabīradāsa was born in Saṃ. 1455 = A.D. 1398 and died in Saṃ. 1505 = A.D. 1448. He wrote in Hindī the following two works on jyotiṣa.

1. *Bāragrantha*. Manuscript:

NPS 49 E of 1935–37. Copied in Saṃ. 1747 = A.D. 1690. Copy at the Kāśī Hindū Viśvavidyālaya kā Pustakālaya in Vārāṇasī.

2. *Svarodaya*. Additional manuscript (see *CESS* A 2, 19b):

NPS 21jha of 1941–43. Property of the Nāgarīpracāriṇī Sabhā at Vārāṇasī.

**KAMALĀKARA*

The AS Bombay manuscript of his *Jātakatilaka* was given the number 297 by mistake in *CESS* A 2, 20b; it should be AS Bombay 353.

**KAMALĀKARA* (*fl.* 1658)

Additional manuscripts of his *Siddhāntatattvaviveka* (see *CESS* A 2, 21a–23a):

Benares (1963) 36925. 30ff. Incomplete (with a vāsanā = *Śeṣavāśanā*?). No author mentioned.

**KAMALĀKARA BHAṬṬA* (*fl.* 1612)

Additional information on the manuscripts of his *Kālanirṇaya* (see *CESS* A 2, 23a–23b):

*AS Bombay 744. 59ff. Incomplete (*Sarvaśāstrārtha*). From Bhāu Dājī.
*Florence 120. 32ff. (*Tithinirṇaya*).

Additional manuscripts of his *Śāntikamalākara* (ese *CESS* A 2, 23b):

Jammu and Kashmir 4645. 5ff. Copied in Saṃ. 1741 = A.D. 1684. Incomplete.
AS Bengal 2184 (G 1935). 318ff. Copied by Yajñeśvara Dīkṣita Bhāgavata on 10 śuklapakṣa of Jyeṣṭha in Śaka 1680 = *ca.* 15 June 1758.
IO 1759 (160b). 206ff. Copied in A.D. 1801. From H. T. Colebrooke.
IO 1758 (178). 199ff. Copied in A.D. 1806. From H. T. Colebrooke.
AS Bombay 730. 47ff. Copied in Śaka 1770 = A.D. 1848. Incomplete (jananaśānti).
Jammu and Kashmir 4778. 235ff. Copied in Saṃ. 1928 = A.D. 1871.
Anup 2225. 46ff.
Anup 2226. 262ff. Property of Anantabhaṭṭa the son of Kamalākara.
Anup 2227. 211ff. (ff. 7 and 10–17 missing).
Anup 2228. 207ff. (ff. 112–142 missing).
*AS Bombay 729. 357ff. From Bhāu Dājī.
AS Bombay 731. 111ff. Incomplete. From Bhāu Dājī.
AS Bombay 732. 94ff. Incomplete (śatacaṇḍīsahasracaṇḍīprayoga).
Baroda 343. 414ff. (ff. 55–65 missing) (*Śāntiratna*).
Baroda 2286. 3ff. (vyatīpātādiśānti).
Baroda 9390. Ff. 3–181. Incomplete (*Śāntiratna*).
Berlin 1244 (Chambers 490). 189ff.
BORI 251 of 1884/87. 35ff. From Gujarāt.
BORI 306 of 1884/87. 169ff. (ff. 156 and 168 double). From Mahārāṣṭra.
BORI 170 of 1895/1902. 363ff.
Calcutta Sanskrit College (Smṛti) 364. 37ff. (*Śāntikaumudī*).
Calcutta Sanskrit College (Smṛti) 366. Ff. 77–140. Incomplete (*Śāntiratna*).

KAMALĀKĀNTA JHĀ (*fl.* 1938)

Īśvarīdatta, the astrologer of the adhipati of Kucavihāra, was the father of Yadunātha Śarman, who spent five years at the court of Rameśa, the lord of Mithilā, and then went to the court of Viśvanātha, the lord of Chatrapura. Yadunātha had five sons: Aniruddha of Surapurī, Luṭṭī, Devakānta, Kamalākānta, and Sūryakānta. Kamalākānta, the pupil of Durgādatta, taught at the Śyāmābhavana Saṃskṛta Vidyālaya in Māṇḍūkīyā in Darabhaṅga, Mithilā, and wrote Sanskrit, *Vimalā*, and Hindī, *Saralā*, ṭīkās on the *Praśnabhūṣaṇa* of Jīvanātha Jhā (*fl. ca.* 1846/

1900) as well as a pariśiṣṭa which he completed on Wednesday 15 śuklapakṣa of Mārgaśīrṣa in Saṃ. 1995 = 7 December 1938. These were published with the mūla as *HSS* 131, Banārasa 1941; 2nd ed., Banārasa 1954.

KAMALĀKĀNTA ŚUKLA (*fl.* 1968/1969)

Author of a *Bṛhadavakahaḍācakra*, which contains an example dated Saṃ. 2026, Śaka 1890 = A.D. 1968/69. Together with the author's Hindī vyākhyā, *Bālabodhinī*, this was edited by Avadhavihārī Tripāṭhī, *VSG* 154, Vārāṇasī 1970.

**KARAVINDASVĀMIN*

His *Śulbapradīpikā* (see *CESS* A 2, 24a) was edited by Satya Prakash and Ram Swarup Sharma, New Delhi 1968.

**KARKA*

An inconclusive discussion of the age of his *Karkabhāṣya* (see *CESS* A 2, 24a) is found in B. R. Kulkarni [A 3. 1943].

**KALYĀṆAVARMAN* (*fl. ca.* 800)

Additional manuscripts of his *Sārāvalī* (see *CESS* A 2, 26a–29a):

GJRI 1124/236. 38ff. Incomplete.
GJRI 3250/462. Ff. 1–88 and 90–100. Incomplete (ends in adhyāya 52).
Kurukṣetra 1246 (19587).
Kurukṣetra 1247 (19659).
LDI (VDS) 1317 (9730/2). Ff. 2v–4. Incomplete (adhyāya 35). No author mentioned.

**KAVICŪḌĀMAṆI*

Additional manuscript of his *Sūryasiddhāntanavanīta* (see *CESS* A 2, 29b):

Benares (1963) 34653. 6ff. Copied in Saṃ. 1906 = A.D. 1849. Incomplete (candrasūryagrahaṇādhikāra). Ascribed to Cakravartin.

**KAVICŪḌĀMAṆI* (*fl. ca.* 1620)

Additional manuscripts of his *Jyotiṣakalpataru* (see *CESS* A 2, 29a–29b):

Bharatpur S 10. No author mentioned.
Dharwar 700(690). 139ff. No author mentioned.
Kurukṣetra 357(19630). With a *Rogāvalī*.

KAVIPATI

Author of a *Tattvapañcāśikā*. Manuscript:

Kurukṣetra 364 (58).

**KAVĪNDRA KṚṢṆA* (*fl. ca.* 1625/75)

Additional manuscripts of his *Padyapañcāśikā* = *Tattvapradīpajātaka* (see *CESS* A 2, 30a):

Benares (1963) 34667. 7ff. Copied in Saṃ. 1855 = A.D. 1798. No author mentioned.
Bharatpur S 3. No author mentioned.

**KAVĪNDRĀCĀRYA SARASVATĪ* (*fl. ca.* 1600/75)

Author (see *CESS* A 2, 30a) of a *Samarasāra* in Hindī. Manuscript:

NPS 39 of 1904. Copied in Saṃ. 1833 = A.D. 1776. Property of the Mahārāja Banārasa kā Pustakālaya at Rāmanagara, Vārāṇasī.

See also D. R. Chanana [A 3. 1969] and M. P. Paradkar [A 3. 1969].

KĀNHA DVIJA (*fl.* 1878)

Author of a *Jyotissārāvalī* in Hindī in Saṃ. 1935 = A.D. 1878. Manuscript:

NPS 29 of Saṃ. 2004–2006. Property of Paṇḍita Rāmabakasa Miśra of Udayīpura, Pilakichā, Jaunapura.

**KĀMADHA*

This is the abbreviation for the *Kāmadhenupaddhati* of Jayarāma Bhaṭṭa, not the name of an author as in *CESS* A 2, 31a; see NCC, vol. 3, p. 351.

**KĀLIDĀSA* (*fl.* eighteenth century?)

His *Uttarakālāmṛta* (see *CESS* A 2, 34b) was edited with his own Hindī vyākhyā by Jagannātha Bhasīna, Dillī Saṃ. 2028 = A.D. 1971.

KĀŚĪDĀSA

Author of a *Jyotiṣa* in Hindī. Manuscript:

NPS 226 of 1926–28. Copied in Saṃ. 1784 = A.D. 1727. Property of Paṇḍita Śivakaṇṭha Dūbe of Devadārupura, Khīrī.

**KĀŚĪNĀTHA*

Additional manuscripts of his *Praśnapradīpa* (see *CESS* A 2, 35b–36b):

GJRI 978/90. 7ff. Incomplete.
GJRI 980/92. Ff. 11–13. Incomplete.
GJRI 981/93. 2ff. Incomplete.
GJRI 1144/256. 7ff. Maithilī.
GJRI 3178/390. 17ff.
Kurukṣetra 649 (19634)
Kurukṣetra 650 (19868). No author mentioned.
Kurukṣetra 651 (19771).
Kurukṣetra 652 (50131).

KĀŚĪNĀTHA

Additional manuscripts of his *Lagnacandrikā* (see *CESS* A 2, 36b–39a):

GJRI 1070/182. Ff. 24–30 and 33–38. Copied in Saṃ. 1877 = A.D. 1820. Incomplete.
GJRI 1172/284. Ff. 6–22. Incomplete.
GJRI 2987/320. Ff. 1–8 and 10–30. Maithilī. Incomplete.
GJRI 3213/425. 64ff. Incomplete.

The *Lagnacandrikā* with the bhāṣāṭīkā of Rāma Vihāri Sukula was edited by Śivadayālu Pāṇḍeya, 13th ed., Lakhanaū 1968.

KĀŚĪNĀTHA

Additional manuscripts of his *Śīghrabodha* (see *CESS* A 2, 39a–44a):

GJRI 1106/218. Ff. 2–45. Copied in Saṃ. 1702 = A.D. 1645. Incomplete.
Leningrad (1914) 303 (Ind. II 97). 11ff. Copied on Monday 12 śuklapakṣa of Phālguna in Saṃ. 1777 = 27 February 1721 Julian.
GJRI 1108/220. 35ff. Copied in Saṃ. 1793 = A.D. 1736.
AS Bengal 2758 (G 6352). 18ff. Copied on Sunday 7 kṛṣṇapakṣa of Pauṣa in Saṃ. 1814 = 29 January 1758.
WHMRL B. 5. f. Ff. 2–3, 8–12, 14–19, 22–30, 35–37, 40–43, 45–58, 60–62, 64, and 66–68. Copied on Sunday 13 śuklapakṣa of Āśvina in Saṃ. 1845, Śaka 1705 = 12 October 1788. Incomplete.
GJRI 1102/214. 36ff. Copied in Saṃ. 1846 = A.D. 1789.
AS Bengal 2762 (G 9353). 30ff. Copied on 5 śuklapakṣa of Kārttika in Saṃ. 1847 = *ca.* 10 November 1790.
GJRI 3232/444. 29ff. Copied in Saṃ. 1870 = A.D. 1813.
AS Bengal 2759 (G 9861). 40ff. Copied on 7 śuklapakṣa of Mārgaśīrṣa in Saṃ. 1872 = *ca.* 7 December 1815.
GJRI 1100/212. Ff. 9–29. Copied in Saṃ. 1879 = A.D. 1822. Incomplete.
GJRI 1105/217. Ff. 48 and 50–53. Copied in Saṃ. 1883 = A.D. 1826. Incomplete.
AS Bengal 2764 (G 9620). 51ff. Copied in Saṃ. 1892, Śaka 1757 = A.D. 1835.
*WHMRL G. 3. f. 24ff. Copied by Devacanda on a Tuesday in Āśvina in Saṃ. 1907 = A.D. 1850.
AS Bengal 2760 (G 4305). 19ff.
AS Bengal 2761 (G 7781). 39ff.
AS Bengal 2763 (G 9254). 11ff. Incomplete (vivāhaprakaraṇa).
GJRI 1101/213. 13ff. Incomplete (ends in prakaraṇa 2).
GJRI 1103/215. 11ff. Incomplete.
GJRI 1104/216. 28ff. Maithilī. Incomplete.
GJRI 1107/219. Ff. 2–12. Maithilī. Incomplete.
GJRI 1176/288. Ff. 4, 9–11, and 14. Incomplete.
GJRI 3221/433. 29ff. Maithilī. Incomplete. (*Laghubodhasaṅgraha*).
GJRI 3229/441. 9ff. Maithilī. Incomplete.
GJRI 3230/442. 13ff. Incomplete.
GJRI 3231/443. 20ff. Incomplete.
GJRI 3233/445. 37ff. Incomplete.
GJRI 3234/446. 16ff. Incomplete.
GJRI 3235/447. 26ff. Incomplete.
GJRI 3236/448. 31ff.
GJRI 3237/449. 5ff. Maithilī. Incomplete.
GJRI 3265/477. 16ff. Maithilī. Incomplete.
GJRI 3266/478. 14ff. Maithilī. Incomplete.
Kurukṣetra 1114 (19589).
Kurukṣetra 1115 (19863).
Kurukṣetra 1116 (50363).

KĀŚĪNĀTHA BHAṬṬĀCĀRYA

The manuscript of his *Muhūrtamuktāvalī* is (see *CESS* A 2, 44a):

Kunte A 19. 5ff. Copied in A.D. 1819. Property of Pandit Gulāb Sinha of Delhi.

KĀŚĪNĀTHA BHAṬṬA (*fl.* seventeenth or eighteenth century)

Additional manuscript of his *Kālanirṇayadīpikā* (see *CESS* A 2, 44a):

Benares (1956) 13978. 8ff.

Additional manuscripts of his *Tithinirṇayadīpikā* (see *CESS* A 2, 44b):

Benares (1956) 13911. 13ff. Incomplete.
Benares (1956) 13912. Ff. 1–11, 14–21, and 23–27. Incomplete.

KĀŚĪNĀTHA UPĀDHYĀYA (d. 1805).

Bhāskara Upādhyāya (or Pādhye) of Golavali in the Ratnagiri district of Koṅkaṇa was the father of Nāro, the father of Ananta, the father of Kāśī Upādhyāya, the father of Yajñeśvara and Ananta. Ananta moved to Pāṇḍuraṅga (Pandharpur on the Bhīmā) where, by his wife Annapūrṇa, he became father of Kāśīnātha or Bābā and of Viṭṭhala (d. *ca.* 1825); Ananta died in Śaka 1696 = A.D. 1774. Kāśīnātha wrote a *Dharmasindhusāra* in Śaka 1712 = A.D. 1790, and died in Śaka 1727 = A.D. 1805. See P. V. Kane [1930/62] vol. 1, pp. 463–465. Manuscripts:

Baroda 1192. 84ff. Incomplete (ends in adhyāya 2).
Tanjore D 18153 = Tanjore BL 394. 24ff. Incomplete (pt. I).
Tanjore D 18154 = Tanjore BL 395. 24ff. Incomplete (pt. I).

Tanjore D 18155 = Tanjore BL 396. 24ff. Incomplete (pt. I).
Tanjore D 18156 = Tanjore BL 397(1). 41ff. Incomplete (pt. I).
Tanjore D 18157 = Tanjore TS 217. 30ff. Incomplete (pt. I).
Tanjore D 18158 = Tanjore SK 154. 22ff. Incomplete (pt. I).
Tanjore D 18159 = Tanjore BL 397(2). 92ff. Incomplete (pt. II).
Tanjore D 18160 = Tanjore TS 218. 68ff. Incomplete (pt. II).
Tanjore D 18161 = Tanjore SK 155. 58ff. Incomplete (pt. II).
Tanjore D 18162 = Tanjore BL 397(3). 355ff. Incomplete (pt. III).
Tanjore D 18163 = Tanjore TS 219. 254ff. Incomplete (pt. III).
Tanjore D 18164 = Tanjore SK 156 + 157. 205ff. Incomplete (pt. III).

The *Dharmasindhusāra* has been published:

at Bombay in Śaka 1772 = A.D. 1850 (IO 20. K. 13);
at Poona in Śaka 1782 = A.D. 1860 (BM);
at Puṇya in Śaka 1783 = A.D. 1861 (BM and IO 13. E. 16), 2nd ed. Poona 1870 (IO 17. B. 2);
at Bombay in Saṃ. 1926 = A.D. 1869 (IO 14. B. 7);
at Poona in 1870 (IO 13. E. 35);
at Ratnagiri in 1872 (IO 24. D. 9);
at Bombay in Śaka 1796 = A.D. 1874 (IO 24. D. 26);
with the Marāṭhī bhāṣāntara of Bāpuśāstri Moghe, at Bombay in Saṃ. 1931 = A.D. 1874 (IO 26. G. 7);
at Bombay in 1879 (IO 13. E. 27);
at Poona in 1882 (IO 13. E. 4);
with the Hindī translation of Ravidatta Śāstrī, at Mumbaī in Saṃ. 1948 = A.D. 1891 (BM 14033. bb. 39);
at Bangalore in 1892 (BM 14028. d. 46);
at Bombay in Saṃ. 1964 = A.D. 1907 (IO 22. I. 12);
with a Marāṭhī bhāṣāntara, edited by Yajñeśvara Gopāla Dīkṣita, Puṇeṃ 1911 (BM 14027. d. 4 and IO 23. I. 17);
with the Marāṭhī bhāṣāntara of Lakṣmaṇa Nārāyaṇa Jośī, Poona [1925] (IO San. D. 403); and
with the Hindī ṭīkā, *Dharmadīpikā*, of Vaśiṣṭhadatta Miśra, and the ṭippaṇī, *Sudhā*, of Sudāmā Miśra Śāstrī, as *KSS* 183, Vārāṇasī 1968.

KĀŚĪNĀTHA VĀSUDEVA ABHYAÑKARA (*fl.* 1944/1962)

The son of MM. Vāsudeva Abhyañkar, Professor at Ferguson College in Poona, and great-grandson of Bhāskara Śāstrin Abhyañkara, who served the Marāṭha court at Saptarṣipattana (Sātārā), Kāśīnātha Abhyañkara wrote several articles on jyotiḥśāstra (see bibliography) and served first at the Rājakīya Pāṭhaśālā in Ahmadabad, later at the Bhandarkar Oriental Research Institute in Poona. He finished a commentary, *Marīci*, on I 1–III 3 of the *Upadeśasūtra* of Jaimini at Ahmadabad on Monday 15 śuklapakṣa of Kārttika in Saṃ. 2002 = 19 November 1945. This was published in his *The Upadeśa Sūtra of Jaimini*, Ahmedabad 1951.

KĀŚĪRĀJA (*fl.* 1832)

The son of Mahārāja Cetasiṃha and a resident of Kāśī, Kāśīrāja wrote a *Muṣṭikapraśna* in Hindī. Manuscript:

NPS 189 B of 1929–31. Copied in Saṃ. 1802 (read 1902?) = A.D. 1845 (?). Property of Paṇḍita Rāmabhajana Miśra of Behadarakalāṃ, Saṇḍīlā, Haradoī.

KĀŚĪRĀMA

Author of a Hindī ṭīkā on the *Upadeśasūtra* of Jaimini. Manuscript:

NPS 110 B of 1932–34. Property of Paṇḍita Gaṇeśaprasāda Vyāsa of Madāna, Mainapurī.

KĀŚĪRĀMA (*fl.* 1613)

A Pāṭhaka Brāhmaṇa residing in Kāśī, Kāśīrāma wrote a *Lagnasundarī* in Hindī in Saṃ. 1670 = A.D. 1613. Manuscript:

NPS 110 A of 1932–34. Copied in Saṃ. 1971 = A.D. 1914. Property of Lāla Mukuṭavihārīlāla Guptā of Kaṭarābājāra, Śikohābāda, Mainapurī.

**KĀŚĪRĀMA VĀCASPATI BHAṬṬĀCĀRYA* (*fl. ca.* 1650/1700)

Additional manuscripts of his *Malamāsatattvaṭīkā* (see *CESS* A 2, 45a–45b):

AS Bengal 1966 (G 1561). 68ff. Bengālī. Copied in Śaka 1723 = A.D. 1801.
Calcutta Sanskrit College (Smṛti) 105. 90ff. Bengālī. Copied in Śaka 1756 = A.D. 1834.
Benares (1956) 14208. Ff. 1–32 and 32b–36. Bengālī. Incomplete.

**KĀŚĪRĀMA PĀṬHAKA* (*fl.* 1907)

Besides the ṭīkā on the *Vivāhavṛndāvana* (see *CESS* A 2, 45b) Kāśīrāma wrote a Hindī ṭīkā on the *Laghujātaka* of Varāhamihira (*fl. ca.* 550); this was published at Bambaī in Saṃ. 1993, Śaka 1858 = A.D. 1936.

KUTUB KHĀN

Author of a *Ratnajātaka*. Manuscript:

Kunte A 16. 10ff. Property of Pandit Jwālā Datta of Gujrānwāla.

KŪRMA

A resident of Pāranera, which is said to be Pārāśarapura, Kūrma wrote a *Dharmanibandha*. A section of this is the *Tithinirṇaya*. Manuscript:

AS Bengal 2228 (G 5451). 16ff. Copied by Dhanañjaya, the son of Śūdra Vīradeva, on Tuesday 10 kṛṣṇapakṣa of Pauṣa in NS 872, Śaka 1673 = 1 December 1752 Julian.

The last verse is:

pārāśarapuraṃ cāsti pāranera iti smṛtam/
tatrasthena ca kūrmeṇa racitaṃ dharmabandhanam//

KṚPĀRĀMA (*fl.* 1715)

A Nāgara Brāhmaṇa and a protégé of Savāī Jayasiṃha (1686/1743), Kṛpārāma wrote a *Samayabodha* in Hindī in Saṃ. 1772 = A.D. 1715. Manuscripts:

NPS 156 of 1909–11. Property of Bālagovinda Halavāī of Navābagañja, Bārābaṅkī.
NPS 245 B of 1926–28. Property of Bābū Jayamaṅgalarāya of Gājīpura.

**KṚPĀRĀMA* (*fl.* 1735)

A Kāyastha resident in Sāhajahāṃpura, Kṛpārāma wrote the Hindī version of his *Jyotiṣasāra* (see *CESS* A 2, 47b–48a) in Saṃ. 1792 = A.D. 1735. Additional manuscript:

NPS 182 of 1906–08. Copied in Saṃ 1909 = A.D. 1852. Property of the Bijāvaranarésa kā Pustakālaya of Bijāvara.

**KṚPĀRĀMA MIŚRA* (*fl.* 1792)

Benares 35298 in *CESS* A 2, 48b is an error for Benares 35289.

**KṚṢṆA*

Additional manuscripts of his *Triṃśadyogāvalī* (see *CESS* A 2, 51a):

BORI 894 of 1891/95. 36ff. Incomplete (*Yogāvalī* of Śrīkṛṣṇa).

**KṚṢṆA*

Additional information on the manuscript of his *Prabhā* (see *CESS* A 2, 51b):

*AS Bengal 2649 (G 8109). 223ff. Bengālī. Copied by Candraśekhara Śarman in Śaka 1608 = A.D. 1686.

The first verse is:

praṇamyāsārasaṃsārapārāvāratariṃ harim/
kṛṣṇācāryo vitanute dīpikāyāḥ prabhām imām//

KṚṢṆA

Author of a vyākhyā on the *Bṛhajjātaka* of Varāhamihira (*fl. ca.* 550). Manuscript:

Baroda 13350. 178ff. Nandināgarī.

**KṚṢṆA BHAṬṬA*

Additional manuscripts of his *Kālacandrikā* (see *CESS* A 2, 52a):

*Oudh III (1873) IX 1. 68 pp. Copied in A.D. 1792. Property of Paṇḍit Chhoṭe Lāla of Oonao Zillah.
Benares (1956) 12521. 30ff.
Benares 1956) 13839. 28ff.
*Oudh (1879) IX 11. 76 pp. Property of Paṇḍit Śyām Lāl of Lucknow Zila.

**KṚṢṆA BHAṬṬA*

Additional manuscripts of his *Cūḍāratna* (see *CESS* A 2, 52a–52b):

AS Bengal 2639 (G 8190). 40ff. Copied on Thursday 8 kṛṣṇapakṣa of Pauṣa in Śaka 1482 = 9 January 1561. Incomplete (vivāhapaṭala). No author mentioned.
Ānandāśrama 4264.
Ānandāśrama 4266.

KṚṢṆA MIŚRA

Author of a *Joginī daśā vicāra* in Hindī. Manuscript:

NPS 124 A of 1932–34. Copied in Saṃ. 1844 = A.D. 1787. Property of Paṇḍita Bāṅkelāla of Śikohābāda, Mainapurī.

He also wrote a *Praśnavicāra* in Hindī. Manuscript:

NPS 124 B of 1932–34. Property of Paṇḍita Bāṅkelāla of Tāḍhūpura, Śikohābāda, Mainapurī. (Kṛṣṇajū Miśra).

KṚṢṆA (*fl.* 1686)

The son of Nīlakaṇṭha, Kṛṣṇa (or Śrīkṛṣṇa) wrote a commentary, *Marīci*, on the *Muhūrtacintāmaṇi* of Rāma (*fl.* 1600) in Śaka 1608 = A.D. 1686. Manuscripts:

AS Bengal 2714 (G 6440A). Ff. 1–22 and 24–57. Incomplete.
AS Bengal 2716 (G 6440B). 36ff. Incomplete (ends in prakaraṇa 11).

Verses 1 and 3 are:

athaikadantaṃ harijīvanaṃ ca
śrīnīlakaṇṭhaṃ pitaraṃ ca natvā/
karomi kaṇṭhābharaṇāya yogyaṃ
muhūrtacintāmaṇim apy anargham//
ato ᵓyam udyogabharo ᵓsti jātaḥ

śrīkṛṣṇanāmnaḥ prathitānvayasya/
tad atra sujñāḥ kramasaṃskṛtaṃ taṃ
mātsaryam utsārya vilokayantu//

The date of composition is given in the verse:

gajābhrāstimite śāke divākaravinodataḥ/
gurupādābjaniratah śrīkṛṣṇo vyalikhat kramāt//

This Kṛṣṇa is probably identical with Kṛṣṇa, the son of Nīlakaṇṭha, the son of Śivadāsa, who was the pupil of Keśava and revised the *Śīghrabodha* of *Kāśīnātha* (*fl.* before 1559) in seven prakaraṇas:

1. nakṣatra.
2. śubhāśubha.
3. gocaragarbhādhānādisaṃskāra.
4. vivāha.
5. miśra.
6. yātrāvastuveśmapraveśa.
7. name missing.

Manuscript:

*AS Bengal 2765 (G 6395). 19ff. Copied in Saṃ. 1846 = A.D. 1789. See *CESS* A 2, 41a.

The first verse is:

śrīguruṃ keśavaṃ natvā śrīkṛṣṇena kramādimāḥ/
kāśīnāthoktayo muktā bhūṣyante ratnamālayā//

The colophon begins: iti śrīmacchivadāsātmajanīlakaṇṭhatanujaśrīkṛṣṇena kṛtakrame śīghrabodhe.

KṚṢṆADATTA JHĀ (*fl.* 1804)

Bhībhāīnātha was the father of Bavue, Nena, Kailū, and Mukunda; and Bavue was the father of Sone, Nandalāla and Kṛṣṇadatta. This last wrote a ṭīkā, *Subodhinī*, on the *Pañcasvarāḥ* of Prajāpatidāsa at Īśapura in Śaka 1726 = A.D. 1804; this was published with the mūla and the *Saralā* of Govinda Śarman (*fl.* 1940) at Banārasa, 2nd ed. Saṃ. 1998 = A.D. 1941.

**KṚṢṆAMITRA* (= *KṚṢṆAMIŚRA*)

Additional information on manuscripts of his *Kālamārtaṇḍa* (see *CESS* A 2, 58b):

*AS Bengal 2769 (G 2908). 18ff. Copied in Saṃ. 1857 = A.D. 1800. Purchased on Thursday 12 kṛṣṇapakṣa of Māgha in Saṃ. 1893 = 2 March 1837 from Pāṇḍe Iṭāra Rāmajī Yāvana, a resident of Iskandare Vīrapuragrāma.
*AS Bengal 2770 (G 10127). 7ff. Copied on Friday 12 śuklapakṣa of Bhādrapada in Saṃ. 1885 = 19 September 1828. Incomplete (tithinirṇaya).
Benares (1956) 13262. 10ff.
*Oudh IX (1877) IX 4. 20 pp. (*Tithinirṇayamārtaṇḍa*). Property of Paṇḍita Śarayūprasāda of Fyzābād Zillah.
*Oudh XX (1888) IX 96. 28 pp. Property of Rāma Svarūpa of Gonda Zila.

KṚṢṆASIṂHA

Author of a *Svapnādhyāya* in Hindī. Manuscript:

NPS 224 of 1923–25. Copied in Saṃ. 1892 = A.D. 1835. Property of Ṭhākura Maheśasiṃha Kohalī of Becaīsiṃha kā Puravā, Kesaragañja, Baharāīca.

KṚṢṆĀNANDA

Author of a ṭīkā on an *Āyurdāyagaṇanā*. Manuscript:

Rajshahi, Varendra Res. Soc. 669. See NCC, vol. 5, p. 12.

**KṚṢṆĀNANDA SARASVATĪ*

Additional manuscripts of his *Jaiminisūtraṭīkā:* (see *CESS* A 2, 61b–62a):

Jaipur (II). 96ff. Copied in Saṃ. 1779 = A.D. 1721.
Benares (1963) 36194. Ff. 1–2, 5–66, and 69–88, and 2ff. Incomplete.
Jaipur (II). 91ff.

**KEDĀRADATTA JOŚĪ* (1961/1968)

Author (see *CESS* A2, 62a–62b) also of a *Gaṇitapraveśikā*, published at Dillī-Vārāṇasī-Paṭanā in 1967, and of a *Jyautiṣa meṃ svaravijñāna kā mahattva*, published at Dillī-Paṭanā-Vārāṇasī in 1968.

**KEVALARĀMA PAÑCĀNANA* (*fl.* 1728/1762)

Additional manuscripts of his *Grahacāra* (see *CESS* A 2, 63b):

Calcutta, Saratkumar Ray 407. See NCC, vol. 6, p. 246.
Sūcīpattra 16. See NCC.

Additional manuscript of his *Dṛkpakṣasāraṇī*, composed for Savāī Jayasiṃha (1686/1743) (see *CESS* A 2, 63b):

Calcutta Sanskrit College 55. 26ff.

**KEŚAVA*

The AS Bengal manuscript of his *Vyavahārasāra* (*sic;* the title as given in *CESS* A 2, 64a is wrong) calls him Keśavārka; this work may, then, be the *Brahmatulyasāra* of Keśavārka listed in *CESS* A 2,

75a. There are ten prakaraṇas in the *Vyavahārasāra*:

1. tithi.
2. vāra.
3. yoga.
4. nakṣatra.
5. karaṇa.
6. saṅkrānti.
7. gocara.
8. rāśikūṭa.
9. vivāhatyājya.
10. vivāha.

**KEŚAVA SOMAYĀJIN BHĀRADVĀJA*

Additional information on the manuscripts of his *Nakṣatreṣṭiprayoga* (see *CESS* A 2, 65b):

*AS Bengal 633 (G 247). 27ff. Copied on Tuesday 4 kṛṣṇapakṣa of Mārgaśīrṣa in Saṃ. 1882 = 27 December 1825.
*AS Bengal 634 (G 10439). 19ff. A note on f. 19v connects Keśava with Mālavīya Bhāradvājī Rāmacandra Somayājin of Naimiṣa (*fl.* 1447/1449).
Benares (1953) 3377. 42ff.
Benares (1953) 3500. 6ff. (*Nakṣatreṣṭayaḥ* of Keśava Svāmin).

The last verse is:

nakṣatreṣṭīḥ prāha baudhāyanas tu
tatra cāpastambasūtroktamārge/
bālānāṃ tadbodhanārthaṃ jagāda
bhāradvājaḥ keśavaḥ somayājī//

**KEŚAVA* (*fl.* 1496/1507)

Additional manuscript of his *Grahakautuka* (see *CESS* A 2, 66a):

Dāhilakṣmī XXXIII 42. See NCC, vol. 6. p. 244.

Additional manuscripts of his *Jātakapaddhati* (see *CESS* A 2, 66b–70b):

GJRI 924/36. 6ff. Copied in Saṃ. 1922 = A.D. 1865.
PrSB 967 (Göttingen, Mu II 15). Ff. 2–4 and 31–51. Śāradā. Copied Wednesday 6 śuklapakṣa of Phālguna in (Saptarṣi) Saṃ. ⟨49⟩ 58 = 14 March 1883. With the udāharaṇa of Viśvanātha.
GJRI 898/10. 6ff. Maithilī.
GJRI 899/11. 5ff. Maithilī. Incomplete.
GJRI 900/12. Ff. 3–4. Maithilī. Incomplete.
GJRI 923/35. 7ff. Incomplete.
GJRI 3125/337. 8ff. Maithilī. Incomplete.
Kurukṣetra 187 (50366).
Kurukṣetra 334 (19543).

Additional information on manuscripts of his *Muhūrtatattva* (see *CESS* A 2, 72a–73b):

*Oxford CS c. 315 (ix). 14ff. Incomplete.

**KEŚAVA KAVĪNDRA* (*fl. ca.* 1550?)

Additional manuscripts of his *Saṅkhyāparimāṇanibandha* (see *CESS* A 2, 64b–65a):

Mithila I 410 D. 56ff. Maithilī. Copied in Śaka 1657 = A.D. 1735. Property of Pandit Premdhar Jhā of Ujan, Jhañjhārpur, Darbhanga.
Mithila I 410 F. 35ff. Maithilī. Copied in Śaka 1750 = A.D. 1828. Property of Pandit Jībanāth Jhā of Lagamā, Biraul, Darbhanga.
Mithila I 410 A. 31ff. Maithilī. Copied by Rāmadatta Śarman of Rahuāgrāma on Monday 10 śuklapakṣa of Śrāvaṇa in Śaka 1801, Sāl. Saṃ. 1287 = 28 July 1879. Property of the Rāj Library, Darbhanga.
Mithila I 410 I. 46ff. Maithilī. Copied in Śaka 1818 = A.D. 1896. Property of Pandit Bālakṛṣṇa Jhā of Nanaur, Tamuria, Darbhanga.
Mithila I 410 E. 28ff. Maithilī. Copied in Śaka 1835 = A.D. 1913. Property of Pandit Kapileśvar Jhā of Sakhabad, Manīgāchī, Darbhanga.
IO 5513 (1348b). 6ff. Bengālī. Incomplete. From H. T. Colebrooke.
Mithila I 410 32ff. Maithilī. Property of Pandit Rabināth Jhā, Professor at M. M. Vidyālaya and resident of Andhraṭharhi, Darbhanga.
Mithila I 410 B. 33ff. Maithilī. Incomplete. Property of the Rāj Library, Darbhanga.
Mithila I 410 C. 36ff. Maithilī. Property of the Rāj Library, Darbhanga.
Mithila I 410 G. 40ff. Maithilī. Property of Pandit Śrīkānt Jhā of Naḍuār, Jhañjhārpur, Darbhanga.
Mithila I 410 H. 45ff. Maithilī. Property of Pandit MM. Rājināth Miśra of Saurāth, Madhubani, Darbhanga.
Mithila I 410 J. 20ff. Maithilī. Incomplete. Property of Pandit Rāghava Jhā of Andaulī, Mādhavapur, Darbhanga.
Mithila I 410 K. 41ff. Maithilī. Property of Pandit Rāghava Jhā of Andaulī, Mādhavapur, Darbhanga.
Mithila I 410 L. 41ff. Maithilī. Property of Pandit Tārānāth Jhā of Dharmapur, Jhañjhārpur, Darbanga.
Mithila I 410 M. 41ff. Maithilī. Property of Pandit Gaṇānand Jhā of Lohnā, Jhañjhārpur, Darbhanga.
Mithila I 410 N. 24ff. Maithilī. Property of Pandit Dāmodar Jhā of Andhrāṭhārhi, Darbhanga.

Keśava may be identical with the Kāyastha Mazumdar who administered Tirabhukti from about 1546 to 1557.

KEŚAVAPRASĀDA DŪBE (or *DVIVEDIN*) (*fl.* 1840/1873)

The son of Paramasukha and a resident of Āgarā, Keśavaprasāda wrote the following works in Hindī on jyotiḥśāstra.

1. An *Aṅgasphuraṇa* in Saṃ. 1926 = A.D. 1869

Manuscript:

NSP 193 A of 1929–31. Copied in Saṃ. 1931 = A.D. 1874. Property of Paṇḍita Kāśīrāma Jyotiṣī of Rijaura, Eṭā.

2. A. *Jyotiṣasāra* in Saṃ. 1930 = A.D. 1873. Manuscripts:

NPS 193 D of 1929–31. Copied in Saṃ. 1933 = A.D. 1876. Property of Lāla Jayanārāyaṇa of Nagalārājā, Naukheḍā, Eṭā.
NPS 193 E of 1929–31. Copied in Saṃ. 1936 = A.D. 1879. Property of Paṇḍita Śiva Śarmā of Nagarādhīra, Sarāya Agata, Eṭā.
NPS 230 A of 1926–28. Copied in Saṃ. 1939 = A.D. 1882. Property of Rāyalāla of Ramuāpura, Dhauraharā, Khīrī.
NPS 230 B of 1926–28. Copied in Saṃ. 1939 = A.D. 1882. Property of Paṇḍita Manīlāla Tivārī of Gaṅgāputra, Miśrikha, Sītāpura.
NPS 193 C of 1929–31. Copied in Saṃ. 1939 = A.D. 1882. Property of Paṇḍita Rāmakumāra Miśra of Basīṭha, Kāsagañja, Eṭā.

3. A *Mayūracitra* in Saṃ. 1926 = A.D. 1869. Manuscripts:

NPS 230 C of 1926–28. Copied in Saṃ. 1929 = A.D. 1872. Property of Paṇḍita Rāmanātha Pujārī of Bisavāṃ, Sītāpura.
NPS 230 D of 1926–28. Copied in Saṃ. 1931 = A.D. 1874. Property of Paṇḍita Baladevaprasāda Tivārī of Antā, Kakavana, Kānapura.

4. A *Horā yā śakunagamana*. Manuscript:

NPS 193 B of 1929–31. Copied in Saṃ. 1930 = A.D. 1873. Property of Ṭhākura Khañjanasiṃha of Sikandarāmaū, Alīgaḍha.

KEŚAVĀNANDA ŚARMAN

The son of Bhīmadatta, the son of Śivarāma of the Ḍabarālajāti, Keśavānanda, a resident of Timalīgrāma, wrote a Hindī ṭīkā, *Subodhinī*, on the *Jātakacandrikā* of Jayadeva (*fl.* 1750); this was published with the mūla at Bambaī in 1958; reprinted Bambaī 1963.

**KEŚAVĀRKA* (*fl.* thirteenth or fourteenth century)

Additional manuscripts of his *Vivāhavṛndāvana* (see *CESS* A 2, 75a–77a):

Leningrad (1914) 301 (Ind. I 15). 12ff.
Leningrad (1914) 302 (Ind. V 94). 1f. Incomplete (adhyāya 5).

Another edition of the *Vivāhavṛndāvana* was published with the *Vivāhadīpikā* of Gaṇeśa (b. 1507) at Mumbaī in Saṃ. 1966, Śaka 1831 = A.D. 1909.

KOKĀ PAṆḌITA

Author of a *Sāmudrikanārīdūṣaṇa* in Hindī. Manuscripts:

NPS 199 A of 1929–31. Copied in Saṃ. 1710 = A.D. 1653. Property of Paṇḍita Gaṅgārāma Gauḍa of Jalālī, Alīgaḍha.
NPS 199 C of 1929–31. Copied in Saṃ. 1890 = A.D. 1833. Property of Paṇḍita Bābūrāma, adhyāpaka at Rāmanagara, Āvāgaḍha, Eṭā.

KONERI

Author of a *Koneriyantra*. Manuscript:

GJRI 902/1. 2ff.

**KAUṬILYA* (*fl.* third century B.C.?)

T. R. Trautmann, *Kauṭilya and the Arthaśāstra*, Leiden 1971, pp. 174–184, adduces convincing evidence that the present form of book II, in which the astronomical material (see *CESS* A 2, 78b) appears, dates to *ca.* A.D. 150. It remains true, however, that the astronomy itself is older by many (perhaps six) centuries. See also G. Dwiwedi [A 3. 1969].

KṢEMAÑKARA MIŚRA

Author of a *Tithinirṇayasāra*, possibly identical with the *Tithinirṇaya* of Kṣemarāma (*fl.* 1720). Manuscript:

Kurukṣetra 403 (50679).

**KṢEMARĀMA* (*fl.* 1720)

Additional information on manuscripts of his *Tithinirṇaya* (see *CESS* A 2, 79a–79b):

*Florence 121. 42ff.

Cf. Kṣemañkara Miśra.

**KHAḌGASENA* (*fl.* 1651/1656)

Mānūsiṃha was the father of Lūṇarāja and Ṭhākurasīdāsa, Lūṇarāja the father of Khaḍgasena, who studied under Caturabhoja Bairāgī of Āgarā and resided in Nāranaula in Bāgaḍadeśa, Pañjāba. Additional manuscripts of his *Trilokadarpaṇa* (see *CESS* A 2, 79b):

*BORI 598 of 1875/76. 99ff. Copied in Saṃ. 1798 = A.D. 1741. From Jepur.
NPS 208 of 1923–25. Copied in Saṃ. 1920 = A.D. 1863. Property of the Jaina Mandira (Baḍā) at Bārābaṅkī.
NPS 19kha of Saṃ. 2010–2012. Copied in Saṃ. 1930 = A.D. 1873. Property of the Digambara Jaina Pañcāyatī Mandira at Ābūpurā, Mujaphpharanagara.

NPS 19ka of Saṃ. 2010–2012. Property of the Ādināthajī kā Mandira at Ābūpurā, Mujaphpharanagara.

*NABBĀBA KHĀNAKHĀNĀ (1556/1627)

Author of a *Trayatriṃśayogāvalī*. Manuscript:

SOI 2541 = SOI Cat. I: 1490–2541. 14ff.

Additional editions of his *Kheṭakauṭuka* (see *CESS* A 2, 79b–80a) are: with a Hindī translation, pt. 1, Lakhanaū 1899 (BM 14053. b. 38), and with a Hindī translation, Baṃbaī Saṃ. 1958 = A.D. 1901 (BM 14053. c. 68. (2)).

KHUŚĀLA DŪBE

Author of a *Jātaka* in Hindī. Manuscript:

NPS 238 A of 1926–28. Property of Vāsudevasahāya of Mādhogañja, Pratāpagaḍha.

KHUŚYĀLA KAVI

Author of a *Ṣaḍṛtusaṅkrāntivicāra;* see NCC, vol 5, p. 187.

KHUSĀLA KAVI

Author of a *Bhuvanasārasaṅgraha* in Hindī. Manuscript:

NPS 46 of Saṃ. 2004–2006. Copied in Saṃ. 1893 = A.D. 1836. Property of Hariharadatta Dūbe of Baharā, Tiyarā, Jaunapura.

GAÑGĀDĀSA

Author of a *Tithiprabandha* in Hindī; *cf.* the *Tithiprakāśa* of Gaṅgādāsa Trivedin. Manuscript:

NPS 70ka of Saṃ. 2001–2003. Property of the Nāgarīpracāriṇī Sabhā in Vārāṇasī.

*GAÑGĀDĀSA TRIVEDIN (or DVIVEDIN)

Additional information on the manuscripts of his *Tithiprakāśa* (see *CESS* A 2, 80b):

*AS Bengal 2771 (G 6461). 4ff. Copied on Wednesday 11 kṛṣṇapakṣa of Phālguna in Saṃ. 1751 = 27 February 1695. Property of Narasiṃha Pāṭhaka of Vijayapuragrāma. (Dvivedin).
AS Bengal 2772 (G 2935). 25ff. With a ṭīkā, the *Tithiprakāśaprakāśikā*.
Darbhanga 78 (Dh 64(d)). 6ff. Maithilī. Incomplete. No author mentioned.
Mithila I 162. 5ff. Maithilī. Property of Pandit Sureśa Miśra of Saurāth, Madhubani, Darbhanga. (Dvivedin).
*PUL II 3537 was mistakenly recorded as 3539 in *CESS* A 2.

There is a vyākhyā by Cakrapāṇi Pāṭhaka.

GAÑGĀDĀSA (or *GAÑGĀRĀMA*) *MIŚRA* (*fl. ca.* 1750)

The father of Chatrasāla Miśra (*fl.* 1787) and a resident of Canderī, Gaṅgādāsa *alias* Gaṅgārāma wrote a *Ramalasāra* in Hindī. Manuscript:

NPS 115 of 1923–25. Copied in Saṃ. 1913 = A.D. 1856. Property of Mahantinī Lakṣmaṇadāsī, kuṭī of Bābā Jhāmadāsa of Jagesaragañja, Sulatānapura.

*GAÑGĀDHARA

Additional manuscripts of his *Parāśaratulya* (see *CESS* A 2, 80b):

Kotah 158. 8 pp. No author mentioned.

*GAÑGĀDHARA (*fl.* 1420)

Additional manuscripts of his *Amṛtasāgarī* (see *CESS* A 2, 81a–82a):

Oudh XX (1888) VIII 83. 112pp. Copied in A.D. 1683 (*Gaṇitāmṛtasāraṇī* attributed to Divākara). Property of Paṇḍita Pratāpa Nārāyaṇa of Allahabad Zila.

LDI (KS) 1023 (10674). 60ff.

*GAÑGĀDHARA MIŚRA (*fl.* 1929/41)

Author (see *CESS* A 2, 85b) of a ṭīkā, *Ādarśatala*, on the *Pratibhābodhaka* of Sudhākara Dvivedin (*fl.* 1879/1907); this was published with the mūla at Banārasa in 1942. Gaṅgādhara also wrote a ṭippaṇī on and pariśiṣṭa to the *Kṛtyasārasamuccaya* of Amṛtanātha Jhā Śarman (b. 1755) in Śaka 1859 = A.D. 1937; these were edited with the mūla by Kṛṣṇamohana Śāstrin as *KSS* 129, Banārasa 1953. From this we learn that the father of his father Haṃsarāja was named Śekharadatta.

GAÑGĀPRASĀDA (*fl.* 1958)

A resident of Murāra, Gaṅgāprasāda was co-author with Haradeva Śarman Trivedin of the first khaṇḍa of a *Vyāpāra ratna* in Hindī; this was published at Dillī in 1958.

*GAÑGĀRĀMA

Additional manuscripts of his *Yuddhajayotsava* (see *CESS* A 2, 86a–86b):

GJRI 1061/173. Ff. 1–16 and 18–20. Copied in Saṃ. 1799 = A.D. 1742. Incomplete.
GJRI 1060/172. 19ff.
GJRI 1062/174. Ff. 2–4, 6–12, and 14–20. Incomplete.

GAÑGĀRĀMA

Author of a *Śakunaśāstra* in Hindī. Manuscript:

GJRI 1095/207. 15ff. Copied in Saṃ. 1890 = A.D. 1833.

GAÑGĀRĀMA MIŚRA

Author of a *Cintāmaṇi praśna* in Hindī. Manuscript:

NPS 118 of 1923–25. Copied in Saṃ. 1935 = A.D. 1878. Property of Alakhī Bābā of Rādhākuṇḍa, Baharāïca.

**GAÑGĀRĀMA DVIVEDA* (*fl.* 1718)

Additional manuscripts of his *Ratnadyota* (see *CESS* A 2, 86b–87a):

AS Bengal 2774 (G 9789). 16ff. Copied in Saṃ. 1793 = A.D. 1736.
*AS Bengal 2773 (G 6350). Ff. 1–29 and 31–44. Copied for Nandakiśora, Yugalakiśora, and Devakīnandana on Sunday 2 śuklapakṣa of Vaiśākha in Saṃ. 1866, Śaka 1731 = 16 April 1809.
GJRI 1064/176. 1f. Incomplete. No author mentioned.

The date on which he completed his work, Sunday 8 kṛṣṇapakṣa of Mārgaśīrṣa in Saṃ. 1775 (the date is irregular) (*not* 1053, which is Śaka 975), is given in the following verses at the end:

śarādrisaptendumite hi varṣe
śrīmārgaśīrṣe ᵓpy asite ᵓṣṭamīṣu/
vāre ᵓrkasaṃjñe hanumatpureṣu
vyalīlikhad grantham anāntarāc ca//
vaṃśīdharātmajaḥ śrīmān gaṅgārāmākhyavid dvijaḥ/
tasya putro bhadramaṇir loke satkīrtikārakaḥ//

**GAṆAPATI*

Additional manuscripts of his *Grahaśāntipaddhati* (see *CESS* A 2, 87b):

BORI 97 of 1892/95. 84ff.
Florence 133. 31ff.

He may possibly be identical with Gaṇapati Rāvala (*fl.* 1686).

**GAṆAPATI*

Additional manuscripts of his *Ratnadīpaka* (see *CESS* A 2, 88a–89a):

BORI 561 of 1895/1902. 14ff. Incomplete. Ascribed to Nāmadeva.
GJRI 1065/177. 16ff. Incomplete (dvādaśabhāvaphala).

**GAṆAPATI RĀVALA* (*fl.* 1686)

Additional manuscripts of his *Muhūrtagaṇapati* (see *CESS* A 2, 89b–92a):

Benares (1956) 13683. 15ff. Copied in Saṃ. 1742 = A.D. 1686. (*Parvanirṇaya*)
Baroda 10548. 27ff. Copied in Saṃ. 1806 = A.D. 1749 (*Parvanirṇaya*).
Baroda 9222. 17ff. Copied in Śaka 1712 = A.D. 1790 (*Parvanirṇaya*).
Baroda 558. 18ff Copied in Śaka 1764 = A.D. 1842 (*Parvanirṇaya*).
AS Bengal 2727 (G 524) = *Mitra, Not. 1296. This includes a second copy. 13ff. Incomplete (ends in prakaraṇa 7).
AS Bengal 2726 (G 9601). 34ff. Incomplete.
Bharatpur S. 27. No author mentioned.
Dharwar 692 (200). 69ff.
GJRI 1021/133. 64ff. Incomplete (I, 1–15 missing).
GJRI 1022/134. 58ff. Incomplete.
Kotah 277. 116pp. No author mentioned.
Kurukṣetra 809 (19839).
Kurukṣetra 810 (50176).

**GAṆAPATIDEVA ŚĀSTRIN* (*fl.* 1930/1961).

Author (see *CESS* A 2, 92b) of a Hindī vyākhyā, *Bhāvaprabodhinī*, on the *Camatkāracintāmaṇi* of Nārāyaṇa Bhaṭṭa; this was published as *HSS* 45, Banārasa 1935; 2nd ed., Banārasa 1948; 3rd ed., Vārāṇasī 1963.

GAṆARĀMA ṚṢI

Author of a *Sagunauṭī* in Hindī. Manuscript:

NPS 75 of Saṃ. 2001–2003. Copied in Saṃ. 1920 = A.D. 1863. Property of Hanumatadatta Tripāṭhī, sanātana dharmopadeśaka at Ismāïlagañja, Ilāhābāda.

GAṆEŚA

Author of a *Jātakadīpikā*. Manuscript:

LDI (VDS) 1299 (9714). 8ff. Copied by Rāmacandr a

**GAṆEŚA* (b. 1507)

Additional manuscripts of his *Grahalāghava* (see *CESS* A 2, 94a–100a):

Viśvabhāratī 147. Copied in Śaka 1751 = A.D. 1829. See NCC, vol. 6, pp. 258–259.
Allahabad Municipal Mus. 87 and 88. See NCC.
BORI 509 of 1895/1902. 106ff. With the *Harṣakaumudī* of Nṛsiṃha.
Cocanada, Telugu Academy 1158. See NCC.
GJRI 907/19. 13ff. Maithilī. Incomplete.
GJRI 908/20. 14ff. Maithilī.
GOML Madras R 981a. Ff. 1–4. Telugu. Incomplete (adhyāyas 1–3). With the udāharaṇa of Viśvanātha. Purchased in 1913/14 from P. Ādinārāyaṇāvadhāni of Pedakallepalli.
IM Calcutta 1450; 3426; 6667; 6925A; 8975; 9040; 9131; 9137; and 9320. See NCC.

Jodhpur 462. See NCC.
Kurukṣetra 282 (50085).
Kurukṣetra 283 (50087).
Kurukṣetra 285 (50054). With an udāharaṇa.
LDI (VDS) 1294 (9856). 18ff. With the *Harṣakaumudī* of Nṛsiṃha.
Leningrad (1914) 296 (Ind. V 92). Ff. 1 and 3–16.
Mysore (1955) 5163. 57ff. Grantha. With the ṭīkā of Viśvanātha. No author mentioned.
Nagpur, Deo Coll. 132. See NCC.
NS Press 241. See NCC.
Osmania University 137/5/b. 24ff. Incomplete. No author mentioned.
Poona, Bhāratīya Itihāsa Saṃśodhaka Maṇḍala 48; 49; thi 846; thi 344; thi 347; thi 353; and vi 125/25. See NCC.
Poona, Fergusson College, Mandlik Library Suppl. 209 and 211. See NCC.
Rajapur 27; 46; 55; and 721. See NCC.
Satara, Khuperkar I. xxi. 4. See NCC.
Śṛṅgeri 165 and 281. See NCC.
Udaipur, Nathdwara 184, 19; 184, 20–21; and 184, 24. See NCC.
Viśvabhāratī 115; 129; and 2971(e). See NCC.
Waltair, Andhra Univ. 520. 1. G. 19. See NCC.

Additional manuscripts of his *Tithicintāmaṇi* (see *CESS* A 2, 100b–103a):

Benares (1956) 13195. Ff. 5, 7, 9, 11, 13, 15, 17–39, and 41–83. Incomplete. No author mentioned.
Bharatpur S 36. No author mentioned.
Jaipur (II). 8ff. Ascribed to Nandarāma Miśra.

Additional manuscripts of his *Buddhivilāsinī* (see *CESS* A 2, 103a–104a):

Benares (1963) 37333. 14ff. Incomplete. No author mentioned.
Kurukṣetra 682 (50357).

Additional manuscripts of his *Bṛhattithicintāmaṇi* (see *CESS* A 2, 104a–104b):

BORI 901 of 1884/87. 260ff. (ff. 1 and 8 missing). Copied in Śaka 1682 = A.D. 1760. From Mahārāṣṭra. No author mentioned.
BORI 871 of 1887/91. 175ff. From Mahārāṣṭra. No author mentioned.

Additional manuscripts of his *Vivāhadīpikā* (see *CESS* A 2, 104b–106a):

Dharwar 696 (686). 70ff. Copied in Śaka 1780 = A.D. 1858.
AS Bengal 2694 (G 6418B). Ff. 74–77. (lagnaśuddhi).
AS Bengal 2695 (G 6418A). Ff. 9–73. Incomplete.

Another edition of the *Vivāhadīpikā* was published at Mumbaī in Saṃ. 1966, Śaka 1831 = A.D. 1909.

**GAṆEŚA* (*fl. ca.* 1550/1600)

Originally dated *ca.* 1600, Gaṇeśa's floruit must be extended backwards by about 50 years in light of the date of his cousin Jñānarāja (*fl.* 1503).

Additional manuscripts of his *Tājikabhūṣaṇa* (see *CESS* A 2, 107a–109a):

*Paris BN 212 P (Sans. dév. 317). F. 1 (= Paris BN 1005 BB), 6–16, 20–22, and 26–29. Copied by Jaganātha, the son of Gokala of Pijareta, on Monday 11 śuklapakṣa of Kārttika in Saṃ. 1745, Śaka 1611 = 14 October 1689. Incomplete. Acquired May 1842.
Florence 297. 22ff. Copied in Saṃ. 1765 = A.D. 1708. A few verses after the *Tājikasāra* of Haribhadra. No author mentioned.
Bharatpur S 8. No author mentioned.
Kotah 243. 19pp. No author mentioned.

**GAṆEŚA* (*fl.* 1613)

Additional manuscripts of his *Jātakālaṅkāra* (see *CESS* A 2, 110a–114a):

GJRI 931/43. 24ff. Copied in Saṃ. 1765 = A.D. 1708.
GJRI 3131/343. 13ff. Copied in Saṃ. 1894 = A.D. 1837.
WHMRL G. 38. g. 6ff. Copied by Jātirāma Brāhmaṇa Ṣaḍaṅkavidyārthin of the Chivevaṃśa on 2 kṛṣṇapakṣa of Bhādrapada in Saṃ. 1898 = *ca.* 1 September 1841.
GJRI 933/45. 17ff. Maithilī. Copied in Śaka 1769 = A.D. 1847.
GJRI 934/46. 15ff. Maithilī. Incomplete (ends at 3, 28).
GJRI 1182/294. 2ff. Maithilī. Incomplete.
GJRI 3132/344. 16ff. Maithilī. Incomplete (ends in adhyāya 7).
GJRI 3133/345. 12ff. Incomplete (ends with adhyāya 6)
GJRI 3134/346. 3ff. Incomplete (ends in adhyāya 3).
GJRI 3135/347. 19ff. Maithilī.
GJRI 932/44. 46ff. With the ṭīkā of Haribhānu.
Kurukṣetra 337 (19866).
Kurukṣetra 338 (50369).
Kurukṣetra 339 (50453).
LDI (VDS) 1300 (9723). 7ff.

**GAṆEŚA* (*fl.* 1681)

The verses quoted below from his *Tithimañjarī* (see *CESS* A 2, 93a) show that its epoch was Śaka 1603 = A.D. 1681. Verses 1–2a are:

namaskṛtya bhavānīṃ ca jagadutpattikāriṇīm/
kapakṣasaṃmitāṃ vakṣye gaṇeśas tithimañjarīm//
śāko vihīno °gnikhabhūpamānaiḥ.

Verse 30 at the end gives his genealogy:

śrīgauḍajñātivaryaḥ prathitaguṇagaṇaḥ somanātho dvijanmā
jātaḥ śāṇḍilyagotre śrutipathanipuṇas tatsuto lālabhaṭṭaḥ/
tatsūnuḥ khyātakīrtir budhajanamahitaḥ śrīmahādevanāmā
tatputro jñānabhaṭṭo dvijavaratilakas tasya sūnur gaṇeśaḥ//

Additional information on the manuscripts:

*Florence 266. 30ff. Copied by Ratneśvara, the son of Paṇḍya Divākara, in Saṃ. 1797, Śaka 1662 = A.D. 1740.

GAṆEŚA (*fl.* 1825)

Author of a *Guṇanidhi sāra* in Hindī in Saṃ. 1882 = A.D. 1825. Manuscript:

NPS 32 A of 1906–08. Copied in Saṃ. 1887 = A.D. 1830. Property of Lālā Vidyādhara of Horīpura, Datiyā.

GAṆEŚADATTA (*fl.* 1790)

A resident of Rājagaḍha, Gaṇeśadatta wrote a *Muhūrtamuktāvalī* in Hindī in Saṃ. 1874 = A.D. 1790. Manuscript:

NPS 61 of 1932–34. Copied in Saṃ. 1847 = A.D. 1790. Property of the Sarvopakāraka Nāgarī Pustakālaya at Achanerā, Āgarā.

**GAṆEŚADATTA PĀṬHAKA* (*fl.* 1962/1971)

The son of Baladevadatta Pāṭhaka, a Sarayūparīṇa Brāhmaṇa residing at Piyarīkalāṃ, Vārāṇasī, Gaṇeśadatta (see *CESS* A 2, 114a) finished a Saṃskṛta and Hindī ṭīkā, *Subodhinī*, on the *Narapatijayacaryā* of Narapati (*fl.* 1176) on Sunday 9 śuklapakṣa of Vaiśākha in Śaka 1894 = 2 May 1971. This was published as *KSS* 205, Vārāṇasī 1971. He also wrote an anvaya on the *Camatkāracintāmaṇi* of Nārāyaṇa Bhaṭṭa, which was published at Benares in 1966.

GAṆEŚABHAṬṬA

Author of a *Tithinirṇaya*. Manuscript:

Benares (1956) 13472. 13ff.

**GADĀDHARA*

Additional information about the manuscripts of his *Grahayāgapaddhati* composed at Gaḍhānagara (see *CESS* A 2, 114b):

*Berlin 1250 (Chambers 665). 41ff. Copied in Saṃ. 1651 = A.D. 1594.

*Mithila I 121. 27ff. Maithilī. Property of Pandit MM. Rajināth Miśra of Saurath, Madhubani, Darbhanga.

*Mithila I 122. 41ff. Maithilī. Property of Pandit Sadānand Jhā of Andhraṭhārhi, Darbhanga.

The last two verses are:

iti śrīgaḍhānagare śrīgadādhareṇātiprayāsena/
nānāpurāṇanibandhād ākarāc ca saṃgṛhya//
mūlavākyāni vicarya grahapaddhatiḥ kṛtā lakṣahomasyāpi/
paddhatiḥ saiva kiṃcidviśeṣas tu sadbhir ūhyam//

**GADĀDHARA*

Additional manuscripts of his ṭīkā on the *Bhuvanadīpaka* of Padmaprabha Sūri (*fl.* 1165) (see *CESS* A 2, 114b):

IM Calcutta 1601. See NCC, vol. 6, p. 254.
IM Calcutta 1602. Incomplete. See NCC.

**GADĀDHARA RĀJAGURU* (*fl. ca.* 1725/1750)

Additional manuscript of his *Kālasāra* (see *CESS* A 2, 115a–115b):

AS Bengal 2220 (G 4080). 171ff. Oriyā.

**GARGA*

Additional manuscript of his *Gargayātrā* (see *CESS* A 2, 116a):

Leningrad (1914) 304 (Ind. II 98). 3ff. Copied by Kṛṣṇa, the son of Sadāśiva, at Kāśī on 5 śuklapakṣa of adhika Śrāvaṇa in Saṃ. 1844 = *ca.* 20 July 1787.

**GARGA*

Additional manuscripts of his *Gargasaṃhitā* (see *CESS* A 2, 116a–120a).

1. (A 2, 116a–117b):

*Paris BN 245.1 (Sanscrit bengali 184). 207pp. Bengālī. Copied from a manuscript copied by Kumārānunanda in Śaka 1460 = A.D. 1538. From Guérin.

5. (A 2, 118b):

Madras Univ. R.K.S. 317(b). See NCC, vol. 6, p. 18.
Trivandrum Palace Library 876 G. See NCC.

6. (A 2, 118b–119a):

AS Bengal 2622 (G 2141) I = *Mitra, Not. 3227.
Baroda 2323. 6ff. (jyeṣṭhānakṣatrajananaśāntividhiprayoga).
*GOML Madras D 3252. 6pp. (*Utpātaśānti*).

GOML Madras D 3278. 7pp. Nandināgarī. (kuhūśāntikalpa).
GOML Madras D 3316. 7pp. Nandināgarī. (jyeṣṭhānakṣatraśānti).
GOML Madras D 3356. 4pp. Telugu. (nālaveṣṭanaśānti).
GOML Madras D 3377. 2pp. Nandināgarī. (pūrvāṣāḍhānakṣatrajātaśānti).
GOML Madras D 3378. 5pp. Grantha. (pūrvāṣādhādinakṣatrajananaśānti).
GOML Madras D 3406. 4pp. Telugu. (roganakṣatraśānti).
GOML Madras D 3407. 6pp. Telugu. (roganakṣatraśānti).
GOML Madras D 3424. 3pp. Nandināgarī. (viśākhānakṣatraśānti).
GOML Madras D 3425. 2pp. Grantha. (viṣaghaṭikājananaśānti).
Kerala 2803 (3944 A 2). 15 granthas. Grantha. (ekanakṣatraśānti).
Kurukṣetra 354 (50702). (jyeṣṭhāśānti).

7. (A 2, 119b):

Udaipur 524. Copied in Saṃ. 1746 = A.D. 1689.
Kotah 249. 238pp.
PrSB 961 (Göttingen Mu I 26 (B)). 1f. Śāradā. (*Vṛddhagārgya*).

8j. *Ekāṃśayogaprakaraṇa*. Manuscript:

Benares (1963) 34617. 5ff.

8k. *Sudarśanakalpasārasamuccaya*. Manuscript:

Kerala ———. (3208 A). See NCC.

**GARGA*

Additional manuscript of his *Pallīśaraṭavidhāna* (see *CESS* A 2, 120b), here entitled *Saraṭapallīpatanaśānti* of Vṛddhagārgya:

GOML Madras D 3456. 6pp. Telugu.

GARGA

Author of a *Yantrapraśna; cf.* the *Gargapraśna* of Garga. Manuscript:

GJRI 2954/310. 1f. (In Marāṭhī).

**GARGA*

Additional manuscripts of his *Lokamanoramā* (see *CESS* A 2, 120b–122b):

GJRI 989/101. 2ff. Copied in Saṃ. 1704 = A.D. 1647.
GJRI 1084/196. 7ff. Copied in Saṃ. 1839 = A.D. 1772.
GJRI 984/96. 2ff. Copied in Saṃ. 1874 = A.D. 1817.
GJRI 982/94. 10ff. Copied in Saṃ. 1880 = A.D. 1823. With a ṭīkā.
LDI (VDS) 1293 (9736). 19ff. Copied by Becara Badara at Rājanagara under Jagatavallabha Pārśvanātha Cintāmaṇi in Saṃ. 1903 = A.D. 1846.
Benares (1963) 37487. Ff. 1–2, 2b–3, 1f., ff. 4–5, 1f., ff. 6–17, and 1f. Incomplete. With the *Śivālikhita*.
GJRI 903/15. 4ff. With a ṭīkā.
GJRI 983/95. 4ff.
GJRI 3179/391. 7ff. Maithilī. Incomplete.
GJRI 3180/392. 2ff. Maithilī.
Kotah 300. 2pp.
Kurukṣetra 216 (50059).
PUL II 3663. 5ff.

GARGA

Author of a *Sārasaṅgraha*. Manuscript:

AS Bengal 2635 (G 4300). 42ff. Copied on Wednesday 13 śuklapakṣa of Pauṣa in Saṃ. 1944 = 28 December 1887.

The colophon begins: iti śrīgargācāryakṛtasārasaṃgrahe.

**GARGA (fl. ca.* 900)

Additional manuscripts of his *Pāśakevalī* (see *CESS* A 2, 122b–126a):

LDI (KS) 1038 (10672). 7ff. Copied in Saṃ. 1718 = A.D. 1661. (*Upadeśamālaśakunāvalī*).
Benares (1963) 37533. 8ff. Copied in Saṃ. 1787 = A.D. 1730. No author mentioned. (*Marutpraśnajñāna*).
Jaipur (II). 10ff. Copied in Saṃ. 1796 = A.D. 1739. (*Pāśākevalī*).
AS Bengal Vern. 375 (G 6946). 5ff. Copied in Saṃ. 1851 = A.D. 1794. No author mentioned. (*Pāśākeralī* in Hindī).
NPS 22 of Saṃ. 2010–2012. Copied in Saṃ. 1943 = A.D. 1886. Property of Bābūrāma Mistrī of Khaṭīkāna, Mujaphpharanagara. (*Kevalī* in Hindī).
Baroda 9770. 4ff. (*Praśnapāśāvalī*).
Florence 480. 12ff. (*Pāśakāvalī*).
GJRI 2985/318. 5ff. Maithilī. (*Pāśakeralī*).
Jodhpur 1828. (*Kaivalyaśākuna*). See NCC, vol. 5, p. 79.
Kathmandu (1960) 226 (I 522). 8ff. Nevārī. (*Pāśakeralī*).
Kurukṣetra 621 (50585). No author mentioned. (*Pāśakeralī*).
Kurukṣetra 622 (19639). (*Pāśāvalī = Pāśakevalī*).
LDI (KS) 1037 (10536). 5ff. Copied by Paṇḍita Dhiravijaya Gaṇi, the pupil of Dhanavijaya Gaṇi, at Jīrṇadurga. No author mentioned. (*Pāśākevalī*).
*Paris BN (Senart) 166 (Sanscrit 1716). 9ff. (*Pāśākevalī*).
*Paris BN (Senart) 250 (Sanscrit 1557). 11ff. (*Śakunāvalī*).

Poona, Bhāratīya Itihāsa Saṃśodhaka Maṇḍala vi. 290. See NCC, vol. 6, p. 17.

GIRADHARA

Author of a *Śakunāvalī* in Hindī. Manuscripts:

NPS 76 of Saṃ. 2001–2003. Property of Bholānātha (Bhorelāla) Jyotiṣī of Dhātā, Phatehapura.
Udaipur, Nathdwara 207, 8. Ascribed to Giridharaji. See NCC, vol. 6, p. 20.

GIRIDHARA GOSVĀMIN

The son of Viṭṭhalanātha Gosvāmin and a resident of Braja, Giridhara wrote a *Muhūrtamuktāvalī* in Hindī. Manuscript:

NPS 168 A of 1906–08. Property of Rāmaneta Mantrī of Rājya Ṭīkamagaḍha.

GIRIDHARA PAṆḌITA

The son of Muktāmaṇi Paṇḍita, Giridhara wrote a *Laghusaṅgraha* of which the first section deals with kāla. Manuscript:

AS Bengal 2215 (G 10388). 28ff. Incomplete.

The second verse is:

manubhṛgupramukhair munibhiḥ kṛtāḥ
kati no tantracayā nigamādṛtāḥ/
tad avalokitum aprabhur ādarād
giridharaḥ kurute laghusaṅgraham//

The colophon begins: iti śrīmuktāmaṇipaṇḍitātmajagiridharapaṇḍitaviracite.

GIRIDHARA MIŚRA

See Vedāṅgarāya.

**GIRIDHĀRIN MIŚRA*

There are manuscripts of both his *Āyurdāyavicāra* and his *Lagnavāda* (see *CESS* A 2, 127a) in Mithilā; see NCC, vol. 6, p. 22.

GUṆAVIṢṆU

Author of a vyākhyā on a *Navagrahamantra*. Manuscript:

AS Bengal 848 (G 3597). Ff. 8–11. Bengālī. Copied by Rāmasundara Śarman.

**GUṆARATNA SŪRI* (*fl. ca.* 1375)

Additional manuscripts of his avacūrṇi on the *Kṣetrasamāsa* of Somatilaka Sūri (*fl.* 1298/1367) (see *CESS* A 2, 127a–127b):

LDI 3012 (3668). 17ff. Copied in Saṃ. 1480 = A.D. 1423.
*BORI 590 of 1895/98. 14ff. Copied in Saṃ. 1511 = A.D. 1454.
LDI 3011 (4564). 26ff. Copied by Kālidāsa Vipra of Nalapadranagara in Saṃ. 1565 = A.D. 1508.
LDI 3008 (5642). 29ff. Copied by Harṣarāja Gaṇi in the saṅghāḍā of Mahāmahopādhyāya Dharmasāgara Gaṇi, the pupil of Paṇḍita Vicārasāgara Gaṇi, at Khayarapurāgrāma under Vijayadāna Sūri in Saṃ. 1612 = A.D. 1555.
LDI 3007 (6872). 19ff. Copied by Kalyāṇakuśala, the pupil of Rājakuśala Gaṇi, at Sāraṅgapura in Mālvā under Hīravijaya Sūri of the Tapā Gaccha in Saṃ. 1641 = A.D. 1584.
LDI 3009 (8080). 16ff. (f. 1 missing). Copied by a pupil of Ānandahaṃsa Gaṇi for Harṣavimala, the pupil of Paṇḍita Ānandavijaya Gaṇi. Incomplete.
LDI 3010 (2254). 11ff.
LDI 3013 (5686). 14ff.
LDI (KS) 506 (10103). 84ff.
LDI (KS) 507 (10819). 20ff.
LDI (KS) 508 (10832). 12ff.
LDI (VDS) 502 (9817). 23ff. (ff. 1–22 missing). Incomplete.
Paris BN (Senart) 70 (Sanscrit 1576). 15ff.

**GUṆĀKARA* (*fl.* between 1100 and 1400)

Additional manuscript of his *Horāmakaranda* (see *CESS* A 2, 127b–128b):

Udaipur 547. Copied in Saṃ. 1720 = A.D. 1663.

NORI GURULIṄGA ŚĀSTRIN (*fl.* 1901)

Author of an Āndhraṭīkā on the *Muhūrtamārtaṇḍa* of Nārāyaṇa (*fl.* 1571/1572), published at Madras in 1901 (BM 14053.ccc.38 and IO 1913).

**GURUSEVAKA MIŚRA*

Additional manuscripts of his *Gaṇakapuṣpaśirovataṃsa* (see *CESS* A 2, 129a):

*WHMRL G. 93. k 20ff. Copied by Pūjya Vajīrā Riṣa, pupil of Pūjya Suddhā Riṣajī, at Paṭṭīnagara on Wednesday 13 śuklapakṣa of Māgha in Saṃ. 1921 = 8 February 1875.
Chani 340. See NCC, vol. 5, p. 235.

GULĀBADĀSA (*fl.* 1745)

Author of a Hindī ṭīkā on the *Śīghrabodha* of Kāśīnātha in Saṃ. 1802 = A.D. 1745. Manuscripts:

NPS 68 of 1932–34. Copied in Saṃ. 1823 = A.D. 1766. Property of Ṭhākura Lokamānasiṃha of Akabarapura, Mustaphābāda, Mainapurī.
NPS 130 of 1929–31. Copied in Saṃ. 1823 = A.D. 1766. Property of Umādatta, adhyāpaka at Cāū, Phirojābāda, Āgarā.

GOKULACANDA

The son of the hakīma Rāmacanda and a resident of Mathurā, Gokulacanda wrote a *Sagunaparīkṣā* in Hindī. Manuscript:

NPS 127 of 1929–31. Copied in Saṃ. 1927 = A.D. 1870. Property of Lālā Dilasukharāya of Nagarābhagata, Paṭiyārī, Eṭā.

**GOKULAJIT TRIPĀṬHIN* (*fl.* 1632)

According to NCC, vol. 6, p. 111, Gokulajit, the son of Harijit and the brother of Gopīnātha, Śaṅkarajit, and Śyāmajit, flourished during the reign of Shāh Jahān (1628/1658), and wrote his *Saṅkṣepatithinirṇayasāra* (see *CESS* A 2, 129a–129b) for Kalyāṇamalla, rājā of Iladurga, in A.D. 1632. See also NCC, vol. 3, p. 257.

**GOKULANĀTHA UPĀDHYĀYA* (*fl. ca.* 1675/1740)

The son of Umā and Pītāmbara of the Phaṇadahakula, Gokulanātha was patronized by Fateh Shāh of Garhwal (d. 1699) and Mādhavasiṃha of Mithilā (*fl.* 1700/39) (see NCC, vol. 6, pp. 112–114). Additional manuscripts of his *Māsamīmāṃsā* (see *CESS* A 2, 129b):

Mithila I 293. 23ff. Maithilī. Copied by Rajanīnātha on Monday 10 kṛṣṇapakṣa of Bhādrapada in Śaka 1687 = 9 September 1765. Property of Pandit Maṇīśvar Jhā of Lālaganj, Jhañjhārpur, Darbhanga.
Mithila I 293 G. 17ff. Maithilī. Copied in Śaka 1765 = A.D. 1843. Property of Pandit Sureśa Miśra of Saurāṭh, Madhubanī, Darbhanga.
Mithila I 293 E. 16ff. Maithilī. Copied in Sāl. San. 1295 = *ca.* A.D. 1887. Property of the Śrī Chitradhar Library of Tabhaka, Dalsingh Sarai, Darbhanga.
GJRI 3539/177. 13ff. Maithilī.
Mithila I 293 A. 20ff. Maithilī. Property of Pandit Rabināth Jhā, Professor at M. R. Vidyālaya and resident of Andhraṭhārhi, Darbhanga.
Mithila I 293 B. 8ff. Maithilī. Incomplete. Property of Pandit Mahīdhar Miśra of Lālabāg, Darbhanga.
Mithila I 293 C. 13ff. Maithilī. Property of Pandit Śaktināth Jhā of Ujan, Jhañjhārpur, Darbhanga.
Mithila I 293 D. 20ff. Maithilī. Property of Pandit Manohar Ṭhākur of Tabhaka, Dalsingh Sarai, Darbhanga.
Mithila I 293 F. 10ff. Maithilī. Property of Pandit Balbhadra Jhā of Jogiārā, Darbhanga.

**GOPĀLA*

Additional manuscripts of his *Gopālaratnākara* (see *CESS* A 2, 130a):

Cocanada, Telugu Academy 4530. Incomplete. See NCC, vol. 6, p. 148.
GOML Madras D 13651. Ff. 1–36. Telugu. Incomplete (48 verses). With an Āndhraṭīkā.
GOML Madras D 13652. Ff. 29–35. Grantha. Incomplete (dvādaśabhāva).
Hiersemann. No author mentioned.
Mysore (1911 + 1922) 3186. 39ff. No author mentioned.
Oppert I 1227. Property of Vaṅkīpuram Śrīnivāsācāryār of Tiruvallūr, Chingleput.
Oppert I 1368. Property of Śrīnivāsa Rāghavācāryār of Uttaramallūr, Chingleput.
Oppert I 3839. Property of the Śaṅkarācārya Maṭha at Kumbhaghoṇam, Tanjore.
Oppert I 7097. Property of A. Pappulu Lakṣmaṇaśāstrulu of Vijayanagaram, Vizagapatam.
Oppert II 1960. 13pp. Telugu. Property of Veṅkaṭeśvarajosya of Siddhavaṭa, Kaḍapa.
Oppert II 2090. 350pp. Telugu. Property of Kandālla Veṅkaṭācārya of Śiṅgamāla, Pullampeṭa, Kaḍapa.
Oppert II 5252. 2 copies. Property of Piccudīkṣitar of Akhilāṇḍapuram, Tanjore.
Oppert II 7436. Property of Veṅkaṭarāmaśāstrī of Pillūr, Māyavaram, Tanjore.

GOPĀLA

Author of a *Grahacūḍāmaṇisāriṇī*. Manuscript:

Poona, Fergusson College, Mandlik Library, p. 74. See NCC, vol. 6, pp. 132 and 247.

GOPĀLA

Author of a *Ramalaśāstra* in Hindī. Manuscript:

NPS 52 A of 1920–22. Copied in Saṃ. 1921 = A.D. 1864. No owner mentioned.

**GĀRGYA GOPĀLA*

For additional manuscripts of his *Rahasyaprakāśa* (see *CESS* A 2, 130b) see NCC, vol. 6, p. 147.

**GOPĀLA BHAṬṬA*

Additional information on manuscripts of his *Kālakaumudī* (see *CESS* A 2, 130b):

*Oudh XVII (1884) IX 25. 340pp. Property of Paṇḍita Chandū Lāla of Partabgarh Zila.
*Oudh XVIII (1885) IX 21. 450pp. Property of Śivadīna Rāma of Rae Bareli Zila.

GOPĀLA MIŚRA

Author of a *Vivāhavṛndāvana*. Manuscript:

Kurukṣetra 990 (72).

*GOPĀLA NYĀYAPAÑCĀNANA (*fl. ca.* 1600)*

Additional manuscripts of his *Tithinirṇaya* (see *CESS* A 2, 131a):

Mithila I 152 F. 29ff. Bengālī. Copied in Śaka 1620 (?) = A.D. 1698 (?). Property of Pandit Dāmodar Jhā of Sahapur, Pandaul, Darbhanga.
AS Bengal 2105 (G 3644) I. 22ff. Bengālī. Copied in Śaka 1640 = A.D. 1718.
Mithila I 152. 18ff. Maithilī. Copied in Śaka 1722 = A.D. 1800. Property of Pandit Rabināth Jhā, Professor at M. R. Vidyālaya and resident of Andhraṭhārhi, Darbhanga.
Mithila I 152 A. 25ff. Maithilī. Copied in Sāl. San. 1245 = *ca.* A.D. 1837. Property of Pandit Chaturānand Jhā of Baḍasām, Madhepur, Darbhanga.
Benares (1956) 12921. Ff. 73–100. Bengālī. Incomplete.
Benares (1956) 13167. Ff. 2–21 and 21b–25. Bengālī. Incomplete (*Kālanirṇaya*).
Benares (1956) 14080. 24ff. Incomplete.
Darbhanga 62 (Dh 49(a)). Ff. 1–21. Maithilī. No author mentioned.
Mithila I 151. 26ff. Maithilī. Property of the Rāj Library at Darbhanga.
Mithila I 152 B. 22ff. Bengālī. Property of Pandit Dāmodar Jhā of Thārhī, Andhrathārhī. Darbhanga.
Mithila I 152 C. 28ff. Bengālī. Property of Pandit Tārānātha Jhā of Dharmapur, Jhañjhārpur, Darbhanga.
Mithila I 152 D. 20ff. Maithilī. Property of Pandit MM. Rājināth Miśra of Saurāth, Madhubani, Darbhanga.
Mithila I 152 E. 32ff. Maithilī. Property of Pandit Umākānt Jhā of Tarauni, Sakri, Darbhanga.

Additional manuscripts of his *Saṅkrāntinirṇaya* (see *CESS* A 2, 131a–131b):

Mithila I 409 B. $4\frac{1}{2}$ff. Maithilī. Copied in Śaka 1711 = A.D. 1789. Property of Pandit Rabināth Jhā of Andhrāṭharhī, Darbhanga.
AS Bengal 2107 (G 3645). Ff. 42v–48. Bengālī.
AS Bengal 2108 (G 3895). 8ff. Bengālī.
Benares (1956) 12922. 7ff. Bengālī. Incomplete.
Mithila I 409. 6ff. Bengālī. Property of Dāmodar Jhā of Andhrāṭharhi, Darbhanga.
Mithila I 409 A. 12ff. Maithilī. Incomplete. Property of Babu Karpūr Jhā of Andhrāṭharhi, Darbhanga.
Mithila I 409 C. 6ff. Maithilī. Property of Pandit Manohar Ṭhākur of Tabhaka, Dalsingh Sarai, Darbhanga.
Udaipur 573.
Viśvabhāratī 574(a). See NCC, vol. 6, p. 144.

GOPĀLA JANA (fl. 1776)

A resident of Maū Rānīpura, Jhāṃsī, Gopāla wrote a *Samarasāra* in Hindī in Saṃ. 1833 = A.D. 1776. Manuscript:

NPS 3 of 1904. Property of the Mahārāja Banārasa kā Pustakālaya at Rāmanagara, Vārāṇasī.

GOPĀLA (fl. 1864)

Author of a *Nārāyaṇaśakunāvalī* in Hindī in Saṃ. 1921 = A.D. 1864. Manuscript:

NPS 52 B of 1920–22. Property of Paṇḍita Devīdayāla Miśra of Ṭhākuradvārā, Khajuhā, Phatehapura.

**GOPĀLA ŚĀSTRIN NENE (fl. 1932/1936)*

He completed the *Varṣakṛtyadīpaka* (see *CESS* A 2, 132a) on Thursday 5 śuklapakṣa of Māgha in Saṃ. 1988 = 10 February 1932. There was a second edition of *KSS* 96 published at Vārāṇasī in 1967.

GOPĀLADEVA

Author of an *Alaṅkārayānaka*. Manuscript:

Radh. 33. See NCC, vol. 6, p. 142.

**GOPĪNĀTHA*

Additional manuscripts of his *Budhavallabhā* (see *CESS* A 2, 132a–132b):

AS Bengal 2654 (G 6394). 32ff. Copied on 3 śuklapakṣa of Vaiśākha in Saṃ. 1690 = *ca.* 30 April 1633.
Pingree 12. 34pp. Copied by Viśveśvara Datta from VVRI 2617 (2317 in *CESS* A 2, 132b is an error) in A.D. 1960.
Leningrad (1914) 293 (Ind. II 93). 42ff.

GOPĪNĀTHA PAṆḌITA

Author of a *Saṅkṣepagrahayajña*. Manuscript:

IM Calcutta 3228. See NCC, vol. 6, p. 164.

GOPĪNĀTHA ŚARMAN (b. 1847)

Born at Viṣṇupura in Kāmarūpa in A.D. 1847 and educated at Navadvīpa and Benares, Gopīnātha wrote a *Daivajñabhāskara;* see *JUG* 15, 1, 1964, 87 and NCC, vol. 6, p. 165.

**GOPĪRĀJA*

Additional manuscript of his *Tithitaraṅgiṇī* (see *CESS* A 2, 133a).

IM Calcutta 1334. See NCC, vol. 6, p. 166.

**GOPEŚA KUMĀRA OJHĀ (fl. 1956/1971)*

Author (see *CESS* A 2, 133b–134a) of the second khaṇḍa of the *Vyāpāra ratna* in Hindī, published at Dillī in 1958; of a Hindī ṭīkā, *Bhāvārthabodhinī*, on the *Phaladīpikā* of Mantreśvara, published at Dillī-

Vārāṇasī-Paṭanā in 1969; of a Hindī ṭīkā, *Candrikā*, on the *Jātakādeśamārga* of Putumana Somayājin, published at Dillī-Vārāṇasī-Paṭanā in 1971; and of a *Triphalā*, which consists of Hindī ṭīkās on the *Suślokaśataka* of Miṭṭhana, on the rājayogādhyāya from a *Śatamañjarī*, and on the *Veḍājātaka* of Naracandropādhyāya (*fl.* 1266/1267), published at Dillī-Vārāṇasī-Paṭanā in 1971. The third edition of his *Sugamajyotiṣapraveśikā* was published at Dillī-Vārāṇasī-Paṭanā in 1970.

**GOBHILA*

Additional manuscripts of his *Navagrahaśānti* (see *CESS* A 2, 134a):

Berlin 325. (Chambers 404). 12ff. Copied by Gopālajīka, the son of Cintāmaṇi Dvivedin, at Ḍhākāgrāma in Bañgāladeśa in Saṃ. 1711 = A.D. 1654.
Anup 2192. 13ff. Copied by Boharā Harinātha at Karaṇapura in Saṃ. 1739 = A.D. 1682.
Baroda 9098. 15ff. Copied in Saṃ. 1804 = A.D. 1747.
Baroda 5809. 16ff. Copied in Saṃ. 1899 = A.D. 1842.
Baroda 4609(a). Ff. 2v-3. (śānti).
Baroda 5879. 13ff.
Baroda 8047. 15ff.
BORI 207 of 1880/81. 11ff.
BORI 249 of 1887/91. 7ff. From Gujarāt.
IM Calcutta 2189. See NCC, vol. 6, p. 171.
PL, Buhler I D 185. 4ff (*Grahasthāpana*). Property of Bholānātha Śāstrī of Ahamadābāda.

There is also a *Gobhilagrahayajña*. Manuscripts:

IM Calcutta 1957 and 6089 (Incomplete). See NCC, vol. 6, p. 171.

MĀLAVĪYA GOVARDHANA SŪRI

Author of a *Tithikalpadruma*. Manuscript:

AS Bengal 2788 (G 5804). 28ff.

The colophon begins: iti śrīmālavīyagovardhanasūriviracite.

**GOVARDHANA* (*fl.* 1544?)

Additional information on manuscripts of his *Padmakośa* (see *CESS* A 2, 134b–135b):

*Oxford CS c. 315 (vi). 6ff. Copied for Krapārāma on Thursday 7 śuklapakṣa of Vaiśākha in Saṃ. 1786 = 24 April 1729. No author mentioned.
Kerala 6718 (8958). 180 granthas. Malayālam. No author mentioned.

GOVINDA

Author of a ṭīkā on a *Camatkāracintāmaṇi*. Manuscript:

Gwalior, Mātṛbhūmi 84. Incomplete. See NCC, vol. 6, pp. 190 and 387.

This may be the *Cintāmaṇi* of Govinda. Manuscript:

Dāhilakṣmī XIX 11. See NCC, p. 190.

GOVINDA

The son of Sadāśiva Miśra, Govinda wrote the *Vākyaratnākara* in five paricchedas:

1. prātarmadhyāhnādikṛtya.
2. śuddhi.
3. adhikāra.
4. śrāddha.
5. tithinirṇaya.

The last may possibly be the *Tithinirṇaya* of Govindabhaṭṭa (see *CESS* A 2, 142b). Manuscripts:

AS Bengal 2799 (G 1720). 29ff. Copied by Śivavakasa on Wednesday 2 kṛṣṇapakṣa of Kārttika in Saṃ. 1859, Śaka 1724 = 10 November 1802.
Mithila I 412. 26ff. Maithilī. Copied by Giridhārin. Property of Pandit Rāghava Jhā of Andauli, Mādhavapur, Darbhanga.

The last verse is:

putro ᵓkarod gaṇakamiśrasadāśivasya
prātaḥ prabodhasamayāt tithinirṇayāntam/
ratnākarākhyaguṇapūritasañgrahe ᵓsmiṃs
tattuṣṭaye bhavatu cakrisamudraputryoḥ//

The colophon begins: iti śrīgovindaviracite.

GOVINDA

Author of a ṭīkā on a *Śiśubodhinī*. Manuscript:

Mithilā. See NCC, vol. 6, p. 191.

GOVINDA UPĀDHYĀYA

Author of a *Malamāse niṣedhavicāra*. Manuscripts:

Mithila I 292. 7ff. Maithilī. Property of the Rāj Library, Darbhanga.
Mithila I 292 A. 15ff. Maithilī. Property of Pandit Gaurīkānt Jhā of Devahī, Tamuria, Darbhanga.

The colophon is: śrīgovindopādhyāyasya kṛtir iyam.

**GOVINDA KAVĪŚVARA*

Additional manuscript of his *Saṃvitprakāśa* (see *CESS* A 2, 136b–137a):

Śāstrī, Not. 1911. 323. 16ff. Copied in Saṃ. 1873 = A.D. 1816. Property of Paṇḍita Gañgādeo of Harapurā, Gopālagañja, Chāprā.

GOVINDA DĪKṢITA

Apparently the author of a vyākhyā entitled *Govindadīkṣitīya* on his own *Jātakacandrikā* (see *CESS* A 2, 137a). Manuscripts:

Baroda 13364(b). 30ff. Nandināgarī. Incomplete (ends in adhyāya 2).
Baroda 13382(b). 8ff. Nandināgarī.

*GOVINDA PAṆḌITA (fl. 1598?)

Additional manuscripts of his *Jyotiṣaratnasaṅgraha* (see *CESS* A 2, 137b):

IM Calcutta 5357. Incomplete. See NCC, vol. 6, p. 200.
Kunte B 82. 29ff. Property of Pandit Dilarām of Gujrānwāla.

*GOVINDA (b. 2 October 1569)

Additional manuscript of his *Rasālā* (see *CESS* A 2, 137b–138b):

Kerala 6715 (1707). 1500 granthas. Incomplete.

Additional manuscripts of his *Pīyūṣadhārā* (see *CESS* A 2, 138b–141a):

AS Bengal 2717 (G 8624). Ff. 1–2, 4–22, and 26–35. Copied by Śivalāla Gujarāthī on 9 śuklapakṣa of Vaiśākha in Saṃ. 1860 = *ca.* 29 April 1803. Incomplete (vināyakādiśānti). Formerly property of Vaijanāthabhaṭṭa Maunī.
Dharwar 694 (684). 59ff. Copied in Śaka 1755 = A.D. 1833.
AS Bengal 2715 (G 6489). 60ff., 58ff., 22ff., 31ff., 85ff., 110ff., 3ff., 5ff., 4ff., 5ff., 77ff., and 18ff.
Kurukṣetra 811 (19527).
Kurukṣetra 812 (19528).
Leningrad (1914) 308 (Ind. V 97). Ff. 2–82, 108ff., 32ff., 30ff., 108ff., 135ff., 20ff., 127ff., 38ff., and ff. 1 and 3–27.
N-W P I (1874) 92. 120ff. Ascribed to Nīlakaṇṭha. Property of Rāmakṛṣṇa of Benares.

GOVINDABHAṬṬA

Additional manuscript of his *Tithinirṇaya* (see *CESS* A 2, 142b):

Alwar 1326.

GOVINDAŚARMAN

A member of the Daśaputrakula, Govindaśarman wrote a *Malamāsanirūpaṇa*. Manuscript:

IM Calcutta 3135. Incomplete. See NCC, vol. 6, p. 207.

GOVINDAŚARMAN (*fl.* 1940)

Lakṣmaṇa of Pūrāgrāma in Baliyāpura, Mithilā, was the father of Rāma, Gaṇeśa, Maheśa, and Vindhyeśvarīprasāda; Maheśa was the father of Govindaśarman, who was the pupil of Raghunātha and who wrote a ṭīkā, *Saralā*, on the *Pañcasvarāḥ* of Prajāpatidāsa in Saṃ. 1997 = A.D. 1940. This was published with the mūla and the *Subodhinī* of Kṛṣṇadatta Jhā (*fl.* 1804) at Banārasa, 2nd ed. Saṃ. 1998 = A.D. 1941.

GOVINDASVĀMIN (*fl. ca.* 800/850)

Concerning his *Mahābhāskarīyabhāṣya* (see *CESS* A 2, 143b–144a) see also R. C. Gupta [A3. 1971].

GOVINDĀNANDA KAVIKAṄKAṆA (*fl.* 1510/1535)

Additional manuscript of his *Arthāratnaprabhā* (see *CESS* A 2, 144a–144b):

Viśvabhāratī 670. See NCC, vol. 1, rev. ed., p. 386.

Additional manuscripts of his *Arthakaumudī* (see *CESS* A 2, 144b–145a):

AS Bengal 2646 (G 3580). 127ff. Bengālī. Copied in Śaka 1544 = A.D. 1522.
AS Bengal 2647 (G 5601A). 30ff. Uḍiya. Incomplete (ends in adhyāya 2).
AS Bengal 2648 (G 5603B). In tripāṭha form. Uḍiya. Incomplete (to end of adhyāya 2).

Additional manuscripts of his *Varṣakriyākaumudī* (see *CESS* A 2, 145a):

AS Bengal 2691 (G 3557). 113ff. Bengālī. Copied in Śaka 1533, Malla 919 = A.D. 1611. Incomplete (dānakriyākaumudī).
AS Bengal 2692 (G 687) = *Mitra, Not. 1530.
IO 1654 (411). 393ff. (f. 162 missing). Bengālī. From H. T. Colebrooke.

GOSVĀMIN YĀJA

Gosvāmin was the son of Nṛsiṃha. Additional manuscripts of his *Tithisiddhivallī* (see *CESS* A 2, 145a):

RORI Cat. III 12860. 6ff. Copied by Rādhekṛṣṇa Natthūrāma Ojhā in Saṃ. 1913 = A.D. 1856. Ascribed to Jaya Gosvāmin, the son of Nṛsiṃha.
IM Calcutta 1312. See NCC, vol. 6, p. 217.

GAUTAMA

Additional manuscript of his *Gautamajātaka* (see *CESS* A 2, 145a–145b):

Benares (1963) 34455. 2ff. Copied in Saṃ. 1873, Śaka 1738 = A.D. 1816. No author mentioned.

GAUTAMA

Presumed author of a *Gautamabhāṣā*. Manuscript:

Mithilā. See NCC, vol. 6, p. 230.

GAUTAMA

Author of a *Praśnamālikā*, which may be identical with the *Śakunāvali* of Gautama (see *CESS* A 2, 145b). Manuscript:

GJRI 985/97. 2ff.

Another manuscript of the *Śakunāvali* is:

Udaipur, Nathdwara 188, 2–5. See NCC, vol. 6, p. 224.

GAUTAMA

Author of a *Vyatīpātavaidhṛtirajasvalāśānti*. Manuscripts:

Adyar Cat. 19 E 22. 6ff. Telugu.
GOML Madras D 3433. 7pp. Nandināgarī (from a *Gautamasaṃhitā*).

GAUTAMA SVĀMIN

Alleged author of an *Horājñāna*; see NCC, vol. 6, p. 231.

GAURĪKĀNTA CAKRAVARTIN

Author of a ṭīkā on a *Sarvatobhadracakra*. Manuscript:

Śāstrī, Not. 1900. 401. 5ff. Bengālī. Property of Paṇḍita Rāmatāraṇa Ṭhākura of Kāṭhālpāḍā *via* Naihāṭi.

**GHAṬĪGOPA*

Additional manuscripts of his *Āryabhaṭīyavyākhyā* (see *CESS* A 2, 147a–147b):

Kerala C 638 (C 157). 94pp. Malayālam. Incomplete (the Daśagītikā is missing).
Kerala C 651 (C 736). 49pp.

The next to the last verse is:

ghaṭīgopābhidhānasya vāṅmanaḥkāyavṛttibhiḥ/
yat kṛtaṃ padmanābhasya pūjā tad akhilaṃ bhavet//

GHANARĀMA (*fl.* 1699)

A Kāyastha, Ghanarāma wrote a Hindī translation of the *Līlāvatī* of Bhāskara (*b.* 1114) in Saṃ. 1756 = A.D. 1699 for Udyotasiṃha, the rājā of Ochaḍā. Manuscript:

NPS 35 of 1906–08. Property of the Dayitānareśa kā Pustakālaya in Dayitā.

GHANAŚYĀMA

Author of *Yātrāmaṅgala*. Manuscript:

Sūcīpattra 18. See NCC, vol. 6, p. 275.

GHANAŚYĀMA VYĀSA (*fl.* 1870)

Author of a *Jyotiṣa kī lāvanī* in Hindī in Saṃ. 1927 = A.D. 1870. Manuscript:

NPS 135 of 1926–28. Copied in Saṃ. 1939 = A.D. 1882. Property of Paṇḍita Śivakaṇṭha Bājapeyī of Jaitīpura, Unnāva.

GHANAŚYĀMARĀYA

Author of a *Svapnaparīkṣā* = *Svapnārthacintāmaṇi* in Hindī, allegedly in Saṃ. 1928 = A.D. 1871 though that date is later than the earliest manuscript. Manuscripts:

NPS 134 A of 1926–28. Copied in Saṃ. 1910 = A.D. 1853. Property of Paṇḍita Śivakaṇṭha Tivārī of Baragadiyā, Sītāpura.
NPS 134 B of 1926–28. Copied in Saṃ. 1930 = A.D. 1873. Property of Rāyalāla of Ramuāṃpura, Dauraharā, Khīrī.
NPS 134 C of 1926–28. Copied in Saṃ. 1934 = A.D. 1877. Property of Paṇḍita Śrīkṛṣṇa Dūbe of Śivadattapura, Baratāla, Sītāpura.

CAKRACŪḌĀMAṆI

Alleged author of a ṭīkā on the *Siddhāntaśiromaṇi* of Bhāskara (b. 1114). Manuscript:

N-W P V (1880) A 23. 62ff. Property of Pandit Mākhana Misra of Muttra.

CAKRACŪḌĀMAṆI (*fl. ca.* 1620)

See Kavicūḍāmaṇi (*fl. ca.* 1620).

CAKRADHARA

The son of Vāmana or Vāmadeva, Cakradhara wrote a *Yantracintāmaṇi* or *Sadyantracintāmaṇi* in 4 adhyāyas:

1. yantropakaraṇasādhana.
2. tripraśnādhikāra.
3. grahānayanādhikāra.
4. prakīrṇādhyāya.

There is a vivaraṇa on this by Cakradhara himself, and a commentary, *Yantradīpikā*, by Rāma (*fl.* 1625). See S. B. Dikshit [1896] 352. The manuscripts of the *Yantracintāmaṇi* are:

Benares (1963) 35769. 41ff. Copied in Śaka 1556 = A.D. 1634. With the *Dīpikā* of Rāma.
AS Bengal 6904 (G 1707). 16ff. Copied by Jyotirvid Indrajit on Wednesday 8 kṛṣṇapakṣa of Vaiśākha

in Saṃ. 1729 = 8 May 1672. With the *Dīpikā* of Rāma.
PUL II 3544. 4ff. Copied in Saṃ. 1730 = A.D. 1673. Incomplete (turīyayantra).
Benares (1963) 35324. 2ff. Copied in Saṃ. 1785 = A.D. 1728.
BORI 408 of 1884/86. 28ff. Copied in Saṃ. 1795 = A.D. 1738. With a ṭīkā.
Baroda 3394. Ff. 7–29. Copied in Saṃ. 1826 = A.D. 1769. With the *Dīpikā* of Rāma. Incomplete.
VVRI 1062. 11ff. Copied in Saṃ. 1829 = A.D. 1772. With a vyākhyā.
Benares (1963) 36994. 7ff. Copied in Śaka 1706 = A.D. 1784. With a bhāṣya.
RORI Cat. II 5317. 16ff. Copied by Manasārāma in Saṃ. 1845 = A.D. 1788. With the *Dīpikā* of Rāma.
RORI Cat. II 6108. 29ff. Copied by Ānandakṛṣṇa in Saṃ. 1854 = A.D. 1797. With the *Dīpikā* of Rāma.
RORI Cat. II 5619. 13ff. Copied by Vrajavāsī Miśra at Kāśī in Saṃ. 1895 = A.D. 1837. With the *Dīpikā* of Rāma.
AS Bengal 6903 (G 1604). 25ff. Copied on 1 śuklapakṣa of Caitra in Saṃ. 1899 = 10 April 1842. With the *Dīpikā* of Rāma.
ABSP 1179. 21ff. Copied in Saṃ. 1903, Śaka 1768 = A.D. 1846. With the *Dīpikā* of Rāma.
RORI Cat. II 6885. 36ff. Copied by Lakṣmīcanda Lālā in Saṃ. 1903 = A.D. 1846. With the *Dīpikā* of Rāma.
Baroda 9191. 25ff. Copied in Saṃ. 1918 = A.D. 1861. With the *Dīpikā* of Rāma.
Alwar 1913. With the *Dīpikā* of Rāma. 3 copies.
Ānandāśrama 3456.
AS Bengal 6905 (G 1355). 11ff. With the *Dīpikā* of Rāma.
AS Bengal 6906 (G 1763). 24ff. With the *Dīpikā* of Rāma.
Baroda 3259. 20ff. With the *Dīpikā* of Rāma.
Baroda 9267. 17ff. With the *Dīpikā* of Rāma.
Benares (1963) 35498 = Benares (1909–1910) 1925. 8ff. With his own vivaraṇa.
BM 465 (Add. 14,365k). 8ff. With his own vivaraṇa. From Major T. B. Jervis. See SATE 12.
Bombay U 375. 15ff. With his own vivaraṇa and the *Pratodayantra* of Gaṇeśa.
BORI 847 of 1884/87. 12ff. With his own vivaraṇa. From Gujarāt.
BORI 974 of 1886/92. 6ff.
BORI 874 of 1887/91. 49ff. With the *Dīpikā* of Rāma. From Mahārāṣṭra.
CP, Kielhorn XXIII 123. 11ff. Property of Javāhara Śāstrī of Chāndā.
CP, Kielhorn XXIII 124. 28ff. With the *Dīpikā* of Rāma. Property of Balīrāma Subhājī of Chāndā.
IO 2909 (1989). 16ff. With his own vivaraṇa and the *Pratodayantra* of Gaṇeśa. From Dr. John Taylor.
Jammu and Kashmir 1922. 8ff. With the *Dīpikā* of Rāma.
Jammu and Kashmir 2826. 77ff. With the *Dīpikā* of Rāma.
Mysore (1922) 4440. 17ff.
Nagpur 1663 (1230). 3ff. From Nasik.
Nagpur 1664 (1546). 7ff. From Nasik.
Nagpur 1665 (1548). Ff. 8–11. From Nasik.
Oxford 1535 (Sansk. d. 203) = Hultzsch 320. 21ff. With the *Dīpikā* of Rāma.
PUL II 3829. 29ff. With the *Dīpikā* of Rāma.
RORI Cat. III 15456. 48ff. With the *Dīpikā* of Rāma.
SOI 9416. With the *Cābukayantra* of Gaṇeśa.

The *Yantracintāmaṇi* was published with Cakradhara's vivṛti, Rāma's *Yantradīpikā*, and his own Hindī bhāṣānuvāda, by Bhāgīrathīprasāda Śarman at Benares in 1883 (IO 996); and edited with Hindī and Saṃskṛta ṭīkās by Sundaradeva Śarman, Mathurā 1898 (BM 14053. c. 56. (4)). The last verse is:

āsīd agrajarājavanditapadaḥ śrīvāmano viśruto
jyotiḥśāstramahārṇavāmṛtakaras
tatsūktiratnākaraḥ/
tatsūnuḥ kṣitipālamaulivilasadratnaṃ
grahajñāgraṇīś
cakre cakradharaḥ kṛtī savivṛtiṃ
sadyantracintāmaṇim//

The manuscripts of his vivaraṇa are:

Benares (1963) 35341 = Benares (1878) 115 = Benares (1869) XXIV 12. 10ff. Copied in Saṃ. 1732 = A.D. 1675.
BORI 43A of 1898/99. 13ff. Copied in Saṃ. 1822 = A.D. 1765.
Benares (1963) 37086. 10ff. Copied in Saṃ. 1858 = A.D. 1801.
Benares (1963) 35498 = Benares (1909–1910) 1925. 8ff.
Benares (1963) 37049. 2ff.
BM 465 (Add. 14,365k). 8ff. From Major T. B. Jervis. See SATE 12.
Bombay U 375. 15ff.
BORI 847 of 1884/87. 12ff. From Gujarāt.
IO 2909 (1989). 16ff. With the *Pratodayantra* of Gaṇeśa. From Dr. John Taylor.
N-W P II (1878) B 12. 14ff. Property of Mākhanji of Mathurā.

Verse 1 is:

vijānatāṃ golamodo ᵓsti gamyaṃ
tasmāt pareṣāṃ sugamaṃ yato naḥ/
sadyantracintāmaṇināmadheyaṃ
nijapraṇītaṃ vivṛṇomi yantram//

The colophon begins: iti śrīvāmadevasutatantrajñasiṃhacakradharaviracitaṃ.

PAṆḌITA CAKRADHARA (*fl.* 1920)

The son of Paṇḍita Lakṣmīdhara and a resident of Devaprayāga in Gaḍhavāla, Cakradhara completed his bhāṣāṭīkā and udāharaṇa on the *Pañcāṅgamañjūṣā* of Mukundarāma (*fl.* 1910) and his *Sūtrapañcāṅgasāraṇī* on Friday 10 śuklapakṣa of Āśvina in Śaka 1842 = 22 October 1920. These were published with the *Pañcāṅgamañjūṣā* at Mumbayī in 1922.

The colophon begins: iti śrīgaḍhavāladeśāntargataśrīdevaprayāgakṣetranivāsiśrīmatpaṇḍitalakṣmīdharātmajaśrīmatpaṇḍitacakradharajyotirvitkṛtāyāṃ.

CAKRAPĀṆI

Author of a *Kīrtivardhana* in 3 sections:

1. melāpakathana.
2. varṣamāsatithikanyāśuddhikathana.
3. lagnaśuddhikathana.

Manuscript:

AS Bengal 2784 (G 6405). Ff. 1–13 and 15–16.

The colophon begins: iti śrīcakrapāṇiviracite.

CAKRAPĀṆI

The son of Satyarūpā and Satyadhara, Cakrapāṇi wrote two works on jyotiḥśāstra.

1. *Jātakendu*. Manuscript:

Kathmandu (1960) 121 (I 1538). 31ff.

The first verse is:

natvā śrīmatpādapadmaṃ murārer
brahmeśādyaiḥ saṃsmṛtaṃ cittapīṭhe/
daivajñānāṃ tuṣṭaye cakrapāṇis
triskandhajño jātakenduṃ vidhatte//

The colophon begins: iti śrīmatsatyadharapaṇḍitātmajaśrīcakrapāṇiviracito.

2. *Praśnatattva*. Manuscripts:

RORI Cat. III 10996(6) 19ff. Copied in Saṃ. 1857 = A.D. 1800.
Kathmandu (1960) 230 (III 344). 15ff. Copied by Maheśvararāja on 2 kṛṣṇapakṣa of Pauṣa in Nep. Saṃ. 949 = 10 January 1830.
Benares (1963) 36420 = Benares (1903) 1070. 25ff. Copied in Saṃ. 1891, Śaka 1756 = A.D. 1834.
Poleman 4713 (U Penn. 698). 12ff. Copied in Saṃ. 1897, Śaka 1762 = A.D. 1840.
PUL II 3653. 15ff. Copied in Saṃ. 1919 = A.D. 1862.
Alwar 1848. 2 copies.
Benares (1963) 34312. Ff. 1–4 and 1f. Incomplete.
CP, Hiralal 3167. Property of Devnāth of Ḍoṅgargaon, Bhaṇḍārā.
Jammu and Kashmir 2920. 37ff.
Mithila 185. 10ff. Maithilī. Property of Pandit Lakṣmīvallabha Jhā of Bhakharaini, Madhepur, Darbhanga.
RORI Cat. II 5811. 16ff.
VVRI 2573. 10ff.

The second verse is:

cakrapāṇir iti satyadharasya
khyāta ātmaja ihācyutabhaktaḥ/
prārthitaḥ sa kurute bahuśiṣyaiḥ
praśnatattvam atilāghavam ādyam//

The last verse is:

śrīmatsatyadharādyaṃ
suṣuve sādhvīha satyarūpākhyā/
upakṛtaye śiṣyānāṃ
cakrapāṇinā tena racitaṃ hi//

CAKRAPĀṆI

The son of Kāmarāja, the son of Vāsudeva, a resident of Balālasaṃjñanagara, Cakrapāṇi wrote two works on jyotiḥśāstra. He is quoted by Mahādeva (*fl.* 1520) in his *Jayalakṣmī*.

1. *Jyotirbhāskara*. Manuscript:

Mitra, Not. 2825. 67ff. Copied in Śaka 1686 = A.D. 1764. Property of Rājā Rāmacānd of Naśīpur, Zillā Murshidābād.

The first verse is:

natvā girijayā sārdhaṃ giriśaṃ jagatāṃ gurum/
jyotirbhāskarasaṃjño ᵓsau kriyate cakrapāṇinā//

The colophon begins: iti śrīmahāmahopādhyāyaśrīcakrapāṇikṛto.

2. *Vijayakalpalatā*. Manuscripts:

Anup 5139. 46ff. Copied by Makaranda Vaiṣṇava at Govindasaṃnidhi on the bank of the Kaḷindī at Vṛndāvana in Saṃ. 1711 = A.D. 1654. Formerly the property of Haricaraṇa Miśra.
Jaipur (II). 79ff. Copied in Saṃ. 1717 = A.D. 1660.
Alwar 1964.
Anup 5140. 38ff. Incomplete.
BORI 209 of 1883/84. 23ff. From Gujarāt.
GJRI 1092/204. 11ff. Incomplete (ends at III 182).
Jaipur (II). 21ff. (*Vijąyabhairavakalpalatā*).
Oxford 1587 (Sansk. f. 18) = Hultzsch 330. 20ff. (17ff. according to Hultzsch).
RORI Cat. III 11329. 52ff. (ff. 35–39 missing). Incomplete.
SOI 6031 = SOI (List) 390.

CAKRAPĀṆI

Author of a *Tithinirṇaya*. Manuscript:

Rajputana, p. 35. From Udaipur.

See Cakrapāṇi Pāṭhaka.

CAKRAPĀṆI

Author of 2 sets of astronomical tables.

1. *Pañcāṅgasāraṇī*. Manuscript:

Kathmandu (1960) 218 (I 1177). 11ff. Nevārī.

2. *Sūryagrahaṇasāraṇī*. Manuscript:

Kathmandu (1960) 498 (I 1177). 40ff. Nevārī.

CAKRAPĀṆI

Author of a *Muhūrtamālā*. Manuscripts:

Anup 4992. 10ff.
Jaipur (II). 8ff.

CAKRAPĀṆI

Author of a *Līlāvatī* in Hindī. Manuscript:

NPS 108 of Saṃ. 2001–2003. Property of the Nāgarīpracāriṇī Sabhā (Yājñika Saṅgraha), Vārāṇasī.

CAKRAPĀṆI PĀṬHAKA

Author of a vyākhyā on the *Tithiprakāśa* of Gaṅgādāsa Trivedin. Manuscripts:

Mithila I 163. 16ff. Maithilī. Copied on Friday 11 kṛṣṇapakṣa of Kārttika in Śaka 1700 = 13 November 1778. No owner mentioned.
Mithila I 163 A. 7ff. Maithilī. Incomplete. Property of Babu Chandradhārī Singh of Rauti Deaurhī, Madhubani, Darbhanga.

The last verse is:

samyak samālokya sato nibandhān
prītyai janānām atilālasānām/
svalpākṣarair eva tithiprakāśaṃ
vyākhyātavān pāṭhakacakrapāṇiḥ//

CAKRAPĀṆI MIŚRA

Author of a *Vyavahārādarśa* in 11 ucchvāsas. Manuscripts:

BORI 247 of 1884/87. 56ff. Copied on Sunday 2 śuklapakṣa of Jyeṣṭha in Saṃ. 1806 = 7 May 1749.
Munich 364. 136pp. Copied from BORI 247 of 1884/87.

The colophon begins: iti śrīmiśracakrapāṇiviracite.

CAKRAPĀṆI PĀṬHAKA (fl. 1874)

Author of a Hindī ṭīkā on the *Muhūrtacintāmaṇi* of Rāma (*fl.* 1600), published at [Benares] in 1874 (IO 465).

CAKRAVARTIN

See Kavicūḍāmaṇi.

CAKRAVARTIN

Author of a *Bhāsvatīpaddhati*, apparently based on the *Bhāsvatī* of Śatānanda (*fl.* 1099). *Cf.* also Cakravipradāsa. Manuscript:

BORI 842 of 1887/91. 29ff. (f. 1 missing). Copied in Saṃ. 1710 = A.D. 1653. From Gujarāt.

CAKRAVIPRADĀSA

Alleged author of a ṭīkā on the *Bhāsvatī* of Śatānanda (*fl.* 1099). Manuscript:

Oudh XIII (1881) VIII 15. 40pp. Property of Mahanta Nanda Gopāla of Lucknow Zila.

CAÑGADEVA (fl. ca. 1200/1220)

The earliest known member of Cañgadeva's family was Trivikrama of the Śāṇḍilyagotra; his son, Bhāskara Bhaṭṭa, was given the title of Vidyāpati by Bhojarāja, the Paramāra Mahārāja of Dhārā (*ca.* 995/1056); and Bhāskara's descendents in order were: Govinda, Prabhākara, Manoratha, Maheśvara (*fl.* 1114), Bhāskara (b. 1114), Lakṣmīdhara, who served at the court of the Yādava Jaitrapāla. Maheśvara's younger son was Śrīpati, whose son, Gaṇapati, was the father of Anantadeva (*fl.* 1222). Lakṣmīdhara's son was Cañgadeva, who is mentioned (in an inscription at Pāṭṇā near Chalisgaon in Khandesh recording the endowment by Soïdeva the Nikumba on 9 August 1207 of a maṭha for the study of the works of Bhāskara (b. 1114)) as the astrologer of Siṅghaṇa, the Yādava ruler of Devagiri from 1209/10, and the founder of the maṭha. See Bhau Daji [1865]; F. Kielhorn [1888/92]; and S. B. Dikshit [1896] 247–248.

Verses 17–24 of the inscription are:

śāṇḍilyavaṃśe kavicakravartī
trivikramo ᵓbhūt tanayo ᵓsya jātaḥ/
yo bhojarājena kṛtābhidhāno
vidyāpatir bhāskarabhaṭṭanāmā//
tasmād govindasarvajño jāto govindasaṃnibhaḥ/
prabhākaraḥ sutas tasmāt prabhākara ivāparaḥ//
tasmān manoratho jātaḥ satāṃ pūrṇamanorathaḥ/
śrīmān maheśvarācāryas tato ᵓjani kavīśvaraḥ//
tatsūnuḥ kavivṛndavanditapadaḥ sadvedavidyālatākandaḥ kaṃsaripuprasāditapadaḥ
sarvajñavidyāsadaḥ/
yacchiṣyaiḥ saha ko ᵓpi no vivadituṃ dakṣo vivādī kvacic
chrīmān bhāskarakovidaḥ samabhavat
satkīrtipuṇyānvitaḥ//
lakṣmīdharākhyo ᵓkhilasūrimukhyo
vedārthavit tārkikacakravartī/
kratukriyākāṇḍavicārasāra-
viśārado bhāskaranandano ᵓbhut//
sarvaśāstrārthadakṣo ᵓyam iti matvā purād ataḥ/

jaitrapālena yo nītaḥ kṛtaś ca vibudhāgraṇīḥ//
tasmāt sutaḥ siṃghaṇacakravarti-
daivajñavaryo ᵓjani caṅgadevaḥ/
śrībhāskarācāryanibaddhaśāstra-
vistārahetoḥ kurute maṭhaṃ yaḥ//
bhāskararacitagranthāḥ
siddhāntaśiromaṇipramukhāḥ/
tadvaṃśyakṛtāś cānye
vyākhyeyā manmaṭhe niyamāt//

CAṆḌIKA

Author of a *Muhūrtarāja*. Manuscripts:

Tanjore D 11572 = Tanjore BL 4290. 12ff. Incomplete.
Tanjore D 11573 = Tanjore BL 4295. 27ff. Incomplete.
Tanjore D 11574 = Tanjore BL 4296a. 14ff. Incomplete.

The first verse is:

gaṇeśaṃ giraṃ padmajanmācyuteśān
gurūṃś caṇḍikaś caṇḍikāṃ gotradevīm/
grahān sūryamukhyān munīn arthasiddhyai
namaskṛtya kurve muhūrtādhirājam//

CAṆḌĪCARAṆA SMṚTIBHŪṢAṆA (*fl.* 1883/1901)

Author of a ṭīkā on the *Tithitattva* of Raghunandana (*fl.* 1520/1570), 2nd ed., Calcutta 1901 (BM 14033. bb. 45. (3) and NL Calcutta 180. Jc. 90. 76), 3rd ed., Calcutta 1906 (IO 21. E. 5 and NL Calcutta 180. Jc. 90. 90). He also wrote with Bhūtanātha Vidyāratna a Bengālī bhāṣānuvāda of the *Śuddhidīpikā* of Śrīnivāsa, published at Calcutta in 1883 (IO 9. D. 2), 2nd ed. Calcutta 1901 (BM 14053. ccc. 33).

CAṆḌĪDATTA (*fl.* 1874)

Author of a Hindī ṭīkā on the *Śīghrabodha* of Kāśīnātha, published at Lucknow in 1874 (IO 920).

CAṆḌĪDĀSA

The son of Rāghava, Caṇḍīdāsa wrote a ṭīkā on the *Karaṇakutūhala* (1183) of Bhāskara (b. 1114). Manuscripts:

AS Bengal 6840 (G 7749). 49ff. Copied by Dhanasundara, the pupil of Bhaṭṭāraka Śrīkakka Sūri at Bīkānera in Saṃ. 1716 = A.D. 1658.
Florence 262. 46ff. Incomplete.
RORI Cat. II 7885. 31ff. Copied by Naiṇasāgara at Ākolāgrāma.

The first verse is:

natvā devaṃ trinetraṃ prakaṭitavibhavaṃ
sarvakartāram ekaṃ
śrautasmārtakriyāyāṃ nipuṇataramate
rāghavasyātmajo ᵓham/
jñātvā siddhāntatattvaṃ suguruvacanataḥ
śiṣyaśikṣārtham etac
caṇḍīdāsaḥ subodhaṃ karaṇam atitarāṃ
bhāskarīyaṃ karomi//

CAṆḌŪ JYOTIṢĪ

A resident of Jodhapura, Caṇḍū wrote a *Caṇḍavāṇī sāriṇī*. Manuscript:

RORI Cat. II 9534. 34ff.

CAṆḌŪ (*fl.* 1769/1841)

Author of pañcāṅgas for various years from Saṃ. 1826 = A.D. 1769 to Saṃ. 1898 = A.D. 1841. Manuscripts:

RJ 3019 (vol. 4, p. 285). 23 years.

CAṆḌEŚVARA

Alleged author of a *Gaurījātaka*. Manuscript:

WHMRL Q. 23. k.

CAṆḌEŚVARA

Author of a *Caṇḍeśvarajātaka*. Manuscripts:

BORI 307 of 1882/83. 26ff. Copied in Saṃ. 1814 = A.D. 1757. (aṣṭakavarga).
AS Bengal 6978 (G 7893). 13ff. Copied by Nandarāma on Friday 8 śuklapakṣa of Phālguna in Saṃ. 1882 = 17 March 1826. (aṣṭakavarga).
Anup 4554. 11ff.

CAṆḌEŚVARA

Alleged author of a *Jñānapradīpa*. Manuscript:

Oudh VIII (1876) VIII 16. 136pp. Property of Kṛṣṇadatta Śakadvīpī of Bārābanki Zillah.

CAṆḌEŚVARA (*fl.* 1185)

A vājapeyasomayājin from Mithilā, Caṇḍeśvara wrote a bhāṣya on the *Sūryasiddhānta* in which he uses as an example Tuesday 2 śuklapakṣa of Caitra in Śaka 1107, Kali 4286 = 5 March 1185. He refers to the commentary on the *Sūryasiddhānta* by Mallikārjuna (*fl.* 1178). Manuscripts:

AS Bombay 293. 64ff. Copied by Jyotirvittilaka Nārada, the son of Bhīma, the son of Govardhana, at Kāpikāsthāna in Saṃ. 1452, Śaka 1318 = A.D. 1395/96. Incomplete (adhyāyas 11–13). From Bhāu Dājī.
Kathmandu (1960) 505 (I 1165). Ff. 11–224. Maithilī. Copied by Kāmbhūśarman of Ratnapuranagara on Monday 7 śuklapakṣa of Phālguna in Lakṣmaṇa Saṃ. 392 = *ca.* 14 February 1502. Incomplete.

Kathmandu (1960) 504 (IV). 214ff. Nevārī. Copied by Jayakīrtirāja at Bhaktapattana on Tuesday 14 śuklapakṣa of Āṣāḍha in Nep. Saṃ. 665 = 23 June 1545 during the reign of Jayaprāṇamalladeva (Prāṇamalla ruled Bhatgaon from 1519 to 1547).
Baroda 3351. 59ff. Copied in Saṃ. 1716 = A.D. 1659.
Jaipur (II). 17ff. Copied in Saṃ. 1763 = A.D. 1706.
AS Bengal 6941 (G 10758). 166ff. Nevārī. Copied on 11 śuklapakṣa of Phālguna in Nep. Saṃ. 830 = 27 February 1710.
PL, Buhler IV E *448. 73ff. Copied in Saṃ. 1786 = A.D. 1729. Property of Bālakṛṣṇa Jośī of Ahmadābād. Buhler notes another copy.
Alwar 2025 = Rajputana, p. 57.
Benares (1963) 36079 = Benares (1910–1911) 2008. Ff. 43–78 and 123–126. Incomplete.
BORI 886 of 1884/87. 98ff. Incomplete. From Gujarāt.
BORI 600 of 1895/1902. 170ff. Incomplete.

The colophon begins: iti śrīmaithilavājapeyasomayājiśrīcaṇḍeśvarācāryaviracita.

Caṇḍeśvara apparently also wrote in 35 adhyāyas a *Praśnavidyā*, sometimes known as the *Praśnacaṇḍeśvara*, though this title is also borne by other works, notably those of Baudhācārya and of Rāmakṛṣṇa. Though the first verse of the *Praśnavidyā* is identical with that of the *Sūryasiddhāntabhāṣya:*

namas te paramātmaikarūpāya paramātmane/
svecchāvabhāsitāśeṣadehābhinnāya śambhave//

the colophon and the next to the last verse cited below fit in better with the Caṇḍeśvara (*fl.* 1314) discussed below. The question of authorship must, then, for the moment be left open. The manuscripts of the *Praśnavidyā* are:

Anup 4556. 44ff. Copied by Narasiṃha, the son of Trivikrama Śukla of the Sārasvatajñāti, at Āgarā in Mālava in Saṃ. 1620 = A.D. 1563 during the reign of Akbar (1556/1605). Property of Anūpasiṃha (1674/98).
Anup 4555. 113ff. Copied by Mālīkacarā at Ādūṇī in Saṃ. 1749 = A.D. 1692.
Poleman 5135 (U Penn 1881). 4ff. Copied Saṃ. 1812 = A.D. 1755. No author mentioned.
Benares (1963) 36464. 65ff. Copied in Saṃ. 1821 (read 1817), Śaka 1682 = A.D. 1760.
Poleman 4714 (U Penn 1835). Ff. 21–104. Copied in Saṃ. 1857 = A.D. 1800. Incomplete (begins at end of adhyāya 9).
BORI 164 of A 1883/84. 69ff. Copied in Saṃ. 1869 = A.D. 1812.
Śāstrī, Not. 1907. 193. 17ff. Copied by Haladhara Śarman in Śaka 1744 = A.D. 1822. Property of Paṇḍit Śrīpati Bhaṭṭācaryya of Khunvera, Garvetagram, Zilla Midnapur.
Oudh (1876–1878) VIII 1. 152pp. Copied in A.D. 1835. Property of Mannālāla of Tirwā, Lucknow Zila.
Alwar 1847.
AS Bengal 7154 (G 8118). 55ff. Bengālī.
Benares (1963) 35072. Ff. 9, 11, 14–26, 30–34, 36–38, 40–41, and 43, and 2ff. Incomplete. No author mentioned.
Benares (1963) 36465. 25ff. Incomplete.
Benares (1963) 37455 = Benares (1909) 1825. 16ff. Incomplete.
CP, Hiralal 3160. Property of Kuñjrām of Aḍbhār, Bilāspur.
Florence 308. 33ff. Incomplete (ends in adhyāya 11).
Kathmandu (1960) 92 (I 245). 10ff. Incomplete.
Kurukṣetra 648 (19799).
LDI 7358 (7056). 42ff.
Oudh VII (1875) VIII 11. 160pp. Property of Jānakīprasāda of Bārābānki Zillah.
Oudh VIII (1876) VIII 10. 33pp. Property of Raghunātha of Bārābanki Zillah.
Oudh XIX (1887) VIII 5. 228pp. Property of Gaṅgādhara Vājapeyin of Unao Zila.
Oxford 1549 (Sansk. d. 186) = Hultzsch 282. 63ff.
PUL II 3662. Ff. 2–15 (ff. 1 and 4 missing). Incomplete (to adhyāya 8).
RORI Cat. III 10996(10). 75ff.
SOI 5990 = SOI (List) 370.
VVRI 2542. 13ff. Incomplete. No author mentioned.
WHMRL N. 155. No author mentioned.

The next to the last verse is:

candrasya vittāgālato vivektā
taṅgole (?) rakṣāstutakarmakartā/
mantreṇa sarvaṃ gālataṃ bravīti
caṇḍeśvaro mantrakṛtām variṣṭhaḥ//

The colophon begins: iti mahārājādhirājaśrīcaṇḍeśvarācāryaviracitāyāṃ.

Another text ascribed to Caṇḍeśvara with a similar initial verse is the *Tithinirṇaya*, which begins:

svecchavibhāvitaḥ śeṣabhedabhinnāya śambhave/
kālau vidhibaladarśapūrṇamāsyādikarmasu//

Manuscripts:

AS Bengal 2663 (G 6557). 26ff. Incomplete.
Śāstrī, Rep (1901–1906), p. 4. (*Kālanirṇaya*).

The colophon begins: iti tithinirṇaye caṇḍeśvarācāryaviracite.

CAṆḌEŚVARA ṬHAKKURA (*fl.* 1314)

Karmāditya, the son of Harāditya, the son of Viṣṇu, had two sons, Devāditya and Bhavāditya. Devāditya, a minister to the Mahārāja of Tīrabhukti, was the father of Vīreśvara, Dhīreśvara, Gaṇeśvara, Jaṭeśvara, Haradatta, Lakṣmīdatta, and Śubhadatta.

Vīreśvara became Mahāsandhivigrahika, and this position was inherited by his son, Caṇḍeśvara, who served the Kārṇāṭa lord of Mithilā, Harisiṃhadeva (*ca.* 1303/1324); Caṇḍeśvara claims to have conquered the king of Nepāla and to have had himself weighed in gold on the bank of the Vāgvatī in Śaka 1236 = A.D. 1314 in two verses of his *Vivādaratnākara:*

śrīcaṇḍeśvaramantriṇā matimatānena prasannātmanā
nepālākhilabhūmipālajayinā
dharmendudugdhābdhinā/
vāgvatyāḥ saritas taṭe suradhunīsāmyaṃ dadhatyāḥ
śucau
mārge māsi yathoktapuṇyasamaye dattas
tulāpuruṣaḥ//

. . . .

rasaguṇabhujacandraiḥ saṃmite śākavarṣe
sahasi dhavalapakṣe vāgvatīsindhutīre/
gadita tulitam uccair ātmanā svarṇarāśiṃ
nidhir akhilaguṇānām uttamaḥ somanāthaḥ//

See P. V. Kane [1930/62] vol. 1, pp. 366–372; U. Mishra [A 3. 1930]; B. Bhattacharya [1941], [1944/45], and [1965/67] 17–37; and U. Thakur [A 3. 1969].

Caṇḍeśvara's major work was the *Smṛtiratnākara* in seven sections, of which the first, the *Kṛtyaratnākara* in 22 taraṅgas, is of interest to us. Manuscripts:

AS Bengal 2662 (G 3604). Ff. 1–13, 16, 40–61, and 100–160. Bengālī. Copied at Vavambhauligrāma on Friday 9 kṛṣṇapakṣa of Vaiśākha I in Lakṣmaṇa Saṃ. 392 = 24 April 1500. Incomplete.
Paris BN 788 (Sanscrit Bengali 150). 247ff. Bengālī. Copied in A.D. 1570.
Dacca 1055 C. See NCC vol. 4, p. 278.
IO 1387 (989). 258ff. Bengālī. From H. T. Colebrooke.

The *Kṛtyaratnākara* was edited by Kamalakṛṣṇa Smṛtitīrtha, *BI* 237, Calcutta 1921–1925.

Verse 15 is:

nepālaṃ giridurgamaṃ tu javanād unmūlya
tadbhūpatīn
sarvān rāghavavaṃśajān aripos
tulyapratāpānalaiḥ/
devaṃ viśvavarapradaṃ paśupatiṃ saṃspṛśya
yo ᵓpupūjat
keṣāṃ naiṣa dharātale stutipadaṃ
mantrīndracaṇḍeśvaraḥ//

The colophon begins: iti saprakriyamahāsāndhivigrahikaṭhakkuraśrīvīreśvarātmajasaprakriyamahāsāndhivigrahikaṭhakkuraśrīcaṇḍeśvaraviracite.

Caṇḍeśvara also wrote a *Kṛtyacintāmaṇi* on jyotiḥśāstra in dharmaśāstra. Manuscripts:

Kathmandu (1960) 41 (I 1304). 356ff. Nevārī. Copied in Nep. Saṃ. 709 = A.D. 1589.
IO 1621 (1274b and 1492b). 129ff. Bengālī. Copied in A.D. 1806. From H. T. Colebrooke.
CP, Hiralal 989. Property of Kāśīdatt of Khairāgaṛh, Khairāgaṛh State.
Kathmandu (1960) 40 (I 1311). 341ff.
Kathmandu (1960) 42 (I 1047). 7ff. Nevārī. Incomplete (svapnaphalakathanaprakāśa).
Mithilā. See NCC, vol. 4, p. 275.

The last one and a half verses are:

jyotirjñāś ca mahītale sumanasaḥ sarve kṛtārthāḥ
kṛtāḥ
śrīcaṇḍeśvaramantriṇā viracitā yatkṛtyacintāmaṇiḥ//
nepālādyā vipakṣā raṇabhuvi vijitā yena
yadbhūriyajñaiḥ
santuṣṭaḥ svargaloke harir api mahitā yena
jīveśvarādyāḥ/
śrīmanmantrīśacaṇḍeśvara iti vidito yo
mahīkalpavṛkṣas
tasyācandrārkam eṣā kṛtir iha vimalā rājatāṃ
viśvavandyā//

CATURAVIJAYA GAṆI

The pupil of Muktivijaya, Caturavijaya wrote a stabaka in Old Rājasthānī on the *Muhūrtacintāmaṇi* of Rāma (*fl.* 1600). Manuscripts:

RORI Cat. III 10883. 66ff. Copied by Ṛddhivijaya at Āgarā in Saṃ. 1808 = A.D. 1751 during the reign of Ahammada Śāha (1748/54).
RORI Cat. II 6293. 102ff. (ff. 1–8 missing). Copied in Saṃ. 1880 = A.D. 1827.
RORI Cat. II 4272. 97ff.

CATURTHĪLĀLA ŚARMAN (*fl.* 1917)

Author of a *Muhūrtaprakāśa* on which he wrote a Hindī ṭīkā, *Caturthīlālī*, published at Bombay in Saṃ. 1974 = A.D. 1917 (IO 28. K. 5).

MAHĀPAṆḌITA CATURBHUJA

Author of an *Adbhutasāgarasāra*. Manuscripts:

Mithila 8 A. 45ff. Maithilī. Copied in Śaka 1709 = A.D. 1787. Property of Pandit Ravinātha Jhā, professor at M. R. Vidyālaya and resident of Andhrā Ṭhārhi, Darbhanga.
Mithila 8. 38ff. Maithilī. Copied by Bhāī Jīvaśarman on Wednesday 7 śuklapakṣa of Vaiśākha in Śaka 1789, Sāl. San. 1275 = 8 May 1867. Property of Pandit Vāsudeva Jhā of Bārāhi, Nowhaṭṭā, Bhagalpur.
Mithila 8 B. 28ff. Maithilī. Incomplete. Property of Pandit Gaṅgādhara Jhā of Jonkī, Deodhā, Darbhanga.

Mitra, Not. 1930. 96ff. Maithilī. Property of Paṇḍita Yāgeśvara Pāṭhaka of Mujonā, Tājapura, Darbhāṅgā.

The colophon begins: iti mahāpaṇḍitaśrīcaturbhujaviracita.

CATURBHUJA

Author of a ṭīkā or udāharaṇa on a *Paddhatibhūṣaṇa*. Manuscripts:

BORI 527 of 1895/1902. 54ff. Incomplete.
SOI 6006 = SOI (List) 382. Ascribed to Caturbhuja Murāri Vyāsa.
SOI 8166. No author mentioned.

CATURBHUJA

Author of a ṭīkā on a *Sṛṣṭikaraṇa*. Manuscript:

BORI 207 of A 1883/84. 30ff.

CATURBHUJA MIŚRA

Author of a ṭīkā or vivaraṇa on the *Jyotiṣaratnamālā* of Śrīpati (*fl.* 1040). Manuscripts:

Alwar 1793.
Benares (1963) 35064. Ff. 1–51 and 63–82. Incomplete.

CANDRA NṚPATI

Author of a *Lagnadarpaṇa*. Manuscript:

Paris BN 969 (Sanscrit Bengali 202) III = Guérin 52. Bengālī. Copied in A.D. 1840.

MUNI CANDRA SŪRI

Author of two works in Prākṛta.

1. *Kālavicāraśataka*. Manuscript:

LDI (NC) 2040/1. 2ff.

2. *Candrasūryamaṇḍalavicāra*. Manuscripts:

LDI (NC) 2041/1. 1f.
LDI (NC) 2041/2. 2ff.

CANDRA SŪRI (*fl. ca.* 1150)

See Śrīcandra Sūri (*fl. ca.* 1150).

CANDRA (*fl.* 1771)

Author of a *Candraprakāśa* in Hindī in Saṃ. 1828 = A.D. 1771. Manuscript:

NPS 145 of 1906–08. Copied in Saṃ. 1886 = A.D. 1829. Property of Lālā Vidyādhara of Haripurā, Datiyā.

CANDRAKARA

Author of a *Granthasaṅgraha*. Manuscript:

Mithilā. See NCC, vol. 6, p. 345.

CANDRAKĀNTA TARKĀLAÑKĀRA (1836/1909)

Professor of philosophy at the Calcutta Sanskrit College from 1883 to 1887, and a resident of Serapura, MM. Candrakānta wrote a ṭippaṇa on the *Kālanirṇaya* of Mādhava (*fl. ca.* 1350/75) at Calcutta in Śaka 1809 = A.D. 1887. This was published with the *Kālanirṇaya* as *BI* 101, Calcutta 1890, and at Kalyāṇa-Muṃbaī in Śaka 1838, Saṃ. 1973 = A.D. 1916.

CANDRAKĪRTI

Alleged author of a ṭīkā, *Kārtabuddhivilāsinī*, on a *Sārasvata*. Manuscript:

N-W P V (1880) A 24. 223ff. Property of Pandit Mākhana Misra of Muttra.

CANDRACŪḌA BHAṬṬA PAURĀṆIKA (*fl.* 1610)

The son of Umaṇa Bhaṭṭa (or Umāpati Bhaṭṭa), the son of Dharmā Bhaṭṭa, Candracūḍa wrote a *Kālasiddhānta* = *Nirṇayasiddhānta* in Saṃ. 1667 = A.D. 1610. Manuscripts:

Kerala 3481 (7618). 1250 granthas. Copied in Śaka 1656 = A.D. 1734.
N-W P VII (1882) Dharmaśāstra 2 = N-W P VIII (1884) Dharmaśāstra 10. 80ff. Copied in Saṃ. 1800 = A.D. 1743. Property of Paṇḍita Bābūśāstrī Tailaṅga of Benares.
VVRI 3646. 43ff. Copied in Śaka 1694 = A.D. 1772.
CP, Kielhorn XIX 39. 463ff. Copied in Śaka 1709 = A.D. 1787. (*Kāladivākara*). Property of Dādā Āvaḷe of Chāndā.
Jammu and Kashmir 4102. 43ff. Copied in Saṃ. 1941 = A.D. 1884.
Adyar Index 1256 = Adyar Cat. 39 E 3. 130ff.
Alwar 1294.
Benares (1956) 13238. Ff. 1–2, 4–36, and 38–63. Incomplete.
Benares (1963) 35489 = Benares (1903) 1174. 80ff. According to Benares (1903) copied in Saṃ. 1882 = A.D. 1825. Incomplete.
BORI 528 of 1883/84. 27ff. Incomplete. From Mahārāṣṭra.
CP, Hiralal 845. Property of Dīnānāth of Singharī, Bilāspur.
CP, Hiralal 846. Property of Mohanlāl of Ratanpur, Bilāspur.
CP, Hiralal 847. Property of Govind Sundar Śāstrī of Piñjaḍ, Akolā.

CP, Hiralal 848. Property of Bājīrāv Śāstrī of Murtizāpur, Akolā.
CP, Hiralal 849. Property of Divākar Bhaṭṭ of Multāī, Betūl.
CP, Hiralal 850. Property of Prahlād Bhaṭṭ Lothe of Girolī, Akolā.
CP, Hiralal 851. Property of Sadāśiv Almulvār of Gaḍborī, Chāndā.
CP, Kielhorn XIX 48. 66ff. Property of Nārāyaṇabhaṭṭa of Chāndā.
Kerala 3480 (4625). 1250 granthas.
NS Press 113. See NCC, vol. 4, p. 38.
N-W P I (1874) Law 241 = N-W P I (1874) Law 285. 46ff. Property of Pūrṇānanda Jyotiṣī of Benares.

There is also a Gujarātī ṭīkā on the *Nirṇayasiddhānta* attributed to Candracūḍa with a query. Manuscript:

Baroda 1598. 13ff.

CANDRADATTA PANTA (b. 1915)

A resident of Kāśī, Candradatta wrote a *Varṣacandraprakāśa* in Hindī, which was published at Dillī-Vārāṇasī-Paṭanā in 1964; second ed., Dillī 1971; a *Candrahastavijñāna* in Hindī, which was published at Dillī-Vārāṇasī-Paṭanā in 1966; and a *Praśnacandraprakāśa* in Hindī, which was published at Dillī-Vārāṇasī-Paṭanā in 1970.

CANDRAPRABHA

Author of a *Candronmīlana*. Manuscripts:

PUL II 3380. 58ff. Copied in Saṃ. 1728 = A.D. 1671.
Baroda 3118. 41ff. (f. 22 missing).
PL, Buhler IV E 93. 19ff. Property of Mayāśaṅkara Jānī of Ahmadābād. Buhler notes another copy.
PUL II 3381. F. 2v. Incomplete (paṭala 4).
PUL II 3382. 33ff. With his own *Dīpikā*.

Candraprabha also wrote his own commentary, the *Dīpikā*, on the *Candronmīlana*. Manuscripts:

AS Bengal 7021 (G 10302). Ff. 13–24. Copied on Friday 13 kṛṣṇapakṣa of Śrāvaṇa in Saṃ. 1882 = 12 August 1825. Incomplete (ends with paṭala 27). No author mentioned.
BORI 810 of 1884/87. 37ff. No author mentioned.
Munich 368c. Ff. 7–24. Incomplete (begins with sūtra 18 of paṭala 10, ends in paṭala 27). No author mentioned.
PUL II 3382. 33ff.

CANDRAPRABHA (*fl.* 1398)

See Devānanda (*fl.* 1398).

VIPRA CANDRABHĀNU

Author of a ṭīkā on the *Gaurījātaka*. Manuscript:

Jaipur (II).

CANDRABHĀNU (*fl.* 1766)

Candrabhānu completed a ṭīkā, *Subodhajananī*, on the *Śīghrabodha* of Kāśīnātha on Friday 1 śuklapakṣa of Vaiśākha in Saṃ. 1823 = 9 May 1766. Manuscripts:

VVRI 2492. 77ff. Copied in Saṃ. 1823 = A.D. 1766.
Goṇḍal 399. 87ff. Copied in Saṃ. 1868 = A.D. 1811. Incomplete.
Alwar 1978.

CANDRABHĀLAMAṆI ŚĀSTRIN (*fl.* 1924)

Author of a *Jyautiṣadaivajñaprabhā* = *Muhūrtakalikā*, published at Ayodhyā in 1924 (IO San. D. 966 (a)).

CANDRAMA

A resident of Aḷiyapura, Candrama wrote a *Lokasvarūpa* in Kannaḍa in 125 verses. Manuscripts:

GOML Madras (Kannaḍa) D 408. 10ff. Karṇāṭakī. Copied by Kāpettina Dharanappa Heggade of Kārkala for the Rājā of Mangalore.
GOML Madras (Kannaḍa) D 409. 28pp. Karṇāṭakī.

KOVIDA CANDRAMAṆI (*fl.* 1720)

A protégé of Udyotasiṃha and Pṛthvīsiṃha, mahārājas of Oḍachā, Candramaṇi wrote the following works in Hindī on jyotiṣa:

1. *Muhūrtadarpaṇa*. Manuscript:

NPS 64 of 1929–31. Copied in Saṃ. 1839 = A.D. 1782. Property of Paṇḍita Śāligrāma Dūbe of Nandagavāṃ, Jaitapurakalāṃ, Āgarā.

2. *Ramalavicāra*. Manuscript:

NPS 243 of 1926–28. Copied in Saṃ. 1933 = A.D. 1866. Property of Gaṅgāviṣṇu Jyotiṣī of Banthara, Unnāva.

CANDRAŚEKHARA

Author of a *Praśnacuḍāmaṇi*. Manuscript:

Śāstrī, Not. 1911. 166. 17ff. Bengālī. Property of Babu Vaikuṇṭhanāth Chakravarti of Khunverā, Garvetā, Zilla Medinīpur.

CANDRAŚEKHARA PAṬANĀYAKA

Author of an udāharaṇa on the *Līlāvatī* of Bhāskara (b. 1114). Manuscript:

CP, Kielhorn XXIII 142. 169ff. Copied in Saṃ. 1874 = A.D. 1817. Property of Vaikuṇṭhamiśra of Sammalpur.

CANDRAŚEKHARA PAṬNAIKA

Author of a *Jātakaratnākara* with two ṭīkās, *Taraṇī* and *Prakāśikā*. Manuscript:

Cuttack 3. See NCC, vol. 6, p. 369.

CANDRAŚEKHARA ŚARMAN

A member of the Vārendra kula and a resident of Navadvīpa, Candraśekhara wrote a *Smṛtidurgabhañjana* in 4 adhyāyas; see S. C. Banerji [1957] 195–196. Manuscripts:

Mitra, Not. 937. 84ff. Bengālī. Copied in Śaka 1729 = A.D. 1807. Formerly property of Harinārāyaṇa Śarman. Property of Vrajanātha Vidyāratna of Navadvīpa.
AS Bengal 2819 (G 5139). 4ff., 3ff., and 13ff. Bengālī. Copied by Rādhākānta Śarman. Incomplete (adhyāyas 1–3).
Benares (1956) 12939. 15ff. Bengālī. Incomplete (*Tithidurgabhañjana*).
Benares (1956) 14076. 6ff. Bengālī. (*Tithidurgabhañjana*).
Calcutta Sanskrit College (Smṛti) 384. 29ff. Bengālī. Incomplete.
Calcutta Sanskrit College (II) 37 (8/163). 14ff. Bengālī. Incomplete (adhyāya 1).
Dacca 2293. See S. C. Banerji.
Mitra, Not. 339. 9ff. Bengālī. Incomplete (adhyāya 1). Property of Rājā Satīśacandra of Krishnanagar.
Mitra, Not. 4055. 68ff. Bengālī. Property of Akṣayacandra Bhaṭṭācārya of Pāntā, Zilla Varddhamāna.

The colophon begins: vārendrakulasaṃbhūtanavadvīpanivāsiśrīcandraśekharaśarmaviracite.

CANDRAŚEKHARA VĀCASPATI
(*fl. ca.* 1750/1800)

The grandson of Vidyābhūṣaṇa and the cousin of Jagannātha Tarkapañcānana (1695/1806), Candraśekhara wrote among other works a *Smṛtisārasaṅgraha*. Manuscripts:

Calcutta Sanskrit College (Smṛti) 203. 119ff. Bengālī. Copied on 15 Bhādrapada of Śaka 1730 = *ca.* 4 September 1808.
AS Bengal 2074 (G 3693). 69ff Bengālī.
Calcutta Sanskrit College (Smṛti) 204. 80ff. Bengālī.
IO 1490 (482). 117ff Bengālī. From H. T. Colebrooke.
Mitra, Not. 272. 103ff. Bengālī. Property of Bābū Yatīndra Mohana Ṭhākura of Calcutta.

The first verse is:

śivaṃ natvā smṛtes tattve kriyate sārasaṅgrahaḥ/
śrīvācaspatidhīreṇa vaidhakṛtyapravarttaye//

The colophon begins: iti śrīcandraśekharavācaspatibhaṭṭācāryakṛtaḥ.

CANDRAŚEKHARA SIṂHA (1835/1904)

MM. Sāmanta Candraśekhara Siṃha wrote a *Siddhāntadarpaṇa* in 5 adhikāras and a pariśiṣṭa in Orissa. This was edited by Jogeś Chandra Rāy, Calcutta 1899. For his lunar theory see P. C. Sengupta [1932a] 17–18.

CANDRAŚEKHARA JHĀ (*fl.* 1924)

The son of Veṇī and the pupil of Muralīdhara Jhā (*fl.* 1908/16), Candraśekhara, a resident of Mānecaukagrāma, Mojapharapura, was a professor of jyautiṣaśāstra at the Yugala Kiśora Rūīyā Saṃskṛta Pāṭhaśālā in Kāśī. He completed in Śaka 1846 = A.D. 1924 a ṭīkā, *Vyaktavāsanā*, on the *Līlāvatī* of Bhāskara (b. 1114). This was published at Benares in 1924.

CANDRAŚEKHARA GOPĀLAJĪ ṬHAKKURA
(*fl.* 1952/59)

Author of a *Jyotiṣavijñāna* in Gujarātī, published at Amadābād in 1952, 2nd ed. 1954, and 3rd ed. 1959, and of a *Daśāphaladarpaṇa* in Gujarātī, published at Muṃbaī in 1957.

CANDRAŚEKHARA PĀṬHAKA (*fl.* 1966)

Author of a Hindī ṭīkā on the *Śīghrabodha* of Kāśīnātha, published at Benares in 1966.

CANDRASIṂHA

Author of a *Hāyanaphala*. Manuscript:

Baroda 3362. 19ff. (f. 18 missing).

CANDRASENA

Author of a *Kevalajñānahorā*. Manuscripts:

Arrah II 11. See NCC, vol. 5, p. 50.
Bhuleśvara, Bombay, Pannalal Jain Sarasvati Bhavan 2347. See NCC and Velankar, p. 96.
Mudvidri, Bhandar of Cārukīrti Bhaṭṭāraka 24. See NCC and Velankar.
Mysore and Coorg 2875. 3000 granthas. Property of the Jaina Maṭha at Sravana Belgola.
Sravan Belgula, Bhandar of Bhattarakaji 152. See NCC and Velankar.
Sravan Belgula, Bhandar of Daurbali Jinadas 22. See Velankar.

CANDRASENA

Author of a *Cūḍāmaṇisāra*. Manuscript:

Mysore, p. 9. See NCC, vol. 6, p. 373.

CANDRĀYAṆA

Alleged author of:

1. *Tithikalpavṛkṣa*. Manuscript:

Jaipur (II). 1f.

2. *Sūryasiddhāntasāraṇī*. Manuscript:

Jaipur (II). 2ff.

CARAṆADĀSA

Author of a *Jñānasvarodaya* or *Svarodayasāra* (*Sarodhāsāra*) in Hindī. Manuscripts:

BORI 730 of 1895/1902. 8ff. Copied in Saṃ. 1827 = A.D. 1770.
LDI (MPC) P/7177. 7ff. Copied in Saṃ. 1869 = A.D. 1812.
Vidyābhūṣaṇa 11. 71ff. Copied by Brāhmaṇa Kanīrāma at Bāsanā on Saturday 30 Kārttika in Saṃ. 1884 = 17 November 1827.
LDI (MPC) P/7185. Ff. 2–9. Copied in Saṃ. 1886 = A.D. 1829. Incomplete.
RJ 395 (vol. 2, p. 36). 27ff. Copied in Saṃ. 1895 = A.D. 1838. Incomplete. Property of Lūṇakaraṇajī Pāṇḍyā of Jayapura.
RORI (Rājasthānī) 1759. 14ff. Copied in Saṃ. 1902 = A.D. 1845.
GJRI 1177/289. 32ff. Copied in Saṃ. 1933 = A.D. 1876.
Benares (1963) 34566. Ff. 1–39 and 39b–40, 2ff., ff. 41–44, and ff. 1–42.
LDI (LDC) 1221. 10ff.
SOI 798 = SOI Cat. I: 1408–798.
SOI 3281 = SOI Cat. II: 1135–3281. 11ff.
SOI 11506.

CĀÑGADEVA

Author of an *Uttarapañcaviṃsīpatrikā*. Manuscript:

Poona, Bhāratīya Itihāsa Saṃśodhaka Maṇḍala 102/1. See NCC, vol. 2, p. 305.

CĀÑGADEVA

Author of a *Praśnapradīpa*. Manuscript:

Mithila 191. 2ff. Maithilī. Copied by Gonū Śarman in Śaka 1783, Sāl. San. 1269 = A.D. 1861.

The first verse is:

praśnaparāyaṇagranthaṃ vighnarājena nirmitam/
cāñgadevakṛtaṃ bhaktyā tvatprasādāt karomy aham//

CĀṆAKYA

Cāṇakya is the name of the authority on arthaśāstra often called Kauṭilya (*fl.* third century B.C./second century A.D.); he is quoted by Kalyāṇavarman (*fl. ca.* 800) in *Sārāvalī* 7, 3; 46, 6; and 47, 45, and by ps.-Pṛthuyaśas in *Horāsāra* 18, 67–68. There is also attributed to him an *Uḍujātaka*. Manuscript:

GOML Madras D 13625. 20pp.

The second verse is:

navagrahadaśābhuktiphalabodhaprakāśakam/
cāṇakyaḥ sarvasārajño vakṣye ᵓham uḍujātakam//

CĀMUṆḌARĀYA

Author of a *Sāmudrikalakṣaṇa*. Manuscript:

Arrah, Digambara Bhandar, Kaṇṇaḍa 924. See Velankar, p. 433.

CĀRITRA MUNI

Author of an avacūri on the *Sañgrahaṇīratna* of Śrīcandra Sūri (*fl. ca.* 1150); see Velankar, p. 410.

CIKKA RĀJĀ (*fl.* 1672/1704)

The rājā of Mysore from 1672 to 1704, Cikka Rājā is said to have written a *Śakunanimitta* in Kannaḍa. Manuscript:

Mackenzie, Hala Kanara Philology 9 (p. 341).

CICCHU DAIVAJÑA

Author of a *Praśnasāra*. Manuscript:

CP, Kielhorn XXIII 88. 7ff. Copied in Sam. 1824 = A.D. 1767. Property of Govindarāma Bhaḍajī of Sāgar.

CIṬṬARĀMA (*fl.* 1857)

The grandson of Rajādārāma (?) of Lavapura, Ciṭṭarāma wrote a pañcāñga for Saṃ. 1914 = A.D. 1857 at Sudhāsarapura. Manuscript:

Leningrad (1914) 323 (Ind. III 23). 15ff.

Verses 2–3 are:

āsīl lavapure vidvān rajādārāmaviśrutaḥ/
tatpautraciṭṭarāmeṇa nirmitā tithipatrikā//
vedacandrāñkacandrābde sudhāsarapure vare/
paropakṛtaye hy eṣā dvijānāṃ vṛttikāriṇī//

CITTARASIṂHA (*fl.* 1861)

An Assistant Police Inspector in Gopālagañja, Cittarasiṃha wrote a *Jyotiṣasāranavīnasañgraha* in Hindī in Saṃ. 1918 = A.D. 1861. Manuscript:

NPS 18 of 1935–37. Property of Paṇḍita Rāmakṛṣṇa Tivārī of Phaphūṃda, Iṭāvā.

CITRAGUPTA (*fl. bef. ca.* 750)

Author of a *Jātaka* cited by Kalyāṇavarman (*fl. ca* 800) in *Sārāvalī* 54, 12.

CITRABHĀNU (*fl.* 1530)

The pupil of Gārgya Nīlakaṇṭha (*b. ca.* 14 June 1444) and the teacher of Śaṅkara Vāriyar (*fl.* 1556), Citrabhānu wrote a *Karaṇāmṛta* whose epoch is given as Kali 4608 = A.D. 1507 in the second verse, but whose chronogram in the last verse is 1,691,513 or *ca.* 30 March 1530. See K. K. Raja [1963] 153–154. Manuscripts:

Kerala 3051 (C. 1380 A) = Kerala C 663 A. 17ff. Malayālam. Formerly property of Tuppan Tuppan Nambūri of Punnorkoḍu Manakkal.
Kerala 3052 (C. 1380 B) = Kerala C 663 B. 85ff. Malayālam. With a vyākhyā. Formerly property of Tuppan Tuppan Nambūri of Punnorkoḍu Manakkal.
Kerala 3053 (T. 734). 700 granthas. With a vyākhyā.

Verse 2 is:

kalyabdo ᵓṣṭābhrasaḍvedahīno ᵓbdacaturaṃśayuk/
dināni ca vināḍyaḥ syur abdārdhaṃ nāḍikā api//

The last verse is

buddhyonmathyoddhṛtaṃ yatnāt tantrābdheś citrabhānunā/
tad etat kālatattvajñā gṛhṇantu karaṇāmṛtam//

CIDAMBARA GAṆEŚA (*fl.* 1907/1915)

Author, with Veṇīmādhava Kṛṣṇa, of pañcāṅgas for Śaka 1829–1833 and 1835–1837 = A.D. 1907–1911 and 1913–1915, published at Dhāravāḍa in 1907–1915 (BM 14096. a. 8. (1–8)).

CIDĀNANDA (*fl.* 1850)

Also known as Karpūracanda and Karpūravijaya, Cidānanda composed a *Svarodayaśāstra* in Gujarātī (sometimes identified as Hindī); the date of composition is variously given as Saṃ. 1907 = A.D. 1850 and Saṃ. 1975 = A.D. 1918, but the existence of manuscripts copied before 1918 seems to decide decisively in favor of the earlier date. Manuscripts:

RORI (Rājasthānī) 2510. 21ff. Copied in Saṃ. 1911 = A.D. 1854.
BORI 912 of 1892/95. 12ff. Copied in Saṃ. 1917 = A.D. 1860.
LDI (LDC) 2749. 16ff. Copied in Saṃ. 1925 = A.D. 1868.
LDI (MPC) P/8497. 7ff. Copied in Saṃ. 1939 = A.D. 1882.
LDI (DJSC) 420. 33ff. Copied in Saṃ. 1944 = A.D. 1887.
LDI (DJSC) 67. 14ff.
LDI (LDC) 4569/2. 22ff. With an artha.
LDI (LDC) 5092. 20ff.
LDI (LDC) 5792. 8ff.

CINTĀMAṆI

Author of a *Camatkāracintāmaṇi;* there is a stabaka in Old Gujarātī. Manuscripts:

Oudh XX (1888) VIII 65. 16pp. Copied in A.D. 1596. Property of Paṇḍita Pratāpa Nārāyaṇa of Allahabad Zila.
LDI 6752 (7697). 11ff. Copied in Saṃ. 1754 = A.D. 1697. With the stabaka.
LDI 6750 (6833). 12ff. Copied by Devendravijaya, the pupil of Sabhārañjanapaṇḍita Amṛtavijaya Gaṇi, in Saṃ. 1796 = A.D. 1739. With the stabaka.
LDI 6751 (3035). 12ff. Copied by Paṇḍita Tejavijaya Gaṇi in Saṃ. 1829 = A.D. 1772. With the stabaka.
LDI (AKC) 726. 16ff. Copied in Saṃ. 1869 = A.D. 1812.
LDI 6754 (7338). 15ff. Copied by Paṇḍita Ṛṣabhavijaya Gaṇi, who was with Raṅgavijaya, at Prahlādanapura in Saṃ. 1872 = A.D. 1815. With the stabaka.
Oudh (1879) VIII 9. 22pp. Copied in A.D. 1818. (*Bhāvacintāmaṇi*). Property of Śyāma Lāla of Lucknow Zila.
LDI 6747 (1028). 22ff. (f. 1 missing). Copied by Muni Hemavijaya at Nāgorīśālā (Ahmadabad) in Saṃ. 1883 = A.D. 1826. With the stabaka.
LDI 6749 (7414). 15ff Copied in Saṃ. 1895 = A.D. 1838. With the stabaka.
LDI 6748 (4277). 11ff. Incomplete (bhāvādhyāya). With the stabaka.
LDI 6753 (7193). 14ff. With the stabaka.
LDI 6755 (6986). 16ff. Incomplete. With the stabaka.
Oudh XX (1888) VIII 85. 110pp. (*sic* !). Property of Paṇḍita Pratāpa Nārāyaṇa of Allahabad Zila.

CINTĀMAṆI

The pupil of Cūḍāmaṇi, Cintāmaṇi wrote a *Ramalotkarṣa* also known as *Ramalapraśnatantra*, *Ramalasaṅgraha*, *Ramalaśāstra*, *Ramalacintāmaṇi*, *Prastāracintāmaṇi*, etc; it contains a saṃjñātantra and a praśnatantra. S. B. Dikshit [1896] 489, on the basis of an unidentified manuscript at Ānandāśrama dated Śaka 1653 = A.D. 1731, dates him before Śaka 1600 = A.D. 1678. Manuscripts:

Baroda 3398. 28ff. Copied in Saṃ. 1783 = A.D. 1726.
Benares (1963) 37368. 14ff. Copied in Saṃ. 1785 = A.D. 1728.
Benares (1963) 37559. 10ff. Copied in Saṃ. 1789, Śaka 1654 = A.D. 1732.

GVS 2912 (2673). 36ff. Copied on Thursday 1 kṛṣṇapakṣa of Āṣāḍha I in Saṃ. 1800 = 14 June 1744.
Baroda 7347. 23ff. Copied in Saṃ. 1810 = A.D. 1753. Incomplete.
BORI 352 of 1882/83. 26ff. Copied in Saṃ. 1811 = A.D. 1754. From Gujarāt.
Calcutta Sanskrit College 113. 29ff. Copied in Saṃ. 1816 = A.D. 1759.
Bombay U Desai 1490. Ff. 1–26 and 28–47. Copied in Saṃ. 1818 = A.D. 1761.
Goṇḍal 331. 3ff. Copied in Saṃ. 1821 = A.D. 1764. Incomplete.
Benares (1963) 37668 = Benaras (1897–1901) 622. 36ff. Copied in Saṃ. 1825 = A.D. 1768.
PUL II 3855. 19ff. Copied in Saṃ. 1828 = A.D. 1771.
BORI 896 of 1891/95. 34ff. Copied in Śaka 1706 = A.D. 1784. Incomplete.
Probstain 14. 24ff. Copied in Saṃ. 1842 = A.D. 1785.
BORI 412 of 1895/1902. 18ff. Copied in Śaka 1710 = A.D. 1788.
BORI 413 of 1895/1902. 38ff. Copied in Śaka 1710 = A.D. 1788.
Benares (1963) 37565 = Benares (1878) 52 = Benares (1869) XI 5. 22ff. Copied in Saṃ. 1846, Śaka 1711 = A.D. 1789.
Nagpur 1743 (989). 17ff. Copied in Śaka 1713 = A.D. 1791. From Nasik.
Goṇḍal 329. 20ff. Copied by Bhīmajī, the son of Viśrāma of the Pokaraṇajñāti, on Monday 1 śuklapakṣa of Kārttika I in Saṃ. 1849, Śaka 1714 = 15 October 1792.
Benares (1963) 37650. 1f. and ff. 1, 5–6, and 6b–34. Copied in Śaka 1718 = A.D. 1796. Incomplete.
Kathmandu (1960) 231 (III 17). 25ff. Copied by Kṛṣṇa Gujarāti Moḍha Cātravedī in Saṃ. 1857, Śaka 1721 = A.D. 1800. Incomplete (praśnatantra).
Benares (1963) 37593. 32ff. Copied in Saṃ. 1862 = A.D. 1805.
Goṇḍal 330. 36ff. Copied on Friday 1 kṛṣṇapakṣa of Māgha in Śaka 1729 = 11 February 1808.
Oudh XI (1878) VIII 2. 62pp. Copied in A.D. 1811. Incomplete (praśnatantra). Property of Rājā Rāmanātha of Faizābād Zila.
AS Bombay 391 = AS Bombay (Indraji) 102. 24ff. Copied in Saṃ. 1877 = A.D. 1820.
Benares (1963) 37605. 30ff. and 5ff. Copied in Saṃ. 1882 = A.D. 1825.
RORI Cat. I 3714. 25ff. Copied by Kamalasāgara at Nāgapura in Saṃ. 1882 = A.D. 1825. Incomplete (praśnatantra).
Benares (1963) 37558 = Benares (1903) 1130. 26ff. Copied in Saṃ. 1883 = A.D. 1826.
Mithila 298. 32ff. Maithilī. Copied by Śivaprasāda Kāyastha on Thursday 11 kṛṣṇapakṣa of Āṣāḍha in Saṃ. 1889 (incorrect data). Property of Pandit Bikal Jhā of Lalganj, Jhanjharpur, Darbhanga.
Baroda 1256. 22ff. Copied in Saṃ. 1890 = A.D. 1833.
Nagpur 1742 (2358). 21ff. Copied in Śaka 1762 = A.D. 1840. From Nagpur.
AS Bengal 7261 (G 7777). 21ff. Copied at Devīpura on Friday 6 śuklapakṣa of Vaiśākha in Saṃ. 1901 = 24 May 1844 (?).
Benares (1963) 36797. 29ff. Copied in Śaka 1773 = A.D. 1851. Incomplete (praśnatantra).
Mithila 298 B. 32ff. Maithilī. Copied by Vacanū on Thursday in the middle of Pauṣa in Śaka 1777 = *ca.* 23 January 1856. Property of Pandit Mahīdhara Miśra of Lalabag, Darbhanga.
Baroda 2030. 52ff. Copied in Saṃ. 1919 = A.D. 1862.
Baroda 9198. 74ff. Copied in Saṃ. 1925 = A.D. 1868.
Alwar 1849. 2 copies.
Alwar 1926.
Alwar 1927.
Anup 5044. 5ff.
AS Bombay 390. 8ff. From Bhāu Dājī.
Baroda 1406. 59ff.
Baroda 3267. 30ff. With a Gujarātī ṭīkā. Incomplete.
Baroda 5622. 71ff.
Baroda 8906. 56ff. (*Prastāracintāmaṇi*).
Baroda 9294. 33ff. Incomplete (saṃjñātantra).
Baroda 13395. 88ff. Nandināgarī.
Benares (1963) 37367 = Benares (1905) 1494. 14ff. Incomplete. (praśnatantra).
Benares (1963) 37478 = Benares (1903) 1153. 25ff. Incomplete.
Benares (1963) 37482. Ff. 1–3 and 1f. Incomplete.
Benares (1963) 37561. Ff. 6–18. Incomplete.
Benares (1963) 37617. Ff. 1–2 and 1–8. Incomplete.
Benares (1963) 37636. 41ff. and 1f.
Benares (1963) 37667. 4ff. Incomplete. Probably identical with Benares (1897–1901) 621. 11ff. Copied in Saṃ. 1837 = A.D. 1780.
Benares (1963) 37669. Ff. 7–9 and 11–35. Incomplete.
Benares (1963) 37670 = Benares (1878) 53 = Benares (1869) XI 6. 15ff. Incomplete.
Bombay U Desai 1491. 13ff. Incomplete (saṃjñātantra).
Bombay U Desai 1492. 14ff. Incomplete (praśnatantra).
Bombay U Desai 1493. 12ff. Incomplete (saṃjñātantra).
Bombay U Desai 1494. 17ff. Incomplete (saṃjñātantra).
BORI 720 of 1883/84. 10ff. From Mahārāṣṭra.
CP, Hiralal 4529. Property of Govindbhaṭṭ of Jubbulpore.
CP, Hiralal 4530. Property of the Bhonsalā Rājas of Nāgpur.
CP, Hiralal 4533. Property of Śrīkṛishṇa Pāṇḍuraṅg of Bālāpur, Akolā.
CP, Kielhorn 132. 17ff. Property of Javāhara Śāstrī of Chāndā.
DC 132. 17ff.
IO 3132 (92c). 24ff. From H. T. Colebrooke.
Jammu and Kashmir 2863. 15ff.

Jammu and Kashmir 2951. 24ff.
LDI 7441 (7705). 21ff. (f. 1 missing). Incomplete.
LDI (KC) K/912. 31ff.
LDI (KS) 1035 (10940). 31ff. Copied by Jñānasāgara at Vikramapura.
LDI (LDC) 312. 62ff.
Mithila 297. 7ff. Maithilī. Incomplete (praśnatantra). Property of Pandit Rāmakṛṣṇa Chaudharī of Ekama, Supaul, Bhagalpur.
Mithila 298 A. 10ff. Maithilī. Incomplete. Property of Pandit Sādhu Jhā of Yamathari, Jhanjharpur, Darbhanga.
Mithila 298 C. 8ff. Maithilī. Property of Pandit Rāmakṛṣṇa Chaudharī of Ekama, Supaul, Bhagalpur.
Oudh III (1873) VIII 20. 52pp. Property of Paṇḍit Bhālacandra of Oonao Zila.
Oudh (1879) VIII 12. 60pp. Property of Śyāma Lāla of Lucknow Zila.
Oxford 1550 (Sansk. d. 195) = Hultzsch 302. Ff. 10–45.
PL, Buhler IV E 400. 23ff. Property of ——— of Khambhāliyām. Buhler notes 4 other copies.
PL, Buhler IV E 409. 22ff. Property of Tātyā Bhaṭṭa of Mulhera.
PrSB 969 (Göttingen Mu I 23(A)). Ff. 1–2, 7–14, and 11–28. Śāradā and Devanāgarī. Incomplete.
PrSB 970 (or. oct. 738). Ff. 1–2, 4–17, and 1–19. Incomplete. Now at Marburg.
PUL II 3856. 13ff. (f. 3 missing). Incomplete (saṃjñātantra).
RJ 3079 (vol. 4, p. 290). 15ff. Incomplete.
RORI Cat. II 4759. 10ff. Incomplete (saṃjñātantra).
RORI Cat. II 4760. 10ff. Incomplete (praśnatantra).
RORI Cat. II 9615. 70ff. (ff. 1–2, 7–8, 11, 20, 22–25, 47, 49–51, and 69 missing). Copied by Gasarāma at Karavāḍa.
RORI Cat. III 13981. 12ff. (f. 2 missing). Incomplete.
SOI 3628 = SOI Cat. II: 1097–3628. 25ff., 12ff., and 28ff.
SOI 3931 = SOI Cat. II: 1096–3931. 16ff.
VVRI 1587. 19ff.
WHMRL F. 39. c. Ff. 8–25. Incomplete.

The *Ramalacintāmaṇi* is alleged to have been published at Kāśī (Mysore GOL C 238 and C 273).

Verse 2 is:

vilokya yavanaśāstraṃ rāmalaṃ praśnasaṅgraham/
cintāmaṇiḥ karoty evaṃ ramalotkarṣam adbhutam//.

The colophon begins: iti śrīdaivajñacūḍāmaṇiśrīmanmahārājavanditapādāmbujaśiṣyajanānandadāyisarvavidyākuśalasarvaśāstreṣu kṛtaśramaśrīcintāmaṇipaṇḍitavaryair viracite.

CINTĀMAṆI

Author of a ṭīkā in Tamil on the *Sarvārthacintāmaṇi* of Veṅkaṭeśa: vol. 1 was published at Ce<u>nn</u>ai in 1967.

CINTĀMAṆI (*fl. ca.* 1530)

The son of Jñānarāja (*fl.* 1503), Cintāmaṇi composed a ṭīkā, *Grahagaṇitacintāmaṇi*, on his father's *Siddhāntasundara*. Manuscripts:

Anup 5337. 59ff. Copied in Saṃ. 1725 = A.D. 1668.
Anup 5336. Ff. 2–172. Formerly property of the Jyotiṣarāja = Vīrasiṃha (b. 1613).
Anup 5338. 58ff. Incomplete.
Anup 5339. 31ff. Incomplete (adhikāra I).
AS Bombay 291. 50ff. Incomplete (madhyagatisādhana and part of sphuṭagatisādhana). From Bhāu Dājī.
Benares (1869) XXI 1. 10ff. Incomplete (golādhyāya).
Benares (1963) 34970. 45ff. Incomplete.
Benares (1963) 35318 = Benares (1878) 96 = Benares (1869) XVIII 8. Ff. 12–78. Incomplete.
BORI 26 of 1869/70. 58ff. Incomplete (adhikāra 1 of golādhyāya).
Jammu and Kashmir 3091. 83ff. Incomplete (*Grahagaṇitamaṇi*).
PL, Buhler IV E 529. 55ff. Incomplete. Property of Khuśāla Bhaṭṭa of Ahmadābād.
SOI 9400.
SOI 9401. Incomplete (golādhyāya).
SOI 9402. Incomplete (gaṇitādhyāya).

The colophon begins: iti śrīmatsakalasiddhāntavāsanāvicāracaturapracuratarāparaśāstrarahasyābhijñadaivajñajñānarājagrathitasiddhāntasundaravāsanābhāṣye sujanavicakṣaṇaikabhūṣye jñānādhirājasūnupaṇḍitacintāmaṇiviracite.

CINTĀMAṆI (*fl.* 1633)

See Rājarṣi (*fl.* 1633).

CINTĀMAṆI (*fl.* 1661)

The son of Govinda (b. 2 October 1569), the son of Nīlakaṇṭha (*fl.* 1569/87), the son of Ananta (*fl. ca.* 1575), the son of Cintāmaṇi of the Gārgyagotra, Cintāmaṇi completed a ṭīkā, *Saṃmaticintāmaṇi*, on the *Muhūrtamālā* of Raghunātha (*fl.* 1660) at Kāśī on 15 śuklapakṣa of Śrāvaṇa in Saṃ. 1718 = 31 July 1661 during the regn of Aurangzīb (1658/1707). Manuscripts:

Benares (1963) 37217. Ff. 1–55, 55b–61, 63–122, 1–21, 143–234, and 234b–300. Copied in Saṃ. 1813 = A.D. 1756. Incomplete.
AS Bengal 2746 (G 6328). 300ff.
Baroda 111. 21ff. Incomplete (vāstuprakaraṇa).
Baroda 124. 48ff. (ff. 23–26 and 31–38 missing). Incomplete (saṃskāraprakaraṇa).
Baroda 5346. 90ff. Incomplete.
Baroda 9209. 28ff. Incomplete (saṃjñāprakaraṇa). No author mentioned.
Baroda 9241. 54ff. Incomplete (tyājyātyājyaprakaraṇa and prakīrṇa).

Benares (1963) 35306. Ff. 1–62, 64–157, 159–174, and 174b–187.
Kavīndrācārya 808. No author mentioned.
PL, Buhler IV E 356. 140ff. Property of Harirāmaśāstrī of Aṅkaleśvara.

Verses 3–8 at the end are:

āsīd gargasya vaṃśe gaṇakakulamaṇir jyotiṣāṃ saṃprakāśair
vidvadvṛndāravindodghaṭanadinamaṇir bhaṭṭacintāmaṇir yaḥ/
niḥśeṣaśrotranetravyatikaravilasanmānasaḥ svarbhramantaṃ
sūtraprotā trinetrodbhavamukhacarasvarṇagolaṃ vyakārṣīt//
tasmāc chrīmān ananto vidhur iva jaladher āvir āsīd asīmā-
bhyāsān mīmāṃsakānāṃ sadasi sadṛśatāṃ ko ᵓpi lebhe na yena/
vedāntanyāyavidyāśrutimukhanigamajñānavijñānatattvo
nityaṃ satyaprakṛtyā kalim akṛtakṛtaṃ yaś ca tasmai namo ᵓstu//
yasyodurānanda (?) nibandhakartā
kartā punas tājikanīlakaṇṭhyāḥ/
sa nīlakaṇṭhas tata āvir āsīd
asīmaśabdārṇavapāradṛśvā//
sa nīlakaṇṭhāc chitikaṇṭhapuryāṃ
govindaśarmājani dharmakarmā/
yaḥ śrījahāṅgīrasabhāsu x x
⟨mā⟩dhuryaśauryaś ca mauhūrtikatārakāsu (?)//
muhūrtacintāmaṇinīlakaṇṭhī-
siddhāntasabhyābharaṇādikānām/
ṭīkāṃ bahūnām api durghaṭānāṃ
bodhāya cakre ᵓlpadhiyāṃ budhānām//
rāmāṅghrisannidhisahādhyayanātimitra-
daivajñavaryaraghunāthakavipraṇītām/
govindaśarmatanayo ᵓtra muhūrtamālāṃ
cintāmaṇir guṇimaṇir viśadīkaroti//

The date of composition is given in the verse:

dhṛtighanamitagatavikrama-
śāke rājye ᵓvaraṅgajevasya/
nabhasi sasitapañcadaśyāṃ
saṃmaticintāmaṇiḥ kṛtaḥ kāśyām//

CINTĀMAṆI DĪKṢITA (1736/1811)

The son of Lakṣmī and Vināyaka Somayājin of the Vatsagotra, a resident of Cittapūrṇa (Cipaḷūṇa) in Śūrpārakṣetra, Cintāmaṇi was born in Śaka 1658 = A.D. 1736 and died in Śaka 1733 = A.D. 1811. He is said to have composed a *Sūryasiddhāntasāraṇī* (see S. B. Dikshit [1896] 297). He also wrote in Śaka 1713 = A.D. 1791 at Saptarṣi (Sātārā), presumably under the Marāṭha Śāhu (1777/1810) and the Peshwa Madho Rao II (1774/1795), a *Golānanda* in 148 verses. There is a ṭīkā by his grandson Yajñeśvara (*fl. ca.* 1800). Manuscripts:

BORI 40 of 1907/15. 12ff. Copied in Śaka 1737 = A.D. 1815.
BORI 41 of 1907/15. 7ff. Copied in Śaka 1737 = A.D. 1815. No author mentioned.
Baroda 9178. 32ff. With a vyākhyā.
Bombay U 341. 9ff.
BORI 43 of 1907/15. 30ff. With an upapattikaṭīkā. No author mentioned.
Kavīndrācārya 849. With a ṭīkā. No author mentioned.
SOI 9978.

Verse 2 is:

lakṣmīvināyakau natvā tadākhyau pitarāv api/
brūte cintāmaṇir yantraṃ golānandākhyam adbhutam//

Verses 44–46 at the end are:

śrīśūrpārakṣetre
ᵓsti mahad yac cittapūrṇanāma nagaram/
tadvāsivātsyajyotir-
vidvināyakasomayājitanujena//
cintāmaṇinoktam etat
pitṛprasādāptagolavidyena/
samprati vasatā saptarṣau
kṛṣṇātaraṅgotthavāyubhiḥ pūte//
śrīśālivāhanaśake
viśvaghanair unmite ᵓjanīdam etat tu/
pravibhāvayanti gaṇake
ye prauḍhasabhāsv api yaśasvinaḥ syus te//

CINTĀMAṆI RAGHUNĀTHA ĀCĀRYA
(b. 17 March 1828)

Cintāmaṇi was born at Madras on 2 śuklapakṣa of Caitra in Śaka 1750 = 17 March 1828. At the age of 17 he became First Assistant at the Madras Observatory, where he cataloged stars from 1867 till 1878; he died on 5 February 1880. He was made a Fellow of the Royal Astronomical Society in 1872. Among his writings are a *Jyotiṣacintāmaṇi* in Tamil with a Sanskrit translation, published at Madras in 1874, and a *Śukragrastasūryoparāga*, published at Poona in 1874 (IO 2346). See S. B. Dikshit [1896] 304–305.

CINTĀMAṆI PURUṢOTTAMA PURANDARE VASAĪKAR (*fl.* 1892)

Author of a pañcāṅga for Śaka 1814 = A.D. 1892, published at Mumbaī in 1892 (BM 14096. a. 7. (2)).

CIRAÑJĪVA MIŚRA

The son of Pūrṇānanda Miśra, the son of Kṛṣṇa, a resident of Argala, Cirañjīva of Mathurā wrote a *Śaraccandrodaya*. Manuscripts:

Mithila 362. 62ff. Copied by Khajaisiṃgha Māthuravāsī Āgareka on Monday 2 kṛṣṇapakṣa of Pauṣa in Saṃ. 1818, Śaka 1683 = 11 January 1762. Property of Pandit Rāmacandra Jhā of Mahinathapur, Deodha, Darbhanga.
Benares (1963) 35011 and 35012. Ff. 1–29 and 30–56. Copied in Saṃ. 1841 = A.D. 1784.
Bombay U Desai 1436. 72ff. Incomplete (ends at VII(?) 5).

Verses 1–2 at the end are:

āsīt sūryasutopakaṇṭhanagare vidvadvaraiḥ pūrite
śobhāḍhye ᵓrgalasaṃjñake haripuraḥ kṛṣṇābhidhāno dvijaḥ/
tarkālaṅkṛtaśabdaśāstracaturo jyotirvidām agraṇīḥ
pūrṇānanda iti prathām adhigatas tatsūnur āsīd bhuvi//
cirañjīvakas tatsuto māthuro ᵓbhūt
kavīndro budho jyotiṣāṃ vākpravīṇaḥ/
śaraccandrapūrvodayas tena tene
mude kairavāṇāṃ budhānāṃ bhavāya//

The colophon begins: iti śrīmiśrapūrṇānandātmajamiśraciranjīvaviracite.

CIRAÑJĪVA BHAṬṬA (*fl.* 1647)

The son of Rāghavendra, the son of Kāśīnātha, Cirañjīva wrote under the patronage, and often under the name, of Kṛpārāma (*fl. ca.* 1600/1650), who ruled a territory near Agra, and his son, Yaśovanta Siṃha. His works include:

1. A ṭīkā on the *Jyotiṣkedāra* of Kṛpāśaṅkara (*fl.* 1627). Manuscript:

BORI 913 of 1886/92. 28ff. Copied in Saṃ. 1773 = A.D. 1716.

2. A vyākhyā, *Rāmaprakāśa* or *Kālatattvārṇavasetu*, on a *Kālatattvārṇava*, written in 1647 at Indurakhī in Gwalior; see NCC, vol. 4, p. 282. NCC, vol. 4, p. 21 suggests that this was written by Cirañjīva's father, Rāghavendra. Manuscripts:

Benares (1956) 12124. Ff. 1–82, 82b–166, and 166b–351. Copied in Saṃ. 1704 = A.D. 1647. (*Rāmaprakāśa* = *Kālanirṇayasetu* of Kṛpārāma).
IO 1600–1602 (909–911). Ff. 1–248, 249–474, and 475–737. Bengālī. Copied by the son of Jīvana in Śaka 1721 = A.D. 1799. From H. T. Colebrooke.
Kerala 3409 (1572). 14000 granthas. Grantha. Ascribed to Kṛpārāma.

3. A vyākhyā, *Rāmaprakāśa*, on the *Kālanirṇayadīpikā* of Rāmacandra (*fl. ca.* 1400). The manuscript is not clear about the authorship, mentioning only Cirañjīva's father, Rāghavendra, the son of Kāśīnātha; to this Rāghavendra is the work attributed in NCC, vol. 4, p. 29. Manuscript:

IO 1664–1666 (885, 886, 887). Ff. 1–179, 180–380, and 381–558. Copied in Saṃ. 1862 = A.D. 1805. From H. T. Colebrooke.

CIRAÑJĪVA BHAṬṬĀCĀRYA (*fl. ca.* 1725?)

The son of Śatāvadhāna Bhaṭṭācārya, Cirañjīva, a native of Navadvīpa and resident of Benares, wrote a *Tājikaratnākara* or *Tājikaratna* under the patronage of Yaśovanta Siṃha, who is said to have been a Naib Nazim of Dacca in the early eighteenth century. He may be identical with Cirañjīva Bhaṭṭa (*fl.* 1647). Manuscripts:

Benares (1963) 34850. Ff. 2–49 and 52–78. Copied in Saṃ. 1795 = A.D. 1738. Incomplete.
Benares (1963) 36813. Ff. 1–28, 34–60, and 62. Copied in Saṃ. 1804 = A.D. 1747. Incomplete.
Alwar 1805. Ascribed to Ratnākara, son of Śatavadhana.
AS Bengal 7098 (G 6339). 21ff. Incomplete (ends at IV 48).
Benares (1963) 37073 = Benares (1911–1912) 2075. 5ff. Incomplete.
Kerala 6720 (9705). 1200 granthas. Incomplete.
Oudh XXII (1890) VIII 13. 6pp. Property of Kedāranātha of Āgrā Zila (*Jātakaratnākara* of Ratnākara).

Near the beginning is the verse:

dṛṣṭvā tājakabhūṣaṇaṃ gaṇapater gauḍasya cūḍāmaṇeḥ
sāraṃ kalpataros tathaiva gaditaṃ śrīnīlakaṇṭhasya ca/
anyeṣāṃ kṛtināṃ kṛtāni bahuśaḥ saṃvīkṣya niṣkṛṣya ca
śreyo yad bahusaṃmataṃ matam iha brūmaḥ svapadyena tat//

At the end of I is the verse:

dvaitādvaitamatādinirṇayavidhiprodbuddhabuddhiḥ śruto
bhaṭṭācāryaśatāvadhāna iti yo gauḍodbhavo ᵓbhūt kariḥ/
nānāśāstravidā tadātmajaciranjīvena yan nirmitaṃ
divyaṃ tājakaratnam asya samabhūt pūrṇeyam anidaprabhā//

The colophon begins: iti śrīciranjīvabhaṭṭācāryadaivajñaratnākarodāhṛte.

CIRAÑJĪVA MAITHILA (*fl.* 1921)

Author of a Hindī translation, *Hitaprabhā*, of the *Laghujātaka* of Varāhamihira (*fl. ca.* 550), published at Darabhaṅgā in Saṃ. 1978 = A.D. 1921 (BM 14053. b. 37. (3)).

CUNNĪRĀMA (*fl.* 1837)

Author of a *Karaṇabhūṣaṇa* in Śaka 1759 = A.D. 1837. There are 5 kiraṇas:

1. sūryacandraspaṣṭīkaraṇa.
2. tārāgrahaspaṣṭīkaraṇa.
3. upakaraṇa.
4. candragrahaṇānayana.
5. sūryagrahaṇānayana.

Manuscript:

AS Bengal 6832 (G 10055). 16ff. Copied on Thursday 4 kṛṣṇapakṣa of Āṣāḍha in Saṃ. 1894 = 20 July 1837.

Verse 1 is:

śāko ᵓṅkabāṇādriśaśāṅkahīnaḥ
syād abdasaṅghīyam athārkanighnaḥ/
māsair yuto ᵓdhaḥ sagajāt śataghnād
ato ᵓbdhinārācaradhrāṃśayuk syāt//

The colophon begins: iti śrīcāturvedacunnīrāmakṛte.

CUNNĪLĀLA

Author of a *Varṣapaddhati*. Manuscript:

VVRI 1346. 65ff.

CŪḌĀMAṆI

Alleged author of a *Jyotiḥsārasamuccaya*. Manuscript:

PUL II 3474. 17ff. Incomplete (adhyāyas 8–9).

CŪḌĀMAṆI

Author of a *Nakṣatraśakunāvalī* in Rājasthānī and Gujarātī. Manuscript:

LDI (LDC) 3345/15. Ff. 163–164.

CŪḌĀMAṆI

The teacher of Cintāmaṇi, Cūḍāmaṇi wrote a *Ramalaśāstra*. Manuscript:

AS Bengal 7260 (G 5563). 9ff.

The colophon begins: iti śrīdaivajñacūḍāmaṇiviracite.

CŪḌĀMAṆI CAKRAVARTIN

Author of a *Makarandasādhanaprakriyā*, apparently based on the *Makaranda* of Makaranda (*fl.* 1478). Manuscript:

RORI Cat. II 6275. 12ff. Copied in Saṃ. 1890 = A.D. 1833.

CŪḌĀMAṆI (*fl.* before *ca.* 800)

An authority cited by Kalyāṇavarman (*fl. ca.* 800) in *Sārāvalī* V 20.

CŪḌĀMAṆI (*fl. ca.* 1620)

See Kavicūḍāmaṇi (*fl. ca.* 1620).

COLA

Author of a vyākhyā on the *Pārāśarīhorā* = *Uḍudāyapradīpa* of Parāśara. Manuscript:

PUL II 3633. 131ff. Grantha.

COLA VIPAŚCIT

The son of Ārya Sūrya, Cola, also known as Colarāja, Colapparāja, Cola Sūri, and Cola Kavi, wrote a vyākhyā, *Gaṇakopakāriṇī*, on the *Sūryasiddhānta*. Manuscripts:

GOML Madras R 1706. 102ff. Copied in 1915/16 from a manuscript belonging to Uppulūri Veṅkaṭakṛṣṇammagāru of Kottapalli, Godāvari District. Incomplete (adhyāyas 1–11).
GOML Madras R 3351. 106ff. Copied by Vāsudevaśarman, the son of Vināyakagopālaśarman, in 1920/21 from a manuscript belonging to the Raja of Chirakkal, Malabar. Incomplete (ends in adhyāya 13).
Adyar Index 7223.
Baroda 13368(a). 60ff. Nandināgarī. Incomplete.
Baroda 13379(a). 67ff. Nandināgarī. Incomplete.
Benares (1963) 35401. 49ff. Incomplete (adhyāyas 1–4).
GOML Madras D 13508. 266pp. Grantha. Incomplete (ends in adhyāya 13).
GOML Madras D 13509. 135pp. Grantha. Incomplete (ends at 13, 16).
Mysore (1911 + 1922) 2084. 95ff.
Mysore (1911 + 1922) 2565. Ff. 52–139. Incomplete (adhyāyas 4–14).
Mysore (1911 + 1922) 2598. 82ff. Incomplete (adhyāyas 1–13).
Mysore (1911 + 1922) B 572. 109ff. Incomplete (adhyāyas 1–12).
Oppert II 4592. (*Colarājīya*). Property of the Śaṅkarācāryasvāmimaṭha of Śṛṅgeri, Cikkamogulūr Division.
Oppert II 6268. (*Colapparājīya*). Property of Narasiṃhācārya of Kumbhaghoṇam, Tanjore District.
PUL II 4080. 67ff. Telugu.
PUL II 4081. 96ff. Grantha. Incomplete (adhyāyas 1–6).
PUL II 4082. Ff. 132–170. Grantha. Incomplete (adhyāyas 7–11).

The second verse is:

āryasūryatanūjena vidvatpādābjasevinā/
triskandhārthavidā samyañ nāmnā colena sūriṇā//

The last verse in adhyāya 1 is:

prajñodagraviśiṣṭaśiṣyanivahaślāghyopadeśakriyā-
pratyakṣīkṛtaviprakīrṇaviṣayaskandhatrayīmarmaṇā/
colākhyena vipaścitā viracite śrīsūryasiddhāntika-
vyākhyāne gaṇakopakāracature ᵓdhyāyo ᵓyam ādyo gataḥ//

CAUṆḌAPĀYANA

Author of a *Yāgakālanirṇaya*. Manuscripts:

Hultzsch 1. 436. 56ff. Telugu. Property of Goṭṭimukkula Vīrarāghava Somayāji of Brāhmaṇakrāka.
Hultzsch 1. 606. 17ff. Telugu. Property of Vedam Veṅkaṭasubrahmaṇya Somayāji of Allūr.
Hultzsch 1. 652. 44ff. Telugu. Property of Iṅguva Vīrarāghava Somayāji of Kareḍu.

CAUTHAMALA

Author of a *Kevalī* in Hindī. Manuscript:

NPS 19 of Dillī 1931. Copied in Saṃ. 1852 = A.D. 1795. Property of Svāmin Ravidatta Śarman of Narelā, Dillī.

CYAVANA

One of the legendary founders of jyotiḥśāstra (see, e.g., *Nāradasaṃhitā* 1, 3 and S. Dvivedin [1892] 1), Cyavana is first referred to by Varāhamihira (*fl. ca.* 550) in *Bṛhadyātrā* 29, 3. The existence of the following manuscript of a *Cyavanasiddhānta* is doubtful:

Kavīndrācārya 865.

CHAGANALĀLA (*fl.* 1868)

Author of a pañcāṅga for Saṃ. 1925 = A.D. 1868, published at Agra in 1868 (IO 2650).

CHAJA MAL

Author of a bhāṣāṭīkā on the *Ṣaṭpañcāśikā* of Pṛthuyaśas (*fl. ca.* 575). Manuscript:

Kunte A 25. 18ff. Property of Paṇḍita Jvālā Datta of Gujrānwāla.

CHATRASĀLA MIŚRA (*fl.* 1787)

The son of Gaṅgādāsa (or Gaṅgārāma) Miśra (*fl. ca.* 1750), Chatrasāla was the senāpati of Durjanasiṃha, the rājā of Canderī, and flourished in Saṃ. 1844 = A.D. 1787. He wrote a *Śakunaparīkṣā* in Hindī. Manuscript:

NPS 21 B of 1906–08. Property of the Ṭīkamagaḍhanareśa kā Pustakālaya in Ṭīkamagaḍha.

He also wrote a *Svapnaparīkṣā* in Hindī. Manuscript:

NPS 21 C of 1906–08. Copied in Saṃ. 1849 = A.D. 1792. Property of Lālā Kundanalāla of Bijāvara.

CHADŪRĀMA = CHIDDŪRĀMA (*fl.* 1813)

The son of Dharaṇīdhara and the pupil of Rāmacaraṇa (?), Chadūrāma, a resident of Siddhapurī, wrote a *Lagnasundarī* in Hindī in Saṃ. 1870 = A.D. 1813. Manuscripts:

NPS 67 B of 1929–31. Copied in Saṃ. 1893 = A.D. 1836. Property of Paṇḍita Keśavarāma of Śamaśābāda, Āgarā.
NPS 102 of Saṃ. 2004–2006. Copied in Saṃ. 1917 = A.D. 1860. Property of Rāmaharṣa of Goḍavā, Kaitholā, Pratāpagaḍha.
NPS 67 A of 1929–31. Copied in Saṃ. 1931 = A.D. 1874. Property of Paṇḍita Hariprasāda Ācārya of Āṃnavalakheḍā, Āgarā.
NPS 43 of 1912–14. Copied in Saṃ. 1941 = A.D. 1884. Property of Paṇḍita Brajarāja, pradhānādhyāpaka at Jvālāpura, Sahāranapura.
NPS 78 of 1923–25. Property of Paṇḍita Śivaśaṅkara of Bībīpura, Jaitapura, Bārābaṅkī.
NPS 67 C of 1929–31. Property of Paṇḍita Jānakīprasāda of Bamaraulī Kaṭārā, Āgarā.

CHALĀRI

Alleged author of a *Saṅkṣiptatithinirṇaya; cf.* the *Kālataraṅga* of Chalāri Nṛsiṃha (*fl.* 1198). Manuscript:

Benares (1956) 13649. 14ff.

CHĀJURĀŪ

Author of a *Jyotiṣkedāra*. Manuscript:

Kunte A 21. 29ff. Property of Paṇḍita Gulāb Sinha of Delhi.

CHĀJŪRĀMA DVIVEDIN (*fl.* 1735)

A resident of Koṭā, Chājūrāma wrote a *Tājikasāra* in Hindī in Saṃ. 1792 = A.D. 1735. Manuscript:

NPS 43 of 1932–34. Copied in Saṃ. 1792 = A.D. 1735. Property of Rādheśyāma Dvivedin of Svāmīghāṭa, Mathurā.

JAGAJĪVANA DĀSA GUPTA (*fl.* 1968/1973)

Author of a *Daśāphalavicāra* and a *Saṅkṣiptagocaraphalavicāra* in Hindī, published with his own Hindī ṭīkās at Dillī-Vārāṇasī-Paṭanā in 1968, and of a *Jyotiṣarahasya* in Hindī, of which vol. 1 was published at Vārāṇasī in [1968], vol. 2 at Dillī-Paṭanā-Vārāṇasī in 1973.

JAGAJJYOTIRMALLA (*fl.* 1613/37)

The son of Trailokyamalla (1560/1613), the son of Vipramalla or Viśvamalla (1547/60), the son of Prāṇamalla (1519/47), the son of Bhuvanamalla (1505?/47?), the son of Rāyamalla (1482/1505) (all of the preceding were rājās of Bhaktapura or Bhatgaon in Nepal), the son of the nṛpati Jayayakṣamalla of the race of Raghu, Jagajjyotirmalla ruled Bhatgaon from 1613 to 1637, and composed a commentary, *Svarodayadīpikā*, on the *Narapatijayacaryā* of Narapati (*fl.* 1175), which was completed on 10 śuklapakṣa of Āśvina in Śaka 1536 = *ca.* 13 September 1613. Manuscript:

Kathmandu (1960) 199 (I 1186). 202ff. Maithilī. Copied for Jagajjyotirmalla by Śrīvaṃśa Maṇiśarman on Saturday 14 śuklapakṣa of Jyeṣṭha in Śaka 1536, Lakṣmaṇa Saṃvat 494 = 11 June 1614.

The author's genealogy is given in the following verses at the beginning:

āsīd viśvaviśobhinirmalayaśorāśau raghor anvaye/
vikhyāto jayayakṣamallanṛpatir dātāvadātāśayaḥ//3//

. . . .

putrās trayas tasya nṛpasya jātā
dākṣiṇyadānādiguṇāvadātāḥ/
jyāyān abhūt teṣu balatpratāpaḥ
śrīrāyamallaḥ sphuradugracāpaḥ//6//

. . . .

tataḥ samajani sphurattarabhujoṣmadarpānalo
dayāvinayadānavān bhuvanamallanāmā nṛpaḥ/
x x x x x x x x x x x x x x x x x
paripālayan pramuditāḥ prakāmaṃ prajāḥ//8//
śrīprāṇamallo nṛpatis tato ᵓbhūd
akhaṇḍadormaṇḍalabāṇavarṣaḥ/
audāryagāmbhīryadaśāṅgarājya-
payodhivṛddhau sakalaḥ sudhāṃśuḥ//9//
śrīvipramallo nṛpatis tato ᵓbhūd
abhūtapūrvaprabalapratāpaḥ/
yaḥ pūrṇacandro janatānurāge
tyāge hariścandra ivāparo yaḥ//10//
tasmāt trailokyamallaḥ samajani rajanījānijetā yaśobhiḥ
sāhityanyāyaśāstrāgamavarakavitāraṇyasañcārasiṃhaḥ/
dātā bhoktāvadātāśayagatir anaghaś caṇḍikāpādasevā-
paṇḍībhūtādhijātaḥ śivacaraṇasarojanmacintādvirephaḥ//11//
tatputro dānakarṇo jayati jayajagajjyotimallo narendro
jyotiḥsāhityaśāstrasmṛtivividhakalāmbhodhipāraṅgamajñaḥ/
nārīṇām apy arīṇāṃ sapadi mukhavidhuṃ yasya dṛṣṭvātikaṣṭhād
vaivarṇya x x mūrchāprabhṛtibahuvidhā hanta bhāvā bhavanti//12//

The date is given in the following verse:

āśvinaśukladaśamyāṃ śāke ṣaḍdahanabāṇavidhuvalite/
śatadaśadaṇḍakasamaye deyam udāharaṇam asmābhiḥ//

JAGATKĪRTI BHAṬṬĀRAKA

A resident of Saṅgrāmapura, Jagatkīrti wrote a ṭīkā on a *Candronmīlana*. Manuscript:

RJ 1648 (vol. 2, p. 270). 69ff. Copied in Saṃ. 1754 = A.D. 1697. Property of Baḍā Terahapanthiyoṃ of Jayapura.

JAGADĪŚA JHĀ (*fl.* 1911)

The son of Khakhanu, the son of Būlana, Jagadīśa, a resident of Rāmabhadrapura, taught at the Lakṣmīśvarī Pradhāna Saṃskṛta Pāṭhaśālā in Ānandapura, Darabhaṅga. His pupil, Kuśeśvara Śarman Kumara, published his *Vitribhalagnabhramaṇa* in 26 verses at Anandpur on Wednesday 1 śuklapakṣa of Śrāvaṇa in Śaka 1833 = 26 July 1911.

JAGADĪŚAPRASĀDA TRIPĀṬHIN (*fl.* 1899)

Author of a Hindī ṭīkā on the *Jātakapaddhati* of Keśava (*fl.* 1496/1507), published at Bombay in Saṃ. 1956 = A.D. 1899 (BM 14053. d. 63); 2nd edition, Bombay Saṃ. 1981 = A.D. 1924 (IO San. D. 707).

JAGADĪŚVARA

Author of a *Jātakacandrikā*. Manuscript:

Assam (1930) 26. 67ff. Copied in Śaka 1820 = A.D. 1898.

JAGADEVA

Author to whom is attributed a *Praśnacintāmaṇi*; Jagadeva is perhaps an error for Bhojadeva (*fl. ca.* 995/1056). Manuscript:

GVS 2844 (1755). Ff. 2–6. Incomplete.

JAGADDEVA (*fl. ca.* 1175)

The son of the Mahattama Durlabharāja (*fl.* 1160), the son of Narasiṃha, the son of Rājapāla, the son of Āhilla or Jāhilla of the Prāgvaṭavaṃśa, who was a minister to the Caulukya monarch Bhīmadeva (*ca.* 1031/1065), Jagaddeva, a resident of Gujarāt, is said to have finished his father's *Sāmudrikatilaka*, and also wrote a *Svapnacintāmaṇi* in 2 adhikāras: śubhasvapna and duḥsvapna. Manuscripts:

Baroda 619. 20ff. (ff. 3–6 missing). Copied in Saṃ. 1519 = A.D. 1462. Incomplete.

RORI Cat. II 9952. 16ff. (f. 1 missing). Copied by Mādhava on Tuesday 13 śuklapakṣa of Pauṣa in Saṃ. 1650 = 25 December 1593. Incomplete.
PL, Buhler IV E *454. 17ff. Copied in Saṃ. 1675 = A.D. 1618. Property of Bālambhaṭṭa of Surata. Buhler notes 3 other copies.
IO 3136 (2345b). 29ff. (ff. 8–12, 16–17, and 19–20 missing). Copied by Sāmi Harisaṃkaragiri on Sunday 2 kṛṣṇapakṣa of Jyeṣṭha in Saṃ. 1687 = 16 May 1630. Incomplete.
Anup 5184. Ff. 72–92. Incomplete (śubhasvapna).
Anup 5362 = Bikaner 738. 19ff.
AS Bengal 7347 (G 8217). 7ff.
Baroda 2168(b). 5ff. Incomplete.
Baroda 9202. 22ff.
Baroda 12976. 15ff.
Benares (1963) 37377. Ff. 2–5, 10–22, and 24–27. Incomplete.
Benares (1963) 37378. 19ff.
Benares (1963) 37566. Ff. 3–20. Incomplete.
Bombay U Desai 1510. 22ff.
BORI 1022 of 1886/92. 12ff.
CP, Hiralal 6734. Property of Gopāl Jaikṛishṇa of Kuṭāsā, Akolā.
GVS —— (889). 9ff. No author mentioned.
GVS 2979 (2298). Ff. 1–15 and 17–30. Incomplete.
Jaipur (II). 35ff.
Jammu and Kashmir (2) 760. 46ff.
Kavīndrācārya 837.
LDI 7387 (337/1). 8ff.
LDI 7388 (2709). 14ff. Incomplete.
LDI (KC) K/951. 10ff.
LDI (KS) 1031 (10979). 10ff. Incomplete.
Oudh XX (1888) VIII 17. 30pp. Property of Paṇḍita Pratāpa Nārāyaṇa of Allahabad Zila.
PUL II 4088. 19ff.
SOI 2565/2.
SOI 6060.
Tokyo U 488. 40ff. Nevārī.

The *Svapnacintāmaṇi* has been published with a Marāṭhī anuvāda at Bombay in 1848 (IO 11.D.3.); by Janārdana Hari Āṭhalye with a Marāṭhī ṭīkā at Ratnagiri in 1873 (IO 1599); by Śeṣācala Śāstri with an Āndhra tātparya in Telugu characters at Madras in 1911 (BM 14055.d.13); and, edited from BORI 1022 of 1886/92 and IO 2345b, by J. von Negelein, Giessen 1912.

The colophon begins: iti śrīmahattamadurlabharājātmajajagaddevaviracite.

JAGADDHARA (*fl.* thirteenth century?)

The son of Vidyādhara, the son of Śrīdhara, the son of Anantakaṇṭha of the Bhāradvājagotra and the Mādhyandinaśākhā of the Yajurveda, the astrologer Jagaddhara, formerly a resident of Thihāra (or Takāri), Vapabhūmi (Vipra), Madhyadeśa, received grants in Orissa upon his immigration to Paṭavāḍapaṭaka, Koṇṭaravaṅga, Dakṣiṇatosala, from the Bhañja monarchs Yaśabhañja and Vīrabhañja Yuvarāja. See Binayak Misra, *Dynasties of Mediaeval Orissa*, Calcutta 1933, pp. 50–51.

JAGADDHARA ŚARMAN ŚROTRIYOPĀDHYĀYA (*fl.* 1903)

Author of pariśiṣṭas to the *Varṣakṛtya* of Rudradhara Śarman, published at Kāśī in Śaka 1825 = A.D. 1903 (BM 14033.bbb.24 and IO San. C. 218); pt. 1 was published at Darbhanga in 1927 (IO San. D. 1089 (a)).

JAGADBANDHU SIṂHA (*fl.* 1908/1915)

Author of a *Jyotiṣārṇava*, published with an Utkala bhāṣānuvāda, pt. 1, Cuttack 1908 (IO San. B. 505 (m)), and pt. 2, Puri 1915 (IO San. C. 39 (b)).

JAGADRĀMA

The son of Gaṅgārāma, Jagadrāma wrote a *Śiśusaukhya*. Manuscript:

Anup 5200 = Bikaner 645. 25ff. Copied at Bīkānera in Saṃ. 1739 = A.D. 1682. (Bikaner, through some confusion, calls this the *Jātakapaddhati* in 8ff.).

The colophon begins: iti śrījyotirvidgaṅgārāmātmajajyotirvidjagadrāmaviracitaṃ.

JAGANNĀTHA

The son of Govinda, Jagannātha wrote a *Jyotiṣaśāstra*. Manuscript:

GVS 2822 (3834). Ff. 11–31. Copied in Tuesday 4 śuklapakṣa of Mārgaśīrṣa in Saṃ. 1621 = 7 November 1564. Incomplete.

JAGANNĀTHA

Alleged author of a *Padmakośa;* see Govardhana.

JAGANNĀTHA

The son of Mohana and the pupil of Sukhānanda, Jagannātha wrote a *Bhāvarahasya*. Manuscripts:

BORI 544 of 1895/1902. 10ff. Incomplete (grahabhāvādhyāya).
Leipzig 1105. 5ff. Incomplete (to 2, 5).
VVRI 4620. 10ff. Incomplete (adhyāyas 1–3).

Verses 2–3 are:

natvā gurusukhānandaṃ vidyāyāṃ ca bṛhaspatim/
yasya smaraṇamātreṇa bhāvarahasyaṃ kathitam//
jambūmārge śaivamārgānurakto
bhūdevānām agraṇī mohanākhyaḥ/
tatputraḥ syāc chrījagannāthanāmnā
cakre horābhāvacakraṃ prakāśya//

JAGANNĀTHA

Author of a *Muhūrtadīpaka*. Manuscript:

LDI (LDC) 698. 29ff. Copied in Saṃ. 1893 = A.D. 1836.

JAGANNĀTHA TRIPĀṬHIN

The son of Nātha Tripāṭhin, Jagannātha wrote a *Ratnahāra* in 7 prakaraṇas. Manuscripts:

Mithila 293. 9ff. Maithilī. Copied by Śivanātha at Parihārapūragrāma on Thursday 7 kṛṣṇapakṣa of Bhādrapada in Śaka 1714 = 6 September 1792. Property of Pandit Mahīdhara Miśra of Lalbag, Darbhanga.
Jaipur (II).
VVRI 6924. 17ff. Incomplete.
WHMRL G. 60. d. Ff. 20v–57.

The first verse is:

gaṇeśaṃ hariṃ bhāratīṃ bhānum īśaṃ
jagannāthanāthātmajo ᵓham praṇamya/
game praśnasūtau vivāhe munīnāṃ
matenānvitaṃ ratnahāraṃ karomi//

The colophon begins: iti śrītripāṭhināthātmajaśrītripāṭhijagannāthaviracite.

TĀTĀ JAGANNĀTHA SŪRI

Author of a *Lokacandrikā* in 4 adhyāyas, on which he wrote his own Telugu ṭīkā. The adhyāyas are:

1. bhāvasādhana.
2. dṛṣṭisādhana.
3. balasādhana.
4. āyurdāya.

Manuscript:

GOML Madras (Telugu) R 733. 20ff. Copied in 1919/20 from a manuscript belonging to Koṭikalapūḍi Śivarāmadāsa Pantulugāru of Bobbili. Incomplete (jātakabhāga).

JAGANNĀTHA SAMRĀṬ (*fl. ca.* 1720/1740)

Traditionally said to have been discovered by Jayasiṃha I Mirzā (1605/1667) of Amber during a campaign against the Marāṭha chief Śivājī (1627/1680) in 1664/1665, at which time he was supposed to have been twenty years old, Jagannātha in fact is known only from his translations from the Arabic made for Jayasiṃha II Savāī (1699/1743) in the eighteenth century. See S. Dvivedin [1892] 102–110.

1. *Rekhāgaṇita*, a Sanskrit translation of Naṣīr al-Dīn al-Ṭūsī's (1201/1274) Arabic recension of Euclid's *Elements* in 15 adhyāyas; see L. Wilkinson [1837] and L. J. Rocher [1953/54]. Manuscripts:

Benares (1963) 35605 = Benares (1878) 122 = Benares (1869) XXVIII 1. Ff. 2, 1–32, 34–45, 56–68, 79–187, and 220–292. Copied by Lokamaṇi for the Samrāṭ on Sunday 4 śuklapakṣa of Āṣāḍha in Saṃ. 1784 = 11 June 1727. See vol. 1, appendix, and vol. 2, appendix I of the edition.
Jaipur (II). 244ff. Copied in Saṃ. 1785, Śaka 1650 = A.D. 1728.
Goṇḍal 337. 4ff. Copied in Saṃ. 1862 = A.D. 1805. Incomplete. No author mentioned.
Calcutta Sanskrit College 117. 315ff. Copied in Saṃ. 1878 = A.D. 1821.
Oxford 797 (Wilson 425). 172ff. (ff. 1–8 missing). Copied in A.D. 1821. Incomplete.
Baroda, Bāl Gaṅgādhar Śāstrī Jāmbhekar. Copied by Indrajit Śarman, the son of Jyeṣṭhārāma, a resident of Prabhāsapattana, and a teacher at the Amarelī Gurjara Śālā, on 5 śuklapakṣa of Kārttika in Saṃ. 1886 = 31 October 1829.
RORI Cat. II 5594. 264ff. Copied by Badrīnātha Gauḍa Brāhmaṇa on 12 kṛṣṇapakṣa of Bhādrapada in Saṃ. 1920 = 8 October 1863.
Baroda, Harilāl Harṣadharāi Dhruva. 144ff. Copied for Rāo Bahādur Justice Janārdan Sakhārām Gādgil from Bāl Gaṅgādhar's manuscript on 5 śuklapakṣa of Āṣāḍha in Saṃ. 1946 = 2 July 1889.
Bombay, Kamalāśaṅkara Prāṇaśaṅkara Trivedin. Copied from Jammu and Kashmir 2829 in 1899. Incomplete (adhyāyas 10–12).
Ānandāśrama 3693. See vol. 2, appendix II of the edition.
Baroda 12765. Ff. 4–15. Incomplete (adhyāya 1).
Baroda, Harilāl Harṣadharāi Dhruva. Pp. 1–70 and 1–65. Incomplete (adhyāyas 1–3 and 10–15). From Pandit Durgāprasāda Dviveda of Jaipur.
Baroda, Harilāl Harṣadharāi Dhruva. 85ff. Incomplete (adhyāyas 1–5). From Pandit Durgāprasāda Dviveda of Jaipur.
Benares (1963) 35707. Ff. 2–6. Incomplete (adhyāya 1).
Benares (1963) 35708. Ff. 41–165. Incomplete (adhyāyas 7–15).
Benares (1963) 36798. 4ff. Incomplete. No author mentioned.
Benares (1963) 36936 = Benares (1878) 118. Ff. 1–2 and 1–218.
BORI 514 of 1892/95. 54ff. Incomplete. No author mentioned.
Calcutta Sanskrit College 116. 258ff.
CP, Hiralal 4835. Property of Govindprasād Śāstrī of Jubbulpore. No author mentioned.
IO 2882 (252b). 66ff. From H. T. Colebrooke.
Jaipur (II). 135ff.
Jaipur (II). 24ff.
Jammu and Kashmir 2829. 192ff.
Kerala ———. Copied from Benares 36936.
Kurukṣetra 925 (19553).

N-W P VIII (1884) 11. 271ff. Property of Lāla Sītārāma, B.A., of Benares.
Paris BN 245.6 (Sans. beng. 184). Pp. 42–93. Bengālī. Incomplete. From Guérin.
Paris BN 304.5. (Sans. beng. 187). Pp. 50–127. Bengālī. Incomplete. From Guérin.
SOI 4747 = SOI (List) 1 = SOI Cat. II: 1101–4747. Ff. 17–273.
SOI 9428.
SOI 10051. Incomplete (adhyāyas 1–4).
Vṛndāvana, Āryasamājagurukula (see V. Raghavan in *JOR Madras* 26, 1956–57, 78).

The *Rekhāgaṇita* was edited by Harilāl Dhruva, the son of Harṣadarāya, and Kamalāśaṅkara Trivedin, the son of Prāṇaśaṅkara, 2 vols., *Bombay SS* 61–62, Bombay 1901–1902, on the basis of Benares 35605 and 36936, the 3 Dhruva manuscripts, and the Trivedin manuscript:

Verses 3–9 are:

śrīgovindasamāhvayādivibudhān vṛndāṭavīnirgatān
yas tatraiva nirākulaṃ śucimanobhāvaḥ svabhaktyānayat/
mlecchān mānasamunnatān svatarasā nirjitya bhūmaṇḍale
jīyāc chrījayasiṃhadevanṛpatiḥ śrīrājarājeśvaraḥ//
karaṃ janārdanaṃ nāma dūrīkṛtya svatejasā/
bhrājate duḥsaho ᵓrīṇāṃ yathā graiṣmo divākaraḥ//
yeneṣṭaṃ vājapeyādyair mahādānāni ṣoḍaśa/
dattāni dvijavaryebhyo gogrāmagajavājinaḥ//
tasya śrījayasiṃhasya tuṣṭyai racayati sphuṭam/
dvijaḥ samrāḍ jagannātho rekhāgaṇitam uttamam//
apūrvaṃ vihitaṃ śāstraṃ yatra koṇāvabodhanāt/
kṣetreṣu jāyate samyag vyutpattir gaṇite yathā//
śilpaśāstram idaṃ proktaṃ brahmaṇā viśvakarmaṇe/
pāramparyavaśād etad āgataṃ dharaṇītale//
tad vicchinnaṃ mahārājajayasiṃhājñayā punaḥ/
prakāśitaṃ mayā samyag gaṇakānandahetave//

The final verse is:

śrīmadrājādhirājaprabhuvarajayasiṃhasya tuṣṭyai dvijendraḥ
śrīmatsamrāḍ jagannātha iti samabhidhārūḍhitena praṇīte/
granthe ᵓsmin nāmni rekhāgaṇita iti sukoṇāvabodhapradātary
adhyāyo ᵓdhyetṛmohāpaha iha viratiṃ viśvasaṃkhyo gato ᵓyam//

2. *Samrāṭsiddhānta* or *Siddhāntasārakaustubha*, a Sanskrit translation of Naṣīr al-Dīn al-Ṭūsī's Arabic recension of Ptolemy's *Almagest* in 13 adhyāyas with additional notes referring to Ulugh Beg (1394/1449), Jamshīd al-Kāśī (*fl.* 1406/1429), and Muḥammad Shāh, the Mughal emperor (1719/1748); the *Samrāṭsiddhānta* is said to have been composed in A.D. 1732. Manuscripts:

Cambridge R. 15. 138. 51ff., 122ff., and 78ff. Copied in A.D. 1803. Incomplete (ends in adhyāya 13).
Jammu and Kashmir 2792. 411ff. Copied in Saṃ. 1900 = A.D. 1843. Incomplete.
RORI Cat. III 17213. 4ff. Copied by Bihārī Lāla at Jayapura in Saṃ. 1916 = A.D. 1859. Incomplete (yantrādhyāya).
Mithila 392. 304ff. Copied by Hanumānaprasāda Kāestha for Nakalabhaī Cirañjīva Jhā of Mithilā, Librarian of the Sarakāri Kumpanī Pāṭhaśālā, on Wednesday 2 kṛṣṇapakṣa of Āśvina in Saṃ. 1922 = 4 October 1865. Property of Pandit Rudramaṇi Jhā of Mahinathapur, Deodha, Darbhanga.
RORI (Jaipur) IV 77 and 78. Ff. 1–276 and 277–581. Copied in Saṃ. 1955 = A.D. 1898.
Alwar 1994.
Ānandāśrama 4337. Incomplete (adhyāyas 1–2). See S. B. Dikshit [1896] 293.
Baroda 9215(a). 159ff. Incomplete.
Baroda 9215(c). Ff. 222–476. Incomplete (ends in adhyāya 10).
Baroda 10886. 49ff. Incomplete (adhyāyas 1–2).
Baroda 10887. 116ff. Incomplete.
Benares (1963) 35762 = Benares (1878) 125 = Benares (1869) XXIX 1. Ff. 1–24, 31–122, 133–140, 21–23, 124–192, 1–82, and 1–56. With the *Ukara* of Nayanasukhopādhyāya. Incomplete. (Copied in Saṃ. 1859 = A.D. 1802 according to Benares (1878)).
Calcutta Sanskrit College 118a. 118ff. Incomplete.
Calcutta Sanskrit College 119. 288ff. Incomplete.
Calcutta Sanskrit College 151. Ff. 1–106, 227–251, and 326–420. Incomplete.
Calcutta Sanskrit College 152. Ff. 1–150 and 298–325. Incomplete.
Calcutta University 1012. Ff. 1–96 and 99–100. Incomplete.
Calcutta University 1013. Ff. 1–13, 15, and 27–32. Incomplete.
Calcutta University 1014. Ff. 2–28. Incomplete.
Calcutta University 1015. Ff. 1–12. Incomplete.
Jammu and Kashmir 2857. 186ff. Incomplete.
N-W P X (1886) A 33. 4ff. Incomplete (parvasambhava). Property of Umāśaṅkara Miśra of Azamgarh.
Rajputana, p. 38. At Udaipur. (*Siddhāntabodhaprakāśa*).
RORI Cat. III 11465. 213ff. (ff. 1 and 210–212 missing). Incomplete.
RORI (Jaipur) IV 79. 62ff. Incomplete.
SOI 9427.

The *Samrāṭsiddhānta* was published by Rāmasvarūpa Śarman, 3 vols., New Delhi 1967–1969.

Verses 1–3 = *Rekhāgaṇita* 1–3; verses 4–8 are:

rājādhirājo jayasiṃhadevaḥ
śrīmatsyadeśādhipatiś ca samrāṭ/
śrīrāmapādāmbujasaktacitto

yajvā sadā dānarataḥ suśīlaḥ//
golādiyantreṣu navīnayukti-
pracāradakṣo gaṇitāgamajñaḥ/
satyapriyaḥ satyarataḥ kṛpālus
tigmapratāpo jayati kṣamāyām//
sa dharmapālo gaṇitapravīṇo
jyotirvido golavicāradakṣān/
kārūṃs tathāhūya cakāra vedhaṃ
golādiyantrair dyusadāṃ ca bhānām//
granthaṃ siddhāntasaṃrājaṃ samrāṭ racayati sphuṭam/
tuṣṭyai śrījayasiṃhasya jagannāthasaṃhvayaḥ kṛtī//
arabībhāṣayā grantho mijastīnāmakasthitaḥ/
chvāṇakānāṃ subodhāya gīrvāṇyā prakaṭīkṛtaḥ//

JAGANNĀTHA BHASĪNA (*fl.* 1971)

Retired pradhāna of the Svāmī Rāmatīrtha Mission in Dillī, Jagannātha wrote a Hindī vyākhyā on the *Uttarakālāmṛta* of Kālidāsa, published at Dillī in Saṃ. 2028 = A.D. 1971.

JAGANNĀTHASIṂHA VISENA (*fl.* 1830)

The son of Rājā Devībakhśasiṃha, the Tālukedāra of Dhanagaḍha, Pratāpagaḍha, and a resident of Rāmapura, Ḍerabā, Pratāpagaḍha, Jagannāthasiṃha wrote a *Yuddhajyotiṣa* in Hindī in Saṃ. 1887 = A.D. 1830. Manuscripts:

NPS 77 of 1917–19. Copied in Saṃ. 1891 = A.D. 1834. Property of the Pratāpagaḍhanareśa kā Pustakālaya at Pratāpagaḍha.
NPS 123 of 1909–11. Copied in Saṃ. 1898 = A.D. 1841. Property of Rājā Sāhaba Bahādura of Pratāpagaḍha.
NPS 109ka of Saṃ. 2004–2006. Copied in Saṃ. 1914 = A.D. 1857. Property of Rāya Ambikānāthasiṃha of Nāïna State, Rāyabarelī.
NPS 109kha of Saṃ. 2004–2006. Property of Maṅgalāprasāda Dvivedī of Gogahara, Ḍheṅgura, Pratāpagaḍha.

JAṬĀDHARA (*fl.* 1704)

The son of Vanamālī, the son of Durgamiśra, the son of Uddhava of the Gargagotra, Jaṭādhara wrote a *Phatteśāhaprakāśa*, whose epoch is Śaka 1626 = A.D. 1704, the 48th year of the reign of the Phatteśāha—presumably Aurangzib (1658/1707). See BORI 1883/84, p. 84, and S. B. Dikshit [1896] 292. Manuscript:

BORI 195 of 1883/84. Ff. 6–24. Copied in Saṃ. 1777 = A.D. 1720. Incomplete. From Gujarāt.

JAḌABHARATA

The pupil of Muni Mādhavānanda, Jaḍabharata wrote a *Praśnāvalī*. Manuscript:

Rajputana, p. 47. From Bikaner.

JANAJVĀLĀ (*fl.* 1870)

A resident of Hajaratagañja, Lakhanaū, Janajvālā wrote a ṭīkā in Hindī on the *Praśnamanoramā* of Garga in Saṃ. 1927 = A.D. 1870. Manuscript:

NPS 112 of Saṃ. 2004–2006. Property of the Nāgarīpracāriṇī Sabhā in Vārāṇasī.

JANABHUVĀLA

Author of a *Bhūgolapurāṇa* in Hindī. Manuscript:

NPS 262kha of Saṃ. 2001–2003. Copied in Saṃ. 1862 = A.D. 1805. Property of Ṭhākura Raghunāthasiṃha of Samogarā, Nainī, Ilāhābāda.

JANARĀJA

Author of a ṭīkā on the *Bhuvanadīpaka* of Padmaprabha Sūri. Manuscript:

Benares (1963) 35837. 26ff.

JANĀRDANA

The son of Mukunda, Janārdana wrote a *Jayakaumudī*. Manuscript:

Anup 4604. 46ff. Incomplete.

JANĀRDANA

Author of a *Padyābjamālā*. Manuscript:

BORI 900 of 1884/87. 7ff. Copied in Śaka 1784 = A.D. 1862. From Mahārāṣṭra.

JANĀRDANA BHAṬṬA

Author of a *Bālaviveka* in Hindī. Manuscript:

NPS 267 A of 1906–08. Property of Lālā Vidyādhara of Horīpurā, Datiyā.

JANĀRDANA (*fl.* 1464 or 1599)

The son of Ananta of the Audīcyajñāti, Janārdana wrote a *Vivāhapaṭala* or *Kāmakrīḍāśāstra* in 61 ślokas on Wednesday 8 śuklapakṣa of Phālguna in Saṃ. 1520 = 16 February 1464 or in Śaka 1520 = 21 February 1599. Manuscripts:

PUL II 3946. 8ff. Copied in Śaka 1520 = A.D. 1598/99.
AS Bengal 2679 (G 10329). 10ff. Copied by Jageśvara on Saturday 10 kṛṣṇapakṣa of intercalary Jyeṣṭha.
Baroda 3300. 16ff. Incomplete. Ascribed to Ananta.
Baroda 9761. 14ff.
Dāhilakṣmī XXXV 31. See NCC, vol. 3, p. 346.
PL, Buhler IV E 446. 9ff. Property of Khuśāla Bhaṭṭa of Ahmadābād.

PUL II 3945. 7ff.

Verses 59 and 61 are:

audīcyākhyajñātau ṣaṭkarmā daivajño ᵓnanto ᵓbhūl
lakṣmīkāntaḥ śānto bhaktyā lakṣmīkāntasyāsaktaḥ/
tajjanmā jānākhyaḥ satyavān iṣṭaḥ pitror bhaktas
tenedaṃ kāmakrīḍāśāstraṃ ṣaṣṭiślokaṃ santene//
khanetrabāṇābjavinirmite ᵓtra
varṣe khare phālguni māsi śubhre/
dine ᵓṣṭame jñena yute janākhyaḥ
kṛtvālikhat kautukakṛtyaśāstram//

The colophon begins: ity audīcyajñātīyajanārdanakṛte.

JANĀRDANA BHAṬṬA (*fl.* 1618/1639)

The son of Kṛṣṇabhaṭṭa, Janārdana copied the Oxford manuscript of the *Śīghrasiddhi* of Lakṣmīdhara (*fl.* 1278) between 29 January and 5 February 1639. To this he added a table of yearly parameters of the planets with kṣepakas for 3 March 1618 and 12 verses; see SATE 81–82. Manuscript:

Oxford CS c. 319b. B f. 22v, and C f. 1. Copied by Janārdana in 1639. See SATE 46–47.

JANĀRDANA BHĀSKARA KRAMAVANTA (*fl.* 1858)

Author of a Marāṭhī bhāṣā on the *Jyotiṣasāra* of Śukadeva, published at Muṃbaī in Śaka 1780 = A.D. 1858 (BM); reprinted at Muṃbaī in Śaka 1784 = A.D. 1862 (BM). A Gujarātī translation of the bhāṣā by Sītārāma Rāvajī was published at Muṃbaī in [1864?] (BM).

JANĀRDANA HARI ĀṬHALE (*fl.* 1869/1889)

A resident of Ratnāgiri, Janārdana wrote pañcāṅgas for Śaka 1791–1811 = A.D. 1869–1889: see S. B. Dikshit [1896] 404. He also wrote a Marāṭhī ṭīkā on the *Svapnacintāmaṇi* of Jagaddeva (*fl. ca.* 1175), which was published at Ratnagiri in 1873 (IO 1599).

JANĀRDANA BĀḶĀJĪ MOḌAKA (*fl.* 1888)

Author, with Śaṅkara Bālakṛṣṇa Dīkṣita, of a pañcāṅga for Śaka 1810 = A.D. 1888, published at Ratnāgiri in 1888 (BM 14096.a.3.(4)).

JANAULA

Author of a *Śaniścara kī kathā* in Hindī. Manuscript:

NPS 70 of 1938–40. Copied in Saṃ. 1923 = A.D. 1866. Property of Paṇḍita Ramaṇalāla of Pharaiha, Mathurā.

JANMEJAYA UPADHYA

Author of a *Yoton Jyotiṣṇika*. Manuscript:

Assam (1935/6) 28.

JAMBŪNĀTHA

A resident of Coladeśa, Jambūnātha of the Vādhūlagotra wrote several works on astrology.

1. *Jātakaratna*. Manuscripts:

Tanjore D. 11390 = Tanjore BL 10993(b). Ff. 8–9. Grantha. Incomplete.
Tanjore D 11391 = Tanjore BL 10993(a). 4ff. Grantha. Incomplete.

The first verse is:

parāśarādigranthāṃś ca nanu bhāvārthasaṅgrahān/
ālokya likhyate sārān jambūnāthena dhīmatā//

2. *Jātakasarvasaṅgraha* with his own ṭīkā. Manuscripts:

Kerala 5814 (T. 978). 2800 granthas. Copied in Saṃ. 1694 = A.D. 1637. With the ṭīkā.
Kerala C 685 A (C. 1908A). 23ff. Grantha. With the ṭīkā. Property of Vaṭṭapaḷḷi Maṭham of Śucīndram.

The first verse is:

x x x x x x x x ⟨jambū⟩ nāthena dhīmatā/
horāśāstro x sarvasvaṃ mayā saṅgṛhyate sphuṭam//

The colophon is: iti jātakasarvasaṅgrahe jambūnāthaviracite.

3. *Praśnadīpikā* with his own ṭīkā. Manuscripts:

Adyar List = Adyar Index 3872 = Adyar Cat. 28 M 11. 46ff. Grantha. With the ṭīkā. Incomplete.
PUL II 3654. 86ff. Grantha. With the ṭīkā.

4. *Praśnaratna* or *Praśnāmṛta* in 11 rasas; in this he refers to his *Jātakasarvasaṅgraha* and to his *Praśnasārasamudra*. Manuscripts:

GOML Madras D 13975. 33pp. Grantha. With a ṭippaṇa.
Kerala 10429 (T. 979). 570 granthas.
Kerala C 685 B (C. 1908 B). 29ff. Grantha. Property of Vaṭṭapaḷḷi Maṭham of Śucīndram.
Tanjore D 11502 = Tanjore BL 10999. 10ff. Telugu. Incomplete.
Tanjore D 11503 = Tanjore BL 11051(g). Grantha. Incomplete.

Verse 2 is:

kṛṣṇīyārṇavacandrabhūṣaṇamahālampākaratnāvalī-
praśnābdhīn svadhiyā vimṛśya bahuśas tebhyaḥ kim
apy uddhṛtam/

jambūnāthasamāhvayena viduṣā śrīcoladhātrībhuvānekārthojjvalam alpaśabdamadhuraṃ praśnāmṛtaṃ tāyate//

5. *Praśnasaṅgraha*. Manuscript:

GOML Madras D 17204. 72pp. Telugu. Incomplete.

Verse 1 is:

vādhūlaś colabhūr natvā jambūnātho maheśvaram/
māsābdāhaḥphaloktyarthaṃ kurve
bhāvārthasaṅgraham//

6. *Praśnasārasamudra* in 6 taraṅgas. Manuscripts:

Tanjore D 11509 = Tanjore BL 11012. 64ff. Telugu.
Tanjore D 11510 = Tanjore BL 11013. 114ff. Telugu.

Verse 2 is:

vādhūlakulapadmārko jambūnāthaḥ satāṃ mude/
praśnārṇavaṃ racayati praśnaśāstrāmṛtākaram//

JAMBŪNĀTHA (*fl. ca.* 1475)

See Sundararāja (*fl. ca.* 1475).

JAYA GOSVĀMIN

See Gosvāmin Yāja.

JAYAKṚṢṆA

See Jaikṛṣṇa.

JAYAKṚṢṆA

Author of a *Bālabodhinī*. Manuscripts:

Mithila 212 C. 10ff. Maithilī. Copied in Śaka 1764 = A.D. 1842. Property of Pandit Dharmadatta Miśra of Babhangama, Supaul, Bhagalpur.
Mithila 212 B. 12ff. Maithilī. Copied in Śaka 1765 = A.D. 1843. Property of Babu Puruṣottama Jhā of Babhangama, Supaul, Bhagalpur.
Mithila 212. 10ff. Maithilī. Copied by Śaṅkaradatta Śarman at Rāmanagaragrāma in Śaka 1767, Sāl. San. 1252 = *ca.* A.D. 1845. Property of Pandit Vāsudeva Jhā of Sukpur, Bhagalpur.
Mithila 212 A. 8ff. Maithilī. Copied in Śaka 1803 = A.D. 1881. Property of Pandit Sītārāma Pāṭhaka of Karnapur, Sukpur, Bhagalpur.

The first verse is:

vāgdevatāṃ namaskṛtya kriyate bālabodhinī/
śrīmatā jayakṛṣṇena bālabodhāya kevalam//

GUJARĀTĪ JAYAKṚṢṆADĀSA VEÑKAṬADĀSA (*fl.* 1880)

Author of an Āndhra ṭīkā on the *Kālajñāna* of Kumārasvāmin, published at Madras in 1880 (IO 16. D. 31).

JAYAGOPĀLA PAṆḌITA

Author of a ṭīkā on the *Jātakālaṅkāra* of Gaṇeśa (*fl.* 1613). Manuscripts:

AS Bengal 7047 (G 6424). 16ff. Copied by Gaṇeśadatta on Saturday 8 kṛṣṇapakṣa of Mārgaśīrṣa in Saṃ. 1855 = 29 December 1798. After the colophon is noted: jayagopālanāmapaṇḍiteneyaṃ ṭīkā kṛtā budhaiḥ kṣamasva, and the date Thursday 11 intercalary Vaiśākha of Saṃ. 1869, Śaka 1734 = 21 May 1812.
Benares (1963) 35347 = Benares (1897–1901) 15. 20ff. Copied in Saṃ. 1865 = A.D. 1808, Śaka 1751 = A.D. 1829. One must read either Saṃ. 1885 or Śaka 1731.

JAYADEVA (*fl.* before 1073)

An algebraist cited by Udayadivākara (*fl.* 1073) in his *Sundarī*; see K. S. Shukla [1954a].

JAYADEVA (*fl.* 1671/1675).

The son of Dhāreśvara, the son of Govinda of Śrīpura, Jayadeva wrote a *Tājikamañjarī* in Śaka 1593 = A.D. 1671. Manuscript:

Baroda 3147. 31ff. Copied in Saṃ. 1773 = A.D. 1716. With a vyākhyā.

He also completed a *Praśnanidhi* on Tuesday 2 kṛṣṇapakṣa of Māgha in Saṃ. 1731, Śaka 1596 = 2 February 1675. Manuscripts:

Goṇḍal 189. 35ff. Copied in Saṃ. 1803 = A.D. 1746. With a ṭīkā.
BORI 531 of 1895/1902. 23ff. Copied in Śaka 1722 = A.D. 1800. With a ṭīkā.
PL, Buhler IV E 254. 7ff. Copied in Saṃ. 1907 = A.D. 1850. Property of Nirbhaya Rāma of Mulī. Buhler notes another copy.
Baroda 7702(a). 6ff. Copied in Saṃ. 1918 = A.D. 1861.
Adyar Index 3873 = Adyar Cat. 8 D 35. 14ff.
Baroda 9189. 11ff.
PL, Buhler IV E 255. No ff. given. With a ṭīkā. Property of Tribhuvana Lālajī of Vaḍhavāṇa.

JAYADEVA BHAṬṬA

Author of a *Jātakapaddhati* or *Jātakapaddhatikāmadhenu*. Manuscripts:

AS Bengal 7024 (G 6431). 14ff. Copied on Saturday 14 kṛṣṇapakṣa of Mārgaśīrṣa in Saṃ. 1895, Śaka 1760 = 15 December 1838.
GVS 2802 (4171). Ff. 1–3, 6–8, and 11–14. Copied on Monday 13 śuklapakṣa of Śrāvaṇa in Saṃ. 1900 = 7 August 1843. Incomplete.

The colophon begins: iti śrījayadevabhaṭṭakṛtau.

JAYADEVA ŚARMAN (*fl.* 1750)

Author of a *Jātakacandrikā* in 16 adhyāyas in Śaka 1672 = A.D. 1750. This was published with the *Subodhinī* of Keśavānanda Śarman at Bambaī in 1958; reprinted Bambaī 1963.

JAYANĀRĀYAṆA TARKAPAÑCĀNANA (*fl. ca.* 1898)

A professor at the Calcutta Sanskrit College, Jayanārāyaṇa wrote a *Sūryasaṅkrāntidīpikā*, otherwise known as the *Saṅkrāntidīpikā*. Manuscripts:

Calcutta Sanskrit College (Smṛti) 390. 18ff. Bengālī.
Calcutta Sanskrit College (Smṛti) 391. 25ff.

JAYANTA BHAṬṬA

Author of a ṭīkā, *Bālabodha*, on the *Tattvārthādhigamasūtra* of Umāsvāti (*fl.* first century A.D.). See Velankar, p. 156.

JAYARATNA

Author of a *Jyotiṣasāraprabandha*. Manuscript:

LDI (LDC) 409. 14ff.

JAYARATNA (*fl. ca.* 1725)

A Jaina of the Pūrṇimīya Gaccha and a pupil of Bhāvaratna (*fl.* 1711), Jayaratna wrote a *Jñānaratnāvali*. Manuscripts:

Jammu and Kashmir 4107. 9ff. Copied in Saṃ. 1941 = A.D. 1884 from Alwar 1814.
Alwar 1814.
LDI (LDC) 3713. 2ff.

JAYARĀMA

Author of a *Kṣayamāsanirṇaya*. Manuscript:

Mithilā. See NCC, vol. 5, p. 149.

JAYARĀMA

Author of a *Gaṇitadīpaka*. Manuscript:

Baroda 3099. 1f.

JAYARĀMA

Author of a *Grahagocara*. Manuscripts:

PL, Buhler IV E 64. 7ff. Copied in Saṃ. 1863 = A.D. 1806. Property of Maṇiśaṅkara Jośī of Aṅkaleśvara. Buhler notes 4 other copies.
CP, Hiralal 1530. Property of Tukārām Śaṅkarbhaṭ Joṣī of Ghuikheḍ, Amraoti.

The *Grahagocara* was published with the Gujarātī translation of Gaurīśaṅkara Lalu Mehtā at Amadābāda in Saṃ. 1948 = A.D. 1891 (BM 14053. b. 17. (4)).

JAYARĀMA

Author of a *Tājikakalpalatā*. Manuscripts:

RORI Cat. II 6715. 8ff. Copied at Udayapura in Saṃ. 1768 = A.D. 1711. Incomplete (māsabhāvādhyāya).
RORI Cat. II 5884. 35ff. Copied by Manulāla Vyāsa in Saṃ. 1826 = A.D. 1769.
Baroda 7649. 11ff. Copied in Saṃ. 1914 = A.D. 1857. No author mentioned. Probably identical with PL, Buhler IV E 153. 11ff. No author mentioned. Property of Hariśaṅkara Jośī of Ahmadābād. Incomplete (bhāvādhyāya). Buhler notes another copy.
PL, Buhler IV E 152. 13ff. No author mentioned. Property of Bālakṛṣṇa Jośī of Ahmadābād.

JAYARĀMA

Author of a *Muhūrtālaṅkāra* in at least 17 prakaraṇas. Manuscripts:

Anup 4999. 30ff. Copied in Saṃ. 1711 = A.D. 1654.
Benares (1963) 35931. 64ff. Copied in Saṃ. 1711, Śaka 1576 = A.D. 1654.
BORI 423 of A 1881/82. 30ff. Copied in Saṃ. 1719 = A.D. 1662.
PL, Buhler IV E 367. 41ff. Copied in Saṃ. 1794 = A.D. 1737. Property of Mayāśaṅkara Jānī of Ahmadābād.

JAYARĀMA BHAṬṬA

A resident of Alindra and the son of Śrībhā(?), Jayarāma wrote a *Kāmadhenupaddhati* = *Jātakakāmadhenu;* he may be identical with the author of the *Khecarakaumudī*. Manuscripts:

PL, Buhler IV E 23. 89ff. Copied in Saṃ. 1707 = A.D. 1650. Property of Jagannātha Jośī of Ahmadābād.
BORI 333 of 1879/80. 94ff. Copied in Saṃ. 1716 = A.D. 1659.
BORI 301 of 1882/83. 10ff. Copied in Saṃ. 1726 = A.D. 1669. From Gujarāt. No author mentioned.
IO 3078 (2546). 87ff. Copied at Nalinagara on Thursday 4 kṛṣṇapakṣa of Vaiśākha in Saṃ. 1771, Śaka 1636 = 22 April 1714. Incomplete (fragments interspersed with the *Jātakābharaṇa* of Ḍhuṇḍhirāja). From Gaikawar.
IO 3079 (2457). 71ff. Copied by Bhayarāma Vaṇāśī, a Nāgara Brāhmaṇa, at Ilampura on Sunday 5 śuklapakṣa of Mārgaśīra in Śaka 1650 = 24 November 1728. From Gaikawar.
LDI 6699 (7222). 67ff. Copied by Muni Kesaravardhana at Pāṭaṇamahānagara in Saṃ. 1793 = A.D. 1736.
LDI (LDC) 3683. 59ff. Copied in Saṃ. 1799 = A.D. 1742.
Goṇḍal 22. 10ff. Copied in Saṃ. 1804 = A.D. 1747. No author mentioned.

Florence 281. 5ff. Copied by Tattvahaṃsa Gaṇi at Sūryapurabandira in Saṃ. 1820 = A.D. 1763. No author mentioned.
Benares (1963) 35101. 17ff. Copied in Śaka 1709 = A.D. 1787. Incomplete.
GVS 2766 (5261). 16ff. Copied on Thursday 5 kṛṣṇapakṣa of Āṣāḍha in Saṃ. 1848 = 21 July 1791. No author mentioned.
PUL II 3292. 5ff. Copied in Saṃ. 1854 = A.D. 1797. Incomplete.
LDI 6701 (2717). 12ff. Copied by Mehtā Lakṣmīcandra Kāmeśvara in Saṃ. 1856 = A.D. 1799. No author mentioned.
RORI Cat. II 6094. 76ff. (ff. 67–68 missing). Copied by Jagannātha Vyāsa in Saṃ. 1876 = A.D. 1819. Incomplete.
BORI 525 of 1899/1915. 13ff. Copied in Śaka 1765 = A.D. 1843. (*Jātakakāmadhenu*). No author mentioned.
Goṇḍal 23. 20ff. Copied in Saṃ. 1917 = A.D. 1860. No author mentioned.
Jammu and Kashmir 4103. 9ff. Copied in Saṃ. 1941 = A.D. 1884 from Alwar 1760. Incomplete (dvādaśabhāvaphala from *Jātakakāmadhenu*). No author mentioned.
Adyar Cat. 8 D 39. 26ff. Incomplete (dvādaśabhāvapatiphala of Kāmadha). See NCC, vol. 3, p. 351, and correct *CESS* A 2, 31a. Is this Adyar Index 7623 (*Jātakakāmadhenu*)?
Alwar 1760. (*Jātakakāmadhenu*). No author mentioned.
Baroda 856. 13ff. (f. 3 missing). No author mentioned.
Baroda 7651. 27ff. Incomplete. No author mentioned.
Baroda 13935. 6ff. No author mentioned.
BORI 300 of 1882/83. Ff. 1–5 and 7–10. No author mentioned.
Chani 2838. See NCC.
Goṇḍal 24. No ff. given. Incomplete. No author mentioned.
GVS —— (4589). 1f. No author mentioned.
IM Calcutta 1025. Incomplete. See NCC.
LDI 6702 (5781). 5ff. No author mentioned.
PUL II 3397. 8ff.
SOI 8115. No author mentioned.
SOI 8413. No author mentioned.
SOI 9543. No author mentioned.
SOI 9896. No author mentioned.
Udaipur, Library of Nathdwara 184, 16–17. See NCC.

Verse 6 at the end is:

alindrasaṃstho vijayī guṇāḍhyaḥ
śrībhāsuto ᵓyaṃ jayarāmanāmā/
śrīkāmadhenau janijātakasya
viśeṣatas tadracaṇāṃ cakāra//

JAYARĀMA BHAṬṬA

The son of Śrīmadbhaṭṭa, Jayarāma wrote a *Khecarakaumudī;* he may be identical with the author of the *Kāmadhenupaddhati*. Manuscripts:

VVRI 2462. 8ff. Copied on Tuesday 2 kṛṣṇapakṣa of Śrāvaṇa in Saṃ. 1888 = 23 August 1831.
Benares (1963) 36526. 14ff. Copied in Saṃ. 1907, Śaka 1772 = A.D. 1850. Said to be a part of the *Kāmadhenupaddhati*.
Benares (1963) 34453. Ff. 1–8 and f. 6. Copied in Śaka 1796 = A.D. 1874.
Benares (1963) 34779. 8ff. Copied in Saṃ. 1934 = A.D. 1877.
CP, Hiralal 1130. Property of Vāsudevrāv Golvalkar of Maṇḍlā.
GJRI 3113/325. 17ff. Maithilī.
PL, Buhler IV E 38. 57ff. Property of Bālakṛṣṇa Jośī of Ahmadābād.
Viśvabhāratī 217(1): See NCC, vol. 5, p. 188.

The last verse is:

itthaṃ khecarakaumudī suvipulā jātā budhair vistṛtā
vṛttair dvādaśabhir lasatsphuṭadalair
daivajñahastropamā/
śrīmadbhaṭṭatanayena bhaṭṭajayarāmeṇoditāṃ
daivavic
caināṃ yo nijakaṇṭhagāṃ prakurute bhūpāṅgaṇe
śobhate//

JAYARĀMA BHAṬṬA

The son of Sadāśiva, Jayarāma wrote a *Subodhā*. Manuscripts:

RORI Cat. III 15829(1). 79ff. Copied by Avicala Jośī in Saṃ. 1768 = A.D. 1711. (tithisāriṇī).
RORI Cat. III 15829(2). 15ff. Copied by Satyasāgara in Saṃ. 1800 = A.D. 1743. (tithisāriṇī).
RORI Cat. III 15829 (3). 13ff. Copied in Saṃ. 1865 = A.D. 1808. (pañcāṅgakaraṇasāriṇī).
Adyar Index 7148 = Adyar Cat. 35 C 104. 3ff.

JAYARĀMA (*fl.* 1745)

An Audīcya Brāhmaṇa, Jayarāma wrote a *Ramalāmṛta* at Surata in Saṃ. 1802, Śaka 1667 = A.D. 1745; see S. B. Dikshit [1896] 489. Manuscripts:

Baroda 1260(g). 34ff. Copied in Saṃ. 1890 = A.D. 1833.
Baroda 1266. 34ff. Copied in Saṃ. 1890 = A.D. 1833.
BORI 983 of 1886/92. 18ff.
PL, Buhler IV E 406. 17ff. Property of Bālakṛṣṇa Jośī of Ahmadābād. Buhler notes another copy.

JAYARĀMA JYAUTIṢĪ (d. 1855)

The son of Babuā Jyotirvit, a Mahārāṣṭra Brāhmaṇa, Jayarāma resided in Vārāṇasī, where he was associated with Durgāśaṅkara Pāṭhaka (*fl.* 1837); he was also connected with Lancelot Wilkinson (*fl.* 1834/1837) of Sihora. See S. Dvivedin [1892] 121.

JAYALAKṢMAṆA

Alleged author of a ṭīkā on the *Siddhāntaśiromaṇi* of Bhāskara (b. 1114). Manuscript:

N-W P I (1874) 36. 211ff. in 4 volumes. Property of Rāmeśvara Chaube of Mirzapore.

JAYAVANTAŚIṢYA (*fl.* 1503)

The unnamed pupil of Jñānaśīla Paṇḍita Jayavanta in Saṃ. 1560 = A.D. 1503 wrote a poem of 40 verses on the interpretation of dreams, the *Svapnacatuṣpadī*. Manuscript:

Bombay U 2407. 1f. Incomplete (begins with vs. 29).

JAYAVALLABHA

Author of a *Vidyālagnapaddhati*. Manuscript:

RORI Cat. III 14546. 44ff.

The second verse is:

vivahakadraviradrayādri
gāhāṇaṃ bahukulādriṃdhettūṇaṃ/
raccayaṃ vidyālagnaṃ
vihiṇa jayavallahaṃ nāma//

JAYAVIJAYA

Author of a *Śakunadīpikā* in Gujarātī. Manuscript:

LDI (MPC) P/4868. No ff. given. Copied in Saṃ. 1688 = A.D. 1631.

JAYAŚAÑKARA DEVAŚAÑKARAJĪ ŚARMAN (*fl.* 1969)

Author of a *Prakṛti se varṣā jñāna* in Rājasthānī, published in 2 volumes at Kalakattā in Saṃ. 2026 = A.D. 1969.

JAYAŚĪLA MUNI

Author of a stabaka in Old Gujarātī on the *Sañgrahaṇīratna* of Śrīcandra Sūri. Manuscript:

LDI 3117 (6078). 54ff. Copied for Śrāvikā Premabāī in Saṃ. 1740 = A.D. 1683.

JAYAŚEKHARA

Author of a *Kṣetrasamāsa;* see Velankar, p. 100.

SAVĀĪ JAYASIṂHA (1686/1743)

A Kachwāha (Kacchavaṃśa) Rājput, Jayasiṃha was born at Amber in 1686 to the Mahārāja Viṣṇusiṃha; he succeeded his father as Mahārāja in 1699 and ruled till his death on 2 October 1743. He founded the city of Jaipur in 1728 (see P. D. Pathak [1963/64]), and in the same year is said to have dedicated the Persian *Zīj-i jadīd-i Muḥammad-Shāhī*, probably largely written by Abū al-Khayr Khayr Allāh Khān, to the Mughal emperor Muḥammad Shāh (1719/1748), though the star-catalog is dated A.H. 1138 = A.D. 1725/1726 and the preface was written after 1734; see W. Hunter [1797] and C. A. Storey, *Persian Literature*, vol. 2, pt. 1, London 1958, pp. 93–94. He is best known for constructing the astronomical observatories at Benares, Delhi, Jaipur, Mathurā, and Ujjain; see R. Barker [1777]; J. L. Williams [1793]; S. B. Dikshit [1896] 353–355; A. Ff. Garrett and C. Guleri [1902]; S. Noti [1911]; G. R. Kaye [1918a] and [1920a]; M. F. Soonawala [1940] and [A2. 1952]; and A. P. Stone [1958]. He took an active part in Mughal politics (see J. Tod, *Annals and Antiquities of Rajasthan*, 3 vols., Oxford 1920, vol. 3, pp. 1341–1356; D. C. Sircar [A3. 1936/37]; S. Chandra [A2. 1948]; B. Das Gupta [A2. 1956]; M. L. Sharma [A2. 1969]; and H. C. Tikkimal [A3. 1969]; one of his most noteworthy political acts was the last performance of an aśvamedha in June/July 1742 (see P. K. Gode [A3. 1937] [1937b] [1937/38b] and [A2. 1943]; and V. S. Bhatnagar [A3. 1960]). Besides patronizing Kṛpārāma (*fl.* 1715) Jagannātha Samrāṭ (*fl. ca.* 1720/1740), Kevalarāma Pañcānana (*fl.* 1728/1762), and probably Nayanasukhopādhyāya (*fl. ca.* 1725/1730), Jayasiṃha was responsible for the writing of the following Sanskrit works on astronomy (see also S. B. Dikshit [1896] 292–295 and G. M. Moraes [1951/52]).

1. *Jayavinodasāriṇī*, composed in Śaka 1657 = A.D. 1735; see SATIUS 66b–67a. Manuscripts:

RORI Cat. III 11839. 36ff. Copied by Karuṇākara in Saṃ. 1963 = A.D. 1906. No author mentioned.
Calcutta Sanskrit College 17. 19ff.
Poleman 5107 (Harvard 61). 23ff. See SATIUS 34b.

2. *Yantrarājaracanā*, on the astrolabe. Manuscripts:

Benares (1963) 34439. 11ff. Copied in Saṃ. 1853, Śaka 1718 = A.D. 1796.
Benares (1963) 36972 = Benares (1915–1916) 2521. Ff. 1–3, 5, and 7–18. Incomplete.
Bikaner 759. 2 copies (25ff. and 18ff.).
BORI 180 of A 1883/84. 23ff.
BORI 850 of 1884/87. 6ff. From Gujarāt.
Jammu and Kashmir 2830. 20ff.
Kurukṣetra 846 (19540).
Mithila 274. 8ff. Maithilī. Property of Gañgādhara Jhā of Jonki, Deodha, Darbhanga.
N-W P I (1874) 10. 25ff. With a ṭīkā. Property of Kedāra Nātha of Benares.
Poleman 4715 (Columbia, Smith Indic 73). 35ff.
Poleman 4891 (Columbia, Smith Indic 168). 3ff. Incomplete.
RORI Cat. III 12618. 13ff.

The *Yantrarājaracanā* was edited by Kedarnath [1924] with the translation from A. ff. Garrett and C. Guleri [1902]. It was edited again by Kedāranātha with the *Yantraprabhā* of Śrīnātha and the *Yantrarājaprabhā* of Kedāranātha (*fl.* 1953) as *RPG* 5, Jayapura 1953.

The colophon begins: iti śrīmanmahārājādhirājaśrīsavāījayasiṃhakṛtā.

His genealogy is given in sarga 1 of the *Īśvaravilāsa* of Kṛṣṇabhaṭṭa, edited by Mathurānātha Śāstrī as *RPG* 29, Jayapura 1958.

śrīsūryavaṃśo bhuvanaprakāśakas
tatrāpi puṇyaṃ kathitaṃ raghoḥ kulam/
tatrāpi kīrtiḥ kila mānavaṃśajā
pāvitryam etad bhṛśam uttarottaram//14//
bhāsvadvaṃśavataṃsatāṃ dadhati ye dharmātmanāṃ dhīmatāṃ
dhaireyā dharaṇītale suviditā māndhātṛmukhyā nṛpāḥ/
tasminn eva kule ᵓmale vidhur iva kṣīrāmbudhau pārthivaḥ
pṛthvīrāja iti prasiddha udabhūd yo viṣṇubhaktāgraṇīḥ//16//
tasyābhūt tanayas trivikrama ivāvirbhūtasadvikramaḥ
pṛthvībhārasamūhadhāraṇavidhau śeṣāvatāraḥ svayam/
adhyambāvati yaś ca rājyam akarol labdhaṃ nijaṃ paitṛkaṃ
vikhyāto bhuvi bhāramalla iti sa kṣoṇībhṛtāṃ śekharaḥ//18//
tasya śrībhagavantadāsa uditaḥ putraḥ pavitrakriyaḥ
śūraḥ kṣatriyavaṃśavistaraśirolaṅkārahīrāṅkuraḥ/
indraprasthapurādhirājapuruṣaprotthāpanasthāpana-
svacchandaprasaratpratāpamahimā yo ᵓbhūt prabhūtaḥ svayam//19//
tasyābhūd bhūribhūmīpativinataśiromañjumāṇikyamālā-
sthāne saṃsthāpitājñāmayamadhuravacā medinīmaṇḍalasya/
sākṣād ākhaṇḍalo yaḥ pratidharaṇibhṛtām mānasaṃchedanārthaṃ
hastanyastāsavajraḥ samarabhuvi jayī mānasiṃho mahīndraḥ//20//
tasyābhūd bhāsamāno bhavabhavanabhavadbhūribhāgyaprabhāvo
bhūbhartā bhārabhartā bhuvanabhavikabhṛdbhūyasāṃ vaibhavānām/
bhāvānīte bhavānīpatibhajanabhare bhāvitābhrāntabhavyo
bhūbhūṣā bhāvabhāg bhābhavanam abhibhavan bhūpatīn bhāvasiṃhaḥ//29//
samudbhūto ᵓmuṣmād anupamadhanuṣmān atimahā
mahāsiṃho nāma kṣitivibudhakāmakṣitiruhaḥ/
tapodhyānāsaktaiḥ paramaharibhaktaiḥ sukṛtibhiḥ
śubhāśīrbhiḥ sadyaḥ pratigatavipadyaḥ samabhavat//31//
tasyātmajo narapatir jayasiṃhavarmā
karmāṇi yasya kalayann avaraṅgajebaḥ/
siṃhāsanasthitimadaṃ vijahau samantāt
santāpitaḥ parabalodayibhiḥ pratāpaiḥ//32//
tatputro rāmasiṃhaḥ sakalavasumatībhāgyasaubhāgyabhūmā
bhūyaḥ śyāmāsidhūmānumitaparabalottāpitejohutāśaḥ/
yatsaundaryaprasārair jagati ratipatir bhāvinaṃ mānabhaṅgaṃ
svasyābhijñāya vijñas tanum atanuharakrodhakuṇḍe juhāva//39//
tasya nṛpasya kumāraḥ kumāra iva pārvatīśasya/
śrīkṛṣṇasiṃhanāmā jātaḥ śrīkṛṣṇacaraṇadhṛtacittaḥ //45//

Sarga 2, 1–8 continue with a description of Kṛṣṇasiṃha's son, Viṣṇusiṃha, and sarga 2, 9 through sarga 7 describe the exploits and character of Savāī Jayasiṃha.

JAYĀNANDA

The son of Medhākara, Jayānanda wrote a *Janmapaddhati*. Manuscript:

Jammu and Kashmir 2946. 32ff. Copied in Saṃ. 1873 = A.D. 1816.

JAYĀNANDA

Author of a *Muhūrtadīpa*. Manuscripts:

PL, Buhler IV E 346. 330ff. Copied in Saṃ. 1582 = A.D. 1525. Property of Tātyā Bhaṭṭa of Mulhera. Buhler notes 3 other copies.
Baroda 1326. 26ff. The attribution to Jayānanda is queried.
CP, Hiralal 4254. Property of Rāmacandra Bābāji of Akoṭ, Akolā.
CP, Hiralal 4255. Property of Tukārām Pāṭhak of Yeodā, Amraotī.

JARE

Alleged author of a *Nakṣatranāma*. Manuscript:

N-W P X (1886) A 10. 4ff. Property of Bālābhāū Sapre of Benares.

JALPANĀCĀRYA

Author of a *Janmajālapa*. Manuscript:

PUL II 3871. 2ff. (raudrapatākīcakra).

JASAVIJAYA

See Yaśovijaya.

JĀGEŚVARA

See Yogeśvara.

JĀNAKĪDĀSA

The pupil of Nijānanda, Jānakīdāsa wrote a *Jyotiṣa* in Hindī. Manuscript:

NPS 125ga of Saṃ. 2004–2006. Property of the Nāgarīpracāriṇī Sabhā of Vārāṇasī.

He also wrote a *Bālabodha* in Hindī. Manuscript:

NPS 125gha of Saṃ. 2004–2006. Property of Bhaiyā Hanumataprasādasiṃha of Aṭhadamā Riyāsata, Bastī.

JĀLANDHARANĀTHA

Author of a *Jālandharasvarodaya*. Manuscript:

SOI 3524 = SOI Cat. II: 1007–3524. 27ff. Copied in Saṃ. 1918 = A.D. 1861.

JITĀRI

Indian authority on astrology mentioned by Ibn al-Nadīm (*Fihrist*, p. 271 ed. Flügel, which has the probably erroneous reading j.b.ā.r.y).

JINA

An Indian astrologer frequently cited in Arabic sources of the ninth century (e.g., in al-Ṣaymarī's *Kitāb aṣl al-uṣūl*).

JINAKĪRTI

Author of a bhāṣā ṭīkā, *Bālāvabodha*, on the *Ṣaṭpañcāśikā* of Pṛthuyaśas (*fl. ca.* 575). Manuscript:

WHMRL G. 111. m.

JINACANDRA

Author of a *Navagrahaphalanirṇaya*. Manuscript:

RORI Cat. III 16724. 2ff. Incomplete.

JINACANDRA SŪRI

Author of a *Muhūrtamuktāvalī*. Manuscript:

LDI (LDC) 5377. 10ff. Copied in Saṃ. 1861 = A.D. 1804. With a stabaka.

JINADĀSA

Author of a ṭīkā on the *Kṣetrasamāsa* of Ratnaśekhara. Manuscript:

Baroda 7693. 14ff. Incomplete.

JINANĀTHA

Author of a *Bhāvakutūhala*. The date of the manuscript, if correct, makes an identification with Jīvanātha Jhā (*fl. ca.* 1846–1900) impossible. Manuscript:

Goṇḍal 235. 12ff. Copied in Saṃ. 1855 = A.D. 1798. Incomplete.

JINAPRABHA SŪRI

Author of a *Navagrahapārśvanāthastotra* in 10 verses in Prākṛta. Manuscripts:

Bombay U 2406(61). No ff. given.
Bombay U 2406(65). No ff. given. Incomplete (verse 10).

Verse 10 is:

iya navagraha dhuyagathaṃ
jinappahasūrīhiṃ guṃphitu thavaṇaṃ/
tuhapāsa paḍhaï jo taṃ
asahāvagahā na pīḍaṃti//

To a Jinaprabha Sūri is also attributed a vṛtti on a *Navagrahastotra* (of Bhadrabāhu?); see Velankar, p. 206.

JINAPRABHA SŪRI

Author of a *Sukāladuṣkālajñāna*. Manuscript:

LDI 7293 (2979/1). Ff. 4–6.

JINABHADRA GAṆI KṢAMĀŚRAMAṆA (*fl.* 609)

A famous Śvetāmbara Jaina commentator, Jinabhadra wrote his *Viśeṣāveśyakabhāṣya* in Śaka 531 = A.D. 609 at Valabhī under the Maitraka king Śīlāditya I (*ca.* 590/615). One of his many works is the *Kṣetrasamāsa* or *Bṛhatkṣetrasamāsa*, on which commentaries were written by Haribhadra (*fl.* 1128), Siddha Sūri (*fl.* 1135), Malayagiri (*fl. ca.* 1150), Vijayasiṃha (*fl.* 1158), Devabhadra (?) (*fl.* 1176), Ānanda Sūri (*fl. ca.* 1225), and Devānanda (*fl.* 1398). Manuscripts:

Pattan, Saṅghavī Pāḍā 44. 246ff. Copied on Thursday 7 kṛṣṇapakṣa of Jyeṣṭha in Saṃ. 1274 = 25 May 1217. With the vṛtti of Siddha Sūri.
RAS (Tod) 101. Ff. 738–819. Copied at Pattan on 13 śuklapakṣa of Aśvina in Saṃ. 1332 = 3 October 1275. With the ṭīkā of Malayagiri.
Cambay II 289. Ff. 200–270. Copied at Śrīpattana for Jinavardhana Sūri, who was at the Jinarājasūripaṭṭa in the Kharataragaccha, in Saṃ. 1466 = A.D. 1409. With the ṭīkā of Malayagiri.
RORI Cat. I 421. 23ff. Copied by Manohara Muni at Āgarānagara on 1 śuklapakṣa of Mārgaśīrṣa in Saṃ. 1622 = 22 November 1565 during the reign of Pātasāha Akabara Jalāladi (1556/1605).

LDI (KS) 511 (11092). 34ff. Copied in Saṃ. 1640 = A.D. 1583. (*Laghukṣetrasamāsa*).
RORI Cat. II 5088. 13ff. Copied by Durgadāsa Yati in Saṃ. 1653 = A.D. 1596. With the vṛtti of Haribhadra.
BORI 1138 of 1887/91. 144ff. Copied in Saṃ. 1666 = A.D. 1609. With the ṭīkā of Malayagiri. From Gujarāt.
RORI Cat. I 2115. 16ff. Copied by Muni Devasiṃha in Saṃ. 1684 = A.D. 1627. With a *Bālāvabodha* in Old Rājasthānī.
IO 7514 (1357). Ff. 48–159. Copied by Sā⟨ha⟩ Rahiya, the son of Nāthya of the Vāyacāragotra, at Jesalamerunagara on 5 śuklapakṣa of Āṣāḍha in Saṃ. 1686 = 14 June 1629. With the ṭīkā of Malayagiri. Incomplete. From H. T. Colebrooke.
RORI Cat. II 7403. 9ff. Copied by Sundarahaṃsa Gaṇi at Guḍhā in Saṃ. 1851 = A.D. 1794.
Agra, Vinayadharma Lakṣmī Jñānamandira 1132–1149. See Velankar, p. 98.
Ahmadabad, Bhandar of the Vimala Gaccha Upasraya, Haja Patal's Pole 41 (52). See Velankar.
Ahmadabad, Bhandar of the Vimala Gaccha Upasraya, with Uddyotavimala Gaṇi 15 (21; 23 and 24). See Velankar.
Ahmadabad, Dela Upāśraya Bhandar (ground floor) 56 (1–9) and (first floor) 33 (16; 17; 23; 24; and 29). See Velankar.
AS Bombay 1589. 9ff. From Bhāu Dājī.
AS Bombay 1590. 10ff. From Bhāu Dājī.
AS Bombay 1591. 12ff. From Bhāu Dājī.
Baroda, Hamsavijayaji Maharaj at Kantivijayaji Library 1226. See Velankar.
BORI 16 of 1880/81. 283ff. With the ṭīkā of Malayagiri.
BORI 1137 of 1887/91. 7ff. From Gujarāt. No author mentioned, but see Velankar.
Cambay 42C.
Cambay II 286. Ff. 62–111. With the ṭīkā of Malayagiri.
Chani, Bhandar of Muni Kantavijayaji Maharaj 364. See Velankar.
Florence 589. 14ff. With a vṛtti.
Jaipur, Inner Bhandar of Harisāgara Gaṇi 42. See Velankar.
Jesalmir, Bhandar of Bāṇṭhakī Kundi 173 and 284. See Velankar.
LDI (VDS) 496 (9682). 32ff. With the ṭīkā of Malayagiri. Incomplete.
LDI (VDS) 497 (9537). 16ff. With a *Laghuvṛtti*.
Leumann 112.
Leumann 113.
Leumann 121.
Līṃbaḍī 1463.
Līṃbaḍī 1708.
Mandvi, Anantanātha Mandira of the Kacchi Osval Dasa 92 and 109. See Velankar.
Patan, Bhandar of the Agali Sheri 33 (4) and 53 (17). See Velankar.
Patan, New Sangha Bhandar 16 (7) and 18 (41). See Velankar.
Patan, Sangha Bhandar 76 (147). See Velankar.
Patan, Vadi Pārśvanātha Pustaka Bhandar 18 (15). See Velankar.
RORI Cat. I 1024. 153ff. With the ṭīkā of Malayagiri.
RORI Cat. II 7367. 15ff. With a stabaka in Old Rājasthānī.
RORI Cat. II 9463. 8ff.
Strasbourg 4456 (Sanscr. 371). 18ff.
Strasbourg 4554 (Sanscr. 457). 150ff. With the ṭīkā of Malayagiri.
Surat, Jainānanda Bhandar, Gopipura 42, 187, and 1568. See Velankar.

The *Kṣetrasamāsa* was published with the ṭīkā of Malayagiri at Bhavnagar in Saṃ. 1977 = A.D. 1920 (BM 14101. c. 27; see Velankar and NCC, vol. 5, p. 159).

The last verse is:

taṃ namata bohdajaladhiṃ
guṇamandiram akhalavāgināṃ śreṣṭham/
caraṇaśrīyopagūḍhaṃ
jinabhadragaṇikṣamāśramaṇam//

JINAVARA

Author of a *Sukanāvalī* (*Śakunāvalī*) in Rājasthānī. Manuscript:

RAS (Tod) 148. 40ff. Copied on Friday 7 śuklapakṣa of Caitra in Saṃ. 1875 = 2 April 1819. "from Library of Rawul Moolraj of Jesselmere."

JINENDRA

Author of a *Praśnacintāmaṇisāra* or *Jñānadīpaka* in 73 Prākṛta verses, edited with a Sanskrit version by Jinavijaya Muni in *Jayapāyaḍa nimittaśāstra*, *SJS* 43, Bombay 1958, pp. 87–96.

The colophon begins: iti jinendrakathitaṃ.

JINENDRA BHAṬṬA

Author of a *Mātṛkāśakuna* in 51 verses. Manuscript:

Mithila 259. 4ff. Property of Pandit Ghanaśyāma Jhā of Babhangama, Supaul, Bhagalpur.

Verse 51 is:

idaṃ jñānaṃ jinendreṇa bhāṣitaṃ nānyathā kvacit/
sāgarā yadi śuṣyanti ralanti yadi parvatāḥ/

The colophon begins: śrījinendrabhaṭṭaviracita.

JINEŚVARA

Author of a *Janmapatrīpaddhati*. Manuscript:

RORI Cat. III 12808. 24ff. Copied by Kṛṣṇarāma Tivāḍī in Saṃ. 1940 = A.D. 1883. With a Gujarātī ṭīkā.

JINEŚVARA SŪRI

Author of a vyākhyā on the *Jātakapaddhati* of Keśava (*fl.* 1496/1507). Manuscript:

Baroda 2805. 19ff. Copied in Saṃ. 1940 = A.D. 1883. With a Gujarātī commentary.

JIYĀRĀMA ŚĀSTRĪ (*fl.* 1899)

Author of a Hindī anuvāda of the *Grahalāghava* of Gaṇeśa (b. 1507), edited by Rāmeśvara Bhaṭṭa, Kalyāṇa-Bombay 1899 (BM 14053. ccc. 26).

JĪTĀRĀMA

Author of a bhāṣā in Gujarātī of the *Mahādevī* of Mahādeva (*fl.* 1316). Manuscript:

VVRI 1459. 5ff. Copied in Saṃ. 1786 = A.D. 1729.

JĪMŪTAVĀHANA (*fl.* 1092)

A Brāhmaṇa of the Pāribhadrakula and a resident of Rāḍhā in Bengal, Jīmūtavāhana wrote a vast *Dharmaratna* of which one section, the *Kālaviveka*, was written in Śaka 1013–1014 = A.D. 1091–1092; see P. V. Kane [1930/62], vol. 1, pp. 318–327. Manuscript:

AS Bengal 2653 (G 1568) = Mitra, Not. 1974. 156ff. Bengālī. Property of Ghaṭakasiṃha Vandyaghaṭīya on 4 Āṣāḍha of Śaka 1417 = 25 June 1495.

The *Kālaviveka* was edited by Madhusūdana Smṛtiratna and Pramathanātha Tarkabhūṣaṇa as *BI* 136, Calcutta 1905.

JĪVA

The son of Yājñika Narahari of Gujarāt, Jīva wrote a *Praśnasāra*. Manuscripts:

RORI Cat. II 7715. 3ff. Copied by Vṛjavāsī Sillū in Saṃ. 1906 = A.D. 1849.
Alwar 1862.
AS Bengal 7177 (G 10165). 5ff.
Florence 309(a). 7ff. Ascribed to Jīvapātaka.
Jammu and Kashmir 2926. 5ff.
VVRI 2581. 9ff. Ascribed to Jīvamiśra.

Verse 2 is:

naraharer agnicitas tanayaḥ kaviḥ
sakalapraśnam ṛjuṃ kurute hi saḥ/
nikhilakāvyam iva prakaṭīkṛtaṃ
rasamitaṃ bahuśāstravinirmitam//

The colophon begins: iti śrīyājñikanaraharisutajīvajyotirvitkṛta.

JĪVA

Author of a *Ravistuti* in Gujarātī. Manuscript:

LDI (DJSC) 350/7. 1f. Copied in Saṃ. 1749 = A.D. 1692.

JĪVADATTA

Author of a *Sārapañjikā* in six adhyāyas:

1. vārādinirdeśa.
2. daśakriyādhikāra.
3. yātrādinirdeśa.
4. gṛhanirdeśa.
5. nānākriyāvidhāna.
6. saṅgramādhikāra.

Manuscript:

Kathamandu (1960) 150 (I 1646). 35ff. Nevārī.

JĪVANAKṚṢṆA MUKHOPĀDHYĀYA (*fl.* 1914)

Author of an *Adbhutamayūrapuccha*, published with a Bengālī translation at Cooch Behar in 1914 (BM 14053. c. 71. (1)).

JĪVANĀTHA

Author of an *Āyussādhana*. Manuscript:

Mithilā. See NCC, vol. 2, p. 155.

JĪVANĀTHA

Author of a *Cakrānayanādhyāya*. Manuscript:

Mithilā. See NCC, vol. 6, p. 290.

JĪVANĀTHA

Author of a *Pavanavijaya*. Manuscript:

Anup 4848. 9ff. Copied in Saṃ. 1937 = A.D. 1682.

JĪVANĀTHA

The son of Śrīdatta of the Hariharavaṃśa, Jīvanātha wrote a *Śuddhyaśuddhivicāra* for the Maithila Mahārāja, Chatrasiṃha. Manuscripts:

Mithila 372 B. 12ff. Maithilī. Copied in Śaka 1803 = A.D. 1881. Property of Pandit Suvaṃśa Lāla Jhā of Sukpur, Bhagalpur.
Mithila 372. 23ff. Maithilī. Property of Sītārāma Pāṭhaka of Karnapur, Sukpur, Bhagalpur.
Mithila 372 A. 13ff. (f. 1 missing). Maithilī. Incomplete. Property of Pandit Santoṣi Jhā of Balaha, Sukpur, Bhagalpur.

Verses 1–2 are:

vighneśaṃ gurupādapadmayugalaṃ natvākhilāṃ
paddhatiṃ

saṃvīkṣyāmalamānasaś ca gaṇakastomaikalakṣyaḥ sadā/
śrīdattasya kaves tathā hariharāvaṃśodbhavasyātmajaḥ
śuddhāśuddhavinirṇayaṃ vitanute śrījīvanāthaḥ sudhīḥ//
prodyaddurdharavairidarpadalanaḥ śrīchatrasiṃho nṛpo
vikhyātaḥ suyaśo himāṃśukiraṇaiḥ svacchīkṛtakṣmātalaḥ/
tasyājasramahīpamaulimadhupavyālolapādāmbujasyājñāto mithilādhipasya vibudhair jñeyo mamaiṣa śramaḥ//

The next to the last verse is:

pālīvaṃśaparasparodadhibhavo jyotirvidāṃ viśrutaḥ
prodyacchrīvacanūdharāsuravaraḥ sarvopakārakṣamaḥ/
tasyāhaṃ bhaginīsutaḥ pramuditaḥ śrījīvanāthaḥ sudhīḥ
śuddhāśuddhavivecanaṃ ca kṛtavān modāya vidyāvatām//

JĪVANĀTHA

Author of a *Svarodaya*, *Svaratattvacamatkāra*, or *Ātmaprakāśa*. Manuscripts:

RORI Cat. III 13825(13). Ff. 226–228. Copied by Nayakīrti at Stambhatīrtha in Saṃ. 1584 = A.D. 1527.
Oxford 793 (Walker 213b). Ff. 6–15, Copied by Pītāmbara, the son of Śivadāsa, for Bhaṭa Hariśrama in A.D. 1640.
Dāhilakṣmī XXV 7. Copied in A.D. 1686. Incomplete (*Camatkāracintāmaṇi*). See NCC, vol. 6, p. 386.
SOI 3294 = SOI Cat. II: 1129–3294. 5ff. Copied in Saṃ. 1743, Śaka 1608 = A.D. 1686.
Benares (1963) 37759. 17ff. Copied in Saṃ. 1756 = A.D. 1699.
Anup 5371. 17ff. Copied at Bīkānera from a manuscript belonging to Gaṇeśa Dudhādhārījī in Saṃ. 1859 = A.D. 1802. Ascribed to Jīvananātha.
Baroda 3360. 4ff. Incomplete.
Baroda 4811. Ff. 15–20. (*Ātmaprakāśa*).
Jodhpur 1823. (*Ātmaprakāśasvarodaya*). See NCC, vol. 2, p. 50.
LDI 7426 (1759). F. 41. Incomplete.
LDI (DSC) 9448. 4ff.
LDI (SCC) Sag. 339/1. 6ff.
RORI Cat. I 3702. 13ff.
Tanjore D 11669 = Tanjore BL 4318. 8ff. Incomplete.

Verse 1 is:

camatkāraśivaṃ natvā camatkārāya bhūbhṛtām/
svaratattvacamatkāro jīvanāthena kathyate//

JĪVANĀTHA ŚARMAN

Author of a *Janmapatrikāvidhāna*. Manuscripts:

Benares (1963) 37273. 52ff. and 1f. Maithilī. Copied in Śaka 1776 = A.D. 1854.
Benares (1963) 37274. 50ff. Maithilī. Copied in Śaka 1776 = A.D. 1854.
Benares (1963) 37149. 14ff. Incomplete.
Benares (1963) 37296. 172ff.

The *Janmapatrikāvidhāna* was edited by Harinandana Miśra, pt. 1, Benares Saṃ. 1986 = A.D. 1929 (IO San. 983/i).

JĪVANĀTHA JHĀ (*fl. ca.* 1846/1900)

The son of Śambhunātha, the son of Karuṇākara, Jīvanātha, a Maithila Brāhmaṇa, was the elder brother of Nīlāmbara Jhā (b. 18 July 1823). He wrote a large number of works on jyotiṣa.

1. *Tājikadarpaṇa*. Manuscript:

AS Bengal 7095 (G 10467). 42ff.

The last 2 verses are:

āsīn maithilabhūsuro budhavaro vedāṅgavidyākaraḥ
śrīśrīśrīkaruṇākaraḥ kavikulāny ābhūṣayan tarkavit/
tatputraḥ kṣitipālavanditapadaḥ śrīśambhunāthaḥ kṛtī
śambhudhyānabalena śambhusamatāṃ kāśyām agādādarāt//
tajjena nānāmatam ādareṇa
purātanāṃ x pravilokya ramyam/
śrījīvanāthena vilokanārthaṃ
vidāṃ kṛtas tājikadarpaṇo ᵓyam//

2. *Bhāvakutūhala* in 17 adhyāyas; see Jinanātha. Manuscripts:

Benares (1963) 35394 = Benares (1903) 1296. 38ff. Copied in Saṃ. 1935 = A.D. 1878.
AS Bengal 7215 (G 4800). 60ff. Nevārī.
Benares (1963) 34320. 16ff. and 1f. Maithilī.
Mithila 228. 58ff. Maithilī. Property of Pandit Muktinātha Jhā of Baruary, Parsarma, Bhagalpur.
Mithila 228 A. 14ff. Maithilī. Incomplete. Property of Pandit Mahīdhara Miśra of Lalbag, Darbhanga.

The *Bhāvakutūhala* was published:
at Cawnpor (?) in 1865 (BM):
with the Bengālī translation of Rāmagopāla Jyotirvinoda (= Rāya) at Calcutta in 1896 (IO 1260 and NL Calcutta 180. Kc. 89. 4), 2nd ed. Calcutta 1935 (NL Calcutta 180. Kc. 93. 14);
edited by Jīvānanda Bhaṭṭācārya, 2nd ed., Calcutta 1897 (NL Calcutta 180. Kc. 89. 8);
edited with a Singhalese gloss and notes by B. L. Sarnelis, pt. 1, [Colombo] 1897 (BM 14053. ccc. 4);
edited with the Hindī translation of Nārāyaṇaprasāda by Gajānana Śarman, Bombay Saṃ. 1968 = A.D. 1911 (BM 14053. dd. 19);
and with his own Marāṭhī artha by Lakṣmaṇa Nārāyaṇa Jośī, Puṇeṃ 1931.

The last verse is:

āsīc chrīkaruṇākaro budhavaro vedāṅgavedākaras
tatsūnuḥ kṣitipālavanditapadaḥ śrīśambhunāthaḥ
kṛtī/
vijñavratakṛtādaro gaṇitavij jyotirvidāṃ prītaye
cakre bhāvakutūhalaṃ laghutaraṃ śrījīvanāthaḥ
sudhīḥ//

3. *Pārāśarīvāsanā* or *Tattvadīpika*, a ṭīkā on the *Uḍudāyapradīpa* of Parāśara. Manuscripts:

GJRI 3257/469. 13ff. Maithilī.
Mithila 125. 12ff. Maithilī. Property of Pandit Tarakeśvara Miśra of Tarauni, Sakri, Darbhanga.

The colophon begins: iti śrīmadgaṇakavaryaśambhunāthasutajīvanāthaviracitā.

4. *Praśnabhūṣaṇa* in 17 adhyāyas, published with a Sanskrit ṭīkā, *Vimalā*, and a Hindī ṭīkā, *Saralā*, both by Kamalākānta Jhā (*fl.* 1938), as well as the same scholar's pariśiṣṭa, as *HSS* 131, Banārasa 1941; 2nd ed. Banārasa 1954. The last verse is:

kṛtvā tājikadarpaṇaṃ prathamataḥ śrījīvanāthaḥ
kṛtī
ramyaṃ bhāvakutūhalaṃ ca parataḥ
pārāśarīvāsanām/
vidvaccakramanovinodakaraṇaṃ
chātrānukampāvaśād
anyat praśnavibhūṣaṇaṃ munimataṃ jñātvā paraṃ
nirmame//

5. *Vanamāla* in 6 adhyāyas. Manuscripts:

Mithila 328 A. 5ff. Maithilī. Copied in Śaka 1786 = A.D. 1864. Property of Pandit Janārdana Miśra of Chanaur, Manigachi, Darbhanga.
Benares (1963) 35395 = Benares (1903) 1927. 7ff. Copied in Saṃ. 1935 = A.D. 1878.
Benares (1963) 37325. 4ff. Maithilī.
Mithila 328. 8ff. Maithilī. Property of Pandit Santoṣī Jhā of Balaha, Sukpur, Bhagalpur.

The *Vanamālā* was published with a Telugu translation, at Madras in 1893 (BM 14053. b. 31. (2)); with an Āndhra tātparya, at Madras in 1916 (IO San. B. 162) and at Masulipatam in 1918 (IO San. B. 775(*u*)); and with his own Sanskrit anvaya and Hindī ṭīkā, *Amṛtadhārā*, by Kapileśvara Śāstrin Caudharī (*fl.* 1940/1948) as *HSS* 147, Banārasa 1941. The last verse is:

kṛtvā tājikadarpaṇaṃ prathamataḥ śrījīvanāthaḥ
kṛtī
ramyaṃ bhāvakutūhalaṃ ca parataḥ pārāśrīvāsanām/
cañcatpraśnavibhūṣaṇaṃ viniramādalpākṣarām
arthadāṃ
vidaccakramanovinodajananīṃ kīlālayogāvalīm//

6. *Bhāvaprakāśa*. Manuscript:

Mithila 230. 6ff. Maithilī. Property of Pandit Sītārāma Pāṭhaka of Karnpur, Sukpur, Bhagalpur.

The *Bhāvaprakāśa* was published with his own Hindī ṭīkā, *Bhāvabodhinī*, by Puṣpalāla Jhā as *HSS* 40, Benares 1962. The colophon begins: iti śrīśambhunāthadaivajñātmajaśrījīvanāthadaivajñaviracite.

7. An udāharaṇa on the *Makaranda* of Makaranda (*fl.* 1478). Manuscripts:

Mithila 251 A. 12ff. Maithilī. Copied in Śaka 1810 = A.D. 1888. Property of Pandit Jayānanda Miśra of Parsarma, Bhagalpur.
Mithila 251. 9ff. Maithilī. Copied by Sītārāma Śarman at Kāśī. Property of Pandit Sītārāma Pāṭhaka of Karnpur, Sukpur, Bhagalpur.
Mithila 251 B. 8ff. Maithilī. Property of Pandit Rudrānanda Jhā of Parsarma, Bhagalpur.

Verse 2 is:

makarandoktatithyādeḥ sādhanārtham udāhṛtiḥ/
daivajñānāṃ vinodāya jīvanāthena darśyate//

8. *Vāsturatnāvalī*, completed on Friday 7 kṛṣṇapakṣa of Caitra in Śaka 1766 = 17 April 1846. Manuscripts:

Mithila 344 C. 60ff. Maithilī. Copied in Śaka 1777 = A.D. 1855. Property of Pandit Gopāla Miśra of Tabhaka, Dalsinghsarai, Darbhanga.
Mithila 344. 23ff. Mithilī. Copied by Phekana Śarman at Kāśī on Thursday 9 kṛṣṇapakṣa of Jyeṣṭha in Śaka 1794, Sāl. San. 1279 = 19 June 1873. Property of Pandit Sītārāma Pāṭhaka of Karnapur, Sukpur, Bhagalpur.
Benares (1963) 37324. 34ff. Maithilī.
Mithila 344 A. 30ff. (ff. 7, 10–11, and 25–26 missing). Maithilī. Incomplete. Property of Pandit Suvaṃśalāla Jhā of Sukpur, Bhagalpur.
Mithila 344 B. 4ff. Maithilī. Incomplete. Property of Pandit Śrīnandana Miśra of Kanhauli, Sakri, Darbhanga.

The *Vāsturatnāvalī* was published at Benares in 1883 (BM and IO 459); at Benares in 1888 (IO 267); edited with a Telugu version by N. Guruliṅga Śāstrī, Madras 1897 (BM 14053. ccc. 16 and IO 16. G. 17); edited by Kṛṣṇadatta, Benares 1919 (BM 14055. g. 3 and IO San. D. 235); and edited by Acyutānanda Jhā with his own Saṃskṛta ṭīkā, *Subodhinī*, and a Hindī version, and with his own *Vidhivivekādhyāya*, as *HSS* 152, Banārasa 1941; 2nd ed., Banārasa 1949.

The last verse is:

śāke tarkarasāgacandramilite pakṣe balakṣetare
caitre māsi bhṛgor dine smaratithāv eṣā gatā
pūrṇatām/
nānācāryamataṃ vilokya racitā śrīvāsturatnāvalī
śrīmanmaithilajīvanāthakṛtinā daivajñamodapradā//

9. *Subodhinī*, a ṭīkā on the *Bījagaṇita* of Bhāskara (b. 1114), written for the sons of the Maithila Mahārāja, Lakṣmīśvara Siṃha. It was published with the mūla at Benares in 1885 (IO 6. F. 9), and with the mūla and with his own Saṃskṛta and Hindī ṭīkā, *Vimalā*, by Acyutānanda Jhā as *KSS* 148, Banaras 1949, in a version edited by Lakṣmīnārāyaṇa in the middle of the śuklapakṣa of Jyeṣṭha in Saṃ. 1942 = *ca.* 20 June 1885.

Verses 14–17 at the beginning are:

āsīc chrīkaruṇākaro budhavaro vedāṅgavidyākaras
tatsūnuḥ kṣitipālavanditapadaḥ śrīśambhunāthaḥ kṛtī/
vijñavrātakṛtādaro gaṇitavit siddhāntapāraṅgamaḥ
śambhudhyānabalena śambhusaṃmatāṃ kāśyām agādādarāt//
putrau babhūvatus tasya dvāv ādyo jīvanāthakaḥ/
nīlāmbaraḥ kaniṣṭhaś ca kṛtīśas tapaso balāt//
nīlāmbaro daivavidagragaṇyaḥ
siddhāntapāṇḍityaramāśaraṇyaḥ/
susatkṛtaḥ śrīmithileśamukhyair
ilādhipair vijñajanaprasiddhaḥ//
mithilendravinodāya bījaṭīkā subodhinī/
janānām upakārāya jīvanāthena tanyate//

JĪVARĀMA

Alleged author of a *Koṭacatuṣṭaya*. Manuscript:

CP, Hiralal 1089. Property of Śrīkṛishṇa Pāṇḍuraṅg of Bālāpur, Akola.

JĪVAVIJAYA GAṆI

Author of a stabaka in Old Gujarātī on the *Jambūdvīpaprajñapti*. Manuscript:

RORI Cat. III 13904. 225ff.

JĪVAŚARMAN (*fl.* fourth or fifth century)

An authority on genethlialogy cited by Varāhamihira (*fl. ca.* 550) in *Bṛhajjātaka* 7,9 and 11,1 and *Bṛhadyātrā* 9,1; by Kalyāṇavarman (*fl. ca.* 800) in *Sārāvalī* 35,2 and 39,3 and 19; by Utpala (*fl.* 966/968) on *Bṛhajjātaka* 6,1; 11, 1; and 13,3; and elsewhere. See P. V. Kane [1948/49) 9.

JĪVENDRA

Author of a *Candronmīlana* and of a vyākhyā on the same. Manuscripts:

Benares (1963) 35208. 17ff. Copied in Saṃ. 1824 = A.D. 1767.
Mithila 63. 12ff. Maithilī. With the vyākhyā. Incomplete. Property of Babu Matikānta Jhā of Ekama, Supaul, Bhagalpur.

The first verse of the vyākhyā is:

atratyaśāstramāleṣṭhaṃ vicārya ca punaḥ punaḥ/
candronmīlanavyākhyānaṃ jīvendreṇa vitanyate//

JĪVEŚVARA UPĀDHYĀYA (*fl. ca.* 1280)

Author of a *Ratnaśataka* for Vīreśvara, who was a mantrin and apparently a mahāsandhivigraha (the colophon must be corrupt); Vīreśvara, then, is the father of Caṇḍeśvara Ṭhakkura (*fl.* 1314) and minister of Śaktisiṃhadeva (*ca.* 1276/1296). Manuscripts:

Mithila 289 A. 10ff. Maithilī. Copied in Śaka 1819 = A.D. 1897. Property of Pandit Śrīnandana Miśra of Kanhauli, Sakri, Darbhanga.
Mithila 289. 7ff. (ff. 3–4 missing). Maithilī. Incomplete. Property of Pandit Vāsudeva Jhā of Sukpur, Bhagalpur.

Verse 1 is:

śrīmān vīreśvaro mantrī granthaṃ ratnaśatāhvayam/
jīveśvaram upādhyāyaṃ niyujya kriyate kṛtī//

The colophon in Mithila 289 begins: iti mahāsandhivigraha // kaṇḍatkeralaśrījīveśvarakṛtaṃ.

JESARĀJA

Author of a *Kṣayamāsanirṇaya*. Manuscript:

Benares (1956) 13264. 5ff. Copied in Saṃ. 1790 = A.D. 1733.

JAIKṚṢṆA

Author of a *Ramalanavaratna* in Hindī; this is probably a version of the *Ramalanavaratna* of Paramasukha (*fl.* 1810). Manuscript:

SOI 2598 = SOI Cat. II: 1095–2598. 102ff.

JAINASĀDHU (*fl.* 1635)

See Dhanarāja (*fl.* 1635).

JAINĀCĀRYA (*fl.* 1675/1695)

See Meghavijaya (*fl.* 1675/1695).

JAINENDU

Author of a *Jātakaratnakośa*. Manuscript:

BORI 864 of 1891/95. 25ff. Copied in Saṃ. 1890 = A.D. 1833.

JAIMINI

Alleged author of an *Ārūḍhaśāstra* or *Jñānapradīpikā* edited with a Tamil ṭīkā, *Bhāvaprakāśikā*, of K. Sorṇaśāstrin and V. R. Śrīnivāsācārya, by Kṛṣṇa Śāstrin of Devakota, at Madras in 1899 (BM 14053. ccc. 27).

JAIMINI

Alleged author of the *Upadeśasūtra* in 4 adhyāyas of which each contains 4 pādas. There are commentaries by Nīlakaṇṭha (*Subodhinī* in 1754), Durgāprasāda Dviveda (*Jaiminipadyāmṛta* in 1906; adhyāyas I–II) Vināyaka (*Arthoddyota* in 1911), Rāmayatna Ojhā (1925), Acyutānanda Jhā (1943), Kāśīnātha Vāsudeva Abhyaṅkara (*Marīci* in 1945; to III 3), Annaji (or Anvaji), Kṛṣṇānanda (or Bālakṛṣṇānanda) Sarasvatī, Nṛsiṃha, Parameśvara Yogīndra (*Jyotiṣānanda*), Premanidhi, Malayavarman (*Kāśikā*), Lakṣmaṇa (*Jyotiḥpradīpikā*), Lakṣmīpati, Veṅkaṭeśa (*Bhāvakaumudī*), Vrajarāja Śukla, Somanātha, (*Jyotiṣakalpalatā*), and Haribhānu; see also B. V. Raman [A3. 1950a] and Bhavānīrāma. Manuscripts:

GOML Madras R 371(g). Ff. 57–76v. Telugu. Copied on Monday 3 śuklapakṣa of Jyeṣṭha in Saṃ. 1793 = 31 May 1736 Julian. With an Āndhraṭīkā. Incomplete (I–II only). Purchased in 1911/12 from C. Viśvanātha Śāstrigaḷ of Vizianagaram.

Kathmandu (1960) 122 (I 1209). 82ff. Nevārī. Copied during the reign of Jayaraṇa Jitamalladeva in Bhatgaon (1722/1769) on Sunday 15 kṛṣṇapakṣa of Śrāvaṇa in NS 874 = 18 August 1754. With the ṭīkā of Kṛṣṇānanda Sarasvatī.

BORI 474 of 1892/95. 129ff. Copied in Saṃ. 1821 = A.D. 1764. With the ṭīkā of Kṛṣṇānanda Sarasvatī. Incomplete (ends with II).

Benares (1963) 35674 = Benares (1903) 1285. Ff. 2–6. Copied in Saṃ. 1825 = A.D. 1767. Incomplete.

Benares (1963) 36920. 9ff. Copied in Saṃ. 1852, Śaka 1717 = A.D. 1795.

Baroda 114 (1114?) 12ff. Copied in Saṃ. 1875 = A.D. 1818. This is PL, Buhler IV E 125. 12ff. Copied in Saṃ. 1875 = A.D. 1818. Property of Harirāmaśāstrī of Aṅkaleśvara.

Benares (1963) 34409. 17ff. Copied in Saṃ. 1879 = A.D. 1822. Incomplete (to II 1).

Benares (1963) 34410. 63ff. Copied in Saṃ. 1880 = A.D. 1823. With the *Subodhinī* of Nīlakaṇṭha. Incomplete (to II 4).

BORI 152 of A1883/84. 12ff. Copied in Saṃ. 1882 = A.D. 1825.

SOI 2101 = SOI Cat. I: 1457–2101. 80ff. Copied in Saṃ. 1887 = A.D. 1830. With the *Bhāvakaumudī* of Veṅkaṭeśa.

BORI 475 of 1892/95. 144ff. Copied in Saṃ. 1893 = A.D. 1836. With the ṭīkā of Kṛṣṇānanda Sarasvatī.

PL, Buhler IV E 126. 59ff. Copied in Saṃ. 1893 = A.D. 1836. With the *Bhāvakaumudī* of Veṅkaṭeśa. Property of Maṅgala Śaṅkara of Ahmadābād.

BORI 826 of 1887/91. 15ff. Copied in Saṃ. 1897 = A.D. 1840. From Gujarāt.

RORI Cat. III 15460. 6ff. Copied in Saṃ. 1905 = A.D. 1848. Incomplete (III only; rājayogādhyāya).

Benares (1963) 35186. 54ff. Copied in Saṃ. 1907 = A.D. 1850. With the *Subodhinī* of Nīlakaṇṭha. Incomplete (to II 4).

RORI Cat. II 6290. 27ff. Copied by Bālamukunda Gosvāmin in Saṃ. 1911 = A.D. 1854. With the *Subodhinī* of Nīlakaṇṭha.

PrSB 964 (or. oct. 648). 5ff. Copied by Mīnarāma in Saṃ. 1918 = A.D. 1861. Incomplete (to II 4). Now at Marburg.

PL, Buhler E IV 127. 54ff. Copied in Saṃ. 1923 = A.D. 1866. With the *Subodhinī* of Nīlakaṇṭha. Property of Maṅgala Śaṅkara of Ahmadābād.

RORI Cat. II 5533. 42ff. Copied by Lalitādāsa Vyāsa at Vṛndāvana in Saṃ. 1924 = A.D. 1867. With the *Subodhinī* of Nīlakaṇṭha.

Oudh III (1873) VIII 11. 140pp. Copied in A.D. 1868. With the ṭīkā of Haribhānu. Property of Paṇḍit Bhālacandra of Oonao Zillah.

Benares (1963) 35184. 58ff. Copied in Saṃ. 1927 = A.D. 1870. With a ṭīkā.

Poleman 4833 (Columbia, Smith Indic 171). 8ff. Copied in Saṃ. 1928 = A.D. 1871.

VVRI 4477. 39ff. Copied in Saṃ. 1950 = A.D. 1893. With the *Subodhinī* of Nīlakaṇṭha.

Benares (1963) 34564. 44ff. Copied in Saṃ. 1987 = A.D. 1930. With a ṭīkā in Hindī. Incomplete (to II 4).

Adyar Cat. 21 D 33. 26ff. Grantha. Incomplete (I only; āyurdāya.)

Adyar Cat. 21 F 49. 48ff. Grantha. With the *Jyotiṣānanda* of Parameśvara.

Adyar Cat. 21 F 56. 70ff. Grantha. Incomplete (II only).

Adyar Cat. 22 G 55. 42ff. Grantha. With a ṭīkā.

Adyar Cat. 22 G 56. 42ff. Grantha. With a ṭīkā.

Adyar Cat. 22 G 57. 9ff. Grantha. Incomplete (I 1–2 only).

Adyar Cat. 22 G 58. 47ff. Telugu. With an Āndhraṭīkā. Incomplete (I–III only).

Adyar Cat. 22 G 59. 62ff. Telugu. With the *Jyotiḥpradīpikā* of Lakṣmaṇa.

Adyar Cat. 25 C 27. 5ff. (ff. 1–2 missing). Telegu. Incomplete.

Adyar Cat. 34 I 23. 27ff. Incomplete (I only).

Alwar 1772.

AS Bengal 6955 (G 10462). 83ff. With the *Subodhinī* of Nīlakaṇṭha. Incomplete (ends in III).

Baroda 1110. 4ff. Incomplete (I–II only).

Baroda 1338(e). 9ff. Nandināgarī. Incomplete (ends in II 4).

Baroda 3136. 27ff. With the *Subodhinī* of Nīlakaṇṭha. Incomplete (I–II only).

Baroda 6500. 74ff. Grantha. With the *Jyotiṣānanda* of Parameśvara. Incomplete.

Baroda 13444. Ff. 221(21?)–138. Nandināgarī. With a vyākhyā. Incomplete (I–II only).

Benares (1963) 34371. Ff. 1–10 and 1f. With a ṭīkā. Incomplete.

Benares (1963) 34376. 33ff. With the *Subodhinī* of Nīlakaṇṭha.
Benares (1963) 34383. 8ff. Incomplete.
Benares (1963) 34384. 1f. Incomplete.
Benares (1963) 34806. 23ff. With a ṭīkā. Incomplete.
Benares (1963) 34807. Ff. 7–98. With the ṭīkā of Premanidhi. Incomplete.
Benares (1963) 35182. 27ff. With the ṭīkā of Kṛṣṇānanda Sarasvatī. Incomplete.
Benares (1963) 35183. 6ff. With a ṭīkā. Incomplete.
Benares (1963) 35185. Ff. 1–4 and 6–169. With the ṭīkā of Kṛṣṇānanda Sarasvatī. Incomplete.
Benares (1963) 35204. 15ff. Incomplete.
Benares (1963) 35205. 28ff. With the *Subodhinī* of Nīlakaṇṭha. Incomplete.
Benares (1963) 35214. 14ff. Incomplete (I–II only).
Benares (1963) 35285 = Benares (1897–1901) 553. 41ff. With the *Subodhinī* of Nīlakaṇṭha.
Benares (1963) 36150. 4ff. Incomplete (III 1–3 only).
Benares (1963) 36151. 8ff. Incomplete (I–II only).
Benares (1963) 36215. Ff. 1–7 and 9. Incomplete.
Benares (1963) 36216. 5ff. Incomplete.
Benares (1963) 37069. 1f. Incomplete (III only; ariṣṭādhyāya).
Benares (1963) 37283 = Benares (1878) 177. 7ff. Incomplete.
BORI 531 of 1875/76. 156ff. With the ṭīkā of Kṛṣṇānanda Sarasvatī. From Dilhī.
BORI 909 of 1886/92. 39ff. With the *Subodhinī* of Nīlakaṇṭha.
BORI 910 of 1886/92. 35ff. With the *Subodhinī* of Nīlakaṇṭha.
BORI 473 of 1892/95. 47ff. With the *Bhāvakaumudī* of Veñkaṭeśa.
BORI 406 of 1895/98. 7ff.
BORI 518 of 1895/1902. 73ff. With the ṭīkā of Kṛṣṇānanda Sarasvatī.
CP, Hiralal 1837 and 1838. Property of Govindprasād Śāstrī of Jubbulpore.
CP, Hiralal 1839. Property of Govind Joshi of Jubbulpore.
GJRI 2984/317. 22ff. Maithilī. With the *Subodhinī* of Nīlakaṇṭha. Incomplete (to II 3).
GOML Madras D 13725. Ff. 1–13. Grantha. Incomplete (to IV 2).
GOML Madras D 13726. Ff. 1–3. Telugu. Incomplete (I–II only).
GOML Madras D 13727. Ff. 12–17. Grantha. Incomplete (to III 1).
GOML Madras D 13728. Ff. 105–110. Telugu. Incomplete (I–II only).
GOML Madras D 13729. Ff. 1–11. Telugu. Incomplete.
GOML Madras D 13731. Ff. 85–104. Telugu. With a Karṇāṭakaṭīkā. Incomplete (I–II only).
GOML Madras D 13732. Ff. 32–84. Telugu. With the *Jyotiḥpradīpikā* of Lakṣmaṇa. Incomplete (I only).
GOML Madras D 13733. Ff. 99–108. Telugu. With the *Jyotiḥpradīpikā* of Lakṣmaṇa. Incomplete (to II 1).
GOML Madras D 13734. Ff. 3–24. Telugu. With the *Jyotiḥpradīpikā* of Lakṣmaṇa. Incomplete (I only).
GOML Madras D 13735. Ff. 1–17. Telugu. With the *Jyotiḥpradīpikā* of Lakṣmaṇa. Incomplete (I only).
GOML Madras D 13736. Ff. 39–48. Grantha. With the *Jyotiḥpradīpikā* of Lakṣmaṇa. Incomplete (I only).
GOML Madras D 13737. Ff. 109–112. Telugu. With a vyākhyāna. Incomplete (ends in I 2).
GOML Madras D 16887. 45pp. Telugu. With an Āndhraṭīkā. Incomplete (I–II only).
GOML Madras D 17561. 38pp. Telugu. With an Āndhraṭīkā. Incomplete (I–II only).
GOML Madras D 19228. 94pp. Grantha. With the *Bhāvakaumudī* of Veñkaṭeśa. Incomplete (to II 4).
GOML Madras R 383(a). Ff. 6–10. Telugu. Incomplete (to II 1). Purchased in 1911/12 from C. Viśvanātha Śāstrigal of Vizianagaram.
GOML Madras R 4058(a). Ff. 1–36. Grantha and Tamil. With a vṛtti. Incomplete (ends in II). Presented in 1921/22 by Veñkudīkṣitar of Nañgavaram, Kulittalai, Trichinopoly.
Kathmandu (1960) 33 (III 109). 4ff. Incomplete (to yogādhyāya 3). No author mentioned.
Kathmandu (1960) 123 (I 1209). 11ff. Nevārī.
Kathmadu (1960) 124 (I 1209). 22ff. Nevārī. With the ṭīkā of Kṛṣṇānanda Sarasvatī. Incomplete (I–II only).
Kerala 5907 (2519 X). 40 granthas. Telugu. Incomplete.
Kerala 5908 (3577 B). 175 granthas. Grantha. Incomplete.
Kerala 5909 (9484 A). 60 granthas. Telugu. Incomplete.
Kerala 5910 (C. 2131 B) = Kerala C 686 B. 16ff. Malayālam. Incomplete (I only). Formerly property of Puruṣottaman Śañkaran Nambūrippāḍ of Kunnattunāḍu.
Kerala 5911 (T. 569). 260 granthas. With a vṛtti. Incomplete.
Kerala 5912 (1377). 300 granthas. Grantha. With the *Jyotiḥpradīpikā* of Lakṣmaṇa. Incomplete.
Kerala 5913 (3028 A). 700 granthas. Grantha. With the *Jyotiḥpradīpikā* of Lakṣmaṇa. Incomplete.
Kerala 5914 (3577 C). 700 granthas. Grantha. With the *Jyotiḥpradīpikā* of Lakṣmaṇa. Incomplete.
Kerala 5915 (9484 B). 440 granthas. Telugu. With the *Jyotiḥpradīpikā* of Lakṣmaṇa. Incomplete.
Kerala 5916 (T. 248). 700 granthas. With the *Jyotiḥpradīpikā* of Lakṣmaṇa. Incomplete.
Kerala 5917 (1682). 2400 granthas. With the *Kāśikā* of Malayavarman. Incomplete.
Kerala 5918 (1695). 1100 granthas. With the *Subodhinī* of Nīlakaṇṭha. Incomplete.
Kerala C 691 (C 248). 92ff. With the *Jyotiḥpradīpikā* of Lakṣmaṇa. Incomplete (to I 3).

Kurukṣetra 345 (19720). With the *Subodhinī* of Nīlakaṇṭha.
Kurukṣetra 346 (50697).
Leiden XI 34(b).
Mithila 110. 12ff. Maithilī. Property of Pandit Śaśinātha Miśra of Tarauni, Sakri, Darbhanga.
Mysore 461 (481). No author mentioned.
Mysore (1911 + 1922) 2688. 16ff.
Mysore (1922) 3738. Ff. 30–34.
Mysore (1922) 4542. 32ff. No author mentioned.
N-W P VIII (1884) 5. 17ff. Incomplete (III only). Property of Pandit Devakṛṣṇa Miśra of Benares.
Oppert I 59. 400pp. Grantha. This and other Oppert MSS. are perhaps the mīmāṃsā-work also entitled *Jaiminisūtra*. Property of Narasiṃhācāryār of Ammaṇapākam, Chingleput.
Oppert I 359. 60pp. Grantha. Property of Koṇḍaṅgi Anantācāryār of Kāñcīpuram, Chingleput.
Oppert I 386. 48pp. Grantha. Property of Anantācāryār of Kāñcīpuram, Chingleput.
Oppert I 1240. Property of Vaṅkīpuram Śrīnivāsācāryār of Tiruvallūr, Chingleput.
Oppert I 1833. 25pp. Grantha. Property of Śivasūri Śāstrī of Bhavāni, Coimbatore.
Oppert I 2330. 80pp. Telugu. Property of the Śrī Sarasvatī Bhaṇḍāram Committee of Tiruvallikkeṇi, Madras.
Oppert I 6584. Property of Durbha Rāmaśāstrulu of Maḍḍi, near Padmanābha, Vizagapatam.
Oppert I 7306, Property of the Rāja of Vijayanagaram, Vizagapatam.
Oppert I 7956. With a vyākhyā. Property of Paravastu Veṅkaṭaraṅgācāryār of Viśākhapaṭṭana.
Oppert II 932. Property of Jaḍapatūr Raṅgācāryār of Kāñcīpuram, Chingleput.
Oppert II 2655. 30pp. Grantha. Property of Śaṅkaraśāstrī of Kumāraliṅgam, Uḍumalapeṭa, Koimbatore.
Oppert II 3152. Property of Taḍakamalla Veṅkaṭakṛṣṇarāyar of Tiruvallikeṇī, Madras.
Oppert II 3309. Property of Anantanārāyaṇa Josya and Gurumūrti Josya of Diṇḍukal, Madura.
Oppert II 4604. Property of the Śaṅkarācāryasvāmimaṭha at Śriṅgeri, Cikkamogulūr, Mysore.
Oppert II 6273. Property of Narasiṃhācārya of Kumbhaghoṇam, Tanjore.
Oppert II 6983. Property of Vyāsarājācāri of Kumbhaghoṇam, Tanjore.
Poleman 4832 (Columbia, Smith Indic 170). 10ff.
Poleman 4834 (U Penn. 706). 4ff.
Poleman 4835 (U Penn. 700). 16ff.
PUL II 3439. 20ff.
PUL II 3440. 8ff.
PUL II 3441. 11ff. Incomplete (I–II only).
PUL II 3442. 6ff. Incomplete (I–II only).
PUL II 3443. 6ff. Incomplete (I–II only).
PUL II 3444. 11ff. Incomplete (I–II only).
PUL II 3445. 19ff. Incomplete (to III 3).
RORI Cat. I 1163. 6ff.
RORI Cat. II 8019. 33ff. With the *Subodhinī* of Nīlakaṇṭha. Incomplete.
RORI Cat. III 10987. 54ff. With the *Subodhinī* of Nīlakaṇṭha. Incomplete (to II 4).
RORI Cat. III 15416. 7ff. Incomplete.
SOI 2351 = SOI Cat. I: 1459–2351. 27ff. With a ṭīkā.
SOI 8392. Sith a ṭīkā.
SOI 9494. With the *Subodhinī* of Nīlakaṇṭha.
SOI 9495.
SOI 9497.
SOI 10029. With the *Subodhinī* of Nīlakaṇṭha.
Tanjore D 11331 = Tanjore BL 11064. 9ff. Grantha. Incomplete (I–II only).
Trichinopoly Krishna Iyer 431.
VVRI 2449. 12ff. With a ṭīkā, *Candrikā*. Incomplete.
VVRI 4008. 5ff. With the *Subodhinī* of Nīlakaṇṭha. Incomplete.
VVRI 4051. 16ff. Incomplete (III-IV only).
Weber (A) 35, 13. A copy of the edition lithographed at Benares in 1877.
WHMRL I. 68.
WHMRL I. 85.

The *Upadeśasūtra* has been published:

with the *Subodhinī* of Nīlakaṇṭha at Kāśī in Saṃ. 1931 = A.D. 1874 (BM) and at Kāśī in Saṃ. 1934 = A.D. 1877 (BM);
at Kāśī in Saṃ. 1934 = A.D. 1877 (BM):
with the *Subodhinī* of Nīlakaṇṭha, edited by Rasikamohana Chaṭṭopādhyāya, Kalikātā Saṃ. 1941 = A.D. 1884 (BM 14053. dd. 6 and NL Calcutta 180. Kb. 88. 11 (4));
with the *Subodhinī* of Nīlakaṇṭha at Mumbaī in 1888 (BM);
with the *Subodhinī* of Nīlakaṇṭha at Allahabad in 1888 (IO 3. B. 6) (I–II only);
with his own *Upadeśārthoddyota* by Vināyaka Śāstrī Vetāla at Kāśī in Saṃ. 1968 = A.D. 1911 (BM 14055. d. 11 (1)) (I–II only);
with his own Marāṭhī bhāṣāntara by Viṣṇu Gopāla Navāthe, *Jātakaśiromaṇi* I 8, Bombay 1914 (BM 14055. d. 23. (1) and IO San. C. 160(b)) (*Jaiminīyapaddhati*, I–II only);
with the ṭīkā of Rāmayatna Ojhā, 2nd ed., Benares 1925 (IO San. B. 935(f));
with the *Jaiminipadyāmṛta* of Durgāprasāda Dviveda and the *Jaiminimūlakandalī* of Mādhava, Mumbayī 1925 (I–II only);
with the *Subodhinī* of Nīlakaṇṭha and the Bengālī bhāṣānuvāda of Rādhāvallabha Pāṭhaka, at Calcutta in Śaka 1848 = A.D. 1926 (IO San. B. 990(d) and NL Calcutta 180. Kc. 92. 17);
with his own Saṃskṛta and Hindī ṭīkā, *Vimalā*, by Acyutānanda Jhā as *HSS* 159, Banārasa 1943; 2nd ed., Banārasa 1952 (I–II only);
with an English translation by B. Suryanarayana Rao, Bangalore 1932 (NL Calcutta 180. Kb. 93. 7); rev. by B. V. Raman, Bangalore 1944 (I–II only);

with his own *Marīci* by Kāśīnātha Vāsudeva Abhyaṅkara and an English translation of I–II at Ahmedabad in 1951. Abhyankar has used Kathmandu I 1209; BORI 474 and 475 of 1892/95; BORI 826 of 1887/91; 3 other BORI MSS; and 2 other Kathmandu MSS.

There was an edition in Telugu script with an Āndhraṭīkā published at Madras according to Mysore GOL B 1838.

Closely connected with the *Upadeśasūtra* and also attributed to Jaimini are the *Jaiminisūtrakārikās* in 2 adhyāyas of which each has 4 pādas. Manuscripts:

Benares (1963) 36217 = Benares (1878) 108. 6ff. Copied in Saṃ. 1874 = A.D. 1817.
AS Bengal 6953 (G 5508). 12ff.
Kathmandu (1960) 125 (I 1209). 8ff. Nevārī.
N-W P IX (1884) A 35. 7ff. Property of Pandit Śyāmā Caraṇa of Benares.
Oudh VII (1875) VIII 4. 16pp. Property of Jānakīprasāda of Bārābaṅki Zillah.

The *Kārikās* were published by Durgāprasāda Dviveda in his edition of the *Upadeśasūtra*, Mumbayī 1925, pp. 95–112; and by K. V. Abhyankar in his edition of the *Upadeśasūtra*, Ahmedabad 1951, pp. 167–181. Abhyankar used Kathmandu I 1209.

JAIMINI

Alleged author of a *Jaiminicandrikā*; *cf.* the ṭīkā, *Candrikā*, on the *Upadeśasūtra* of Jaimini. Manuscripts:

Benares (1963) 35213. 20ff. Copied in Saṃ. 1835 = A.D. 1778.
Benares (1963) 34805. 6ff. Incomplete. No author mentioned.
CP, Hiralal 1766. (*Jātakajaimini*). Property of Śrīdev Dīkshit of Maṇḍlā.
SOI 9498. (*Jaiminīyajātakacandrikā*).

JAIMINI

Alleged author of a *Jaiminīyaśakuna*. Manuscript:

SOI 9890.

JAIMINI

Alleged author of a *Dvādaśabhāva* in 8 adhyāyas. Manuscripts:

GOML Madras D 13730. Ff. 69–72. Telugu (*Jaiminisūtra* in margin, *Dvādaśabhāvaphala* at beginning).
Mysore (1911 + 1922) 2621. 32ff. (*Jaiminisūtra* in 8 adhyāyas).
Oppert I 362. No. pp. given. Grantha. Property of Koṇḍaṅgi Anantācāryār of Kāñcīpuram, Chingleput.
Oppert I 3566. 40pp. Grantha. Property of Narasiṃhapuram Rāghavācāryār of Kumbhaghoṇam, Tanjore.
Oppert II 1455. Property of Pattaṅgi Raṅgācāryār of Pillapākkam, Conjeveram, Chingleput.
Oppert II 1628. Property of Śrīraṅgācāryār of Velāmūr, Madhurāntakam, Chingleput.
Oppert II 7596. 185pp. Grantha. Property of the Mahārāja of Pudukoṭa, Tanjore.

JAIMINI

Alleged author of a *Phalaratnamālā* in 6 adhyāyas edited, with the Tamil translation of T. S. Nārāyaṇa Sāmi, by T. S. Vāmana Rāu at Tanjore in 1911 (BM 14055. d. 14 and IO 23. BB. 43).

JORĀVARAMALA (*fl.* 1767)

A Māthura Kāyastha residing in Nāgapura, Jorāvaramala wrote a *Śaniścara kī kathā* in Hindī in Saṃ. 1824 = A.D. 1767. Manuscripts:

NPS 510 A of 1926–28. Copied in Saṃ. 1926 = A.D. 1869. Property of Ṭhākura Tribhuvanasiṃha of Śāhapura, Nerī, Sītāpura.
NPS 510 B of 1926–28. Copied in Saṃ. 1926 = A.D. 1869. Property of Paṇḍita Śivadīna Jośī of Paṭarāsā, Khairābāda, Sītāpura.

JÑĀNACANDRA

The pupil of Sāgaracandra, Jñānacandra wrote a *Kheṭādimañjarī* or *Khecaramañjarī*. Manuscript:

RORI Cat. II 9496. 9ff.

JÑĀNADEVA

Author of a *Nārāyaṇaśakunāvalī* or *Praśnāvalī* preserved in the *Skandapurāṇa*. Manuscripts:

Benares (1963) 37654. Ff. 2–14. Copied in Śaka 1658 = A.D. 1736. Incomplete.
Benares (1963) 37421. 4ff. Copied in Saṃ. 1896 = A.D. 1839.
Alwar 1824.
Anup 4814. 15ff. Copied by Makunda Bhaṭa.
GOML Madras D 13940. Ff. 39–46. Telugu and Grantha.
GOML Madras D 13941. Ff. 114–124. Karṇāṭakī. Incomplete.
PUL II 3674. 10ff.

The first verse is:

athātaḥ sampravakṣyāmi praśnānāṃ śakunāvalim/
jñānadevena kathitā yā svayaṃ bhaktasaṃnidhau//

JÑĀNABHĀSKARA

Alleged author of a *Pāśākevalī*. Manuscript:

RJ 3032 (vol. 4, p. 286). 5ff.

JÑĀNABHĀSKARA

Author of a *Ṣaḍvargaphala* in 6 adhyāyas, in the form of a conversation between Aruṇa and Sūrya; *cf.* the *Sūryāruṇasaṃvāda*. Manuscripts:

Leipzig 553. 9ff. Copied in A.D. 1864.
PL, Buhler IV E 488. 7ff. Property of Uttamarāma Jośī of Ahmadābād.

JÑĀNABHĀSKARA

Alleged author of a *Śakunāvalī*. Manuscript:

PUL II 3964. 23ff. (ff. 5–6 missing). Incomplete.

JÑĀNARĀJA (*fl.* 1503)

The son of Nāganātha and the father of Cintāmaṇi (*fl. ca.* 1530) and Sūryadāsa (*fl.* 1541), Jñānarāja wrote a *Siddhāntasundara* or *Sundarasiddhānta* in Śaka 1425 = A.D. 1503. The work consists of the following chapters:

I grahagaṇitādhyāya.
1. madhyamādhikāra.
2. spaṣṭīkaraṇādhyāya.
3. tripraśnādhyāya.
4. parvasambhūti.
5. candragrahaṇādhikāra.
6. sūryagrahaṇādhikāra.
7. grahodayāstādhikāra.
8. nakṣatracchāyāghaṭīsādhanādhikāra.
9. śṛṅgonnatyadhikāra.
10. grahayogādhyāya.
11. tārāchāyābhadhruvādya.
12. pātādhyāya.

II golādhyāya.
1. bhuvanakośādhikāra.
2. madhyabhuktivāsanādhyāya.
3. chedyake yukti.
4. maṇḍalavarṇana.
5. yantramālā.
6. ṛtuvarṇana.

There is a commentary, *Grahagaṇitacintāmaṇi*, by Jñānarāja's son, Cintāmaṇi (*fl. ca.* 1530). See S. Dvivedin [1892] 56–58 and S. B. Dikshit [1896] 267–271. The latter gives a lineage from one Kāśīnātha Śāstrī dated Śaka 1817 = A.D. 1895 (*cf.* also 273 fn.): Rāma of the Bhāradvājagotra, father of Nīlakaṇṭha, father of Viṣṇu, father of Nīlakaṇṭha, father of Nāganātha, father of Nṛsiṃha, father of Nāganātha and Ḍhuṇḍhirāja (*fl. ca.* 1525); Ḍhuṇḍhirāja was the father of Gaṇeśa (*fl. ca.* 1550/1600), and Nāganātha the father of Jñānarāja (*fl.* 1503), the father of Cintāmaṇi (*fl. ca.* 1530) and of Sūrya (1507/1588), the father of Nāganātha (1558/1615), the (grand)father of Gopāla (1623/1668), the (grand)father of Jñānarāja (b. 1673), the (grand)father of Rāmacandra (d. 1809), the father of Vijñāneśvara (1790/1847), the father of Puruṣottama (1826/1877), the father of Kāśīnātha (b. 1846). The family lived at Pārthapura on the Godāvarī.

Manuscripts of the *Siddhāntasundara* are:

Anup 5335. 24ff. Copied by Govindabhaṭṭa in Śaka 1532 = A.D. 1610. Incomplete.
Rajputana, p. 38. Copied in Śaka 1542 = A.D. 1620. Incomplete (gaṇitādhyāya). At Udaipur.
IO 2901 (2002). 55ff. Copied by Kāśībhaṭṭa on Thursday 11 śuklapakṣa of Śrāvaṇa in Śaka 1574 = 5 August 1652. From Dr. John Taylor.
Baroda 9237. 29ff. Copied in Saṃ. 1716 = A.D. 1659.
Benares (1963) 36907 = Benares (1878) 93 = Benares (1869) XVIII 1. 24ff. Copied in Saṃ. 1721, Śaka 1586 = A.D. 1664.
Jaipur (II). 63ff. Copied in Saṃ. 1763 = A.D. 1706.
PL, Buhler IV E 528. 49ff. Copied in Saṃ. 1804 = A.D. 1747. Property of Khuśāla Bhaṭṭa of Ahmadābād.
BORI 860 of 1887/91. 26ff. Copied in Śaka 1686 = A.D. 1764. Incomplete (pātādhyāya (grahagaṇitādhyāya ?)). From Gujarāt.
IO 2902 (2114b). 37ff. Copied in A.D. 1782. From Gaikawar.
RORI Cat. II 4733. 31ff. Copied by Harisukha Brāhmaṇa on Monday 10 śuklapakṣa of Āśvina in Saṃ. 1843, Śaka 1708 = 2 October 1786.
Benares (1963) 35627 = Benares (1878) 90 = Benares (1869) XVIII 4. Ff. 1–4, 6–7, 11–14, and 16–27. Copied in Saṃ. 1845 = A.D. 1788. Incomplete.
AS Bengal 6935 (G 1435) = Mitra, Not. 1767. 8ff. Maithilī. Copied on Sunday 6 kṛṣṇapakṣa of Śrāvaṇa in Śaka 1712, Sāl. San. 1269 = 21 August 1791. Incomplete (golādhyāya).
AS Bengal 6936 (G 8210). 27ff. Copied on 9 śuklapakṣa of Vaiśākha in Saṃ. 1878 = 9 May 1821. Incomplete (golādhyāya).
AS Bengal 6934 (G 7922). 13ff. Copied on Tuesday 8 kṛṣṇapakṣa of Phālguna in Saṃ. 1889 = 12 March 1833. Incomplete (golādhyāya and grahagaṇitādhyāya 1–11).
Baroda 3345. 51ff. Copied in Saṃ. 1925 = A.D. 1868.
Baroda 11544. 67ff. Copied in Saṃ. 1944 = A.D. 1887. Incomplete (pātādhyāya and golādhyāya).
Alwar 2016. 2 copies.
Ānandāśrama 4350.
Anup 5334. 6ff. Incomplete (golādhyāya).
AS Bombay 289. 15ff. Incomplete (golādhyāya).
AS Bombay 290. 20ff.
AS Bombay 291. 50ff. With the *Grahagaṇitacintāmaṇi* of Cintāmaṇi. Incomplete (grahagaṇitādhyāya 1–2). From Bhāu Dājī.
Benares (1963) 34970. 45ff. With the *Grahagaṇitacintāmaṇi* of Cintāmaṇi. Incomplete.
Benares (1963) 35566 = Benares (1903) 1215. 74ff. No author mentioned.

Benares (1963) 36902. 36ff. This may be identical with Benares (1869) XVIII 6. 22ff.
BM 452 (Add. 14, 365p). 28ff. From Major Thomas Best Jervis. See SATE 13.
BORI 107 of 1866/68. 48ff. No author mentioned.
BORI 219 of A 1882/83. 19ff. Incomplete (part II: golādhyāya or grahagaṇitādhyāya?). Ascribed to Jñānānanda.
BORI 880 of 1884/87. 20ff. Incomplete (golādhyāya). From Gujarāt.
BORI 881 of 1884/87. 8ff. Incomplete (gaṇitādhyāya). From Gujarāt.
BORI 283 of Vishrambag 1. 38ff. No author mentioned.
CP, Kielhorn XXIII 178. 51ff. Property of Javāhara Śāstrī of Chāndā.
Jammu and Kashmir 3091. 83ff. With the *Grahagaṇitacintāmaṇi* of Cintāmaṇi. Incomplete.
Kavīndrācārya 903.
Kotah 127. 57pp.
Lucknow 520. G 39 S (45774).
Lucknow 520. G 39 S (45885).
Lucknow (46271). Is this Jñānarāja's work?
Mithila 417. 10ff. Maithilī. Incomplete (ends in grahagaṇitādhyāya 3). Property of Pandit Vāsudeva Miśra of Champa, Benipatti, Darbhanga.
Oxford CS d. 805(v). 18ff.
SOI 9398. Incomplete (golādhyāya).
SOI 9399. Incomplete (pātādhyāya).
SOI 9906.

Verses 2–4 are:

yannāmākṣararaśmibhis tanugataiḥ kiṃcitkalāvān bhaved
bhaktaḥ svāntaniśākaro hṛtatamās tatroccaradbhiḥ kramāt/
natvā tāṃ bhuvaneśvarīm api guruṃ siddhāntasatsundaraṃ
sujñānandakaraṃ karomi caturajñānādhirājaḥ sphuṭam//
yan nāradāya kathitaṃ caturānanena
jñānaṃ graharkṣagatisaṃsthitirūpam agryam/
śākalyasaṃjñamuninā likhitaṃ nibandhaṃ
padyais tad eva vivṛṇomi savāsanasvaiḥ//
brahmārkenduvaśiṣṭharomakapulastyācāryagargādibhis
tantrāṇy aṣṭakṛtāni teṣu gahanaḥ khecārikarmakramaḥ/
tadratnākaravāsanāvataraṇeḥ siddhāntapotāḥ kṛtāḥ
śrīmadbhojavarāhajiṣṇujacaturvedāryamadbhāskaraiḥ//

The date is given by a verse in the first adhyāya:

sāṅghriśakraśataśodhito
bhavec chālivāhanaśako ᵓbdasañcayaḥ/
saṅguṇaḥ khagaguṇena
yojitaḥ kṣepakeṇa śaradi dhruvo bhavet//

The colophon to the golādhyāya is:

itthaṃ śrīmannāganāthātmajena
prokte tantre jñānarājena ramye/
granthāgārādhārabhūte prabhūte
golādhyāye varṇanaṃ ṣaḍṛtūnām//

Jñānarāja also wrote a *Bījādhyāya* for the *Siddhāntasundara*. Manuscripts:

Benares (1963) 35629. Ff. 16–17. Copied in Saṃ. 1682 = A.D. 1625. Incomplete.
Benares 35626 = Benares (1878) 92 = Benares (1869) XVIII 2. 27ff.
Berlin 833 (or. fol. 231). 21ff. Copied from a manuscript copied by Ekanātha in Śaka 1522 = A.D. 1600.
SOI 9396.
SOI 9397.

JÑĀNAVIMALA SŪRI

Author of a *Pandara Tithini Thoyo* in Old Gujarātī. Manuscript:

LDI (MPC) P/7547. 6ff.

JÑĀNASĀGARA

Author of a *Praśnottaramāṇikyamālā*. Manuscripts:

RJ 3051 (vol. 4, p. 288). 27ff. Copied in Saṃ. 1860 = A.D. 1803.
RJ 3052 (vol. 4, p. 288). 37ff. (f. 1 missing). Copied on 10 kṛṣṇapakṣa of Caitra in Saṃ. 1861 = *ca.* 3 April 1804. Incomplete.

JÑĀNASĀGARA (*fl.* 1408).

The person to whom Guṇaratna Sūri (*fl. ca.* 1375) dedicated his avacūrṇi on the *Kṣetrasamāsa* of Somatilaka Sūri (*fl.* 1298/1367), Jñānasāgara is said to have written an avacūrṇi on the same work in Saṃ. 1465 = A.D. 1408. Like Guṇaratna, Jñānasāgara was a pupil of Devasundara Sūri (b. 1339) of the Tapā Gaccha. See Velankar, p. 99. Manuscripts:

Ahmadabad, Bhandar of the Vimala Gaccha Upasraya, Falusha's Pole 18 (48) and Haja Patel's Pole 41 (47). See Velankar.
BORI 1254 of 1891/95. 10ff.

JÑĀNĀNANDA

Author of a *Ratnapradīpa*. Manuscript:

Oudh XXII (1890) VIII 14. 16pp. Property of Kedāranātha of Āgrā Zillah.

JYEṢṬHADEVA (*fl. ca.* 1500/1575)

The pupil of Dāmodara (*fl. ca.* 1440/1500), the son of Parameśvara (*ca.* 1380/1460), Jyeṣṭhadeva was the teacher of Acyuta Piṣāraṭi (*ca.* 1550–7 July 1621).

He composed the *Yuktibhāṣā* in Malayālam as an exposition of the *Tantrasaṅgraha* (1500) of his fellow-pupil, Nīlakaṇṭha (b. *ca.* 14 June 1444). He was a Nampūri Brāhmaṇa from the Paraññoṭṭu illam in Ālattūr, Kerala, according to an old granthavari at Baroda: pūrvoktadāmodarasya śiṣyaḥ jyeṣṭhadevaḥ/ iddehaṃ paraññoṭṭu nampūriyākunnu/ yuktibhāṣāgranthatte uṇṭākkiyatuṃ iddehaṃ tanne. See K. V. Sarma [1958a] and K. K. Raja [1963] 156–158.

The *Yuktibhāṣā* was edited by R. V. Thampuran and A. R. A. Iyer, vol. 1, Trichur 1948; and by T. Chandrasekharan, Madras 1953. There is also a Saṃskṛta *Gaṇitayuktibhāṣā* closely related to the *Yuktibhāṣā*. The mathematics of the *Yuktibhāṣā* has been studied by C. M. Whish [1830]; K. M. Marar and C. T. Rajagopal [1944] and [1945]; C. T. Rajagopal [1949]; C. T. Rajagopal and A. Venkataraman [1949]; and C. T. Rajagopal and T. V. V. Aiyar [1951] and [1952].

JYOTIRĀJA (*fl.* 1382)

Jyotirāja composed, probably in Nepal in Śaka 1304 = A.D. 1382, a *Jyotirājakaraṇa* in seven chapters:

1. tithyadhikāra.
2. grahādhikāra.
3. tripraśnādhikāra.
4. candragrahaṇādhikāra.
5. sūryagrahaṇādhikāra.
6. sottara.
7. sūryasiddhāntamatameghavikṣepa.

Manuscripts:

Kathmandu (1960) 129 (I 440). 6ff. Nevārī. Copied on Thursday 13 kṛṣṇapakṣa of Jyeṣṭha in Nep. Sam. 538 = 2 June 1418. Incomplete (chapters 2–6).

Kathmandu (1960) 130 (III 440) = Nepal (Regmi), vol. 1, p. 420. 24ff. Nevārī. Copied by Daivajña Jyotirāja at full moon of Caitra in Nep. Sam. 541 = 18 March 1421 during the reign of Jayajyotirmalladeva (*ca.* 1409/1428).

The date is given in the vs. 2a-b:

śāke kṛtābhraviśvonaṃ śāstrābdaṃ taṃ vidhīyate/

Jyotirāja also wrote a *Svarodayadaśā* in Nevārī. Manuscript:

Kathmandu (1960) 524 (III 364) = Nepal (Regmi), vol. 1, p. 441. 35ff. Copied by Daivajña Guṇarāja for Daivajña Manirājabhāra on Sunday 4/5 śuklapakṣa of Vaiśākha in ⟨Nep.⟩ Saṃ. 582 = 5 April 1462 during the reign of Rāyamalladeva; NS 582, however, falls during the reign of Rāyamalla's father, Yakṣamalla (1428/1482).

JYOTIṢARĀJA

Author of a *Praśnavicāra* in Hindī. Manuscript:

NPS 213 of 1926–28. Property of Paṇḍita Rāmasvarūpa Miśra of Arjunapura, Antū, Pratāpagaḍha.

JYOTIṢARĀJA (b. 1613)

See Vīrasiṃha (b. 1613)

JVĀLĀPRASĀDAJĪ MIŚRA (*fl.* 1953)

A resident of Dīnadāra, Murādābāda, Jvālāprasāda wrote a bhāṣāṭīkā on the *Bṛhadyavanajātaka*, which was published at Kalyāṇa-Muṃbaī in 1953.

ṬĪKĀRĀMA AVASTHĪ

The son of Bhavānīprasāda, Ṭīkārāma wrote a Hindī translation of the *Laghujātaka* of Varāhamihira (*fl. ca.* 550). Manuscript:

NPS 324 of 1929-31. Property of Ṭhākura Pratāpasiṃha of Rataulī, Holīpurā, Āgarā.

ṬĪKĀRĀMA DHANAÑJAYA (*fl.* 1931)

The son of Ekadeva Dhanañjaya, Ṭīkārāma, a resident of Khidimagrāma, Gulmī, Nepāla, wrote a ṭīkā, *Manoramā*, on the *Bhāsvatī* of Śatānanda (*fl.* 1099) and a pariśiṣṭa in Saṃ. 1988 = A.D. 1931. This was published at Vārāṇasī [N.D.]. In the final verses Ṭīkārāma claims that he has also written a ṭīkā, *Manoramā*, on the *Līlāvatī* of Bhāskara (b. 1114).

ṬOḌARAMALA (*fl.* 1761)

A Jaina resident of Jayapura, Ṭoḍaramala composed a *Trilokasāra* in Hindī. Manuscripts:

RJ 1801 (vol. 2, p. 284). 303ff. (ff. 1–108 missing). Copied in Saṃ. 1839 = A.D. 1782. Incomplete. Property of Baḍā Terahapanthiyoṃ of Jayapura.

RJ 3371 (vol. 4, p. 321). 289ff. Copied in Saṃ. 1841 = A.D. 1784.

NPS 68ka of Saṃ. 2007–2009. Copied in Saṃ 1880 = A.D. 1823. Property of the Digambara Jaina Mandira (Baḍā Mandira) at Cūḍāvālī Galī, Cauka, Lakhanaū.

RJ 3373 (vol. 4, p. 321). 218ff. Copied by Kālūrāma Sāha, the son of Jaitarāma Sāha, in Saṃ. 1884 = A.D. 1827.

NPS 429 C of 1923–25. Copied in Saṃ. 1901 = A.D. 1844. Property of the Jaina Mandira (Baḍā) at Bārābaṅkī.

RJ 3375 (vol. 4, p. 321). 394ff. Copied by Javāharalāla Suganacanda Sonī of Ajamera in Saṃ. 1969 = A.D. 1912.

RJ 3372 (vol. 4, p. 321). 44ff. Incomplete.

RJ 3374 (vol. 4, p. 321). 125ff.

ṬOḌARAMALLA (*fl.* 1565/1589)

Born at Laharpur in Oudh, Ṭoḍaramalla began his career as a clerk in the Mughal bureaucracy under Akbar (1556/1605). He served in the negotiations with Khān Zamān of Jaunpur in 1565, took part in the siege of Chitor in 1567/1568, investigated the defences of

Surat in 1572, made the revenue settlement of Gujarat in 1574 and entered the central government's finance department, took part as a general in Akbar's campaigns in Bengal in 1574/1576, served as governor of Gujarat in 1576/1577, was in charge of the Bengal mint in 1577, became wazīr in 1577/1578, suppressed the rebellion in Bengal in 1580, and was appointed dīwān in 1582/1583. He died in November of 1589. Between *ca.* 1572 and 1582 Nīlakaṇṭha (*fl.* 1569/1587) published for him the jyotiṣa sections of a vast encyclopedia entitled *Ṭoḍarānanda;* this is often ascribed to Ṭoḍaramalla (or Ṭoḍaravarman) in the manuscripts. See P. V. Kane [1930/62] vol. 1, pp. 421–423, and *Ṭoḍarānanda*, vol. 1, edited by P. L. Vaidya, *Ganga OS* 5, Bikaner 1948.

A part of the *Ṭoḍarānanda* is the *Varṣakṛtyasaukhya*. Manuscript:

Anup 2368. 58ff. Formerly property of Kavīndrācārya Sarasvatī (*fl. ca.* 1600/1675).

Another section is the *Tithinirṇaya*. Manuscript:

Anup 1704 = Bikaner 1035. 61ff. Formerly property of Kavīndrācārya Sarasvatī (*fl. ca.* 1600/1675).

ṬHAKKURA PHERŪ (*fl.* 1315)

The son of Canda of the Dhaṃdhakula and a resident of Kannāṇapura, Ṭhakkura, a Jaina, wrote the following works in Prākṛta (all are edited by Agaracanda and Bhaṃvaralāla Nāhaṭā as *Ratnaparīkṣādisaptagranthasaṅgraha*, *RPG* 44, Jodhpur 1961):

1. *Ratnaparīkṣā* on gems, based on Agastya and Buddhabhaṭṭa, was written at Delhi in Saṃ. 1372 = A.D. 1315 during the reign of Allāvadī or ᶜAlā al-dīn Khaljī (1296/1316); ed. pp. 1–16.

2. *Dravyaparīkṣā* on coins and mints; ed. pp. 17–38.

3. *Dhātūtpatti* on useful substances; ed. pp. 39–44. The manuscript was copied for Purisaḍa, the son of Bhāvadeva, on Monday 8 śuklapakṣa of Phālguna in Saṃ. 1403 = 19 Feburary 1347.

4. *Jyotiṣasāra* on astronomy and astrology in 4 dvāras:

   1. dinaśuddhi.
   2. vyavahāra.
   3. gaṇitapada.
   4. lagnasamuccaya.

Ed., pt. 2, pp. 1–40. The second verse mentions Haribhadra, Naracandra, Padmaprabha Sūri, Yavana, Varāhamihira, Lalla, Parāśara, and Garga. The *Jyotiṣasāra* was also composed in Saṃ. 1372 = A.D. 1315.

5. *Gaṇitasāra* on mathematics; ed., pt. 2, pp. 41–74 See O. Prakash [A2. 1965]. The manuscript was copied on 5 śuklapakṣa of Caitra in Saṃ. 1404 = 16 March 1347.

6. *Vāstusāra* on architecture; ed., pt. 2, pp. 75–103. The *Vāstusāra* was also composed in Saṃ. 1372 = A.D. 1315.

7. *Kharataragacchayugapradhānacatuḥpadikā*; ed., pt. 2, pp. 104–106. The manuscript was copied on 8 śuklapakṣa of Phālguna in Saṃ. 1403 = 19 February 1347.

ṬHĀKURADATTOPĀDHYĀYA

Author of a *Vastuvicāradīpakamaṇi*. Manuscript:

Baroda 13071. 65ff. Copied in A.D. 1922.

ṬHĀKURADĀSA BHAṬṬĀCĀRYA

Author of a *Tithisārasaṅgraha*. Manuscript:

Benares (1956) 14079. 1f.

ṬHĀKURADĀSA BHAṬṬĀCĀRYA (*fl.* 1876)

Author of a *Jyotiṣārthadīpikā*, published with a Bengālī translation at Calcutta in 1876 (IO 8. F. 29), and pt. 1, with a Bengālī translation, at Calcutta in 1911 (BM 14055. d. 12).

ṬHĀKURADĀSA CŪḌĀMAṆI (*fl.* 1911)

Author of a *Jyotiṣadarpaṇa*, published with a Bengālī bhāṣānuvāda at Calcutta in 1911 (IO 19. BB. 22).

ḌHUṆḌHIN

Alleged author of a *Gaurījātaka;* this may be the *Sujātaka* of Ḍhuṇḍhirāja. Manuscript:

VVRI 2387. 15ff.

ḌHUṆḌHIN

Author of a *Māsādinirṇaya*. Manuscript:

BORI 603 of 1882/83. 11ff. From Mahārāṣṭra.

ḌHUṆḌHIRĀJA

The *Anantasudhārasasāriṇī* of Ananta (*fl.* 1525) and the caṣaka on it are ascribed to Ḍhuṇḍhirāja in the following manuscripts:

Benares (1878) 69 = Benares (1869) XIV 8. 13ff. This is Benares (1963) 35420, where the error is corrected.

Benares (1869) XIV 11. 17ff. This is Benares (1963) 35524 = Benares (1878) 72, where the error is corrected.

The mistaken attribution apparently arose from the mention of Ḍhuṇḍhin in the first verse; see *CESS* A 1, 40b, where one must remove Ḍhuṇḍhirāja from the list of commentators.

ḌHUṆḌHIRĀJA

Author of a *Keralabhāṣya* or *Keralavacanāni*. Manuscript:

PUL II 3307. 5ff. Telugu.

The colophon begins: ḍhuṇḍhirājakṛtakeralabhāṣyam.

ḌHUṆḌHIRĀJA

Author of a *Khecarīkalpadruma*. Manuscript:

IM Calcutta 5354. Incomplete. See NCC, vol. 5, p. 188.

ḌHUṆḌHIRĀJA

Author of a *Grahaphalopapatti;* this may be part of the upapatti on the *Makaranda*. Manuscript:

Benares (1963) 35344 = Benares (1878) 109 = Benares (1869) XXIV 5. 6ff.

ḌHUṆḌHIRĀJA

Author of an udāharaṇa on the *Grahalāghava* of Gaṇeśa (b. 1507); perhaps identical with Ḍhuṇḍhirāja (*fl.* 1590). Manuscripts:

Benares (1869) XIV 7. 57ff.
CP, Hiralal 1581. Property of the Bhonsalā Rājas of Nagpur.

ḌHUṆḌHIRĀJA

Author of a ṭīkā, *Vyavahāraprakāśikā*, on the *Bālavivekinī* of Nāhnidatta. Manuscripts:

Anup 4901. 16ff. Copied at Kesurakasava in Saṃ. 1643 = A.D. 1586 during the reign of Rāyasiṃha (1571/1611). Property of Saṃvaladāsa Sāṃgāvata in Saṃ. 1647, 1651, and 1654 = A.D. 1590, 1594, and 1597.
Bombay U Desai 1390. Ff. 10–24. Copied in Saṃ. 1658 = A.D. 1601.
Bombay U Desai 1389. 32ff. Copied in Saṃ. 1818 = A.D. 1761.

The first verse is:

natvā herambam ambāṃ ca ḍhuṇḍhirājena tanyate/
vyākhyā bālavivekinyā vyavahāraprakāśikā//

ḌHUṆḌHIRĀJA

The son of Vināyaka, Ḍhuṇḍhirāja wrote a *Sāyanatattvaviveka* for Rukmāṅgada Dīkṣita. Manuscripts:

Mithila 400. 18ff. Copied on 30 kṛṣṇapakṣa of Bhādrapada in Saṃ. 1920 = *ca.* 11 October 1863. Property of Pandit Śrīnandana Miśra of Kanhauli, Sakri, Darbhanga.
AS Bengal 6831 (G 6368). 32ff. Incomplete (*Āyanatattva*).
SOI 9389. No author mentioned.

The second verse is:

vināyakasuto natvā vināyakapadāmbujam/
vivektuṃ sāyane tattvaṃ ḍhuṇḍhirājaḥ pravarttate//

The colophon begins: iti śrīmatsakalavidvadvṛndapadadvandvapadanāptamahābhāgyodayaśrīrukmāṅgadadīkṣitājñayā ḍhuṇḍhirājaviracitaḥ.

ḌHUṆḌHIRĀJA

Author of a *Sujātaka*, sometimes called *Jātakapaddhati;* its relation to the *Jātakābharaṇa* remains obscure. See Ḍhuṇḍhin. Manuscripts:

GJRI 1130/242. 15ff. Maithilī. Incomplete.
Tanjore D 11400 = Tanjore BL 4199. 21ff.

The first verse is:

śrīmadgurūṇāṃ caraṇāravindaṃ
yatsarvavijñānavidhānarūpam/
praṇamya ḍhuṇḍhir gaṇakaḥ sujātakaṃ
śrīśambhunoktaṃ likhitaṃ vibhuktau//

ḌHUṆḌHIRĀJA (*fl. ca.* 1525)

The son of Nṛsiṃha of Pārthapura on the Godāvarī and the pupil of Jñānarāja, Ḍhuṇḍhirāja is traditionally identified with the nephew of Jñānarāja (*fl.* 1503), who then is his teacher; this makes him also a descendent of Rāma of the Bhāradvājagotra, a resident of Pārthapura in about 1300. Ḍhuṇḍhirāja wrote a popular *Jātakābharaṇa*. See S. Dvivedin [1892] 67–68 and S. B. Dikshit [1896] 273–274. There are many adhyāyas:

1. maṅgala.
2. saṃvatsaraphala.
3. ayanaphala.
4. ṛtuphala.
5. māsaphala.
6. pakṣaphala.
7. tithiphala.
8. vāraphala.
9. nakṣatraphala.
10. navāṃśaphala.
11. yogaphala.
12. karaṇaphala.
13. lagnaphala.
14. ḍiṃbhākhyacakraphala.
15. dvādaśabhāvaphala.
16. ravyādigrahabhāvaphala.

17. dṛṣṭiphala.
18. rāśiphala.
19. cakraphala.
20. gocaraphala.
21. aṣṭakavarga.
22. dvigrahayoga.
23. trigrahayoga.
24. rājayoga.
25. sāmudrika.
26. rājayogabhaṅga.
27. pañcamahāpuruṣayoga.
28. kārakayoga.
29. nābhasayoga.
30. raśmijātaka.
31. grahāṇāṃ dīptādyavasthā.
32. sthānādiyuktagrahaphala.
33. sūryayoga.
34. candrayoga.
35. pravrajya.
36. ariṣṭa.
37. riṣṭabhaṅga.
38. sarvagrahariṣṭabhaṅga.
39. sadasaddaśā.
40. daśāphala.
41. antardaśāphala.
42. naṣṭajātaka.
43. niryāṇa.
44. candrakṛtaniryāṇa.
45. strījātaka.

Manuscripts:

Benares (1963) 36560 = Benares (1878) 30 = Benares (1869) VII 2. Ff. 1–22, 29–32, 32b–36, and 38–94. Copied in Saṃ. 1679 = A.D. 1622. Incomplete.

Berlin 866 (Chambers 280). 135ff. Copied in Saṃ. 1681, Śaka 1546 = A.D. 1624.

DC 181. Ff. 2–14. Copied in Śaka 1547 = A.D. 1625.

Baroda 7383. 31ff. Copied in Śaka 1550 = A.D. 1628. Incomplete.

DC 201. Ff. 2–98. Copied in Śaka 1559 = A.D. 1637.

AS Bombay 359. Ff. 2–114. Copied in Saṃ. 1705 = A.D. 1648. From Bhāu Dājī.

Anup 4650 = Bikaner 643. 130ff. Copied in Saṃ. 1706 = A.D. 1649. Formerly property of Maṇirāma Dīkṣita (*fl. ca.* 1650/1700).

LDI (LDC) 1655. 127ff. Copied in Saṃ. 1717 = A.D. 1660.

Anup 4651. 94ff. Copied at Āṃvāṃ by Gaṅgādhara, the son of Bhīkaṃbhaṭṭa of the Mahārāṣṭrajāti, in Saṃ. 1720 = A.D. 1663. Formerly property of Gaṅgādhara Vāḍholakara.

Baroda 3135. 117ff. Copied in Saṃ. 1725 = A.D. 1668.

Jaipur (II). 133ff. Copied in Saṃ. 1725 = A.D. 1668.

Anup 4530. 1f. Copied by Haṃsarāja in Śaka 1591 = A.D. 1669. Incomplete (grahadānavidhāna).

BORI 342 of 1880/81. 130ff. Copied in Saṃ. 1733 = A.D. 1676.

RJ 2989 (vol. 4, p. 282). 43ff. Copied by Sukhakuśala Gaṇi at Nāgapura on 13 śuklapakṣa of Bhādrapada in Saṃ. 1736 = *ca.* 7 September 1679.

PL, Buhler IV E 120. 116ff. Copied in Saṃ. 1738 = A.D. 1681. Property of Hariśaṅkara Jośī of Ahmadābād. Buhler notes 14 other copies.

Udaipur 520. Copied in Saṃ. 1740 = A.D. 1683.

Berlin 867a (Chambers 320). 64ff. Copied in Saṃ. 1756, Śaka 1620 = A.D. 1699.

GVS 2805 (1570). Ff. 7–93. Copied on Wednesday 2 śuklapakṣa of Pauṣa in Saṃ. 1766, Śaka 1631 = 21 December 1709. Incomplete.

Chāṇī, Ā. Śrī. Vi. Dā. Sū. Sam. Śā. Saṃ. Copied by Yaśovijaya Gaṇi, the pupil of Guṇavijaya Gaṇi, the pupil of Riddhivijaya Gaṇi, the pupil of Vijayaprabha Sūri, at Satyapura on Sunday 12 kṛṣṇapakṣa of Phālguna in Saṃ. 1770, Śaka 1635 = 28 February 1714. See Praśasti (1), p. 286.

IO 3078 (2546). 87ff. Copied at Nalinagara on Thursday 4 kṛṣṇapakṣa of Vaiśākha in Saṃ. 1771, Śaka 1636 = 22 April 1714. Includes extracts from the *Kāmadhenupaddhati* of Jayarāma Bhaṭṭa. From Gaikawar.

Oxford 1575 (Sansk. d. 190) = Hultzsch 286. 136ff. Copied in Saṃ. 1774 = A.D. 1717.

RORI Cat. I 3119. 100ff. Copied by Kuśalā Caturvedī at Manoharapura in Saṃ. 1774 = A.D. 1717. (*Jātakasāra*).

LDI 6739 (2745). 5ff. Copied by Muni Lakṣmīkuśala at Dvīpabandara in Saṃ. 1781 = A.D. 1724. Incomplete (caturaśītiyoga). No author mentioned.

Benares (1963) 37280. Ff. 45–62 and 69–154 and 1f. Copied in Saṃ. 1785 = A.D. 1728. Incomplete.

Baroda 8396. 146ff. Copied in Śaka 1658 = A.D. 1736.

Cāṇasmā, Ni. Vi. Jī. Maṇi. Pu. Copied by Hitavijaya, the pupil of Govindavijaya Gaṇi, on Wednesday 12 kṛṣṇapakṣa of Kārttika in Saṃ. 1797, Śaka 1662 = 5 November 1740 Julian. No author mentioned. See Praśasti (1), p. 322.

Bombay U 495. 177ff. Copied by Rāmaśukla on 5 śuklapakṣa of Vaiśākha in Śaka 1672 = 29 April 1750.

LDI (LDC) 3961. 12ff. Copied in Saṃ. 1809 = A.D. 1752. Incomplete (dvādaśabhāva).

Poleman 4979 (Harvard 102). 71ff. Copied in Saṃ. 1818 = A.D. 1761.

RORI Cat. I 3761. 61ff. Copied by Sujanavijaya at Meḍatā in Saṃ. 1818 = A.D. 1761.

GOML Madras D 13719. 14pp. Copied by Rāma Miśraka on Saturday 3 kṛṣṇapakṣa of Phālguna in Saṃ. 1827 = 2 March 1771. Incomplete (naṣṭajātaka and nirṇayādhyāya).

Mithila 101 C. 53ff. Maithilī. Copied in Śaka 1693 = A.D. 1771. Property of Pandit Yaduvīra Miśra of Khopa, Phulparas, Darbhanga.

Benares (1963) 36356. Ff. 1–26, 29, 36–44, 46, 74–86, 116–124, and 130–134. Copied in Saṃ. 1830 = A.D. 1773. Incomplete.

Mithila 101. 129ff. Copied on Friday 13 śuklapakṣa of Māgha in Saṃ. 1832 = 1 February 1776. Property of Pandit Cirañjīva Jhā of Babhanagama, Supaul, Bhagalpur.
AS Bombay 360. 16ff. Copied in Śaka 1702 = A.D. 1780. Incomplete. From Bhāu Dājī.
RJ 2990 (vol. 4, p. 282). 100ff. Copied by Gaṅgādhara Bhaṭṭa at Nāgapura on 6 śuklapakṣa of Kārttika in Saṃ. 1840 = *ca.* 30 October 1783.
Goṇḍal 116. 114ff. Copied by Vāsudeva of the Udīcyajñāti at Goṇḍalapura on Friday 11 kṛṣṇapakṣa of Āśvina in Saṃ. 1842, Śaka 1707 = 28 October 1785.
LDI (LDC) 4758. 9ff. Copied in Saṃ. 1846 = A.D. 1789. Incomplete (dvādaśaphala). No author mentioned.
RORI Cat. III 15619. 111ff. (ff. 77–96 missing). Copied in Saṃ. 1847 = A.D. 1790.
Leipzig 1028. 114ff. (ff. 18–30 missing). Copied in A.D. 1791. Incomplete.
SOI 1680 = SOI Cat. I: 1387–1680. 126ff. Copied in Saṃ. 1848 = A.D. 1791.
Florence 292. 79ff. Copied by Harinātha in Saṃ. 1849 = A.D. 1792.
Poleman 5117 (McGill, Museum 21). 6ff. Copied in Saṃ. 1850 = A.D. 1793. No author mentioned.
Goṇḍal 117. 119ff. Copied by Ukā, the son of Īśvara Jośī, on Tuesday 2 śuklapakṣa of Pauṣa in Saṃ. 1852 = 11 January 1796.
Berlin 867b (or. fol. 312). 135ff. Copied by Brāhmaṇadakṣiṇī Premacandajī at Argalāpura in Saṃ. 1853 = A.D. 1796.
Mithila 101 B. 37ff. Copied by Śivanātha at the Bali Āśrama in Daḍibhaṅgāgrāma on Wednesday 12 śuklapakṣa of Bhādrapada in Śaka 1718 = 14 September 1796. Property of Pandit Mahīdhara Miśra of Lalabag, Darbhanga.
Osmania University B. 76/4. 149ff. Copied in A.D. 1798.
Baroda 5634. 160ff. Copied in Śaka 1721 = A.D. 1799.
RORI Cat. I 3745. 59ff. Copied by Rāmacandra Yati at Mīrī in Dakṣiṇadeśa in Saṃ. 1856 = A.D. 1799.
PL, Buhler IV E 212. 12ff. Copied in Saṃ. 1857 = A.D. 1800. Incomplete (pañcāṅgaphala). Property of Uttamarāma Jośī of Ahmadābād. Buhler notes another copy.
LDI (LDC) 1602. 139ff. Copied in Saṃ. 1859 = A.D. 1802.
Goṇḍal 118. 152ff. Copied on Sunday 5 śuklapakṣa of Pauṣa in Saṃ. 1860 = 18 December 1803.
RORI Cat. III 18181. 3ff. Copied by Kīrtimalla at Rūpālī in Saṃ. 1861 = A.D. 1804. Incomplete (nirṇayādhyāya only).
Poleman 4981 (Columbia, Smith Indic 36). 101ff. Copied in Saṃ. 1863, Śaka 1728 = A.D. 1806.
GJRI 3128/340. 58ff. Copied in Saṃ. 1865 = A.D. 1808. Incomplete (ends with daśāphalādhyāya).
Benares (1963) 36362. Ff. 1–65 and 65b–99. Copied in Saṃ. 1868 = A.D. 1811.
Oudh III (1873) VIII 2. 186pp. Copied in A.D. 1815. Property of Paṇḍita Chhoṭe Lāla of Oonao Zillah.
Benares (1963) 35316 Ff. 1–85 and 1f. Copied in Saṃ. 1877 = A.D. 1820.
BORI 908 of 1886/92. 97ff. Copied in Saṃ. 1877 = A.D. 1820.
RORI Cat. II 9794. 90ff. Copied by Rāmabala at Kalyāṇapura in Saṃ. 1878 = A.D. 1821.
BORI 213(B) of 1883/84. No ff. given. Copied in Saṃ. 1879 = A.D. 1822.
Poleman 4977 (U Penn. 677). 25ff. Copied by Raghunātha in Śaka 1752 = A.D. 1830.
VVRI 4558. 63ff. Copied in Saṃ. 1888 = A.D. 1831.
Leipzig 1030. 96ff. Copied in A.D. 1832. Incomplete (the manuscript contains excerpts from many works).
RORI Cat. III 12424. 185ff. Copied by Bhavanātha Mehatā in Saṃ. 1891 = A.D. 1834.
LDI (LDC) 636. 21ff. Copied in Saṃ. 1893 = A.D. 1836.
Benares (1963) 36787. 6ff. Copied in Śaka (read Saṃ.) 1896 = A.D. 1839. Incomplete (nirṇayādhyāya only).
LDI 6823 (5032) 80ff. Copied by Bihāri Ṛṣi, the pupil of Vimalacandra Svāmin, at Mālerakoṭalānigama in Saṃ. 1896 = A.D. 1839.
PUL II 3427. 110ff. Copied in Saṃ. 1896 = A.D. 1839.
RORI Cat. III 14096(2). 56ff. Copied by Caturbhuja Raṅgā in Saṃ. 1896 = A.D. 1839.
VVRI 2448. 88ff. Copied in Saṃ. 1897 = A.D. 1840. Incomplete.
Mithila 101 A. 111ff. (ff. 48–67 missing). Maithilī. Copied in Śaka 1766 = A.D. 1844. Incomplete. Property of Pandit Vāsudeva Jhā of Sukpur, Bhagalpur.
RORI Cat. III 11094. 97ff. Copied by Rāmalāla at Śrīnagara in Saṃ. 1904 = A.D. 1847.
PL, Buhler IV E 178. 140ff. Copied in Saṃ. 1906 = A.D. 1849. (*Tājakābharaṇa*). Property of Śivaśaṅkara Jośī of Ahmadābād.
Benares (1963) 34612. Ff. 1–86 and 90–157. Copied in Saṃ. 1907 = A.D. 1850. Incomplete.
PL, Buhler IV E 107. 22ff. Copied in Saṃ. 1908 = A.D. 1851. (*Jātakakaustubha*). Property of Nirbhaya Rāma of Mulī.
RORI Cat. II 7012. 105ff. Copied by Bāladevācārya Puṣkarṇā at Vikramapura in Saṃ. 1909 = A.D. 1852.
RORI Cat. I 40, 10ff. Copied by Gaṇeśa in Saṃ. 1916 = A.D. 1859. (*Jātakasāra*).
Nagpur 722 (1432). 119ff. Copied in Śaka 1783 = A.D. 1861. From Nasik.
Poleman 4982 (Columbia, Smith Indic 50). 110ff. Copied by Vidyādhara in Saṃ. 1918 = A.D. 1861.
GJRI 928/40. Ff. 70–76. Maithilī. Copied in Śaka 1785 = A.D. 1863. Incomplete.
Goṇḍal 119. 147ff. Copied on Thursday 4 kṛṣṇapakṣa of Pauṣa in Saṃ. 1920 = 28 January 1864.
Nagpur 725 (2114). 127ff. Copied in Śaka 1793 = A.D. 1871. From Nagpur.

PUL II 3426. 187ff. Copied in Saṃ. 1929 = A.D. 1872.
Goṇḍal 120. 166ff. Copied by Jyeṣṭhārāma Raghunātha Rāvaḷa at Ṭaṅkāra in Saṃ. 1945 = A.D. 1888.
ABSP 1260. Ff. 57–73. Incomplete.
ABSP 1292. Ff. 1–32. Incomplete.
ABSP 1298. Ff. 1–3. Incomplete (*Jātakaparipāṭīprabandha*).
Adyar Index 2204 =
Adyar Cat. 8 D 69. 256ff.
Adyar Cat. 8 F 75. 120ff. Incomplete (ends with candrarāśiphala).
Alwar 1770.
Ānandāśrama 1987.
Ānandāśrama 2574.
Ānandāśrama 2588.
Ānandāśrama 2605.
Ānandāśrama 4272.
Ānandāśrama 5644.
Ānandāśrama 8235.
Ānandāśrama 8387.
Anup 4652. 105ff. Incomplete.
Anup 4653. 84ff. Incomplete.
Anup 4654. 24ff. Incomplete.
AS Bengal 7034 (G 8705). 36ff. Incomplete (ends with grahadṛṣṭiphalādhyāya).
AS Bengal 7035 (G 7770). 66ff. Incomplete (ends with dīptādigrahaphalādhyāya).
AS Bengal 7122 (G 7925) III. 7ff. Incomplete (nirṇayādhyāya).
Baroda 7650. 122ff.
Baroda 9077. 32ff. Incomplete.
Benares (1963) 34610. 17ff. Incomplete.
Benares (1963) 34770. 10ff. Incomplete (ayanādipañcāṅgaphala). No author mentioned.
Benares (1963) 35032. 41ff. Incomplete.
Benares (1963) 35042. Ff. 6–8, 42–51, and 53–67. Incomplete.
Benares (1963) 35179. 11ff. Incomplete.
Benares (1963) 35211. 28ff. Incomplete.
Benares (1963) 35363 = Benares (1897–1901) 362. 60ff.
Benares (1963) 35732 = Benares (1916–1917) 2713. Ff. 9–19, 32–61, 65–82, 85–97, 99–125, and 129–172. Incomplete.
Benares (1963) 35733 = Benares (1916–1917) 2714. Ff. 1–100 and 102–103. Incomplete.
Benares (1963) 36146. 28ff. Incomplete.
Benares (1963) 36327. Ff. 1–13 and 16–19, ff. 1–4, and 2ff. Incomplete.
Benares (1963) 36328. 10ff. Incomplete (ends with lagnaphala).
Benares (1963) 36357. Ff. 1–111 and 114–115. Incomplete.
Benares (1963) 36358. 48ff. Incomplete.
Benares (1963) 36359 = Benares (1878) 29 = Benares (1869) VII 1. 20ff. Incomplete.
Benares (1963) 36360. Ff. 1–8 and 16–19. Incomplete. No author mentioned.
Benares (1963) 36361. Ff. 1–22, 22b–27, and 27b–28. Incomplete.
Benares (1963) 36789. Ff. 1 and 1–12 and 2ff. Incomplete (dvādaśabhāvanirūpaṇa). No author mentioned.
Benares (1963) 37027. Ff. 1–12 and 7–125.
Benares (1963) 37121. 44ff. Incomplete.
Benares (1963) 37281. Ff. 2–40, 43–44, 44b, 44c–46, and 46b–82, and 1f. Incomplete.
BORI 517 of 1895/1902. 152ff.
BORI 312 of Vishrambag 1. 95ff.
Calcutta Sanskrit College 22. 105ff.
Calcutta Sanskrit College 23. Ff. 7–124.
Cambridge University 246 = Cambridge University Add. 2497. No author mentioned.
CP, Hiralal 1780. Property of Rāmnāth of Jubbulpore.
CP, Hiralal 1781. Property of Govindprasād Śāstrī of Jubbulpore.
CP, Hiralal 1782. Property of Lakṣmīprasād of Jubbulpore.
CP, Hiralal 1783. Property of Madanmohan of Gubrākalā, Jubbulpore.
CP, Hiralal 1784. Property of Murlīdhar of Gubrākalā, Jubbulpore.
CP, Hiralal 1785. Property of Govind Joshī of Jubbulpore.
CP, Hiralal 1786. Property of Śrīdev Dīkshit of Maṇḍlā.
CP, Hiralal 1787. Property of Mādhavrāv of Damoh.
CP, Hiralal 1788. Property of Kanhaiyālāl Guru of Saugor.
CP, Hiralal 1789. Property of the Bhonsalā Rājas of Nagpur.
CP, Hiralal 1790. Property of Tukārām Govind Pāṭhak of Yeodā, Amraotī.
CP, Hiralal 1791. Property of Śivrām of Hoshangābād.
CP, Hiralal 1792. Property of Ajodhyāprasād Brāhman of Seonī.
CP, Hiralal 1793. Property of Vāsudevrāv Golvalkar of Maṇḍlā.
CP, Hiralal 1794. Property of Dālchand Brāhman of Singhansarā, Bilāspur.
CP, Hiralal 1795 and 1796. Property of Chaṇḍīdatt Śāstrī of Menḍhrā, Bilāspur.
CP, Hiralal 1797. Ascribed to Varāhamihira. Property of Rāmkṛṣṇa Pāṇḍe of Haṭṭā, Damoh.
CP, Hiralal 2635. Incomplete (nirṇayādhyāya). Property of Jagmatibāi of Uḍatum, Bilāspur.
CP, Kielhorn XXIII 34. 62ff. Property of Javāhara Śāstrī of Chāndā.
DC (Gorhe) App. 133. Property of Śaṅkara Bālakṛṣṇa Lumpāṭhakī of Puṇatāmbe, Ahmadnagar.
GJRI 927/39. Ff. 1–18, 42–90, and 96–105. Incomplete.
GJRI 929/41. Ff. 1–43, 60–112, and 131–192. Incomplete.
GJRI 930/42. 121ff.

GJRI 3129/341. 65ff. Maithilī. Incomplete.
GJRI 3130/342. 95ff. Incomplete.
GOML Madras D 19374. 34pp. Nandināgarī. Incomplete (lagnaphala to nirṇayādhyāya).
GOML Madras R 1933. 77ff. Telugu. Presented in 1915/16 by Chembrol Rāmasvāmisiddhānti of Vallūr, Godāvarī.
GOML Madras R 4297(a). Ff. 2–99.
GVS 2806 (4299). 18ff. Incomplete (bhāvavicāra).
GVS—(3019). Ff. 3–7, 10–48, 71–86, 91, and 129–183. Incomplete. No author mentioned.
GVS—(4166). No. ff. given.
GVS—(4167). Ff. 1–3, 5–18, and 20–28. No author mentioned.
GVS—(4173). Ff. 1–2, 8–10, and 12–15.
GVS—(4184). Ff. 1–6, 8, 10–14, and 17–19.
IO 3075 (2356). 83ff. Copied by Śrīpati of the Vyāsavaṃśa. From Gaikawar.
IO 3076 (998). 74ff. From H. T. Colebrooke.
IO 3077 (2839). 26ff. Incomplete. From Colin Mackenzie. Probably identical with Mackenzie 18.
IO 6411 (Mackenzie II 41). 130ff. From Colin Mackenzie.
Jaipur (II). 96ff.
Jaipur (II). 4ff. No author mentioned.
Jammu and Kashmir 1190. 18ff. Incomplete.
Jammu and Kashmir 2782. 106ff. Incomplete.
Jammu and Kashmir 2828. 93ff.
Jammu and Kashmir 2936. 134ff. Incomplete.
Jammu and Kashmir 3048. 4ff. Incomplete.
Kathmandu (1960) 113 (I 1167). 44ff. Nevārī. Incomplete.
Kathmandu (1960) 114 (III 261). 33ff. Incomplete.
Kathmandu (1960) 115 (I 1203). 4ff. Nevārī. Incomplete.
Kathmandu (1960) 116 (I 1199). 12ff. Nevārī. Incomplete.
Kathmandu (1960) 117 (I 1195). 40ff. Nevārī.
Kathmandu (1960) 118 (III 331). No ff. given. Incomplete.
Kotah 198. 77pp. No author mentioned.
Kurukṣetra 336 (50632).
LDI 6822 (6667/1). Ff. 10–11, 16–20, and 25–74. Copied by Ratnalābha, the pupil of Pandit Mativardhana, the pupil of Śivalābha Gaṇi, at Bhujanagara. Incomplete.
LDI 6824 (3945). Ff. 9–50 and 53. Copied by Lihā Raṅgiladāsa. Incomplete.
LDI (DJSC) 174. 20ff. No author mentioned.
LDI (LDC) 1365. 74ff.
LDI (LDC) 1757. 17ff.
LDI (LDC) 3329/122. Ff. 274–330.
LDI (LDC) 4692. 13ff.
LDI (LDC) 5625. 50ff.
Leipzig 1029. 76ff. Incomplete (ends with nirṇayādhyāya).
Lucknow 520. J. 35 (4503). No author mentioned.
Madras BE 1547 = Madras BE (Iyer) 299 (1547). Ff. 1–63 and 83–103. No author mentioned.
Mysore (1922) 1110. 57ff. No author mentioned.
Nagpur 720 (62). 87ff. From Khamgaon.
Nagpur 721 (1104). 126ff. From Nasik.
Nagpur 723 (1573). 24ff. From Nasik.
Nagpur 724 (1580). Ff. 2–29. From Nasik.
Nagpur 726 (2631). 91ff. From Nagpur.
N-W P I (1874) 109. 192ff. Property of Jagannātha Jotiṣi of Benares.
N-W P II (1877) A 4. 192ff. Property of Chandra Dhara of Benares.
Oppert II 8218. Property of T. Rāmarow of Tanjore.
Osmania University B. IV/45. 21ff. Telugu. Incomplete.
Osmania University B. IV/48. 48ff. Telugu. Incomplete.
Oudh XX (1888) VIII 41. 240pp. Property of Paṇḍita Pratāpa Nārāyaṇa of Allahabad Zila.
Oudh XX (1888) VIII 120. 166pp. Property of Paṇḍita Pratāpa Nārāyaṇa of Allahabad Zila.
Oudh XX (1888) VIII 121. 32pp. (*Jātakasāra*). Property of Paṇḍita Pratāpa Nārāyaṇa of Allahabad Zila.
Oudh XXII (1890) VIII 29. 160pp. Property of Kedāranātha of Āgrā Zila.
Oxford CS d. 770(v). No author mentioned.
PL, Buhler IV E 14. No ff. given. Incomplete (ṛṇabhaṅgādhyāya). Property of Morārajī of Vaḍhavāṇa.
PL, Buhler IV E 411. No ff. given. Incomplete (rājayogādhyāya). Property of Morārajī of Vaḍhavāṇa.
PL, Buhler IV E 466. No ff. given. Incomplete (śiṣṭādhyāya; read riṣṭādhyāya). Property of Morārajī of Vaḍhavāṇa.
Poleman 4978 (U Penn 710). 140ff. Copied by Keśabhaṭṭa.
Poleman 4980 (Harvard 624). Ff. 1–102, 104–110, and 1f. Incomplete.
Poleman 5118 (U Penn 688). 9ff. Incomplete (rāśisthagrahaphala to mūlādijananaphala).
PrSB 968 (Göttingen Mu II 14(A)). Ff. 7–23. Śāradā. Incomplete.
PUL II 3428. 169ff.
PUL II 3429. 46ff. Incomplete (ends with strījātakādhyāya).
PUL II 3430. 27ff. (ff. 9–20 missing). Incomplete (ends with bhāvaphala).
RORI Cat. I 2945. 64ff. Incomplete.
RORI Cat. II 6429(1). Ff. 2–5 and 7–11. With an artha in Old Rājasthānī. Incomplete.
RORI Cat. II 8161. 4ff. Incomplete.
RORI Cat. II 9210. 113ff. Incomplete.
RORI Cat. III 10546. 36ff. Incomplete (to vs. 631).
RORI Cat. III 15337. 65ff. (ff. 55 and 58–59 missing). Incomplete.
RORI Cat. III 15620. 93ff. Incomplete.

RORI Cat. III 15783. 173ff. (ff. 1–51 and 158 missing). Incomplete.
RORI Cat. III 18209. 6ff. Incomplete.
SOI 4408.
SOI 5037.
SOI 6942 = SOI (List) 607.
SOI 9506.
SOI 10033.
SOI 11088.
Tanjore D 11399 = Tanjore BL 4203. 24ff. Incomplete (ends with antardaśādhyāya).
VVRI 1615. 50ff. Incomplete.
VVRI 2432. 79ff. Incomplete.
VVRI 4771. 21ff. Incomplete.
WHMRL F. 13.
WHMRL F. 14.
WHMRL G. 76. b.
WHMRL G. 110. a.
WHMRL M. 12. e.
WHMRL N. 191. a.
WHMRL O. 9.
WHMRL Q. 23. h.

There are numerous editions of the *Jātakābharaṇa:*
at Muṃbaī in Śaka 1783 = A.D. 1861 (BM and IO 9. F. 27);
at Delhi in Saṃ. 1932 = A.D. 1875 (BM and IO 411);
[NP] in [1876] (IO 12. K. 22);
at Lakhanau in 1879 (BM); 3rd ed., Lucknow 1889 (IO 385);
at Poona in 1883 (IO 920);
ed. Ramaṇīmohana Caṭṭopādhyāya, Calcutta, B.S. 1292 = A.D. 1884 (IO 395), probably identical with the edition of Rasikamohana Caṭṭopādhyāya, Calcutta 1885 (NL Calcutta 180. Kb. 88. 11(2));
with a Bengālī anuvāda in *Śāstrapracāra* 2, Calcutta B.S. 1294 = A.D. 1886/7 (IO 26. G. 20 and NL Calcutta 180. Kb. 88. 12);
at Bombay in Saṃ. 1945 = A.D. 1888 (IO 1493);
with the Hindī ṭīkā of Sūryanārāyaṇa Siddhāntin, Lakhanaū 1900 (BM 14053. ccc. 32);
with the Hindī ṭīkā of Vanamālī Caturvedī, Bombay Saṃ. 1960 = A.D. 1903 (IO 21. G. 3);
with the Hindī ṭīkā, *Śyāmasundarī,* of Paṇḍita Śyāmalāla, Bombay Saṃ. 1962 = A.D. 1905 (IO 18. H. 20);
with the Siṃhalese translation of Mattaka Khemānanda, Colombo 1913 (BM 14055. d. 8. (2));
with the Marāṭhī bhāṣāntara of Mahādeva Bhāskara Goḍabole, Poona 1918 (IO San. D. 131);
ed. Viṃjamāri Tātācārya, in Telugu script, Pemtapāḍu 1929 (IO San. D. 1215(b));
with the Hindī ṭīkā, *Vimalā,* and a pariśiṣṭa by Acyutānanda Jhā, *HSS* 212, Banārasa 1951.

Verses 2–3 are:

udāradhīmandarabhūdhareṇa
pramathya horāgamasindhurājam/
śrīḍhuṇḍhirājaḥ kurute kilārṣyam
āryāsaparyāmalakoktiratnaiḥ//
jñānarājagurupādapañkajaṃ
mānase khalu vicintya bhaktitaḥ/
jātakābharaṇanāma jātakaṃ
jātakajñasukhadaṃ vidhīyate//

The last two verses are:

godāvarītīravirājamānaṃ
pārthābhidhānaṃ puṭabhedanaṃ yat/
sadgolavidyāmalakīrtibhājāṃ
matpūrvajānāṃ vasatisthale yat//
tatraiva daivajñanṛsiṃhasūnur
gajānanārādhanajābhimānaḥ/
śrīḍhuṇḍhirājo racayāṃ babhūva
horāgame ᵓnukramam ādareṇa//

ḌHUṆḌHIRĀJA (*fl.* 1589)

The grandson of Keśava of the Atrigotra, Ḍhuṇḍhirāja wrote a *Grahamaṇi* in Śaka 1511 = A.D. 1589. Manuscript:

AS Bengal 6848 (G 7899) I. 2ff.

The last verse is:

śrīmadatrikulasaṃbhavo dvijaḥ
keśavātmajasutaḥ subuddhimān/
ḍhuṇḍhirāja iti nāmadhārakaḥ
prasphuṭaṃ grahamaṇiṃ cakāra ha//

ḌHUṆḌHIRĀJA (*fl.* 1590)

The son of Rāma, Ḍhuṇḍhirāja wrote a ṭīkā, *Makarandapañcāṅgopapatti,* on the *Makaranda* of Makaranda (*fl.* 1478) in Śaka 1512, Kali 4691 = A.D. 1590. Manuscripts:

SOI 3480 = SOI Cat. II: 1038–3480. Ff. 2–10. Copied in Śaka 1579 = A.D. 1657.
Mithila 247 A. 5ff. Maithilī. Copied in Śaka 1759 = A.D. 1837. Property of Pandit Anantalāl Jhā of Nanaur, Tamuria, Darbhanga.
SOI 3358 = SOI Cat. II: 1062–3358. 8ff. Copied in Saṃ. 1904 = A.D. 1847.
Mithila 247. 3ff. Maithilī. Property of Pandit Bachchā Jhā of Hanuman Nagar, Lohat, Darbhanga.
PUL II 3775. 9ff.

The first verse is:

gaṇeśaṃ giriṃ tātarāmaṃ praṇamya
sudhī ḍhuṇḍhirājas tadāptaprabodhaḥ/
sphuṭīvāsanāṃ mākarandāñkavṛnde
sutithyādipatropajīvye vadāmi//

The colophon begins: iti rāmadaivajñasutaḍhuṇḍhirājaviracitā.

Another part of this may be the *Grahaphalopapatti* of Ḍhuṇḍhirāja.

PAṆḌITA TATTVASUNDARA

Alleged author of a vivṛti on the *Mahādevī* of Mahādeva (*fl.* 1316). Manuscript:

AS Bengal Jaina 6698.

TAMMAṆA

The son of Aśvattha Upādhyāya, Tammaṇa wrote a ṭīkā, *Dīpāvalī*, on the *Vārṣikatantra* of Viddaṇa (or Viddhana). Manuscript:

Anup 5134. 84ff. Incomplete.

TAMMAYAJVĀN

The son of Veṅkaṭāmbā and Malla Yajvān, the son of Malla Yajvān of Śākinīpattana, the son of Honnārya, Tammayajvān or Tammayārya of Paragīpurī wrote a ṭīkā, *Kāmadogdhrī*, on the *Sūryasiddhānta*. Manuscripts:

Baroda 13476. 112ff. Telugu. Copied in Śaka 1740 = A.D. 1818.
GOML Madras R 3906. Ff. 2–214. Grantha. Copied by Gaṅgaya on Thursday 26 Mithuna in Virodhikṛt = July 1851. Purchased from Narasiṃha Śāstrigal of Bhavani, Coimbatore, in 1921/22.
IO 6278 (Burnell 109). 188ff. Grantha. Copied by Subbaya, the pupil of Kṛṣṇa Sūrīndra, from a manuscript belonging to Śāmāvarya of Kumbhaghoṇa on Wednesday 14 kṛṣṇapakṣa of Jyeṣṭha in Kali 4953, Śaka 1774 = 16 June 1852. From A. C. Burnell.
GOML Madras R 5418. 64ff. Copied in A.D. 1928/29 from GOML Madras R 3906.
GOML Madras R 5499. 221ff. Copied in A.D. 1929/30 from GOML Madras R 3906. Incomplete (adhyāyas 3–14).
Baroda 13370. 193ff. Nandināgarī. Incomplete (ends in vaidhṛtādhyāya).
Baroda 13379(b). Ff. 73–168. Nandināgarī.
CP, Hiralal 1531. Incomplete (grahaṇadvayādhikāra). Property of the Bhonsalā Rājās of Nāgpur.
GOML Madras R 6303. 80ff. Telugu. Incomplete (ends in tripraśnādhikāra). Purchased in 1937/38 from E. Śrīnivāsācāryar of Śrīperumbūdūr, Chingleput.
Hultzsch 2. 1068. 24ff. Grantha. Incomplete. Property of the Temple Library at Tiruviḍaimarudūr.
IO 6279 (Mackenzie VIII. 54) = Mackenzie 6. 50ff. Grantha. Incomplete (madhyādhikāra). From Colin Mackenzie.
IO 6280 (Mackenzie VIII. 51) = Mackenzie 51. 100ff. Grantha. Incomplete (somagrahaṇādhyāya to pātādhyāya). From Colin Mackenzie.
IO 6281 (Mackenzie VIII. 84). 43ff. Grantha. Incomplete (golādhyāya). From Colin Mackenzie.
IO 6282 (Mackenzie VIII. 79b). 72ff. Grantha. Incomplete (sphuṭādhyāya). From Colin Mackenzie.
Mysore (1922) 1799. Ff. 42–47. Incomplete.
Musore (1922) 1801. 293ff.
Mysore (1911 + 1922) 3240. 184ff.
Mysore (1922) 3523. Ff. 102–200.
Mysore (1922) 3524. 145ff.
Mysore (1955) 5267. 204ff. Grantha. Incomplete (ends with mānādhyāya). No author mentioned.
Mysore and Coorg 339. Property of Mahādeva Joyisa of Sringeri.
Mysore and Coorg 348. 6000 granthas. Property of Nārāyaṇa Dīkṣita of Bommarasaiyana Agrahara.
Oppert I 1412. 112pp. Grantha. No author mentioned. Property of Aṇṇāsvāmī Śrauti of Bhavāni, Coimbatore.
Oppert I 1413. 89pp. Grantha (*Kāmadogdhrīṭīkā*). No author mentioned. Property of Aṇṇāsvāmī Śrauti of Bhavāni, Coimbatore.
Oppert I 1789 and 1790. Grantha. No author mentioned. Property of Śivasūri Śāstrī of Bhavāni, Coimbatore.
Oppert II 3489. No author mentioned. Property of Gomaṭham Guñjā Narasiṃhācāryār of Melkoṭa, Mysore.
Oppert II 4515. No author mentioned. Property of the Śaṅkarācāryasvāmimaṭha at Śṛṅgeri, Cikkamogulūr, Mysore.
RAS (Whish) 12(2). 246ff. Grantha.

At the beginning are the verses:

śrīhonnāryaḥ sarvatantrasvatantras
tasmāj jātas tādṛśo mallayajvā/
tajjaḥ khyātaḥ sarvasiddhāntavettā
śākinyākhye pattane mallayajvā//
tatputro ᵓhaṃ vedavedāntavedī
jyotirvidyāpāragas tammayajvā/
sūryaṃ natvā sūryatantrasya ṭīkāṃ
honnambāyai kāmadogdhrīṃ karomi//

At the end are the verses:

ghanādrinikaṭe bhāti paścime paragīpurī/
tasyāṃ śrīhonnamāmbāyāḥ prasādī johniyābhidhaḥ//
vidvān śrīsūryasiddhāntādyaṣṭasiddhāntapāragaḥ/
rapītarakulāmbhodhisudhāṃśusadṛśaḥ prabhuḥ//
ṛgvedī vedavedāntaśāstrajño mantrakovidaḥ/
sarvajñasadṛśaḥ prājño vartate sarvabhogabhāk//
tasyātmajo mallayajvā sarvatantrasvatantrakaḥ/
tatputrau mallayajvā ca somanāthādhvarī hy ubhau//
. . . .
tanmadhye mallayajvākhyaveṅkaṭāmbātanūbhavaḥ/
tammayāryas tarkaśāstre vedānte śabdaśāstrayoḥ//
jyautiṣe vedavedāṅgaśrautasmārtākhileṣu ca/
nipuṇaḥ śrīhonnamāmbāvaravāgvibhavaḥ sudhīḥ//
śrīhonnāryasya pautrāc śivagurusadṛśān mallayajvākhyaputrāj
jāto mallādhvarīndrāt paragipuravarasthāyinas tammayāryaḥ/
siddhāntasyārkanāmnaḥ kalitapadavatīṃ kāmadogdhrīṃ suṭīkāṃ

mānādhyāyasya samyag ravigurukṛpayā proktavān ambikāyai//

TAMMAYĀRYA

Author of a *Grahagaṇitabhāskara*. Manuscripts:

Mysore (1955) 5260. 9ff. Telugu.
Mysore (1955) 6165. 6ff. Grantha.
Mysore (1911 + 1922) B 588. 88ff. With a Karṇāṭaṭīkā.
Oppert II 4572. No author mentioned. Property of the Śaṅkarācāryasvāmimaṭha at Śṛṅgeri, Cikkamogulūr, Mysore.

TARKATILAKA (*fl.* 1613)

The son of Dvārakādāsa of the Dakṣavaṃśa, Tarkatilaka of Mathurā finished a ṭīkā on the *Kālamādhava* or *Kālanirṇaya* of Mādhava (*fl.* 1334/1359) for his older brother, Mohana Madhusūdana, on 2 śuklapakṣa of Mādhava in Saṃ. 1760 = 10 April 1613. Manuscripts:

RORI Cat. II 8460. 99ff. Copied by Bhīmajī Josī of Ṭoḍagāḍha for Harivaṃśa, Mahārāja of Būndī (this must be Aniruddha, whose reign began in 1678) in Saṃ. 1751 = A.D. 1694.
BORI 264 of 1886/92. 58ff. (ff. 1–14 missing).
Mitra, Not. 2842. 65ff. Property of the Gopāla Tīrtha Maṭha in Purī.
RORI Cat. II 9986. 71ff. (ff. 59–64 missing). Incomplete.

Verse 2 is:

māthuravipraḥ śrīmān sudarśanaḥ svasya bodhāya/
śrīmādhavapadacintāṃ mohanamiśropadeśataḥ kurute//

At the end are the 2 verses:

khamunirasendumite ᵓbde mādhavaśukladvitīyāyām/
racitaṃ vivaraṇam etan manīṣinā tarkatilakena//
dvārakādāsaputrasya dvārakānāthasevinaḥ/
dakṣavaṃśaprasūtasya kṛtiḥ pracaratāc ciram//

The colophon begins: iti śrīmohanamadhusūdunānujatarkatilakaracitaṃ.

TARKAVĀCASPATI BHAṬṬĀCĀRYA

Author of a *Jyotiḥsāra*. Manuscripts:

AS Bengal 7085 (G 3467). 52ff. Bengālī. Copied by Lakṣmaṇadeva Śarman.
Mithila 115. 55ff. Bengālī. Property of Pandit Gaṅgādhara Jhā of Jonki, Deodha, Darbhanga.

The colophon begins: iti tarkavācaspatibhaṭṭācāryaviracito.

TĀJAKĀCĀRYA or TĀJIKĀCĀRYA

Title of authorities on Tājika (Persian) astrology. Ascribed to such an author are the following works.

1. *Praśnasaṅgraha*. Manuscripts:

Śāstrī, Not. 1904. 134. 46ff. Bengālī. Copied in Śaka 1713 = A.D. 1791. Property of Paṇḍita Kāśīnātha Tarkālaṅkāra of Ākiyādhala, Lohajaṅga, Ḍhākā.
PUL II 3522. 12ff. Bengālī. Copied in Śaka 1721 = A.D. 1799. (*Tājikapraśnakaraṇa*).

The first verse is:

bhāsvantaṃ jagadādhānaṃ praṇamya viduṣāṃ mude/
kurute tājakācāryaḥ praśnānāṃ sārasaṅgraham//

2. *Bhuvanadīpaka;* the author's name, Tājakācārya, may be an error for Tilaka Sūri (*fl.* 1269). Manuscript:

Poleman 4983 (Columbia, Smith Indic 134). Ff. 1–23 and 23b–25. Copied in Saṃ. 1877 = A.D. 1820.

TĀṆḌAVA KAVIRĀJA

Author of a vivaraṇa on a *Mitāṅka*, presumably that of Viśvanātha (*fl.* 1612/1630). Manuscript:

Alwar 1895.

TĀTA MIŚRA

Author of a *Jyotiḥprabodha*. Manuscript:

DC 192. Ff. 17–29.

TĀRAKA (*fl. ca.* 590)

Astrologer consulted at the birth of Harṣa (*ca.* 606/648) who was born, probably at Sthāṇvīśvara, on 12 kṛṣṇapakṣa of Jyeṣṭha in *ca.* 590 to Yaśovatī and Mahārājādhirāja Prabhākaravardhana. See Bāṇa, *Harṣacarita*, ed. A. A. Führer, Bombay 1909, p. 184; ed. S. K. Piḷḷai, *TSS* 187, Trivandrum 1958, pp. 186–187.

TĀHIRA (= *AHAMADA*?) (*fl.* 1598/1621)

The pupil of Ahamada (= Aḥmad) and a resident of Āgarā, Tāhira (= Ẓāhir) wrote a *Sāmudrika* in Hindī in Saṃ. 1678 = A.D. 1621. Manuscript:

NPS 2 of 1917–19. Copied in Saṃ. 1904 = A.D. 1847. Property of Paṇḍita Dayāśaṅkara Pāṭhaka of Maṇḍī Rāmadāsa, Mathurā.

TIPPA (*fl.* 1507)

The son of Gauraṇa, Tippa wrote an *Uparāgadarpaṇa* in which are discussed the eclipses in each of 17 cycles of 60 years from Śaka 1429 = A.D. 1507 to Śaka 2449 = A.D. 2527. Manuscripts:

GOML Madras R 2136(a). Ff. 1–28. Grantha. Copied in 1916/17 from a manuscript belonging to Jayanti Jogannagāru of Haṃsavaram, Tuni, Godāvarī, that was copied by Jayanti Pāpayya on 14 śuklapakṣa of Āṣāḍha in Siddhārtin = 13 July 1859 from a manuscript belonging to Śiṅgarāya Koṇḍayyagāri. Incomplete (begins with Kālayukti of cycle 6 = A.D. 1918).
Lucknow 520. T 532 U (45753).
Oudh VIII (1876) VIII 4. 14pp. Ascribed to Teppaḍa. Property of Rāmanidhi Miśra of Ghāzīpur, Bārābāṅki Zillah.

The second half of the last verse in cycle 6 is:

tippājyotirvidaivaṃ tribhuvanamahite kalpite ᵓnalpabhāsā
ṣaṭṣaṣṭiḥ sūrihṛdyo ᵓgamad ayam uparāgādime darpaṇe ᵓsmin//

Near the end of the last cycle is the verse:

jyotirāgamadhureṇa cetasā
tippanābhidhabudhena kalpite/
atra saptadaśaṣaṣṭirūpite
soparāgamaṇidarpaṇe yayau//

TIMMAṆA

Author of a *Tithinirṇaya*. Manuscripts:

Tanjore D 18588 = Tanjore BL 191a. 5ff.
Tanjore D 18589 = Tanjore BL 192a. 5ff.
Tanjore D 18590 = Tanjore BL 192b. 3ff.

VIṬLAṂPALLI TIMMAṆA ŚĀSTRIN (*fl.* 1910/1917)

Author of a *Tithibhūṣaṇasaṅgraha* with an Āndhra tātparya, published at Viṭlampalli in 1910 (IO 8. K. 28), and of a *Praśnamanoramā* with an Āndhra tātparya, published at Bellary in 1917 (IO San. A. 31(i)).

VELLĀLA TIMMAYYA

Author of a *Pañcāṅgaśiromaṇi*. Manuscripts:

Mysore (1911 + 1922) 2559. Ff. 184–187.
Mysore (1911 + 1922) 2575. 79ff. With a Karṇāṭaṭīkā.

PĀPA TIMMAYYA

Author of a *Lakṣmīnṛsiṃhīyagaṇita*. Manuscripts:

Mysore (1911 + 1922) 2568. 11ff.
Mysore (1911 + 1922) 2588. Ff. 123–134.

TIMMARĀYA

Author of a vyākhyā in Karṇāṭakī on the *Divākarapaddhati* of Divākara (b. 1606). Manuscript:

Mysore (1911 + 1922) 2336. Ff. 33–76.

TIRUKOṬṬINAMBI

The son of Āriyan of Caturvedamaṅgalam, Tirukoṭṭinambi wrote a *Girahaspuṭam* in Tamil. Manuscripts:

GOML Madras (Tamil) R 340. 72ff. Tamil. Restored in 1916/17 from GOML Madras (Tamil) D 2000. With a ṭīkā.
GOML Madras (Tamil) D 2000 = Sastri, Rep. (1896–97) 303. Ff. 152–194. Tamil.

TILAKA SŪRI (*fl.* 1269)

See Siṃhatilaka Sūri (*fl.* 1269).

TĪRTHARĀJA (*fl.* 1750)

A Śākadvīpī Brāhmaṇa and a protégé of Acalasiṃha, the rājā of Alīpura, Bundelakhaṇḍa, Tīrtharāja wrote a *Samarasāra* = *Samayavijaya* in Hindī in Saṃ. 1807 = A.D. 1750. Manuscripts:

NPS 481 A of 1926–28. Copied in Saṃ. 1829 = A.D. 1772. Property of Paṇḍita Avadhabihārī Miśra Pujārī of Kālākāṃkara, Pratāpagaḍha.
NPS 428 of 1923–25. Copied in Saṃ. 1880 = A.D. 1823. Property of Paṇḍita Durgāprasāda of Jū. Jigāniyāṃ, Hajūrapura, Baharāïca.
NPS 481 B of 1926–28. Copied in Saṃ. 1901 = A.D. 1844. Property of Paṇḍita Kālikāprasāda Dūbe of Gauraiyā Rasūlapura, Miśrikha, Sītāpura.
NPS 481 C of 1926–28. Copied in Saṃ. 1909 = A.D. 1852. Property of Ṭhākura Candrikābakhśasiṃha of Khānīpura, Tālābabakhśī, Lakhanaū.
NPS 115 of 1906–08. Copied in Saṃ. 1915 = A.D. 1858. Property of Kāmatāprasāda Dārogā of Ajayagaḍha.
NPS 481 D of 1926–28. Copied in Saṃ. 1932 = A.D. 1875. Property of Ṭhākura Hulāsasiṃha, the jamīndāra of Saṇḍīlā, Macharahaṭṭā, Sītāpura.
NPS 194 A of 1920–22. Property of Paṇḍita Choṭelāla Pahalavāna of Khajuhā, Phatehapura.
NPS 194 B of 1920–22. Property of the Balarāmapuranareśa kā Pustakālaya in Balarāmapura.
NPS 89 of the Dillī Khoja Vivaraṇa, 1931. Property of Paṇḍita Pyārelāla Śarmā of Śāhadarā, Dillī.

TULAJARĀJA (*fl.* 1728/1736)

The youngest son of Dīpāmbā and Ekojī or Vyaṅkajī (Mahārāja 1675/1684), the son of Tukkabāi and Śāhajī Bhonsle (d. 23 January 1664), the son of Mālojī (d. 1619), a noble in the service of the Niẓāmshāhs of Ahmadnagar, Tulajarāja or Tukkoji was the Mahārāja of Tanjore from 1728 to 1736. A learned Sanskrit scholar, he wrote, besides other works, the following:

1. *Inakularājatejonidhi* on gaṇita, jātaka, and saṃhitā. The gaṇita section is in 12 adhyāyas:

1. madhyamagraha.
2. sphuṭa.
3. pāta.
4. upakaraṇa.
5. candragrahaṇa.
6. sūryagrahaṇa.
7. chedyaka.
8. śṛṅgonnati.
9. samāgama.
10. grahayoga.
11. udayāsta.
12. gola.

Manuscripts:

Tanjore D 11323 = Tanjore BL 4263 and 4267. (34ff.) and 95ff. (gaṇita).
Tanjore D 11324 = Tanjore BL 4230. 46ff. Incomplete (jātaka).
Tanjore D 11325 = Tanjore 15395c. Telugu. Incomplete.
Tanjore D 11326 = Tanjore BL 12354. 99ff. Incomplete. (saṃhitā).

Verse 6 at the beginning is:

āsīn mālojirājo dinakarakularāṭ tatsutaḥ śāharājaḥ
putras tasyaikarājaḥ sakalaguṇanidhir
bhosalāmbhodhicandraḥ/
dīpāmbā tasya bhāryā tribhuvanaviditās tatkumārās
trayo ᵓmī
śāhendraśrīśarīphāvanipatitulajakṣoṇapālāvataṃsaḥ//

A somewhat later verse names the amātya, Śivarāya, who may have been responsible for the compilation of this work:

yasyāmātyaḥ śrutīnāṃ smṛtinivahapurāṇetihāsāgamānāṃ
kāvyālaṅkāraśāstroragapativacasāṃ pārago
nītidakṣaḥ/
sarvajñaḥ satyasandho vitaraṇanipuṇaḥ
sarvalokopakartā
brahmajñaḥ kīrtiśālī vilasati śivarāyādhvarīndrāgragaṇyaḥ//

2. *Vākyāmṛta*. Manuscript:

Tanjore D 11327 = Tanjore BL 4628. 71ff. Incomplete.

Verses 10–11 are:

śrīmacchāhajibhūpasūnur avanāv ekaḥ kilaikojirāṭ
dīpāmbāmahiṣīmahīparivṛḍhaḥ sāmrājyasiṃhāsane/
kākutsthā iva tatsutā api tataḥ
saubhrātrasampadyutāḥ
śrīśāhendramukhās trayaḥ samabhavan
kṣmārakṣaṇe tatparāḥ//

śrīmacchāhajibhūmipālaśarabh⟨ojī⟩śakṣamāmaṇḍalādhīśaśrītulajādhipān ajanayad ratnapradīpān yataḥ/
lokasyāsya tamovirodhiviśadaprauḍhaprakāśān ato
dīpāmbety agamat prasiddhim atulām
ekojirājapriyā//

TULAJĀRĀMA ŚARMAN (*fl.* 1923)

Author of a *Praśnadīpikā*, published with a *Praśnasaṅgraha* and a *Śakunāvali* at Surat in Saṃ. 1980 = A.D. 1923 (IO San. B. 519(a)).

TULASĪ SĀDHU

Author of a *Tithiśodaśikā* (*Tithiṣoḍaśikā*?) in Hindī. Manuscript:

LDI (SCC) Sag. 237/10. Ff. 13–14. Copied in Saṃ. 1758 = A.D. 1701.

TULASĪDĀSA

Alleged author of a *Dhruvapraśnāvalī* in Hindī. Manuscript:

NPS 323 N of 1909–11. Property of Paṇḍita Gaṇeśadatta Miśra, Dvitīya Adhyāpaka of the English Branch School in Goṇḍā.

TULASĪDĀSA (1532/1623)

The great Hindī poet, whose verses were used for the purposes of divination (see G. A. Grierson [1926]), is alleged to have written a work (or works) on divination called *Rāmaśakunāvalī* or *Rāmājñājyotiṣa*. Manuscripts:

SOI 3835 = SOI Cat. II: 1100–3835. 24ff. Copied in Saṃ. 1909 = A.D. 1852. (*Rāmājñājyotiṣa*)
Kurukṣetra 915 (50694). (*Rāmājñāśakunāvalī*)
SOI 3280 = SOI Cat. II: 1099–3280. 20ff. (*Rāmājñājyotiṣa*).
SOI 3734 = SOI Cat. II: 1098–3734. 16ff. (*Rāmaśakunāvalī*).

Tulasīdāsa also wrote a *Ratnasāgarajyotiṣa* or *Bṛhaspatikāṇḍa* in Hindī in Saṃ. 1606 = A.D. 1549. Manuscripts:

NPS 143gha of Saṃ. 2004–2006. Copied in Saṃ. 1936 = A.D. 1879. Property of Ṭhākura Rāmakiśunasiṃha of Surerī, Mārikapura, Jaunapura.
NPS 30 of 1903. Property of the Mahārāja Banārasa kā Pustakālaya at Rāmanagara, Vārāṇasī.
NPS 142ga of Saṃ. 2001–2003. Property of Viśvanātha Dūbe of Rekavāreḍīha, Maū Ājamagaḍha.

TŪPHĀNĪ ŚARMAN (*fl.* 1873)

Also known as Vighneśvara, Tūphānī Śarman, a Maithila paṇḍita, finished a compilation known as the *Kṛtitattvasaṅgraha* in Śaka 1795 = A.D. 1873 at Sumohanā in Tirabhukta. This was edited by Rāmacandra Jhā as *KSS* 181, Vārāṇasī 1967.

TEJAḤSIṂHA

The head of the Luṅkāgaccha, Tejaḥsiṃha Ṛṣi Lumpaka wrote a *Pañcaviṃśatikā* on mathematics. Manuscripts:

LDI 7328 (3665/5). Ff. 12–14v. Copied in Saṃ. 1870 = A.D. 1813 (*Gaṇitapañcaviṃśatikā*). With a stabaka in Old Gujarātī.
LDI 7327 (903) 3ff. (*Iṣṭāṅkapañcaviṃśatikā*).

The colophon begins: iti śrīluṅkāgacchādhirājaśrīpūjyaśrī 6 tejasiṃghajīkṛtā.

TEJAḤSIṂHA (*fl.* 1336)

The son of Vikrama of the Prāgvāṭavaṃśa, a minister of the Cālukya monarch Śāraṅgadeva (*ca.* 1276/1296), and the brother of Vijayasiṃha, Tejaḥsiṃha composed a *Daivajñālaṅkṛti* in Saṃ. 1393 = A.D. 1336. Manuscripts:

Kerala 7758 (1728). 500 granthas. Copied in Saṃ. 1582 = A.D. 1525.
PL, Buhler IV E 198. 25ff. Copied in Saṃ. 1618 = A.D. 1561. Property of Motilāla Vaidya of Ahmadābād.
AS Bengal 7131 (G 8406). 22ff. Copied by Prayāgamiśra on Tuesday 9 kṛṣṇapakṣa of Phālguna in Saṃ. 1636 = 8 March 1580.
BORI 327 of 1882/83. 21ff. Copied in Saṃ. 1803 = A.D. 1746. From Gujarāt.
Benares (1963) 34867. 22ff.
Benares (1963) 36135 = Benares (1913–1914) 2359. 22ff. Incomplete. No author mentioned.

Verses 26–30 at the end are:

lakṣmīr yasya pratene svayam acalam ihācandratāraṃ niveśaṃ
yasmin muktāḥ phalanti praguṇataragaṇā doṣapaṅktyā vimuktāḥ/
yasmin viśrāmabhājaḥ paramapṛthutaraśreṇayaḥ sajjanānāṃ
so °yaṃ prāgvāṭavaṃśo jagati vijayate °nalpaśākhāviśālī//
sphūrjaccālukyavaṃśodbhavanṛpatiśirobhūṣaṇībhūtakīrteḥ
śrīmacchāraṅgadevāhvayapuruṣapateḥ pādapadmaprasādāt/
sarvavyāpārapāraṃ samakham apagataḥ sadguṇaughaikapātraṃ
tatra śrīvikramāhvo °jani vijitaripur mantriṇaḥ satyamitram//
mantrī tasmād athādau dhṛtavijayapadāṃ siṃhasaṃjñāṃ dadhāno
jajñe vidvajjanānāṃ hṛdayakumudam uddāyi vākcandrikābhūt/
sāhityanyāyavādapramukhapariṇamatsarvaśāstrābdhipāraṃ
prāptaḥ saukhyaikapātraṃ vinayanayamukhaiḥ sadguṇair gītakīrtiḥ//
tasyoccair mānyabandhus tanujanir ajani khātam ādau ca tejaḥ
prānte siṃheti nāma pradadhad avanataḥ sarvadā sadgurūṇām/
kiṃcillabdhaprabodhaḥ pṛthumatividuṣāṃ pādapadmaprasādāt
snehaukaḥ sajjanānāṃ vinayanayayuto lokadurvākyabhīruḥ//
daivajñālaṅkṛtīti prathitam avitathaṃ saṃjñayā saṃpratene
tenedaṃ vatsarīyaṃ phalam iha sakalaṃ sarvalokopakṛtyai/
hṛjjīvāntarvibhūṣābharaṇagaṇanayā vibhrate ye grahajñāḥ
śaśvad bhūbhṛtsabhāyāḥ śrutidhanagurutāmaitrabhājo °tra te syuḥ//

The date is given in the next, somewhat corrupt verse:

śrībhūbhṛdvikramasya trinidhiśikhidharāsaṃmite °bde tapasye
māse °jyarkṣe kavau x sitamadanadine °trāgataṃ sadguror yat/
pāraṃparyādhṛte °pi svayamanubhavagranthajātārthasya samyak
pūrṇābdīyaṃ phalaṃ sadgrahagaṇitavidāṃ mantrireṇoḥ prasādam//

TEJANĀTHA

A resident of Sapahāṃ Gāṃva, Tejanātha wrote a *Sāmudrika* in Hindī. Manuscript:

NPS 425 of 1923–25. Copied in Saṃ. 1892 = A.D. 1835. Property of Ṭhākura Maheśasiṃha Kohalī of Becaīsiṃha kā Puravā, Kesaragañja, Baharāīca.

TEJASIṂHA ṬHĀKURA (*fl.* 1873)

Author of a *Jñānacandrodaya* in Hindī in Saṃ. 1930 = A.D. 1873. Manuscript:

NPS 477 of 1926–28. Property of Śivanareśasiṃha of Mallāṃpura, Sītāpura.

TEPPADA (*fl.* 1507)

See Tippa (*fl.* 1507).

TOYANĀTHA ŚARMAN (*fl.* 1930)

Author of a pañcāṅga for Nepal for Saṃ. 1987 = A.D. 1930, published at Benares in 1930 (IO San. F. 190(b)).

CATURAGAṆAPATI TRIPURĀRI

Author of a *Tithicakra*. Manuscripts:

Mysore (1911 + 1922) 2559. Ff. 59–62.
Mysore (1911 + 1922) 2649. Ff. 45–49.

TRIPURĀRI (*fl.* 1627)

Author of a *Pañcāṅgaśiromaṇi* in Śaka 1549 = A.D. 1627. Manuscript:

GOML Madras R 457(d). Ff. 137–207. Telugu. With an Āndhraṭīkā. Incomplete (ends in adhikāra 3). Purchased in 1911/12 from M. Rāmakṛṣṇakavi of Vanaparti.

The date is given in verse 2:

ṣaṣṭir navadaśa⟨guṇi⟩taṃ
vyapagatasaṃvatsareṇa saṃmiśram/
navagaganābdhisametaṃ
śakanṛpakālaṃ vijānīyāt//

The colophon begins: iti tripurāriviracite.

TRIVIKRAMA

The son of Gaṅgādhara of the Kauṇḍinyagotra and a resident of Anindamagrāma, Trivikrama wrote a *Kālavidhānapaddhati* in 156 verses. There is a commentary by Śrīdhara and a Siṃhalese translation. Manuscripts:

Kerala 3468 (6058). 7000 granthas. Malayālam. Copied in ME. 953 = A.D. 1778. With the vyākhyā of Śrīdhara.
Kerala 3470 (C 2152) = Kerala C 668. 160ff. Malayālam. Copied in ME 1017 = A.D. 1842. With the vyākhyā of Śrīdhara. Formerly property of Vaittiyappa Pillai Avl. of Munnār.
GOML Madras R 1594(a). Ff. 1–96. Grantha. Copied by Vīrāsvāmin, the son of Avaḻūr Gomaṭham Periyanārāyaṇa Ayyaṅgar, in Śarvajit = A.D. 1887 (?). With a Tamil ṭīkā. Presented in 1915/16 by E. Śrīnivāsarāghavācariyar of Conjeeveram.
GOML Madras R 4469. 276ff. Grantha. Copied in 1924/25 from a manuscript belonging to Śaṅkarasubbā Śāstrigal of Tiruchendur, Tinnevelly. With the vyākhyā of Śrīdhara. Incomplete.
Adyar List. 3. copies. Grantha = Adyar Index 1251 =
Adyar Cat. 22 I 47. 36ff. Grantha.
Adyar Cat. 22 I 48. 8ff. Telugu. Incomplete.
Adyar Cat. 22 I 49. 16ff. Grantha. Incomplete.
Adyar Cat. 26 B 43. 27ff. Grantha. Incomplete.
Adyar Cat. 27 M 27. 356ff. Grantha. With a Tamil ṭīkā. Incomplete.
Adyar Cat. 28 G 15. 32ff. Grantha. Incomplete.
Adyar Cat. 33 I 8. 80ff. Grantha.
Baroda 6845(b). 14ff. Grantha. Incomplete.
Baroda 13358(b). 21ff. Nandināgarī. Incomplete.
Baroda 13366(a). 23ff. Nandināgarī.
Baroda 13376. Ff. 208–279. Nandināgarī. With the vyākhyā of Śrīdhara. Incomplete.
Baroda 13381(c). Ff. 56–76. Nandināgarī.
Baroda 13406. 200ff. Nandināgarī. With the vyākhyā of Śrīdhara. Incomplete.
Baroda 13422(b). 10ff. Nandināgarī.
Baroda 13506. 10ff. Telugu. Incomplete.
BM 201 (Or. 4763). 36ff. Siṃhalese. With a Siṃhalese translation. Incomplete.
BM Or. 6613(19). Siṃhalese. From the Nevill Collection.
BM Or. 6613(20). 49ff. Siṃhalese. From the Nevill Collection.
BM Or. 6613(47). Siṃhalese. From the Nevill Collection.
GOML Madras D 13543. 20ff. Telugu. With the vyākhyā of Śrīdhara. Incomplete.
GOML Madras D 17768. Ff. 116–134. Telugu. Incomplete.
GOML Madras R 1418. 39ff. Grantha. With the vyākhyā of Śrīdhara. Incomplete. Purchased in 1914/15 from Devanāthācāriyar of Rāmeśvaram.
GOML Madras R 2596(a). Ff. 5–39. Telugu. With a Telugu ṭīkā. Presented in 1917/18 by Vāśireḍḍi Candramaulīśvaraprasāda Bahadur, the Zamindar of Muktyala, Kistna.
GOML Madras R 3984. 70ff. Grantha and Tamil. Incomplete. Presented in 1921/22 by Tiruveṅkaṭattayyaṅgār of Sāmavādhyār, Srirangam, Trichinopoly.
Granthappura 872. With the vyākhyā of Śrīdhara. See NCC, vol. 4, p. 35.
IO 6333 (3533g). 15ff. Nandināgarī.
IO 6334 (Burnell 153). 122ff. Grantha. From A. C. Burnell.
IO 6335 (Mackenzie III. 76). 72ff. Telugu. With an Āndhraṭīkā. Incomplete. From Colin Mackenzie.
IO 6336 (Mackenzie V. 16a). 4ff. Karṇāṭakī. With the vyākhyā of Śrīdhara. Incomplete. From Colin Mackenzie.
Kerala 3462 (2619A). 650 granthas. Grantha. Incomplete.
Kerala 3463 (4032C). 400 granthas. Grantha. Incomplete.
Kerala 3464 (5963A). 750 granthas. Grantha. Incomplete.
Kerala 3465 (8967C). 200 granthas. Malayālam. Incomplete.
Kerala 3466 (12060A). 400 granthas. Grantha. Incomplete.
Kerala 3467 (C. 2520G) = Kerala C. 682G. 31ff. Grantha. Incomplete. Formerly property of Brahma Śrī Kāśi Vādhyār of Mahādānapuram.
Kerala 3469 (L. 410). 7000 granthas. Grantha. With the vyākhyā of Śrīdhara.
Kerala 3471 (T. 858). 7000 granthas. With the vyākhyā of Śrīdhara.
Kerala 3472 (916A). 1800 granthas. Grantha. With the vyākhyā of Śrīdhara. Incomplete.
Kerala 3473 (2348). 2800 granthas. Grantha. With the vyākhyā of Śrīdhara. Incomplete.
Kerala 3474 (C. 2014) = Kerala C. 667. 73ff. Grantha. With the vyākhyā of Śrīdhara. Incomplete. Formerly property of Śakti Śāstrī of Āyikuḍi.

Kerala 3475 (12955). 3300 granthas. Grantha and Tamil. With a Tamil ṭīkā.
Kerala 3476 (3592A). 700 granthas. Grantha and Tamil. With a Tamil ṭīkā. Incomplete.
Kerala 3477 (4443C). 1000 granthas. Grantha and Tamil. With a Tamil ṭīkā. Incomplete.
Kerala 3478 (8833). 900 granthas. Grantha and Tamil. With a Tamil ṭīkā. Incomplete.
Kerala C. 669. 874pp. With a vyākhyā.
Mysore (1922) 66. 20ff.
Mysore (1922) 69. 14ff.
Mysore (1922) 222. Ff. 49–57.
Mysore (1922) 465. Ff. 2–194. With the vyākhyā of Śrīdhara.
Mysore (1922) 1818. 13ff.
Mysore (1911 + 1922) 2541. 222ff. With the vyākhyā of Śrīdhara.
Mysore (1922) 4387. 184ff.
Mysore (1911 + 1922) B 759. 215ff. With the vyākhyā of Śrīdhara.
Mysore and Coorg 271. 2000 granthas. With the vyākhyā of Śrīdhara. Property of Nārāyaṇa Dīkṣita of Bommarasaiyana Agrahara.
Oppert I 39. 300pp. Grantha. Property of Narasiṃhācāryār of Ammaṇapākam, Chingleput.
Oppert I 152. 250pp. Grantha. Property of Varadācāryār of Ammaṇapākam, Chingleput.
Oppert I 1214. Property of Vaṅkīpuram Śrīnivāsācāryār of Tiruvallūr, Chingleput.
Oppert I 3555. Grantha. Property of Narasiṃhapuram Rāghavācāryār of Kumbhaghoṇam, Tanjore.
Oppert I 4800. 200pp. Grantha. Property of Appāvājapeya of Tiruvālaṅgāḍu, Tanjore.
Oppert II 1044. Property of Tirupuṭkuli Śrīkṛṣṇa Tātācāryār of Kāñcīpuram, Chingleput.
Oppert II 1437. Property of Pattaṅgi Raṅgācāryār of Pillapākkam, Conjeveram, Chingleput.
Oppert II 3307. Property of Anantanārāyaṇa Josya and Gurumūrti Josya of Diṇḍukal, Madura.
Oppert II 3490. Property of Gomaṭham Guñjā Narasiṃhācāryār of Melkoṭa, Mysore.
Oppert II 6026. Property of Gurusvāmin Śāstrī of Kumbhaghoṇam, Tanjore.
Oppert II 9711. Property of Nārāyaṇopādhyāya of Vedāraṇyam, Tanjore.
Oppert II 10032. Property of Veṅkaṭācala Aiyar of Maṇakkāl, Trichinopoly.
Osmania University 687/b. 36ff. Telugu. Incomplete.
Tanjore D 11351 = Tanjore BL 11080. 20ff. (f. 13 missing). Grantha.
Tanjore D 11352 = Tanjore BL 11028. 19ff. Grantha.
Tanjore D 11353 = Tanjore BL 11029. 12ff. Telugu. Incomplete.

The *Kālavidhānapaddhati* was published with the Drāviḍaṭīkā of Vedānta Rāmānujācārya at Madras in 1916 (IO 8. K. 16), and at Madras in 1922 (NCC, vol. 4, p. 36); neither edition mentions Trivikrama. Verse 155 is:

anindamagrāmanivāsakuṇḍinaḥ
prasiddhagaṅgādharamādanandanaḥ/
trivikramaḥ kālavidhānapaddhatiṃ
cakāra sāṃvatsarikaprasādataḥ//

TRIVIKRAMA

Alleged author of a ṭīkā on the *Jātakābharaṇa*, presumably of Ḍhuṇḍhirāja (*fl. ca.* 1525). Manuscript:

PL, Buhler IV E 121. 31ff. Copied in Saṃ. 1913 = A.D. 1856. Property of Vajeśaṅkara of Dhrāṅgadhrā.

TRIVIKRAMA

Author of a vṛtti on the *Tājikasāra* of Haribhadra (*fl.* 1523). Manuscripts:

Goṇḍal 150. 112ff. Copied by Bhagavāna Hīrajī at Dhorājī on Sunday 11 śuklapakṣa of Vaiśākha in Saṃ. 1860 = 1 May 1803.
VVRI 5049. 219ff. Copied in Saṃ. 1904 = A.D. 1847.

TRIVIKRAMA

The son of Nārāyaṇa and the brother of Jñānakṛṣṇa or Jñānamalla, Trivikrama wrote a *Trivikramaśataka* or *Brahmavyavahāra;* see S. Dvivedin [1892] 85–86. There are commentaries by Gopīnātha (*Budhavallabhā*) and Hṛṣikeśa. Manuscripts:

Baroda 2496. 14ff. Copied in Saṃ. 1541 = A.D. 1484. With the *Budhavallabhā* of Gopīnātha.
Baroda 712. 28ff. Copied in Saṃ. 1596 = A.D. 1539. With the *Budhavallabhā* of Gopīnātha.
Benares (1963) 34953. 12ff. Copied in Saṃ. 1638 = A.D. 1581.
Benares (1963) 36375. 34ff. Copied in Saṃ. 1643 = A.D. 1586. With the *Budhavallabhā* of Gopīnātha.
Jaipur (II). 7ff. Copied in Saṃ. 1646 = A.D. 1589.
Anup 4747. 5ff. Copied in Saṃ. 1655 = A.D. 1598.
AS Bengal 2654 (G 6394). 32ff. Copied on 3 śuklapakṣa of Vaiśākha in Saṃ. 1690 = 30 April 1633 (?). With the *Budhavallabhā* of Gopīnātha.
Anup 4694. 23ff. Copied by Kāratagavarabhau in Saṃ. 1717, Śaka 1580 = A.D. 1658/60 (?). With a ṭīkā.
VVRI 5067. 8ff. Copied for Bhavānīśarman on Sunday 7 śuklapakṣa of Pauṣa in Saṃ. 1729 = 15 December 1672.
Bombay U 366. Ff. 2–8. Copied by Kutūhala in Bhādrapada of Saṃ. 1733 = 28 August–26 September 1676.
LDI (LDC) 6241. 14ff. Copied in Saṃ. 1758 = A.D. 1701.
BORI 167 of A 1883/84. 8ff. Copied in Saṃ. 1759 = A.D. 1702.

Leningrad (1914) 292 (Ind. II 92). 6ff. Copied on Wednesday 8 kṛṣṇapakṣa of Phālguna in Saṃ. 1785 = 12 March 1729 Julian.
VVRI 1651. 11ff. Copied at Mulatrāṇa on Thursday 12 kṛṣṇapakṣa of Vaiśākha in Saṃ. 1788 = 20 May 1731. With the vyākhyā of Hṛṣikeśa. Incomplete.
BORI 419 of 1895/98. 10ff. Copied in Saṃ. 1806 = A.D. 1749.
BORI 416 of 1884/86. 18ff. Copied in Saṃ. 1862 = A.D. 1805. With the *Budhavallabhā* of Gopīnātha.
PL, Buhler IV E 284. 18ff. Copied in Saṃ. 1909 = A.D. 1852. (*Bṛhmaṇa*). Property of Jīvanakuśala Gorajī of Bhuja.
Baroda 3156. 3ff. Copied in Saṃ. 1942 = A.D. 1885.
Pingree 13. 36pp. Copied by Pūrṇacandra Panta from VVRI 1651 on 13 December 1960. With the vyākhyā of Hṛṣikeśa. Incomplete.
Pingree 12. 34pp. Copied by Viśveśvara Datta from VVRI 2617 in A.D. 1960. With the *Budhavallabhā* of Gopīnātha.
Pingree 11. 11pp. Copied by Kamalakānta from VVRI 5067 on 23 March 1961.
Adyar Index 2618 = Adyar Cat. 35 C 22. 70ff. With a vyākhyāna.
Alwar 1812.
Alwar 1813. With the vyākhyā of Hṛṣikeśa.
Anup 4746. 8ff. Formerly property of Maṇirāma Dīkṣita (*fl. ca.* 1650/1700).
Benares (1963) 34513. 6ff.
Benares (1963) 36373. Ff. 2–4. Incomplete.
Benares (1963) 36374 = Benares (1878) 178. Ff. 2–9. Incomplete.
Bombay U 365. 4ff.
Bombay U Desai 1377. 10ff. With the *Budhavallabhā* of Gopīnātha.
BORI 822 of 1884/87. 8ff. From Gujarāt.
BORI 955 of 1886/92. 4ff.
Cambridge University Add. 2132 = Bendall. From Rājasthān.
GJRI 955/67. Ff. 1–2 and 5–7. Incomplete.
GVS — (2230). 4ff. With the *Budhavallabhā* of Gopīnātha.
IO 2884 (1557). 7ff. Bengālī. From H. T. Colebrooke.
Jammu and Kashmir 852. 5ff.
Jammu and Kashmir 3049. 10ff.
Kotah 276. 6pp.
Kurukṣetra 426 (50120).
Leningrad (1914) 293 (Ind. II 93). 42ff. With the *Budhavallabhā* of Gopīnātha.
Oudh (1877–1878) VIII 16. 48pp. Property of Paṇḍita Kṛṣṇadatta of Sītāpur Zila.
PUL II 3547. 10ff.
VVRI 2617. 24ff. With the *Budhavallabhā* of Gopīnātha.
WHMRL Z. 60. c.
Wien (Univ.) 290.

The first verse is:

namaskṛtya paraṃ brahma gaṇakendus trivikramaḥ/
munipraṇītam akhilaṃ vyavahāraṃ pravakṣyati//

Verse 101 is:

nārāyaṇasya tanayo jñānamallānujo dvijaḥ/
trivikramaḥ śataślokair vyavahāram amuṃ
vyadhāt//

TRIVIKRAMA

The son of Mahādeva, Trivikrama wrote a *Siddhāntatattva*. Manuscripts:

RORI Cat. II 5622. 7ff. Copied by Vrajavāsī Sillū at Kāśī in Saṃ. 1895 = A.D. 1838.
Alwar 2003.

TRIVIKRAMA

Author of a *Strījātaka*. Manuscripts:

RORI Cat. III 16057. 14ff. Copied by Nānūrāma Dādhīca in Saṃ. 1933 = A.D. 1876.
Bikaner 734. 37ff.

The last verse is:

trivikrameṇa vacanasya strījātakam anuttamam/
guror ālekhya śuddhaṃ cet kṣamāṃ kurvantu
paṇḍitāḥ//

TRIVIKRAMA (*fl.* 1180)

The teacher of Āmarāja (*fl. ca.* 1200), Trivikrama wrote in Śaka 1102 = A.D. 1180 a *Khaṇḍakhādyakottara* giving additional rules to those in the *Khaṇḍakhādyaka* (665) of Brahmagupta (b. 598). Many verses of this work are quoted by Āmarāja in his *Vāsanābhāṣya*.

TRIVIKRAMA (*fl.* 1713/1737)

The son of Kṛṣṇajit or Kāhnajit, Trivikrama wrote a number of works, apparently at Nalinapura.

1. *Dvādaśabhāvaphala* or *Dvādaśabhāvalikhanānukrama*, completed on Sunday 5 śuklapakṣa of Jyeṣṭha in Saṃ. 1770 = 17 May 1713. Manuscripts:

LDI 6951 (4796). 11ff. Copied in Saṃ. 1770 = A.D. 1713.
LDI 6949 (3525). Ff. 3–16. Copied at Godharāgrāma, Kacchadeśa, in Saṃ. 1771 = A.D. 1714. Incomplete.
RORI Cat. I 1783. 17ff. Copied by Kuśalacanda at Māṇḍavī in Saṃ. 1820 = A.D. 1763.
LDI 6952 (2050). 31ff. Copied by Muni Rāmacandra, the pupil of Muni Trikamacandra, in Saṃ. 1850 = A.D. 1793. (*Dvādaśabhāvavicāra*).
RORI Cat. III 17297. 21ff. Copied on Sunday 5 śuklapakṣa of Āṣāḍha I in Saṃ. 1970 = 6 July 1913 (?).

LDI 6948 (3971). 26ff.
LDI 6950 (8018). 7ff. Incomplete.
LDI 6953 (3927). 2ff. (*Dvādaśabhāvavicāra*).
LDI 6954 (8883). 4ff. Copied by Paṇḍita Harṣavijaya Gaṇi (*Dvādaśabhāvavicāra*).
LDI 6955 (3900). 6ff. (*Dvādaśabhāvavicāra*).

The last verse is:

śrīvikramāt khādrimunīndusaṃmite
śucau site pañcamitārkavāre/
śrīkāhnajitsūnutrivikrameṇa
bhāvakramo ᵓyaṃ likhitaḥ sukhārthaḥ//

2. *Grahasiddhi* or *Grahaśīghrasiddhi*, composed at Nalinapura in Saṃ. 1776 = A.D. 1719. Manuscripts:

RORI Cat. II 8617. 22ff. Copied by Hemasāgara, the pupil of Ratnasāgara, at Nalinanagara in Saṃ 1834 = A.D. 1777. With his own udāharaṇa.
Goṇḍal 70. 27ff. Copied by Vāsudeva, the son of Mādhavajī, the son of Śivarāma Vyāsa of the Udīcyajñāti, at Goṇḍalapura on Saturday 9 kṛṣṇapakṣa of Āṣāḍha I in Saṃ. 1842 = 30 July 1785.
RORI Cat. I 628. 9ff. Copied by Harirāma Mulajī Sārasvata at Mothālā in Saṃ. 1884 = A.D. 1827.
LDI (LDC) 1756. 17ff. Copied in Saṃ. 1920 = A.D. 1863.

3. An udāharaṇa on his own *Grahasiddhi*, composed at Nalinapura in Saṃ. 1794 = A.D. 1737. Manuscript:

RORI Cat. II 8617. 22ff. Copied by Hemasāgara, the pupil of Ratnasāgara, at Nalinanagara in Saṃ. 1834 = A.D. 1777.

4. A set of astronomical tables called the *Bhramaasāraṇī*, perhaps designed to accompany the *Grahasiddhi*. Manuscripts:

RORI Cat. I 596. 161ff. Copied by Kuśalacanda at Māṇḍavībandara in Saṃ. 1818 = A.D. 1761. (*Bhramaṇagrahakoṣṭhakāni*).
RORI Cat. II 4681. 138ff. Copied by Caturavijaya at Pohakaraṇanagara in Saṃ. 1846 = A.D. 1789.
Goṇḍal 252. 24ff. Copied by Morārajī Saradhāra Bhaṭṭa on Wednesday 4 kṛṣṇapakṣa of Bhādrapada in Saṃ. 1858 = 23 September 1801 (?). (*Bhramaṇacandrārkī*).
Benares (1963) 36984 = Benares (1902) 1008. 136ff. Copied in Saṃ. 1904, Śaka 1769 = A.D. 1847. (*Grahasāraṇī*).
RORI Cat. II 9445. 171ff. Copied by Rāvatasundara, the pupil of Motīsundara, at Karṇapura in Saṃ. 1907 = A.D. 1850.
LDI 7303 (1816). Ff. 3–10 and 12–15. (*Sūryacandrasāraṇī*). Incomplete.

5. *Tithisāraṇī* according to the Brāhmapakṣa. Manuscripts:

PL, Buhler IV E 189. 9ff. Copied in Saṃ. 1805 = A.D. 1748. Property of Jīvanakuśala Gorajī of Bhuja.
Goṇḍal 162. 3ff. Copied by Kacarā Govaṃjī Dave in Saṃ. 1860 = A.D. 1803.
Goṇḍal 163. 7ff. Copied by Harikṛṣṇa, the son of Śāma Dave, on Friday 13 kṛṣṇapakṣa of Bhādrapada in Saṃ. 1883 = 29 September 1826.
LDI (LDC) 1642. 6ff. Copied in Saṃ. 1894 = A.D. 1837.
CP, Hiralal 2059. Property of Śaṅkarbhaṭ of Jāvalbūtā, Bulḍānā.
CP, Hiralal 2060. Property of the Balātkār gaṇ Jain Māndir at Kārañjā, Akolā.
CP, Hiralal 2061 and 2062. Property of Śrīnivāsrāv of Ratanpur, Bilāspur.
RORI Cat. I 652. 6ff. With his own ṭīkā.

Verses 1–2 are:

gaṇādhīśaṃ ca devīṃ ca śrīguroś caraṇāmbujam/
natvā ravyādikān kheṭān kṛṣṇajitpramukhān budhān//
tithisāraṇīṃ sugamāṃ brahmapakṣe karomy aham/
yasyāṃ tithyādayaḥ spaṣṭā bhavanti laghukarmaṇā//

6. A ṭīkā on his own *Tithisāraṇī*. Manuscript:

RORI Cat. I 652. 6ff.

It begins: ahaṃ trivikramanāmā gaṇako brahmapakṣe sugamāṃ tithisāriṇīṃ karomi.

TRIVEṆĪPRASĀDA SIṂHA (*fl.* 1955)

A resident of Paṭanā, Triveṇīprasāda wrote a *Grahanakṣatra* in Hindī which was published at Paṭanā in 1955.

TRYAMBAKA

Author of a *Svapnādhyāya*. Manuscript:

Oppert II 2204. 18pp. Telugu. Property of V. Raṅgācāryār of Veṅgamāmbāpuram, Pullampeṭa, Kaḍapa.

TRYAMBAKA BHAṬṬA

Author of a *Grahasāriṇī*. Manuscripts:

BORI 894 of 1886/92. 71ff.
BORI 469 of 1892/95. 71ff.
CP, Hiralal 1573. (*Grahalāghava*). Property of Mādhav Nārāyaṇ Bhope of Warorā, Chāndā.
CP, Hiralal 1587. Property of the Balātkār gaṇ Jain Mandir at Kārañjā, Akolā.
LDI (LDC) 3661. 40ff. (No title given).

TRYAMBAKA (*fl.* 1663/1673)

The son of Viśvanātha (*fl.* 1612/1630), Tryambaka (or Tryambaka Bhaṭṭa) wrote a ṭīkā on the *Viṣṇu-*

karaṇa of Viṣṇu in A.D. 1663. Manuscript:

BORI 193 of A 1883/84. 41ff. Copied in Saṃ. 1864 = A.D. 1807.

He also wrote a *Paddhatikalpavallī* for Anūpasiṃha (*fl.* 1674/1698), the Mahārāja of Bikaner, in Saka 1595 = A.D. 1673/74. Manuscript:

Anup 4827. 9ff. Copied by Tryambaka himself for Anūpasiṃha in Saṃ. 1741 = A.D. 1684.

TRYAMBAKA (*fl. ca.* 1800)

One of the 18 collaborators in writing the *Navagrahapadakāni* for Śarabhoji II of Tanjore (1798/1833); see Acyuta (*fl. ca.* 1800).

TRYAMBAKA GAṆEŚA (*fl.* 1909)

Author, with Śaṅkara Bhālacandra, of a pañcāṅga for Saṃ. 1966, Śaka 1831 = A.D. 1909, published at Gavāliyāra in 1909 (BM 14096. a. 9. (1)).

TRYAMBAKA GO. ḌHAVALE (*fl.* 1956)

Author of a *Jyotirvaibhava* in Marāṭhī, published at Puṇeṃ in 1956.

D. N. RAJE (*fl.* 1950/54)]

Author of a *Jātakarahasya* (Poona 1950), of a *Gṛhiṇījātaka* (Poona 1951), and of a *Jātakanidhi* (Poona 1954).

DATTARĀMA

Author of an *Arghadīpikā*, published with a ṭīkā at Bombay (Mysore GOL B 1624), and of a *Svapnaprakāśikā*, published at Bombay (Mysore GOL B 1653); *cf.* Dattātreya.

DATTARĀMA MĀTHURA (*fl.* 1855)

A resident of Āgarā, Dattarāma wrote a *Ramalanavaratnadarpaṇa* in Hindī in Saṃ. 1912 = A.D. 1855. Manuscript:

NPS 92 D of 1926–28. Copied in Saṃ. 1948 = A.D. 1891. Property of Paṇḍita Śyāmācaraṇa Jyotiṣī, c/o Ādityaprasāda Pāṇḍe of Kaṇaundiyā, Ḍaliyāṃ, Mirajāpura.

DATTĀTREYA

Author of a ṭīkā on the *Lokamanoramā* of Garga. Manuscript:

VVRI 2578. 9ff.

DATTĀTREYA

Author of a *Ghaṭitālaṅkāra*. Manuscript:

CP, Kielhorn XXIII 27. 20ff. Copied in Śaka 1568 = A.D. 1646. Property of Javāhara Śāstrī of Chāndā.

DATTĀTREYA

Author of a *Vivāhabhūṣaṇa*. Manuscript:

CP, Kielhorn XXIII 153. 17ff. Copied in Śaka 1574 = A.D. 1652. Property of Javāhara Śāstrī of Chāndā.

DATTĀTREYA

Author of a *Svapnaprakāśikā*; *cf.* Dattarāma. Manuscript:

Adyar Index 7354.

DATTĀTREYA ŚAÑKARA KEḶAKARA (b. 1933)

Author of a *Praśna jyotiṣa* in Marāṭhī, published at Mumbaī [1970].

DADA RĀJARṢI

See Rājarṣi Bhaṭṭa.

DAYĀNANDA

See Hṛdayānanda.

DAYĀNANDA ŚARMAN = DAYĀNĀTHA JHĀ (*fl.* 1910/54)

The son and pupil of Genālāla, Dayānanda, who was jyautiṣaśāstrapradhānādhyāpaka at the Rājakīya Saṃskṛta Vidyālaya in Mujaphpharapura, wrote a *Bhābhramabodha* in 1910; it was published as *MM* 107, Banārasa 1938. He also wrote a *Vimaṇḍalavakravicāra*, published as *MSVG* 3, Darbhanga 1954.

DAYĀNIDHI KHAḌĪRATNA (*fl.* 1963)

Author of a *Sūkṣmasiddhāntadarpaṇa* published in Oriyā script at Kaṭaka in 1963.

DAYĀPRIYA (*fl.* 1533)

The pupil of Vinayapriya and Tilakapriya, the pupils of Śivavarddhana, the pupil of the five pāṭhakas of Lakṣmīvallabha in the Kharataragaccha, Dayāpriya wrote a *Sārasaṅgraha* at Indraprastha in (Śaka) 1455 = A.D. 1533. Manuscripts:

LDI 7285 (4889). 14ff. Copied in Saṃ. 1755 = A.D. 1698.

Jaipur (II). 17ff. Copied in Saṃ. 1768 = A.D. 1711.

Verses 1–4 at the end are:

śrīmatkharataragacche lakṣmīvallabhapāṭhakāḥ/
paravādimadībhāliṃ pañcāsyā ye ᵓtra viśrutāḥ//
śiṣyās teṣāṃ jaganmukhyāḥ
 śrīmacchrīśivavarddhanāḥ/
vinayapriyas tacchiṣyas tilakapriyasaṃyutaḥ//
tayoḥ satīrthyaḥ sañjātaḥ śrīmān sādhur dayāpriyaḥ/

ittham bālāvabodhe ᵓtra śāstram etad vyacīkḷpat//
śrīīndraprasthe puri bhūvatamse
ṛddhīddhaṛddhāpaṇakaprakīrṇe/
parvendriyābdhīndumite śubhe ᵓbde
māsīha śukre śitapakṣatau ca//

DAYĀRĀMA

Author of a *Sāmudrika* in Hindī. Manuscript:

NPS 154 A of 1906–08. Property of Paṇḍita Mātādīna Khajāñcī of Gaurahāra.

DAYĀLU

Author of a *Muhūrtarāja* in at least six prakaraṇas:

1. saṃvatsara.
2. gocara.
3. saṃskāra.
4. vivāha.
5. yātrā.
6. gṛha.

Manuscript:

AS Bengal 2797 (G 6432). Ff. 2–25. Incomplete.

DAYĀŚAÑKARA

Author of a *Grahadīpikā*. Manuscripts:

Benares (1963) 35914. 8ff. Copied in Saṃ. 1886 = A.D. 1829.
Benares (1963) 37038. 5ff.
Mithilā. See NCC, vol. 6, p. 251.
PL, Buhler IV E 69. 8ff. Property of Jagannātha Jośī of Ahmadābād.

DAYĀŚAÑKARA

Author of a *Tithinirṇaya*. Manuscript:

N-W P I (1874) Law 311. 9ff. Property of Gaṇeśa Rāwa (Rāma) of Benares.

DAYĀŚAÑKARA

The son of Dharaṇīdhara, Dayāśañkara wrote a *Śāñkhāyanagṛhyapradīpa*, of which a part is the *Navagrahamakhaprayoga*. Manuscript:

Adyar List 3072 = Adyar Cat. 34 L 28. 39ff.

DAYĀŚAÑKARA

Alleged author of the following 3 works.

1. A ṭippaṇa on a *Praśnapradīpa*. Manuscript:

N-W P II (1877) B 12. 14ff. Property of Bholā Datta of Benares.

2. A ṭīkā on a *Mallāripaddhati*. Manuscript:

N-W P II (1877) B 11. 23ff. Property of Bholā Datta of Benares.

3. A ṭippaṇa on the *Sūryasiddhānta*. Manuscript:

N-W P II (1877) B 13. 34ff. Property of Bholā Datt of Benares.

DAYĀŚAÑKARA

Author of a ṭīkā on the *Praśnamanoramā* of Garga. Manuscript:

N-W P I (1874) 44. 11ff. Property of Gaṇeśa Rāma of Benares.

DAYĀŚAÑKARA UPĀDHYĀYA (*fl.* 1963)

A resident of Rāmanagara, Vārāṇasī, Dayāśañkara wrote a *Jyautiṣapraśnaphalagaṇanā*, published with his own Hindī vyākhyā, *Vimalā*, as *VSG* 93, Vārāṇasī 1963.

DAYĀSIṂHA GAṆI

The pupil of Jayatilaka Sūri, Dayāsiṃha wrote, under Ratnasiṃha Sūri, a ṭīkā in Old Rājasthānī, the *Bālāvabodha*, on the *Kṣetrasamāsa* of Ratnaśekhara. Manuscripts:

BM (Gujarātī) 14B (Or. 2118B). 121ff. Copied on Tuesday 1 śuklapakṣa of Pauṣa in Saṃ. 1668 = 24 December 1611 Julian.
RORI Cat. I 3493. 90ff. Copied in Saṃ. 1684 = A.D. 1627.
LDI 3046 (6325). 95ff. Copied in Saṃ. 1743 = A.D. 1686.
LDI 3045 (1387). Ff. 2–90. Incomplete.
RORI Cat. II 8814. 112ff.

DAYĀSIṂHA GAṆI (*fl.* 1436)

The pupil of Ratnasiṃha Sūri of the Tapāgaccha, Dayāsiṃha wrote a ṭīkā in Old Gujarātī, the *Bālāvabodha*, on the *Sañgrahaṇī* of Śrīcandra Sūri (*fl. ca.* 1150) completed on Friday 14 śuklapakṣa of Śrāvaṇa in Saṃ. 1493 = 27 July 1436. Manuscripts:

Berlin (Jaina) 753 (or. fol. 1981). 40ff. Copied at Vīragrāma on 7 kṛṣṇapakṣa of Jyeṣṭha in Saṃ. 1511 = *ca.* 17 June 1454.
LDI 3102 (6191). 100ff. (ff. 1–34 missing). Copied in Saṃ. 1577 = A.D. 1520. Incomplete.
LDI 3100 (4374). 36ff. Copied in Saṃ. 1610 = A.D. 1553.
LDI 3104 (2787). 49ff. Copied at Dīvabandara for Śāha Yādava, the son of Śāha Sīdhara, in Saṃ. 1625 = A.D. 1568.
LDI 3103 (4223). 75ff. Copied by Muni Vardhamāna in Saṃ. 1670 = A.D. 1613.

BORI 634 of 1895/98. 38ff. Copied in Saṃ. 1694 = A.D. 1637.
Ahmadabad, Dela Upāśraya Bhandar, ground floor 55 (24–29). See Velankar, p. 410.
AS Bengal Jaina 7412.
BORI 1310 of 1895/1902. See Velankar.
Chani, Bhandar of Kāntivijayajī Mahārāja 897. See Velankar.
LDI 3101 (3407). 35ff. (f. 15 missing). Copied at Pattana. Incomplete.
Limḍī 745, 1237, 1238, and 1570. See Velankar.
Patan, Saṅgha Bhaṇḍāra 56 (2). See Velankar.
Patan, Saṅgha Bhaṇḍāra, Limḍī Pāda 2 (5). See Velankar.

DARŚANAVIJAYA

Jaina author of a *Bṛhaddhāraṇāyantra* edited by Jñānavijaya, Biramgam 1931 (NL Calcutta 180. Kc. 93. 4); 2nd ed., Biramgam 1931 (NL Calcutta 180. Kc. 93. 5).

DALAPATIRĀJA (*fl. ca.* 1511/1512)

The son of Vallabha of the Bhāradvājagotra and the pupil of Sūrya Paṇḍita, Dalapati was the samastakaraṇādhīśvara of Nijāma Sāha, the ruler of Devagiri, who is probably Burhān Niẓām Shāh of Ahmadnagar (1510/1533). He composed an enormous compilation, the *Nṛsiṃhaprasāda*; the Benares manuscript dates some of the 12 sāras of which it consists in Saṃ. 1568 and 1569 = A.D. 1511 and 1512. See P. V. Kane [1930/62] vol. 1, pp. 406–410. One section is the *Kālanirṇayasāra*. Manuscripts:

IO 1476 (401) IV. 75ff. (ff. 12–13 and 28–30 missing). Copied in A.D. 1802. From H. T. Colebrooke.
Anup 1688. 519ff.
DC 6220. 44ff. No author mentioned. From the Kesari Marāṭha Collection.
Jammu and Kashmir 702. 50ff. Incomplete.
N-W P I (1874) Law 53. 60ff. Property of Vāgeśwari Datta of Benares.
VVRI 875. 6ff.

Another section was the *Śāntisāra*. Manuscripts:

Anup 2464. 12ff. Incomplete.
Anup 2465. 52ff.
Jammu and Kashmir 701. 53ff.

DALELAPURĪ

Author of a *Grahabhāvaphala* in Hindī. Manuscript:

NPS 34 of 1938–40. Property of Paṇḍita Ramaṇalāla of Pharaiha, Mathurā.

He also wrote a *Muhūrtacintāmaṇi* in Hindī. Manuscripts:

NPS 19 A of 1935–37. Property of Paṇḍita Jugalakiśora of Jagasaurā, Iṭāvā.
NPS 19 B of 1935–37. Property of Paṇḍita Rāmacandra of Biyāmaū, Balaraī, Iṭāvā.
NPS 19 C of 1935–37. Property of Paṇḍita Kāśīrāmā of Gośapurā, Śikohābāda, Mainapurī.

DAŚAPUTRA

Author of a *Malamāsanirṇaya*. Manuscript:

N-W P X (1886) A 7. 35ff. Property of Bālābhāū Sapre of Benares.

DAŚABALA (*fl.* 1055/58)

The son of Vairocana of the Valabhānvaya and probably a Buddhist, Daśabala wrote two astronomical works in accordance with the Brāhmapakṣa.

1. The *Cintāmaṇisāraṇikā* in 6 prakaraṇas composed in Śaka 977 = A.D. 1055 during the reign of Bhoja (*fl. ca.* 1005–1056). There is a ṭīkā by Mahādeva (*fl.* 1258). Manuscripts:

Rājāpūr Pāṭhaśālā. Copied on Thursday 2 śuklapakṣa of Āṣāḍha in Saṃ. 1558 = 17 June 1501.
Oxford 776 (Walker 190b). Ff. 120–134. Copied on 4 śuklapakṣa of Kārttika in Saṃ. 1596 = 15 October 1539.
CP, Hiralal 2058. Property of Vāsudev Kāle of Mulekheḍi, Bulḍānā.
LDI 6918 (1453). 20ff.
LDI (LDC) 6205/1. 4ff.

The *Cintāmaṇisāraṇikā* was published from the Rājāpūr and Oxford manuscripts by D. D. Kosambi [1952].

The second introductory verse is:

purācāryair etair na khalu vihitāḥ svalpavidhinā
nijagranthe hy antaḥ
sphuṭatithibhayogaprabhṛtayaḥ/
ataḥ paśyan viśvaṃ gaṇitagahanodvignam adhunā
namaskṛtyārkendū diśati tad upāyaṃ daśabalaḥ//

Verse 15 of the tithiprakaraṇa begins: śākaḥ svarahayāṅkonaḥ. And verse 15 of the saṃvatsarānayana is:

śrībhoje caturarṇavāṃ kalayati prauḍhena doṣṇā bhuvaṃ.
vikhyāto bhuvi ratnasaṃbhavalaghur vairocanir vālabhaḥ/
golagranthavidāṃ varo daśabalaḥ saṃsmṛtya śauddhodinaṃ
bodhavyāṃ aṇubuddhibhir vihitavāṃstathyāṃ titheḥ sāraṇīm//

The colophons begin: iti mahākāruṇikabodhisattvadaśabalaviracitāyām.

2. The *Karaṇakamalamārtaṇḍa* in 10 adhikāras composed in Śaka 980 = A.D. 1058; see S. B. Dikshit [1896] 239–240. Manuscript:

BORI 20 of 1870/71. 11ff. From Surat.

Verse 10 of the last adhikāra is:

valabhānvayasañjāto vairocanasutaḥ sudhīḥ/
idaṃ daśabalaḥ śrīmān cakre karaṇam uttamam//

DAŚARATHA

Author of a *Śanistotra* or *Śanaiścarastotra* in eleven verses. Manuscript:

PrSB 155 (or. oct. 739). 2ff. Now in Marburg.

This was published at Calcutta in 1883 (NL Calcutta 180. Nd. 85. 1(3)) and in many collections of stotras (see, e.g., IO, Sanskrit Books, vol. 4, p. 2352). The colophon begins: iti daśarathaproktaṃ.

PAṆḌITA DĀTĀRĀMA

Author of a *Jātakāmṛtaprakaraṇa*. Manuscript:

Chamba 13.

DĀDĀBHĀĪ = DĀDĀBHAṬṬA (*fl.* 1719)

The son of Mādhava Śrīgāṃvakara (or Śrīgrāmakara) (*fl. ca.* 1700) of the Kaśyapagotra and the brother of Nārāyaṇa (*fl. ca.* 1725), Dādābhāī was a Cittapāvana Brāhmaṇa. See S. B. Dikshit [1896] 292 and S. L. Katre [1942b]. He wrote the following works on jyotiṣa.

1. The *Kiraṇāvalī*, a ṭīkā on the *Sūryasiddhānta* composed in Śaka 1641 = A.D. 1719. Manuscripts:

PUL II 4074. 116ff. (ff. 20–30 missing). Copied in Saṃ. 1780 = A.D. 1723.
AS Bengal 6940 (G 6347). Ff. 1–44 and 1–87. Copied in the kṛṣṇapakṣa of Jyeṣṭha in Saṃ. 1849 = A.D. 1792. Incomplete (to the pātādhikāra).
IO 2781 (1122e). 77ff. Copied in A.D. 1800. From H. T. Colebrooke.
Cambridge R. 15. 105. 132ff. Copied in A.D. 1805. Incomplete (adhyāyas I–XI).
VVRI 2388. 12ff. Copied in Saṃ. 1862 = A.D. 1805. Incomplete.
Ānandāśrama 4336.
Ānandāśrama 6586.
BORI 697 of 1883/84. 197ff. From Mahārāṣṭra.
Calcutta Sanskrit College 181. 101ff.
IO 2780 (2261). 86ff.
Kavīndrācārya 893. No author mentioned.
Oxford 772 (Mill 11). 128ff.
Paris BN 304.1 (Sans. beng. 187). Pp. 1–156. Bengālī. From Guérin.
PUL II 4075. 93ff. Incomplete (to mānādhyāya).
RORI Cat. II 4859. 39ff.

The first 2 verses are:

praṇipatya paraṃ brahma sūryāśayamahodadheḥ/
sāracandraṃ samuddhṛtya tanomi kiraṇāvalīm//
cittapāvanajātīyamādhavāṅgabhavaḥ sudhīḥ/
dādābhāī samālocya varāhādikṛtīḥ sphuṭāḥ//

The colophon begins: iti śrīcittapāvanajātīyaśrīgāṃvakaramādhavātmajaśrīdādābhāīkṛte.

2. The *Turīyayantrotpatti*. Manuscripts:

Benares (1963) 35900. 4ff. (*Turīyayantropapatti*)
BORI 821 of 1884/87. 5ff. From Gujarāt.

The colophon begins: iti mādhavatanujadādābhāīkṛtā.

DĀNA

The pupil of Sadāraṅga, Dāna wrote a *Trailokyadīpikācopāī*. Manuscript:

RORI (Rājasthānī) 2162. 4ff.

DĀMODARA

Author of an *Ādeśapraśna* = *Praśnajyautiṣa*. Manuscript:

Kathmandu (1960) 29 (I 1414). 44ff. Nevārī. Incomplete.

This may be part of the *Ādeśasaṅgraha* of Dāmodara (*fl. ca.* 1675/83).

DĀMODARA

Author of an *Iṣṭikāla* according to Gobhila. Manuscript:

AS Bengal 1378 (G 2740) = Mitra, Not. 4089. 1f. Incomplete.

DĀMODARA

Author of a *Kālakaumudī*. Manuscript:

Śāstrī, Rep. (1901–1906), p. 14. Discovered by Kuñja Bihārī in Orissa.

DĀMODARA

Author of a *Gaṇitamanohara*. Manuscript:

Mithilā. See NCC, vol. 5, p. 262.

DĀMODARA

Author of a *Golabandha*. Manuscript:

Benares (1963) 35736. 13ff. Copied in Saṃ. 1724 = A.D. 1667.

DĀMODARA

Author of a *Golādeśa* in 10 chapters:

1. pātālanirūpaṇa.
2. mṛtyulokanirūpaṇa.
3. svarganirūpaṇa.
4. madhyagativāsanā.
5. spaṣṭagativāsanā.
6. tripraśna.
7. grahaṇa.
8. nakṣatrasaṃsthāna.
9. ———.
10. kālanirdeśādeśa.

Manuscripts:

Benares (1963) 35225 = Benares (1903) 1044. 50ff. Copied in Saṃ. 1734 = A.D. 1677.
Kathmandu (1960) 68 (I 1167). 62ff. Copied in NS 830 = A.D. 1710.
Kathmandu (1960) 69 (III 316). 40ff. Copied in Saṃ. 1878 = A.D. 1821.
SOI 9408.

The colophon begins: iti dāmodarakṛtau.

DĀMODARA

Author of a *Jātakakarmapaddhati* or *Dāmodarapaddhati*. Manuscripts:

BORI 105 of 1884/86. 9ff. Incomplete.
Oppert II 4649. Property of the Śaṅkarācāryasvāmimaṭha at Śṛṅgeri, Cikkamogulūr, Mysore.

DĀMODARA

Author of a *Jātakadīdhiti*. Manuscript:

Benares (1963) 35720 = Benares (1912–1913) 2166. 75ff. Incomplete. No author mentioned in Benares (1963).

DĀMODARA

Author of a *Jātakasaṅgraha*. Manuscript:

Oudh XXII (1890) VIII 18. 40pp. Property of Kedāranātha of Āgrā Zila.

DĀMODARA

Author of a *Jātakādeśa*. Manuscripts:

Jammu and Kashmir 4062. 94ff. Copied from Alwar 1769 in Saṃ. 1941 = A.D. 1884.
Alwar 1769.

DĀMODARA

A resident of Vidarbhadeśa, Dāmodara wrote a *Jyotiṣārka*. Manuscripts:

Benares (1963) 36427. 7ff. Copied in Saṃ. 1735 = A.D. 1678.
LDI (LDC) 346. 7ff. (*Jyotiṣa*).

DĀMODARA

Presumed author of the *Dāmodarīya*. Manuscript:

GOML Madras D 13569. 7pp. Telugu. Incomplete.

DĀMODARA

Author of a *Praśnasāra*. Manuscript:

Baroda 3193. 27ff. Copied in Saṃ. 1626 = A.D. 1569.

DĀMODARA

The brother of Keśava, the nephew of Dāmodara, and the grandson of Yajñaśarman of the Bhāradvājagotra, a resident of Chellur in Malabar, Dāmodara wrote a *Muhūrtābharaṇa*. Manuscripts:

GOML Madras R 4442. 54ff. Grantha. Copied in A.D. 1924/25 from a manuscript belonging to the Raja of Chirakkal, Baliapatam, Malabar.
Kerala 13905 (T. 240). 1000 granthas.
Kerala 13906 (TM. 71) 90 granthas. Malayālam. Incomplete.

DĀMODARA

The son of Gaṅgādhara, the son of Devadatta, the son of Mahādeva, the son of Nṛsiṃha, a resident of Jālandhara, Dāmodara wrote a *Yantracintāmaṇi* on tantra. It is sometimes cataloged with jyotiṣa works. Manuscripts:

Nagpur 1666 (1713). 45ff. Copied in Saṃ. 1805 = A.D. 1748. From Nāgpur.
Leningrad (1914) 309 (Ind. II 99). 37ff. Copied by Gaṇeśadāsa of the Kālīyajñāti in Saṃ. 1885 = A.D. 1828.
WHMRL G. 20. i. 50ff. Copied by Śivagirī Gusāṃī on Saturday 10 kṛṣṇapakṣa of Āśvina in Saṃ. 1908 = 18 October 1851. Property of Kṛpāsāgara Pūjajī.
RORI Cat. II 5664. 61ff. Copied in Saṃ. 1918 = A.D. 1861.
Alwar 1912.
BORI 245 of A 1883/84. 29ff.
DC 7545. 34ff.
Paris BN 1005 (Sans. Dév. 331–340) XVI. Incomplete.
Poleman 4984 (Harvard 349). Ff. 1–20, 23–29, 31, 33–72, and 75–77. Incomplete.
Poleman 4985 (Harvard 528). 20ff. Incomplete.
SOI (List) 386.

The *Yantracintāmaṇi* was published at Benares in 1866 (BM), 2nd ed. Kāśī Saṃ. 1935 = A.D. 1878 (BM); at Murādābāda in 1902 (BM 14033. bb. 7 (2)); edited with an Āndhra tātparya by Sūryanārāyaṇa Brahma Somayājin, Madras 1906 (BM

14033. bbb. 6. (2) and IO 3486); and with the Hindī ṭīkā of Baladevaprasādajī Miśra at Bombay in 1929 (IO San. D. 781(g)). Verses 4–9, as found in the WHMRL manuscript with a few obvious corrections, are:

jālandhare pīṭhavare prasiddhe
pratya⟨kṣa⟩rūpo bhuvi vartate yaḥ/
gotre tasmin vedavidyāpravīṇe
yajvā jaiṣī śastikān vedabrāhmān (?)//
tadanvaye paṇḍi⟨ta⟩sannṛsiṃho
jvālāmukho ᵓsau hi mahāprabhāvaḥ/
yāṃ yogamāyām paramārthavidyāṃ
viśeṣapūjyāṃ bhṛguvaṃśajānām//
tasyātmajo ᵓbhūd bhuvi dharmaśīlo
nāmnā mahādeva iti prasiddhaḥ/
naisargavairaprajahuḥ sa satvā
yaṃ prāpya duṣṭāhitara⟨-⟩kāsyaḥ//
tasmād āsīt samativikasa⟨d⟩devadattaḥ kalāvān
mānyo rājñāṃ sadasi viduṣāṃ gadyagaṅgāpravāhaḥ/
uktvacho (?) lāṃ diśi diśi janāḥ kīrtipīṣu sindhuṃ
yasmād yāpi śravaṇapuṭakaiḥ kuñcitākṣāḥ pibanti//
gaṅgādharas tattanayo babhūva
vivekagāṃbhīryaguṇair udāraḥ/
yaṃ prāpya lakṣmī ca sarasvatī ca
tatpā⟨da⟩yugmaṃ sthiratāṃ tanūnām//
dāmodaraḥ sarvakalāpravīṇas
tasmād abhūc chrīgaṇanāthabhaktaḥ/
labdhapratiṣṭho gurudevabhakto
mānyaḥ satāṃ dharmaparāyaṇo ᵓyam//

DĀMODARA

Author of a *Ratnajātaka*. Manuscript:

Alwar 1924.

DĀMODARA

Author of a *Laghukālanirṇaya*. Manuscript:

CP, Kielhorn XIX 41. 19ff. Property of Gaṇapati Śāstrī of Chāndā.

DĀMODARA

Author of a ṭīkā on the *Līlāvatī* of Bhāskara (b. 1114). Manuscript:

PL, Buhler IV E 231. 14ff. Property of Śivaśaṅkara Jośī of Ahmadābād.

This was edited by P. Jhā as *MSVG* 2, Darbhanga 1959.

DĀMODARA

Author of a ṭīkā, *Saṅketamañjarī*, on the Samarasāra of Rāma (*fl.* 1447). Manuscripts:

Benares (1963) 37844 = Benares (1878) 60 = Benares (1869) XIII 4. 19ff. Copied in Śaka 1601, Saṃ. 1736 = A.D. 1678/79.

Benares (1963) 37841 = Benares (1878) 59 = Benares (1869) XIII 3. 15ff. Copied in Saṃ. 1815 = A.D. 1758.

N-W P II (1878) B 13. 35ff. Property of Mākhanji of Mathurā.

DĀMODARA

Author of a *Sāmudrikādeśa* in 6 adhyāyas. Manuscript:

Bombay U Desai 1509. 72ff.

The first verse is:

vighnān aśeṣān vinivārayantaṃ
taṃ ḍhuṇḍhirājaṃ hṛdaye nidhāya/
sāmudrikādeśavaraṃ samagraṃ
dāmodaro ᵓtha prakaṭīkaroti//

Dāmodara also wrote a *Svarādeśa*. Manuscript:

Bombay U Desai 1513. 62ff. Incomplete (ends in adhikāra 4).

These two works may be parts of the *Ādeśasaṅgraha* of Dāmodara (*fl. ca.* 1675/83).

DĀMODARA

Author of a *Sīmantinīmaṅgala*. Manuscript:

Benares (1963) 36324 = Benares (1903) 1042. 10ff. Incomplete (ends with the bhāvaphalādhyāya in the vivāhapaṭala).

DĀMODARA

Author of a vivaraṇa on the *Sūryasiddhānta; cf.* the *Sūryatulya* of Dāmodara (*fl.* 1417). Manuscript:

Kathmandu (1960) 506 (I 992). 24ff. Nevārī. Incomplete.

DĀMODARA

Author of a *Horāpradīpa*. Manuscripts:

BORI 917 of 1891/95. 81ff. Copied in Saṃ. 1774 = A.D. 1717.
Benares (1963) 34529. 34ff. Copied in Saṃ. 1821, Śaka 1686 = A.D. 1764.
Alwar 2032.
Bombay U Desai 1457. 45ff. Incomplete (ends in 94, 6).
BORI 1027 of 1886/92. 10ff.
Kurukṣetra 1350 (19551).
LDI (LDC) 3636. 40ff.
RORI Cat. III 14990. 53ff. (ff. 11 and 22 missing). Incomplete (ends in adhyāya 94).

Verse 3 at the beginning is:

mayayavanavarāhādyair
horoktā karmaṇāṃ manojānām/
jñānapradīpam akhilaṃ
vilokya dāmodaras tanute//

DĀMODARA PAṆḌITA

Author of a ṭīkā, *Bālāvabodha*, on the *Jyotiṣaratnamālā* of Śrīpati (*fl.* 1040). Manuscript:

LDI 6851 (2436). 55ff.

DĀMODARA (*fl.* twelfth, thirteenth, or fourteenth century)

Author of an *Ābdaprabodha* = *Bhojadevasārasaṅgraha*, based in part on the work of Bhojarāja (*fl. ca.* 1005/1056). Manuscripts:

Kathmandu (1960) 15 (I 1692). 77ff. Copied on Friday 2 śuklapakṣa of Phālguna in Śaka 1297 = 22 February 1376 during the reign of Jayārjunadeva (1361/1382).
Kathmandu (1960) 16 (III 226). 101ff. Nevārī. Incomplete.
Kathmandu (1960) 17 (I 1078). 114ff. Copied by Amṛtajīvacandra. Incomplete.
Kathmandu (1960) 18 (I 1206). 93ff. Incomplete.
Kathmandu (1960) 19 (I 297). 98ff. Nevārī. Incomplete.
Kathmandu (1960) 20 (I 619). 79ff. Nevārī. Incomplete.
Oxford Photos 58. 96ff.

The first verse is:

sarvajñam advayam anādim anantam īśaṃ.
mūrdhnābhivandya vacanair vividhair munīnām/
ābdaprabodham udayajñamudānidānaṃ
dāmodaro vyaracayad guṇinaḥ kṣamadhvam//

Verse 4 is:

śrībhojadevanṛpasaṅgrahasarvasāraṃ
sāraṃ ca saṅgrahagatasya varāhasāmyāt/
yogīśvarādibudhasādhumataṃ gṛhītvā
grantho yathāgamakṛto na vikalpanīyaḥ//

DĀMODARA MIŚRA (*fl.* 1387)

The rājaguru of Jhampaṭṭa Nārāyaṇa, a ruler of Kāmarūpa, Dāmodara (see M. Shastri [1954]) wrote a *Smṛtisāgarasāra*, finished on 14 śuklapakṣa of Kumbha in Śaka 1308 = *ca.* 2 February 1387, partially based on his own *Smṛtigaṅgājala*. There are two parts: vrataviveka and śrāddhaviveka, to which M. Shastri [1954] 64 adds a third: antyeṣṭiprakaraṇa. Manuscripts:

Gauhati II. 93 (756) a. Ff. 1–20. Copied by Lakṣmīkānta in Śaka 1622 = A.D. 1700.
Nalbari, Kāmarūpa Sañjīvanīsabhā 146. 15ff. Copied in Śaka 1630 = A.D. 1708.
Gauhati II 76 (610–4). 21ff. Incomplete.

The *Smṛtisāgarasāra* was edited with his own Sanskrit ṭīkā, *Praveśikā*, and Bengālī translation by Ramānātha Gosvāmī as pt. 2 of the *Gaṅgājala*, 2 vols., Gauripur 1930 (Calcutta NL 180. Jc. 93. 26); the whole text was edited from this edition, the Nalbari manuscript, and the first Gauhati manuscript in M. Shastri and P. Caudhuri [1964] 1–76.

Verse 1 of the vrataviveka is:

praṇamya paramātmānam umāṃ ca parameśvarīm/
dāmodaro mahāmiśraḥ kurute sārasaṅgraham//

The last verse in the śrāddhaviveka is:

kumbhe śukle munau granthaṃ mūle ᵓṣṭayutake śake/
cakre trayodaśaśate miśro dāmodaraḥ kṛtī//

Two manuscripts of the *Smṛtigaṅgājala*—one at the Kāmarūpa Sañjīvanīsabhā at Nalbari, the other the property of Paṇḍita Śivanātha Bujar Barua of Datara, Kamrup, — are mentioned on p. 7 of the introduction to M. Shastri and P. Caudhuri [1964]. The concluding verse of the printed *Smṛtigaṅgājala* gives the date of composition as Śaka 1356 = A.D. 1434, and that at the end of the printed antyeṣṭiprakaraṇa of the *Smṛtisāgarasāra* gives the same year; see M. Shastri [1954] 67–68. These dates are at variance with the one given above, and it is not clear how one should reconcile them.

DĀMODARA (*fl.* 1417)

The son and pupil of Padmanābha (*fl. ca.* 1400), the son of Nārmada (or Narmadādeva) (*fl. ca.* 1375), Dāmodara wrote a *Bhaṭatulya* based on the *Āryabhaṭīya* of Āryabhaṭa (b. 476) in Śaka 1339 = A.D. 1417. See S. B. Dikshit [1896] 255–257. Manuscript:

BORI 346 of 1882/83. 23ff. From Gujarāt.

Verse 2 is:

dāmodaraḥ śrīgurupadmanābha-
pādāravindaṃ śirasā praṇamya/
pratyabdaśuddhyāryabhaṭasya tulyaṃ
vidāṃ mude ᵓhaṃ karaṇaṃ karomi//

Verses 16 and 19 at the end are:

śrīnarmadādevasutasya matpituḥ
śrīpadmanābhasya samasya bhāvataḥ/
yasmāt susaṃpannam anugrahād guror
bhūyād ihaitat paṭhanāt pradaṃ śriyaḥ//
sacchiṣye rasakṛt kṛtapraṇatibhiḥ saṃprārthito bījavit/
vaktrāmbhojaraviś cakāra karaṇaṃ dāmodaraḥ satkṛtī//

Dāmodara also wrote a *Sūryatulya* based on the *Sūryasiddhānta; cf.* the *Sūryasiddhāntavivaraṇa* of Dāmodara. Manuscripts:

Anup 5346. 32ff.
IM Calcutta 5356. Incomplete (*Kheṭakarma*). See NCC, vol. 5, p. 188.
Jaipur (II). 25ff.

Finally, he wrote a vṛtti on the *Karaṇaprakāśa* of Brahmadeva (*fl.* 1092). Manuscript:

IO 2915 (2004c). 13ff. Copied by Kāliṅga, the son of Yalla, the son of Nārāyaṇa in *ca.* A.D. 1755. Incomplete (ends in I 11). From Dr. John Taylor.

Verses 3–5 are:

granthārthavic chittibhayād ihārko
ᵓvatīrya bhūmāv akhilaṃ cakāra/
śāstraṃ khilībhūtam ihāvagamya
śrībrahmagupta tvam atha krameṇa//
śrībhāskara tvaṃ ca tataś ca sākṣāt
śrīpadmanābha tvam anāthabandhum/
dayānidhiṃ sarvaguruṃ vareṇyaṃ
śrīpadmanābhaṃ tam ahaṃ namāmi//
tadaṃhrisevābhir avāptavidyo
dāmodaro daivavidāṃ variṣṭhaḥ/
sahopapattyā karaṇaprakāśaṃ
vṛṇoti daivajñamanaḥpratuṣṭyai//

DĀMODARA (*fl.* 1551)

The son of Rāghava, Dāmodara wrote a *Rātrisaṃvitpradīpa* at Jodhapura in Śaka 1473 = A.D. 1551 for Malladeva, the rājā of Mārwār from 1531 to 1562. Manuscript:

Alwar 1937.

DĀMODARA RĀṆABHA (*fl. ca.* 1675/1683)

The son of Jānakī and Raghunātha, a Cittapāvana Brāhmaṇa, and a resident of Kāśī, Dāmodara wrote the following works.

1. *Navaratna*, completed on Thursday 10 kṛṣṇapakṣa of Āśvina in Śaka 1605 = 4 October 1683 Julian. Manuscripts:

Kathmandu (1960) 207 (I 1166). 43ff. Copied by the rājadaivajña Pūrṇānanda, on Monday 8 kṛṣṇapakṣa of Mārgaśira in Śaka 1607, NS 806 = 9 November 1685 Julian.
Oxford Photos 57. A film of Kathmandu I 1166.

Verses 1–4 are:

gaṇeśānaṃ bhāvānīṃ ca śaṅkaraṃ kamalāpatim/
natvā śrībhāskaraṃ pūjyān bāṇāśūnyanṛpair mite//
śālivāhanaśake kāśyām āśvine puṣyabhe gurau/
daśamyāṃ bahule pakṣe nātisaṅkṣiptavistṛtam//
navaratnamayaṃ granthaṃ navadīdhitisaṃyutam/
cittapāvanajātīyo jānakīraghunāthajaḥ//
dāmodaro racayati śiṣyapāṭhakayor mude/
śāstradṛṣṭaṃ lokadṛṣṭam anubhūtaṃ ca kathyate//

2. The *Siddhāntahṛdaya*. Manuscripts:

Anup 5340. 19ff. Copied by Śrīpati in Saṃ. 1735 = A.D. 1678.
Anup 5341. 18ff.
BORI 882 of 1884/87. 24ff. (ff. 2–4 missing). Incomplete. From Gujarāt.

3. A ṭīkā on the *Jñānapradīpa* = *Karmavipāka*, composed at Kāśī in Śaka 1602 = A.D. 1680; this seems also to be called the *Ādeśasaṅgraha* or *Śivamudrā*. *Cf.* the *Ādeśapraśna*, *Sāmudrikādeśa*, and *Svarādeśa* of Dāmodara. Manuscripts:

AS Bengal 6991 (G 10121). 22ff. Incomplete (the *Kālacakrajātaka* in 10 adhyāyas from the *Śivamudrā*).
AS Bengal 6992 (G 6341) I and II. 14ff. Incomplete (*Kālacakrajātaka*).
AS Bengal 7053 (G 6332). 351ff. (Ff. 5–7, 18–61, 70–73, 188–190, 193–245, and 248–257 missing). Incomplete.
Benares (1963) 34937. 35ff. Incomplete.

The verse giving the date is:

śrīmadbhārgavarāmasatkṛtakulajñātiḥ sa dāmodaraḥ
kāśyāṃ netrakhabhūpaśakasamaye jñānapradīpasya tu/
rakṣārthaṃ vimalābhramandirasamaṃ sandīpanaṃ paṇḍitaṃ
prajñācakṣusukhaṃ tanoti bahulārthaṃ spaṣṭaśabdānvitam//

Some colophons begin: iti śrīrāṇabhopanāmakadāmodarakṛtāv ādeśasaṅgrahe śivamudrābhidhāne.

4. A ṭīkā on the *Hastirājavijaya* of Raṇahastin. Manuscript:

Bombay U Desai 1516. 25ff. Incomplete (to 2, 257).

The first verse is:

sītāpatiṃ gaṇapatiṃ bhapatiṃ praṇamya
śrīhastirājavijayasya karoti ṭīkām/
bālāvabodhavidhaye raghunāthaputro
dāmodaro laghutarāṃ viralāṃ sphuṭārthām//

5. A ṭīkā, *Prakāśikā*, on the *Ṣaṭpañcāśikā* of Pṛthuyaśas (*fl. ca.* 575). Manuscripts:

Benares (1963) 36624. 15ff. Copied in Saṃ. 1800 = A.D. 1743.
Nagpur 2340 (2605). 32ff. Copied in Śaka 1698 = A.D. 1776. From Nagpur.
AS Bengal 7363 (G 10027). 15ff. Copied by Rāmeśvara on Thursday 3 śuklapakṣa of Māgha in Saṃ. 1865 = 19 January 1809.

BORI 201 of A 1883/84. 12ff. Copied in Śaka 1745 = A.D. 1823.
AS Bengal 7362 (G 2279). 12ff.
Benares (1963) 36623. 10ff.
Benares (1963) 37024. 13ff.
BORI 523 of 1892/95. 11ff.
GJRI 3242/454. 20ff.
N-W P I (1874) 7 = N-W P I (1874) 82. 25ff. Property of Jagannātha Jotishi of Benares.
N-W P II (1877) B 87. 19ff. Property of Vāgīśvarī Datta of Benares.
Poleman 5023 (U Penn 2604). 10ff.

The next to the last verse is:

jānakīraghunāthābhyāṃ jāto dāmodaraḥ kṛtī/
teneyaṃ racitā ṭīkā supraśnasya prakāśikā//

6. The *Sabhāvinoda*, an encyclopedic work in 10 chapters of which 6 and 7 are devoted respectively to sāmudrika and jyotiṣaśāstra. The *Sabhāvinoda* was composed for Śrīnivāsamalla, the rājā of Lalitapattana in Nepal from 1681 to 1684. See P. K. Gode [A2. 1952].

DĀMODARA RATHA (*fl.* 1920)

Author of a *Vyavahārajyotiṣasārasaṅgraha*, of which pt. 1 was published with an Utkala bhāṣānuvāda at Cuttack in 1920 (IO San. B. 918(i)).

DĀMODARADĀSA

Author of a *Jñānapraśnāvalī* in Hindī. Manuscript:

NPS 87 of 1926–28. Copied in Saṃ. 1916 = A.D. 1859. Property of Paṇḍita Kṛpāśaṅkara Vaidya of Sidhaulī, Sītāpura.

DĀSARĀMA

Author of a *Sūryakāṇḍa* in Hindī. Manuscript:

NPS 157 of Saṃ. 2001–2003. Copied in Saṃ. 1911 = A.D. 1854. Property of Bhāgavata Tivārī of Kurathā, Pīranagara, Gorābājāra, Gājīpura.

DINAKARA

Author of a ṭippaṇa on the *Dhīkoṭida* of Śrīpati (*fl.* 1039/56). Manuscript:

Baroda 1083. 3ff., 9ff., and 7ff. Copied in Saṃ. 1880 = A.D. 1823.

DINAKARA

Author of a *Paribhāṣāprakaraṇa*. Manuscript:

Benares (1963) 37228. 2ff. Incomplete.

DINAKARA BHAṬṬA VIŚVEŚVARA

Author of a *Tithinirṇaya*. Manuscript:

DC 1751. 12ff. Copied in Śaka 1711 = A.D. 1789. From the Dīkṣit (A) Collection.

DINAKARA (*fl.* 1578/1583)

The son of Rāmeśvara (?) and great-grandson of Dunda of the Moḍhajñāti and Kauśikagotra, and a resident of Bārejya or Bāreja on the Brahmamatī or Sabhramatī in Gujarat, Dinakara (see S. B. Dikshit [1896] 277) wrote the following works:

1. The *Candrārkī* on solar and lunar motion written in Śaka 1500 = A.D. 1578; see SATIUS 51b–53a and SATE 101. *Cf.* the *Mahādevīṭīkā* of Divākara (*fl.* 1578). Manuscripts:

Goṇḍal 77. 28ff. Copied on Sunday 1 śuklapakṣa of Bhādrapada in Saṃ. 1737, Śaka 1602 = 15 August 1680 Julian. With a Gurjaraṭīkā.
PL, Buhler IV E 90. 4ff. Copied in Saṃ. 1738 = A.D. 1681. Property of Hariśaṅkara Jośī of Ahmadābād. Buhler notes 12 other copies.
Goṇḍal 80. 6ff. Copied on Thursday 11 kṛṣṇapakṣa of Kārttika in Saṃ. 1745 = 8 November 1688 Julian. Incomplete (ravipañcāṅga).
LDI (LDC) 714. 1f. Copied in Saṃ. 1751 = A.D. 1694.
LDI (LDC) 1411. 2ff. Copied in Saṃ. 1770 = A.D. 1713.
RAS (Tod) 24. 73ff Copied by Muni Ṛṣisenāṣpa, the pupil of Nāthajī, the pupil of Rohitāsajī, the pupil of Bhojarājajī, on Friday 11 śuklapakṣa of Āṣāḍha in Saṃ. 1776 = 18 June 1719 Julian. See SATE 58–59.
LDI 6764 (3106). 12ff. Copied by Ṛṣi Saubhāgya in Saṃ. 1781 = A.D. 1724.
LDI (LDC) 4937. 2ff. Copied in Saṃ. 1785 = A.D. 1728.
Goṇḍal 82. 4ff. Copied in Saṃ. 1814 = A.D. 1757.
Goṇḍal 78. 17ff. Copied on Thursday 9 śuklapakṣa of Vaiśākha I in Saṃ. 1820, Śaka 1686 = 5 May 1763.
Poleman 4827 (Columbia, Smith Indic 180). 2ff. Copied at Rādhanapura on Sunday 7 śuklapakṣa of Phālguna in Sāṃ. 1829, Śaka 1694 = 28 February 1773. See SATIUS 19a.
RORI Cat. II 4870. 3ff. Copied by Śivānanda in Saṃ. 1839 = A.D. 1782.
LDI 6759 (4331). 15ff. Copied by Muni Tīrthavijaya, the pupil of Paṇḍita Vinodavijaya, the pupil of Paṇḍita Amīvijaya, at Nāḍalāīnagara in Saṃ. 1844 = A.D. 1787.
LDI 6763 (7834). 20ff. Copied in Saṃ. 1844 = A.D. 1787.
LDI (LDC) 4815. 23ff. Copied in Saṃ. 1848 = A.D. 1791.

Goṇḍal 79. 12ff. Copied by Vāsudeva Vyāsa, the son of Mādhavajī and a former resident of Khareḍī, at Goṇḍala on 14 kṛṣṇapakṣa of Vaiśākha in Saṃ. 1853 = *ca.* 3 June 1796. With a ṭīkā.
LDI (LDC) 2614. 3ff. Copied in Saṃ. 1856 = A.D. 1799.
Goṇḍal 84. 3ff. Copied on Sunday 9 kṛṣṇapakṣa of Āśvina in Saṃ. 1857 = 12 October 1800.
Benares (1963) 36991. 5ff. Copied in Śaka 1726 = A.D. 1804.
LDI (LDC) 1316. 3ff. Copied in Saṃ. 1868 = A.D. 1811.
RORI Cat. II 9555. 11ff. Copied by Jinasundara at Vikramapura in Saṃ. 1873 = A.D. 1816.
Goṇḍal 85. 2ff. Copied at Bhujanagara in Saṃ. 1878, Śaka 1743 = A.D. 1821.
GVS 2788 (4198). 6ff. Copied at Naḍiāda on Friday 12 kṛṣṇapakṣa of Vaiśākha in Saṃ. 1885 = 9 May 1828.
BORI 510 of 1895/1902. 23ff. Copied in Saṃ. 1904 = A.D. 1847. With a *Jātakapaddhati*.
RORI Cat. I 2584. 4ff. Copied in Saṃ. 1904 = A.D. 1847.
Anup 4566. 1f. Copied by Śaṅkarajati Gusāī in Saṃ. 1906 = A.D. 1849.
RORI Cat. III 15282. 3ff. Copied by Phatehakṛṣṇa in Saṃ. 1910 = A.D. 1853. No author mentioned.
Goṇḍal 83. 3ff. Copied in Saṃ. 1916 = A.D. 1859.
RORI Cat. III 16083. 12ff. Copied by Kuñjalāla Vyāsa at Pohakaraṇa in Saṃ. 1917 = A.D. 1860. No author mentioned.
Goṇḍal 128b. Ff. 1–17. Copied at Rājakoṭa on Friday 14 śuklapakṣa of Phālguna in Saṃ. 1935 = 7 March 1879.
Goṇḍal 81. 4ff. Copied in Saṃ. 1937, Śaka 1802 = A.D. 1880.
Goṇḍal 86. 2ff. Copied on Thursday 4 śuklapakṣa of Māgha in Saṃ. 1970 = 29 January 1914.
Adyar Index 2019 = Adyar Cat. 35 C 104. 6ff.
Baroda 3119. 7ff. With a vṛtti.
Baroda 3120. 3ff.
Baroda 3121. 14ff. (ff. 1–3 missing).
Benares (1963) 35035. Ff. 1–4 and 4–9. No author mentioned.
BORI 445 of A 1881/82. 7ff. Incomplete (māsapraveśasāraṇī).
BORI 308 of 1882/83. 4ff. From Gujarāt.
BORI 315 of Vishrambag 1. 4ff.
Chani, Jaina Śvetāmbara Jñāna Mandira 4055. No author mentioned. See NCC, vol. 6, p. 375.
Dāhilakṣmī XX 2(1). See NCC.
GOML Madras D 14033. 38pp.
GVS—(4203). 4ff. No author mentioned.
GVS—(4491). 3ff. No author mentioned.
GVS—(4577). Ff. 1, 3–4, and 2–3. No author mentioned.
GVS—(5258). 4ff.
IM Calcutta 1123 (no author mentioned) and 1152 (ascribed to Divākara). See NCC.
Jaipur (II). 10ff.
Jhalrapatan, Sri Ailak Pannalal Digambara Jain Sarasvati Bhavan. No author mentioned. See NCC.
Jodhpur 463 and 530 (no author mentioned). See NCC.
Kotah 161. 3pp. (*Candrārkaspaṣṭīkaraṇasāraṇī*). No author mentioned (Kotah 168 in NCC).
LDI 6758 (7401/2). Ff. 5–6. Copied at Siddhapuranagara.
LDI 6760 (4163). 8ff.
LDI 6761 (7031). 7ff.
LDI 6762 (6931). 11ff.
LDI 6765 (6570). 10ff. Incomplete.
LDI 6766 (4356). 24ff. With an Old Gujarātī stabaka.
LDI 6767 (4159). 13ff.
LDI (AKC) 11708/1. Ff. 1–2.
Līṃbaḍī 931 (1376). 6ff. No author mentioned.
Oxford 775 (Walker 208b). 7ff. See SATE 56.
Paris BN 1005 (Sans. Dév. 331–340) VIII.
Poleman 4716 (Harvard 525). 5ff. With an udāharaṇa. See SATIUS 14b.
Poleman 4717, 4923, 4824, and 4823 (Columbia, Smith Indic 190). Ff. 3–6 and 8–17. See SATIUS 14b.
Poleman 4825 (Columbia, Smith Indic 58). 15ff. See SATIUS 19a.
Poleman 4826 (Harvard 934). 14ff. See SATIUS 19a.
Poleman 4883 (Columbia, Smith Indic 34). Ff. 9–11. See SATIUS 24a.
Poleman 4895 (Columbia, Smith Indic 40). 4ff. See SATIUS 25a.
Poleman 4946 (Columbia, Smith Indic MB), XXIV f. 19; XXXVIII 1f.; XXXIX 1f.; LII 2ff.; LXXXVIII 1f.; LXXXIX f. 1; XCI ff. 1–2; and XCII ff. 1–2. See SATIUS 29a–33b.
Poleman 4949 (Columbia, Smith Indic 19). 2ff. See SATIUS 34a.
Poleman 4952 (Columbia, Smith Indic 29). 6ff. See SATIUS 34b.
Poleman 5178 (Columbia, Smith Indic 35). 11ff. See SATIUS 35b.
Poleman 5179 (Columbia, Smith Indic 46). Ff. 2–10. See SATIUS 35b.
RORI Cat. I 224. 2ff.
RORI Cat. I 3253. 1f.
RORI Cat. I 3815. 2ff.
RORI Cat. II 4795 14ff. No author mentioned.
RORI Cat. II 4813. 3ff. (*Candrārkipaddhati*). No author mentioned.
RORI Cat. II 9620. 11ff. (f. 1 missing).
RORI Cat. II 9792. 13ff.
RORI Cat. III 12167(1). 6ff. No author mentioned.
RORI Cat. III 12912. 3ff.
RORI Cat. III 15278 3ff. Incomplete. No author mentioned.
RORI Cat. III 16445. 8ff. No author mentioned.
RORI Cat III 16449(2). 10ff. No author mentioned.
RORI (Rājasthānī) 4746. 11ff. No author mentioned.
RORI (Rājasthānī) 9954. 8ff. No author mentioned.

SOI 9467.
VVRI 2528. 4ff. No author mentioned.

The last verse (38) in some manuscripts reads:

bārejākhye vasan grāme cakre dinakaro mudā/
jātaḥ kauśikagotre ca moḍhajñātisamudbhavaḥ//

In others one finds (verse 35):

śrīmatkauśikagotrajo dvijavaro bārejyasaṃjñe pure
moḍhajñātisamudbhavo dinakaro
daivajñacūḍāmaṇiḥ/
cakre candraravisvakoṣṭakagatau
śrībrahmapakṣāśritau
dṛgpakṣāv api sākṣiṇau ca viśadathy (?) ādhike
prasphuṭām//

2. A commentary on the *Candrārkī*. Manuscripts:

LDI (LDC) 4028. 5ff. Copied in Saṃ. 1751 = A.D. 1694. (vṛtti).
LDI (LDC) 1496. 15ff. Copied in Saṃ. 1782 = A.D. 1725. (vṛtti).
RORI Cat I 2582. 6ff. Copied by Sugaṇapriya in Saṃ 1828 = A.D. 1771. (ṭīkā).
IO 2948 (2541e). 2ff. (ṭippaṇa). See SATE 40.
LDI 6768 (7226/1). Ff. 1v–2 (ṭippaṇī).
PL, Buhler IV E 91. 49ff (ṭīkā). Property of Dharmadāsa of Mulī.
PL, Buhler IV E 92. 22ff. (udāharaṇa). Property of Hariśaṅkara Jośī of Ahmadābād.

The colophon is: iti dinakaraviracitacandrārkīṭippaṇaṃ.

3. The *Kheṭasiddhi*, on the motions of the planets, written in Śaka 1500 = A.D. 1578. See SATE 101–112. Manuscripts:

IO 2947 (2648). Ff. 1–3, 1, 1–12, and 1–75. Copied by Nīlakaṇṭha, the son of Nārāyaṇa Bhaṭṭa, on Friday 8 kṛṣṇapakṣa of Caitra in Saṃ. 1683, Śaka 1559 (read 1549) = 7 April 1626. See SATE 41–42.
Goṇḍal 35. 8ff. Copied at Sihora by Ratneśvara, the son of Divākara, the son of Paṇḍayā Hari, an Udīcya of the Sahasrajñāti and a resident of Siṃhapura, on Saturday 12 śuklapakṣa of Śrāvaṇa in Saṃ. 1793 = 7 August 1736 Julian.
BORI 303 of 1882/83. 6ff. Copied in Saṃ. 1796 = A.D. 1739.
Oudh IV (1874) VIII 1. 12pp. Copied in A.D. 1856. Property of Śivanātha of Unao Zila.
Anup 4503. 83ff. Property of Anūpasiṃha (*fl.* 1674/1698).
Baroda 1081. 5ff.
Jaipur (II). 3ff.
PL, Buhler IV E 45. 84ff. Property of Jayakṛṣṇa of Sudāmāpurī. Buhler notes another copy.
RORI Cat II 4731. 30ff.
RORI Cat. II 8034. 80ff.

Verses 35–36 are:

śrīmadgotre kauśike sāgniko ᵓbhūd
dundākhyo ᵓyaṃ jñātimoḍhe prasūtaḥ/
khyāte grāme brahmamatyāḥ samīpe
bārejyākhye vipravaryair vikīrṇe//
tatpautrajo dinakaraḥ sakalāni kheṭa-
karmāṇi vīkṣya satataṃ hi savāsanāni/
cakre śake khakhatithipramite ca saṃvat
pañcāgnibhūpatimite laghukheṭasiddhim//

4. The *Tithisāraṇī* or *Dinakarasāraṇī* written in Śaka 1505 = A.D. 1583; see SATE 112–114. *Cf.* the *Tithyādicintāmaṇi* of Dinakara (*fl.* 1586). Manuscripts:

RORI Cat. III 15829(6). 10ff. Copied by Avicala Jośī in Saṃ. 1768 = A.D. 1711. No author mentioned.
GVS 2835 (3157). 18ff. Copied on Wednesday 7 śuklapakṣa of Caitra in Saṃ. 1799 = 31 March 1742 Julian.
RORI Cat. I 619. 31ff. Copied by Ratnacandra Muni in Saṃ. 1875 = A.D. 1818.
RORI Cat. III 12758. 5ff. Copied by Jayaśaṅkara Jeṭhārāma Vyāsa in Saṃ. 1923 = A.D. 1866. No author mentioned.
Baroda 3154. 3ff.
Jaipur (II). 18ff.
Poleman 4946 (Columbia, Smith Indic MB) L. 1f. See SATIUS 31a–31b.
RAS (Tod) 36b. 28ff. See SATE 60.
RORI Cat. III 11833. 23ff. Incomplete. No author mentioned.
SOI 5253.

Verse 21 is identical with verse 38 of the first version of the *Candrārkī*.

5. A ṭīkā on the *Grahalāghava* (1520) of Gaṇeśa (b. 1507), composed at Vārejā. Manuscripts:

RORI Cat. I 3788. 23ff. Copied at Rupanagaḍha in Saṃ. 1820 = A.D. 1763.
RORI Cat. III 11029(6). 24ff Copied in Saṃ. 1836 = A.D. 1779. (udāharaṇasāriṇī).

DINAKARA (*fl.* 1586)

The son of Rāmacandra and (adopted ?) son of Śoṣaṇa, a resident of Unnatadurgā (Uparkot, Junāgaḍh, Saurāṣṭra), Dinakara wrote a *Tithyādicintāmaṇi* in Saṃ. 1643 = A.D. 1586; see SATIUS 51a–51b. *Cf.* the *Tithisāraṇī* of Dinakara (*fl.* 1578/83). Manuscripts:

Benares (1963) 37227. 5ff. (*Pañcāṅgasugama*).
Poleman 4718 (Columbia, Smith Indic 53). 2ff. See SATIUS 14b.

Verse 1 is:

śrīsūryapramukhān grahān vidhiharīśān
vighnarājaṃ giraṃ
bhaktyā namya guroḥ padābjayugalaṃ
siddhāntavidvāḍabān/
dṛṣṭvā vai racitaṃ sphuṭaṃ ca sugamaṃ
yāmārdhasādhyaṃ tithi-
pattraṃ yena karomy ahaṃ dinakaras
tithyādicintāmaṇim//

Verses 11–12 are:

śrīmaty unnatadurgānāmni nagare jyotirvidāṃ
bhāskaro
vāyusthāpitavipravaṃśatilakaḥ śrīśoṣaṇākhyo
dvijaḥ/
śrautasmārtavicārasāracaturaḥ śrīśaṅkaropāsakaḥ
kāśīdvāravatīgayātripathigātīrthāśrayaḥ satyavāk//
putras tasya tadaṅghripadmayugajaprāptaprasādaḥ
sudhīr
varṣe rāmayugāṅgabhūparimite śrīvikramārkād
gate/
śrutyādyācyutavāsare dinakaraḥ śrīrāmacandrāṅgajo
vijñas tena kṛto budhaiḥ karuṇayā
tithyādicintāmaṇiḥ//

DINAKARA BHAṬṬA (*fl. ca.* 1600)

The son of Umā and Rāmakṛṣṇa, the son of Nārāyaṇa Bhaṭṭa (b. 1513), the son of Rāmeśvara, Dinakara was the brother of Kamalākara Bhaṭṭa (*fl.* 1612) of Benares. He wrote a *Śāntisāra*. Manuscripts:

BORI 50 of 1902/07. 155ff. Copied in Saṃ. 1663 = A.D. 1606.
Bombay U 1164. 201ff. Copied by Śaṅkara Bhaṭṭa on Saturday 7 kṛṣṇapakṣa of Vaiśākha in Śaka 1616 = 5 May 1694.
Bombay U 1165. 297ff. Copied in Śaka 1688 = A.D. 1766. Incomplete (begins with gaṇḍāntaśāntividhi).
Baroda 1532. 199ff. Copied in Saṃ. 1854 = A.D. 1797. Incomplete.
Calcutta Sanskrit College (Smṛti) 368. 272ff. Copied on 1 śuklapakṣa of Phālguna in Saṃ. 1862 = *ca.* 18 February 1806.
Baroda 1458. 2ff. and 128ff. Copied in Saṃ. 1885 = A.D. 1828.
Oudh VIII (1876) IX 14. 302pp. Copied in A.D. 1860. Property of Paṇḍita Rāmacharaṇa of Bārābānki Zillah.
Anup 2222. 38ff. (*Śāntikarma*).
Anup 2229. 194ff.
AS Bombay 733. Ff. 1–15 and 18–19. Copied for P. H. Jogalekara. Incomplete. From Bhāu Dājī.
Baroda 249. 241ff.
Baroda 1640. 216ff.
Baroda 5020. 222ff.
Baroda 5493. 15ff. Incomplete (rogaśāntiprakaraṇa).
Baroda 10876. 193ff.
Bikaner 981. 207ff.
IO 1754 (2333). 259ff. From Gaikawar.
IO 1755 (2194). 212ff. From Gaikawar.
IO 1756 (522a). 20ff. Incomplete. From H. T. Colebrooke.
IO 1757 (1741)a. Ff. 1–77 and 93–106. Incomplete. From H. T. Colebrooke.
Kerala 6788 (4795 B). 800 granthas. Incomplete (tithinakṣatrayogādiśānti).
Tanjore D 13211 = Tanjore TS 437. 5ff. Incomplete (āśleṣānakṣatrajananaśānti).

The *Śāntisāra* was published at Bombay in 1861 (BM and IO 13. E. 6); at Bombay in 1876 (IO 17. B. 14); at Bombay in 1877 (IO 1. C. 25); and at [NP] in 1887 (IO 14. B. 3). Verse 1 is:

śrīrāmakṛṣṇapitaraṃ natvomāmbāṃ sadāśivam/
rāmaṃ dinakaraśarmā tanute śāntisārakam//

The last verse is:

śrīrāmeśvarasūrisūnur udabhūd yo bhaṭṭanārāyaṇaḥ
kṣauṇīpaṇḍitamānakhaṇḍanajayī śrīrāmakṛṣṇas
tataḥ/
mīmāṃsānayatattvavid dinakaras tasmād abhūt
tatkṛtiḥ
seyaṃ śāntikatantrasāraviṣayā rāmāya dadyān
mudam//

Dinakara also wrote a *Dinakaroddyota*, of which a part is the *Kālakāṇḍa*. Manuscripts:

Anup 2397. 103ff.
Anup 2398. Ff. 1–11, 13, 15–18, 20–54, 56–112, 114–148, 148b–153, and 155–171.
Anup 2399. 3ff. (*Kālanirṇayānukramaṇikā*).
IO 1604 (1217a). 132ff. (ff. 75–82 missing). Incomplete (varṣakṛtya). From H. T. Colebrooke.

DINAKARA (*fl.* 1812/1839)

The son of Ananta of the Śāṇḍilyagotra, Dinakara resided at Poona. See S. B. Dikshit [1896] 298–299. There he wrote the following works on jyotiḥśāstra.

1. *Grahavijñānasāraṇī* in Śaka 1734 = A.D. 1812.

2. *Māsapraveśasāraṇī* in Śaka 1744 = A.D. 1822.

3. *Lagnasāraṇī*.

4. *Krāntisāraṇī* in Śaka 1753 = A.D. 1831.

5. *Candrodayāṅkajāla* in Śaka 1757 = A.D. 1835. Manuscript:

Ānandāśrama 3447.

6. *Dṛkkarmasāraṇī* in Śaka 1758 = A.D. 1836.

7. *Grahaṇāṅkajāla* in Śaka 1755/61 = A.D. 1833/39.

8. A vivṛti on the *Pātasāraṇī* of Gaṇeśa (b. 1507), in Saṃ. 1896, Śaka 1761 = A.D. 1839. Manuscript:

Poleman 4986 (U Penn 697). 9ff.

The first verse is:

natvā vighnaharaṃ pātasāraṇyā vivṛtiṃ sphuṭam/
karomi mandabodhāya hy ahaṃ dinakaraḥ kila//

9. A ṭīkā on the *Yantracintāmaṇi* of Cakradhara.

DIVĀKARA

Author of a *Kṣetrasādhana*. Manuscript:

PUL II 3312. 3ff.

DIVĀKARA

Author of a *Grahayajñadīpa*. Manuscript:

Benares (1953) 3271. 27ff.

DIVĀKARA

Author of a *Jyotirgrantha*. Manuscript:

DC 246.

DIVĀKARA BHAṬṬA (=*DIVĀKARA NANDIN*)

A pupil of Candrakīrti, Divākara wrote a *Laghuvṛtti* on the *Tattvārthādhigamasūtra* of Umāsvāti (*fl.* first century A.D.). See Velankar, p. 156.

DIVĀKARA (*fl.* before 1000)

An authority on astrology mentioned by al-Bīrūnī (b. 973) in his *Fī taḥqīq mā li-ʾl-Hind* (p. 123 ed. Hyderabad; vol. 1, p. 158 trans. Sachau).

DIVĀKARA (*fl.* 1053)

Astrologer at the court of the Śilāhāra monarch Mummuṇi, rājā of Thāṇā, Divākara is mentioned in a grant dated 5 kṛṣṇapakṣa of Āṣāḍha in Śaka 975 = *ca.* 8 July 1053; see G. H. Khare [A2. 1961].

DIVĀKARA (*fl.* 1578)

Author of a ṭīkā in 15 verses on the *Mahādevī* of Mahādeva (*fl.* 1316), written in Śaka 1500 = A.D. 1578; he may be identical with Dinakara (*fl.* 1578) as he uses the *Candrārkī*. Manuscript:

RAS (Tod) 24. 63ff. Copied by Muni Ṛṣiṣenāṣpa, the pupil of Nāthajī, the pupil of Rohitāsajī, the pupil of Bhojarājajī, on Friday 11 śuklapakṣa of Āṣāḍha in Saṃ. 1776 = 18 June 1719 Julian. See SATE 57–58.

The last two pādas of verse 12 are:

evaṃ kṛtaṃ yat tu divākareṇa
vicārya granthān viduṣāṃ hitāya//

DIVĀKARA (b. 1606)

The son of Nṛsiṃha (b. 1586), the son of Kṛṣṇa, the son of Divākara (a pupil of Gaṇeśa [b. 1507]), the son of Bhaṭṭācārya, the son of Rāma of the Bharadvājagotra, a resident of Golagrāma on the bank of the Godāvarī, Divākara was a pupil of his uncle Śiva and a brother of Kamalākara (*fl.* 1658); see S. Dvivedin [1892] 94–98 and S. B. Dikshit [1896] 287. He wrote the following works.

1. A *Jātakamārga* = *Jātakapaddhati* = *Divākarapaddhati* = *Paddhatiprakāśa* = *Padmajātaka*, written in Śaka 1547 = A.D. 1625 at the age of nineteen. Divākara wrote a commentary, the *Gaṇitatattvacintāmaṇi* (1627). There are 8 adhyāyas:

1. bhāva.	5. āyurdāya.
2. dṛṣṭi.	6. antardaśā.
3. bala	7. ariṣṭabhaṅga.
4. iṣṭakaṣṭa	8. (upasaṃhāra)

Manuscripts:

Benares (1963) 36394 = Benares (1878) 165 = Benares (1869) XXXIX 5. 12ff. Alleged to have been copied in Śaka 1547 = A.D. 1625. (*Padmajātaka*).
Anup 4640 = Bikaner 696. 15ff. Copied by Nṛsiṃha in Saṃ. 1699 = A.D. 1642. (*Jātakamārgapadma* = *Paddhatiprakāśa*).
Osmania University B. 46/8. 21ff. Copied in A.D. 1650. With his own *Gaṇitatattvacintāmaṇi*. The catalog's attribution to Keśava (*fl.* 1496/1507), followed in *CESS* A 2, 66b, is evidently wrong.
Oxford 1578 (Sansk. d. 188) = Hultzsch 284. 12ff. Copied in Saṃ. 1707 = A.D. 1650. (*Janmapaddhatiprakāśa*).
RORI Cat. III 15618(2). Ff. 36–50. Copied in Saṃ. 1803 = A.D. 1746. Incomplete (*Jātakamārgapadma*).
RJ 1666 (vol. 2, p. 272). 9ff. Copied in Saṃ. 1830 = A.D. 1773. (*Divākarapaddhati*). Property of Badā Terahapanthiyoṃ of Jayapura.
BORI 867 of 1891/95. 9ff. Copied in Saṃ. 1850 = A.D. 1793. (*Divākarapaddhati*).
RORI Cat. II 4748. 9ff. Copied by Āśārāma Jośī in Saṃ. 1862 = A.D. 1805. (*Paddhatiprakāśa*).
Oudh VII (1875) VIII 6. 10pp. Copied in A.D. 1811. (*Divākarīpaddhati*). Property of Jānakīprasāda of Bārābāṅki Zila.
Poleman 4719 (Columbia, Smith Indic 104). 5ff. Copied on 5 kṛṣṇapakṣa of Māgha in Saṃ. 1890, Śaka 1755 = *ca.* 28 February 1834. (*Paddhatiprakāśa*).
Adyar Index 2158 = Adyar Cat. 21 F 10. 12ff. Telugu. (*Janipaddhatiprakāśa*).
Alwar 1764. (*Jātakapaddhati*).
AS Bombay 357. 14ff. Copied by Narasiṃha, the son of Nāgendra Sūri, at Droṇapura from a manuscript copied by Kāśīpatinandana on Sunday 7 śuklapakṣa of Tapasya (= Phālguna) in Śaka

1600 = 8 January 1679 Julian. (*Jātakapaddhati*). With the *Gaṇitatattvacintāmaṇi*.
Baroda 3161. 3ff. (*Divākarapaddhatiprakāśa*).
Benares (1963) 34302. 6ff. (*Jātakapaddhatiprakāśa*).
Benares (1963) 34796. 10ff. (*Jātakapaddhatiprakāśa*).
Benares (1963) 35796. 10ff. (*Paddhatiprakāśa*).
Benares (1963) 37230. 6ff. (*Jātakamārgapadma* = *Divākarapaddhati*). This is probably identical with Benares (1878) 39. 6ff. (*Janmapaddhati*) and with Benares (1869) VIII 7. 8ff. (*Divākarīyajanmapaddhati*).
BORI 69 of A 1882/83. 6ff. (*Jātakapaddhatiprakāśa*).
GOML Madras D 19287. 7pp. Telugu. Incomplete (to VI 8). (*Janipaddhatiprakāśa*).
Kathmandu (1960) 110 (I 1165). 10ff. Nevārī. (*Jātakamārga*).
Kotah 155. 11pp. (*Paddhatiprakāśa*). No author mentioned.
N-W P X (1886) A 14. 4ff. (*Padmajātaka*). No author mentioned. Property of Bālābhāū Sapre of Benares.
Oppert II 1972. 4pp. Telugu. (*Divākarapaddhati*). Property of Veñkaṭeśvarajosya of Siddhavaṭa, Kaḍapa.
Oudh VIII (1876) VIII 19. 26pp. (*Divākarī*). Property of Devīdatta Śukla of Bārābanki Zila.
Oudh XX (1888) VIII 72. 18pp. (*Rāmavinodaprakāśapaddhati*). Property of Paṇḍita Pratāpa Nārāyaṇa of Allahabad Zila.
Oudh XX (1888) VIII 135. 14pp. (*Janipaddhatiprakāśa*). Property of Paṇḍita Pratāpa Nārāyaṇa of Allahabad Zila.
PL, Buhler IV E 478. 9ff. (*Śrīpatiprakāśa*). Property of Śivaśaṅkara Jośī of Ahmadābād.
PUL II 3400. 10ff. (*Jātakapaddhati*).
PUL II 4013. 9ff. (*Satpaddhati*).
RORI Cat. II 4863. 8ff. (*Paddhatiprakāśa*).
RORI Cat. II 4866. 49ff. (*Paddhatiprakāśa*). With the *Gaṇitatattvacintāmaṇi*.

Verses 1–2 are:

śrīmacchivākhyaṃ gaṇitajñacakra-
cūḍāmaṇiṃ sajjanavṛndavandyam/
vidur vido yaṃ dhiṣaṇena tulyaṃ
taṃ naumi nityaṃ dhiṣaṇāptihetoḥ//
śrīkeśavaśrīpatisundarādi-
praṇītatantrād adhigatya sāram/
prakāśyate sujñadivākareṇa
padmāṃśubhir jātakamārgapadmam//

Verses 99–104 are:

budhavaranarasiṃhanandanena
prabhaṇitasadgaṇitena satpitṛvyāt/
viracitajanipaddhatiprakāśe
viracitam agād idam aṣṭamaḥ prakīrṇaḥ//
godāvarīsaumyataṭasthagola-
grāme bharadvājakulāvataṃsaḥ/
āsīd vidhijñaḥ sakalāgamajñaḥ
kṛṣṇo maheśārcanatatparo yaḥ//
babhūvatus tasya sutau tadādyo
mīmāṃsakādyo gaṇako nṛsiṃhaḥ/
śiromaṇer vārttikam uktiyuktaṃ
yenāmalaṃ bhāṣyam akāri sauram//
vijñātanakṣatranabhogakakṣo
vicāradakṣo vijitāripakṣaḥ/
śivo dvitīyo gaṇako vadānyo
nṛpālamānyo jagato gurur yaḥ//
nṛsiṃhaputreṇa pitṛvyalabdha-
prabodhaleśena divākareṇa/
prakāśitaṃ jātakamārgapadmam
adhyāyarūpāṣṭadalaṃ subodham//
nandenduvarṣeṇa mayā kṛto °yaṃ
grantho raveḥ pādayugaprabhāvāt/
śāke nagāmbhodhiśarendutulye
prācāṃ prabandhān paribhāvya samyak//

2. A ṭīkā, *Gaṇitatattvacintāmaṇi*, on his own *Jātakamārga*, written in Śaka 1549 = A.D. 1627. Manuscripts:

Baroda 11071. 30ff. Copied in Saṃ 1700 = A.D. 1643. Said to be a commentary on the *Varṣagaṇitapaddhati*.
Baroda 3372. 33ff. (f. 1 missing). Copied in Saṃ. 1705 = A.D. 1648. This is PL, Buhler IV E 150. 33ff. Copied n Saṃ. 1705 = A.D. 1648. Property of Khuśāla Bhaṭṭa of Ahmadābād.
Osmania University B. 46/8. 21ff. Copied in A.D. 1650.
BORI 468 of 1892/95. 17ff. (ff. 1–12 missing). Copied in Saṃ. 1791 = A.D. 1734.
BORI 515 of 1899/1915. 29ff. Copied in Saṃ. 1824 = A.D. 1767.
BORI 146 of A 1883/84. 28ff. Copied in Saṃ. 1866 = A.D. 1809.
AS Bengal 7030 (G 6337). 30ff. Copied in Śaka 1776 = A.D. 1854.
Alwar 1738.
AS Bombay 357. 14ff. Copied by Narasiṃha, the son of Nāgendra Sūri, at Droṇapura from a manscript copied by Kāśīpatinandana on Sunday 7 śuklapakṣa of Tapasya (=Phālguna) in Śaka 1600 = 8 January 1679. The post colophon information is identical with that in IO 2001.
Benares. Property of Rājājī Jyotirvid. See S. Dvivedin [1892] 97.
Bombay, Kielhorn XII 3. 40ff. Property of Nānā Dīkshit Maṇerkar of Nargund.
IO 3093 (2001). 25ff. (ff. 13–14 missing). The post-colophonic information is identical with that in AS Bombay 357. From Dr. John Taylor.
N-W P II (1877) A 9. 10ff. Property of Chaṇḍī Datta of Benares.
RORI Cat. II 4866. 49ff.

The first verse is:

kāntaṃ nitāntaṃ śivapādayugmaṃ
citte nidhāyātha divākarākhyaḥ/

sacchiṣyatoṣāya nijapraṇītaṃ
granthaṃ vareṇyaṃ vivarīvarīti//

At the end are the verses:

divākarārādhanalabdhabuddhir
divākarākhyo narasiṃhasūnuḥ/
ramyaṃ nijokter gaṇitasya tattva-
cintāmaṇiṃ saṃracayāṃ babhūva//
yan mayātra samakāri kutracit
tv ekadeśimatakhaṇḍanaṃ varam/
sacchivasya suguror dayābharā-
lokanaprabhavavaibhavaṃ kila//

3. A ṭīkā, *Prauḍhamanoramā*, on the *Jātakapaddhati* of Keśava (*fl.* 1496/1507), written in Śaka 1548 = A.D. 1626. See T. Aufrecht [1891]. Manuscripts:

Poleman 5200 (Columbia, Smith Indic 42). 177ff. Copied by Bhagavanta Daivajña, the son of Moreśvara, the son of Vidyādhara, the son of Raghunātha on Friday 7 śuklapakṣa of Bhādrapada in Śaka 1704 = 13 September 1782.
VVRI 6920. 70ff. Copied in Saṃ. 1867 = A.D. 1810. Incomplete.
Mithila 206. 68ff. Copied in Saṃ. 1923 = A.D. 1866. Property of Paṇḍita Śrīnandana Miśra of Kanhauli, Sakri, Darbhanga.
Alwar 1733.
Baroda 9226. Ff. 3–152.
Benares (1963) 36103. Ff. 1–17 and 19–72. Incomplete.
Benares (1963) 36159 = Benares (1878) 38. 96ff.
Benares (1963) 37039 = Benares (1878) 80 = Benares (1869) XV 8. 7ff. Incomplete.
Benares (1963) 37267. Ff. 1–26 and 28–31. Incomplete.
Benares (1963) 37314. Ff. 31–48. Incomplete.
Mithila 206 A. 101ff. Property of Paṇḍita Rāmacandra Jhā of Mahinathapur, Deodha, Darbhanga.
Mithila 206 B. 91ff. Property of Paṇḍita Gaṅgādhara Jhā of Jonki, Deodha, Darbhanga.
N-W P II (1877) A 1. 120ff. Ascribed to Nṛsiṃha. Property of Chandra Dhara of Benares.
Oxford CS d. 788. 150ff.
PUL II 3416. 103ff.
RORI Cat II 5825. 141ff. (ff. 1–2 missing).
VVRI 2557. 21ff. Incomplete.

The *Prauḍhamanoramā* was published by Vāmanācārya, Benares 1882 (IO 19. C. 42; Mysore GOL B 377, B 1780, and B 1882; and NL Calcutta 180. Kc. 88. 9).

The verses at the end are:

gautamyuttaratīravarttinagare golajñavidvadyute
golagrāmasamāhvaye munibharadvājānvaye
daivavit/
yo ᵓbhūt sujñadivākaro budhavaraḥ
śrīkeśavasyātmajād
daivajñāryagaṇeśasaṃjñakaguror
labdhāvabodhāṃśakaḥ//
tasmān maheśārcanalabdhasaukhyaḥ
śrīkṛṣṇanāmājani vipramukhyaḥ/
kālatrayajñānavatā hi yena
daivajñaśabdo bhuvi sārthako ᵓbhūt//
tasmād abhūtāṃ tanayau tadādyo
mīmāṃsakādyo gaṇako nṛsiṃhaḥ/
śiromaṇer vārttikayuktiyuktaṃ
sauraṃ ca bhāṣyaṃ samakāri yena//
sadvidyārājamānaḥ
sukhadalitaparoddaṇḍacaṇḍābhimānaḥ
siddhāntābhijñasujñadvijavaranibahodgītakīrtiḥ
sumūrtiḥ/
vidyām āsādya sadyo vidadhati
vibudhācāryakasyarddhibhāvaṃ
hṛṣyanto yasya śiṣyāḥ sa jayati jagati śrīśivākhyo
dvitīyaḥ//
śrīmannṛsiṃhasutavaryadivākarākhyaḥ
satpaddhater gaṇakakeśakanirmitāyāḥ/
ṭīkām imāṃ vividhayuktiviśeṣaramyām
ājñāṃ śivasya suguroḥ kṛtavān avāpya//
śāke gajāmbhodhiśarendutulye
siddhāntavitprauḍhamanoramākhyā/
divākaroktā budhakeśavokter
vṛttiḥ sayuktiḥ samagāt samāptim//

4. A vivaraṇa on the *Makaranda* of Makaranda (*fl.* 1478). Manuscripts:

GVS 2864 (874). Ff. 4–8. Copied on Wednesday 5 kṛṣṇapakṣa of Āṣāḍha in Saṃ. 1712 = 11 July 1655 Julian. Incomplete.
AS Bengal 6897 (G 519) = Mitra, Not. 1301. 15ff. Copied by Viśvanātha on 2 śuklapakṣa of Āśvina in Saṃ. 1715 = *ca.* 17 September 1658 Julian.
Baroda 10577. 7ff. Copied in (Saṃ.) 1724 = A.D. 1667 (?).
Florence 295. 8ff. Copied in Saṃ. 1735 = A.D. 1678. (*Jyotiṣamakaranda*).
Benares (1963) 34655. 8ff. Copied in Saṃ. 1777 = A.D. 1720.
Benares (1963) 34936. 10ff. Telugu. Copied in Saṃ. 1806 = A.D. 1749.
Benares (1963) 36815. 10ff. Copied in Saṃ. 1821 = A.D. 1764.
PUL II 3768. 13ff. Copied in Saṃ. 1833 = A.D. 1776.
Benares (1963) 34642. Ff. 6–18. Copied in Saṃ. 1843 = A.D. 1786. Incomplete.
VVRI 2352. 12ff. Copied in Saṃ. 1847 = A.D. 1790.
Paris BN 212 O (Sans. dév. 316). Ff. 1–7 and 11–12 and 2ff. Copied on 2 śuklapakṣa of Māgha in Saṃ. 1848 = 24 January 1792. Acquired May 1842.
Poleman 4721 (Columbia, Smith Indic 79). Ff. 6–7. Copied in Saṃ. 1853, Śaka 1718 = A.D. 1797. Incomplete. See SATIUS 15a.
Poleman 4722 (McGill, Museum 20). 11ff. Copied in Saṃ. 1857 = A.D. 1800.
Florence 296. 11ff. Copied by Kālikādāsa in Saṃ. 1864, Śaka 1729 = A.D. 1807.

Benares (1963) 36134 = Benares (1913–1914) 2358. 7ff. Copied in Saṃ. 1866 = A.D. 1809.
Benares (1963) 35541 = Benares (1897–1901) 911. Ff. 1 and 3–13. Copied in Saṃ. 1875, Śaka 1740 = A.D. 1818. Incomplete.
BORI 496 of 1892/95. 12ff. Copied in Saṃ. 1875 = A.D. 1818.
Benares (1963) 35091 = Benares (1903) 1135. 15ff. Copied in Saṃ. 1877 = A.D. 1820.
Osmania University B. 109/10/a. 9ff. Copied in A.D. 1820.
Benares (1963) 35588. 15ff. Copied in Saṃ. 1878 = A.D. 1821. Incomplete.
Benares (1963) 35531 = Benares (1903) 1279. 11ff. Copied in Saṃ. 1883 = A.D. 1826.
Oudh VII (1875) VIII 7. 24pp. Copied in A.D. 1826. Property of Jānakīprasāda of Bārābānki Zila.
Calcutta Sanskrit College 88. 10ff. Copied in Saṃ. 1884 = A.D. 1827.
Benares (1963) 34347. 7ff. Copied in Saṃ. 1890, Śaka 1755 = A.D. 1833. No author mentioned.
RORI Cat. II 5732. 9ff. Copied by Vrajavāsī Sillu at the Maṇikarṇikāghāṭa in Kāśī in Saṃ. 1894 = A.D. 1837.
Paris BN 957 (Sans. Bengali 189) III = Guérin 30. Copied in A.D. 1840.
Kathmandu (1960) 295 (I 471). 9ff. Copied by Devidatta Śarman Panta on Thursday in the śuklapakṣa of Vaiśākha in Śaka 1765 = 4 or 11 May 1843.
Poleman 4720 (Columbia, Smith Indic 49). 19ff. Copied by Gaphuramaṇi Tripāṭika at Govardhanapura in Kāśī on 14 kṛṣṇapakṣa of Āśvina in Saṃ. 1922 = *ca.* 17 October 1865. See SATIUS 14b.
RORI Cat. II 9059. 12ff. Copied in Saṃ. 1936 = A.D. 1879.
AS Bengal 6893 (G 5512). 15ff. Copied by Sanāthamaṇi at Kāśī on 2 śuklapakṣa of Jyeṣṭha in (Śaka) 1816 = *ca.* 4 June 1894.
ABSP 1115. 15ff. Incomplete.
Alwar 1889. 2 copies.
Baroda 3226. 15ff.
Benares (1963) 34346. Ff. 1–14 and 14b–17.
Benares (1963) 34428. 10ff.
Benares (1963) 34643. 15ff.
Benares (1963) 35715. 23ff.
Benares (1963) 35716. 8ff. Incomplete (to saṃvatsarānayana).
Benares (1963) 35874 = Benares (1878) 74 = Benares (1869) XV 2. 14ff. Incomplete.
Benares (1963) 36192. Ff. 1–4 and 2ff. Incomplete. No author mentioned.
Benares (1963) 37120. 20ff. Incomplete.
Berlin 864 (Chambers 476). 18ff.
BORI 543 of 1875/76. 10ff. From Dilhī. No author mentioned.
BORI 123 of A 1882/83. 11ff.
BORI 171 of A 1883/84. 8ff. Ascribed to Dinakara.
BORI 545 of 1895/1902. 13ff.
Calcutta Sanskrit College 87. 11ff.
Cambridge Univ. Add. 2455. 11ff. See SATE 20.
CP, Hiralal 3733. Property of Śrīdev Dīkshit of Maṇḍlā.
CP, Hiralal 3734. Property of Ajodhyābhaṭṭ of Hardā, Hoshangābād.
IO 2956 (2476c). 11ff. See SATE 38.
Jaipur (II).
Jammu and Kashmir 2794. 9ff.
Jammu and Kashmir 2801. 14ff. Incomplete.
Jammu and Kashmir 2923. 9ff.
Kathmandu (1960) 294 (II 221). 18ff.
Kathmandu (1960) 296 (IV). 6ff.
Kurukṣetra 740 (50132).
N-W P I (1874) 114. 25ff. Ascribed to Dinakara. Property of Pūrṇānanda Jotishi of Benares.
N-W P II (1877) A 13. 16ff. Property of Chaṇḍī Datta of Benares.
Oudh XX (1888) VIII 35. 22pp. and 36. 88pp. Property of Paṇḍita Pratāpa Nārāyaṇa of Allahabad Zila.
Oudh XXII (1890) VIII 7. 80pp. Property of Kedāranātha of Āgrā Zila.
PL, Buhler IV E 315. 15ff. Property of Bālakṛṣṇa Jośī of Ahmadābād.
PUL II 3767. 11ff.
PUL II 3776. 8ff. Incomplete.
RORI Cat. I 3109. 11ff.
RORI Cat. II 4896. 6ff. Ascribed to Nandana.
RORI Cat. III 11826. 9ff.
SOI 2104 = SOI Cat. I: 1460–2104. 13ff.
SOI 3388 = SOI Cat. I: 1061–3388. 53ff.
VVRI 2538. 7ff.

The *Makarandavivaraṇa* was published at Benares in 1869 (BM); at Kāśī in 1880 (BM); and in *Aruṇodaya* I 15, 4–11 at Calcutta in 1890 (BM 14133. g. 16. (pt. 1, no. 15) and NL Calcutta 180. Qa. 89. 1–2).

Verses 1–2 are:

prajñāṃ yataḥ prāpya kṛtapratijñaṃ
spardhāṃ vidhatte prasabhaṃ pratijñam/
ajño ᵓpi taṃ śrīśivanāmadheyaṃ
gurūpamaṃ svīyagurum bhajeyam//
śrīmacchivāt samadhigamya varaprasādaṃ
vṛttāṃśubhir vivaraṇābhinavāravindam/
etad divākaravikāsitam āryavarya-
bhṛṅgā bhajantu makarandapipāsavo ye//

The colophon begins: iti śrīsakalagaṇakasārvabhaumaśrīkṛṣṇadaivajñasutanṛsiṃhasya sutena divākareṇa racitaṃ.

5. A ṭīkā on the *Pātasāraṇī* (1522) of Gaṇeśa (b. 1507). Manuscripts:

AS Bengal 6948 (G 6340) I. Ff. 1–7. Copied on Friday 10 śuklapakṣa of Vaiśākha in Saṃ. 1846 = 1 May 1789. Property of Paṇḍita Rāmeśvara.

Benares (1963) 37298. 8ff. Copied in Saṃ. 1851 = A.D. 1794. With the vivṛti of Viśvanātha. Said to have been composed in Śaka 1688 = A.D. 1766; perhaps one should read Saṃ. 1688 = A.D. 1631, though 1766 could be the date of a previous copy.
PUL II 3626. 4ff.
SOI 10571 (*Pātādhikāra*).

Verse 3 is:

tasmān nṛsiṃhasutavaryadivākarākhyaḥ
śrīmacchivākhyacaraṇāmbujacañcarākaḥ/
niḥsaṃśayārthabahulair vivṛṇomi padyair
bhāvaḥ samastam api sujñagaṇeśasūktaḥ//

6. The *Varṣagaṇitabhūṣaṇa* = *Paddhatibhūṣaṇa* = *Rathoddhatā*. Manuscripts:

Berlin 874 (Chambers 661). 6ff. Copied by Jādavajīka, the son of Vyāsa Mādhavajī of the Ābhyantaranāgarajñāti, for Jajñeśvara, the son of Jāgeśvara Dīkṣita, the son of Bhāïya Dīkṣita, in Saṃ. 1744 = A.D. 1687.
Benares (1963) 37345. 7ff. Copied in Saṃ. 182– = A.D. 1763–1773.
AS Bombay 314. 6ff. Copied in Śaka 1699 = A.D. 1777. From Bhāu Dājī.
LDI (LDC) 1056. 6ff. Copied in Saṃ. 1839 = A.D. 1782. (*Tājika*).
RORI Cat. III 11832. 8ff. Copied by Kamalākara in Saṃ. 1841 = A.D. 1784.
RORI Cat. II 4767. 5ff. Copied in Saṃ. 1847 = A.D. 1790.
Benares (1963) 36024. Ff. 1–3 and 2ff. Perhaps identical with Benares (1903) 1207. 8ff.
Benares (1963) 36097. Ff. 1–2 and 4. Incomplete. No author mentioned.
Berlin 875 (Chambers 794t, i). 5ff. Incomplete (to verse 43).
BORI 518 of 1892/95. 5ff.
CP, Kielhorn XXIII. 146. 11ff. Property of Javāhara Śāstrī of Chāndā.
Jaipur (II). 10ff.
Osmania University Ac/74/3. 12ff.
SOI 4027 = SOI Cat. II: 1107–4027. 9ff. No author mentioned.
SOI 9576. No author mentioned.
WHMRL G. 60. b. 12ff.
WHMRL R. 6.

Verses 1–3 are:

mohāndhakāraughaharaṃ suvṛttaṃ
guror adho ᵓpi sthitibhājam uccam/
gobhiḥ samudbodhitasaddvijendraṃ
śivaṃ guruṃ naumi khagādhirājam//
sujñakṛṣṇatanayo nayārjitaḥ
śrīnṛsiṃha iti yo ᵓtiviśrutaḥ/
vārṣikasya gaṇitasya paddhatiṃ
tatsutaḥ prakurute rathoddhatām//
nṛsiṃhaputreṇa divākareṇa
daivajñatoṣāya vinirmitāyām/
satpaddhatau varṣaphalasya varṣapraveśabhāvādi samāptim āgāt//

7. A ṭīkā, *Mañjubhāṣiṇī*, on the *Varṣagaṇitabhūṣaṇa*. Manuscripts:

BORI 506 of 1895/1902. 14ff. Copied in Saṃ. 1753 = A.D. 1696.
Benares (1963) 37344. 6ff. Copied in Saṃ. 1818, Śaka 1683 = A.D. 1751.
Benares (1963) 37343. 3ff. Incomplete (ends with sahamādhyāya).
Jaipur (II).
Oudh VII (1875) VIII 14. 20pp. Property of Jānakīprasāda of Bārābānki Zila.

DIVĀKARA KĀLA (*fl. ca.* 1625/1650)

The son of Gaṅgā, the daughter of Rāmakṛṣṇa, and of Mahādeva, the son of Rāmeśvara Bhaṭṭa, Divākara was the younger brother of Bālambhaṭṭa and the nephew of Dinakara Bhaṭṭa (*fl. ca.* 1600) and of Kamalākara Bhaṭṭa (*fl.* 1612). He is the author of a *Kālanirṇayacandrikā*. Manuscripts:

Anup 1672 = Bikaner 857. 111ff. Copied in Śaka 1599 = A.D. 1677.
BORI 343 of 1891/95. 77ff. Copied in Saṃ. 1771 = A.D. 1714.
Bombay U 1017. 73ff. Copied by Janārdana Bhaṭṭa Sāgavallīkarajaḍya on Thursday 7 kṛṣṇapakṣa of the intercalary month in Śaka 1690 = 4 August 1768.
Bombay U 1015. 71ff. Copied by Gaṅgādhara, the son of Nārāyaṇabhaṭṭa Nātu, on Wednesday 1 śuklapakṣa of Bhādrapada in Śaka 1702 = 30 August 1780.
CP, Kielhorn XIX 42. 130ff. Copied in Śaka 1702 = A.D. 1780. Property of Gaṇapati Śāstrī of Chāndā.
Bombay U 1016. 126ff. Copied by Nimbābhaṭṭa Śukla, the son of Śivarāma Śukla, at Vīrakṣetra in Gurjaradeśa in Saṃ. 1855 = A.D. 1798.
Adyar Index 1241 = Adyar Cat. 22 I 46. 146ff. Telugu.
Adyar Cat. 24 D 8. 130ff. Grantha. Incomplete.
Anup 1673. 69ff.
Baroda 13630. 59ff.
Benares (1956) 12123. 15ff. (*Tithinirṇaya*).
Benares (1956) 13018. Ff. 1–2, 5–30, and 32–40. Incomplete.
Benares (1956) 13223. Ff. 1–78 and 1–4.
Benares (1956) 13443. 50ff.
Benares (1956) 13903. 2ff. (*Janmatithinirṇaya*).
Bombay, Kielhorn X 21. 120ff. Property of Nānā Dīkshit Maṇerkar of Nargund.
BORI 523 of 1883/84. 14ff. From Mahārāṣṭra.
DC 2581. 37ff. Incomplete. From the Dīkṣit (A) Collection.

Hultzsch 1. 420. 48ff. Telugu. Property of Kesari Yajñayya of Brāhmaṇakrāka.
Kerala 3420 (1715). 2000 granthas.
Kerala 3421 (4975). 250 granthas. Incomplete.
Kurukṣetra 399 (19548). (*Tithinirṇaya*).
Mysore (1922). 2 manuscripts, of which one is incomplete (to pauṣamāsanirṇaya). See NCC, vol. 4, p. 28.
Oppert II 1735. 240pp. Telugu. Property of Upadraṣṭṛ Subbāśāstrī of Kambhālakuṇṭa, Pullampeṭa, Kaḍapa.
Oppert II 1952. 144pp. Telugu. Property of Veñkaṭeśvarajosya of Siddhavaṭṭa, Kaḍapa.
Oppert II 2035. 140pp. Telugu. Property of Kandālla Veñkaṭācārya of Śiñgamāla, Pullampeṭa, Kaḍapa.
Oppert II 2911. Property of Rāja Vellañki Veñkaṭarāmasūryaprakāśa Row of Utukūru, Vissampeṭa, Kṛṣṇa.
Oppert II 3015. Property of Śiṣṭla Sākṣayya of Vissampeṭa, Kṛṣṇa.
Oppert II 9868. Property of Pañcāpageśaśāstrī of Mahādānapuram, Trichinopoly.
SOI Cat. I. See NCC.
Tanjore D 18561 = Tanjore BL 51. 124ff. Incomplete (ends with the janmāṣṭamīnirṇaya).
Tanjore D 18562 = Tanjore BL 52. 118ff. Incomplete.
Tanjore D 18563 = Tanjore BL 53. 93ff. Incomplete.
Tanjore D 18564 = Tanjore BL 9238. 144ff. Grantha. Incomplete.
Tanjore D 18565 = Tanjore TS 540. 70ff. Incomplete.
VVRI 3858. 44ff. Telugu.

Verses 1–2 are:

praṇamya mātaraṃ gañgāṃ bhairavīṃ vanaśañkarīm/
mahādevākhyapitaraṃ śrautasmārtaviśāradam//
divākareṇa sudhiyā sāram uddhṛtya śāstrataḥ/
śiṣṭānāṃ tanyate tuṣṭyai kālanirṇayacandrikā//

The colophon begins: iti śrīmatkālopanāmakabhaṭṭarāmeśvarātmajamahādevadvijavaryasūnubālambhaṭṭānujadivākareṇa.

DIVĀKARA (*fl.* 1683)

Bālakṛṣṇa of the Bhāradvājagotra was the father of Mahādeva, who married Bālā, the daughter of Nīlakaṇṭha Bhaṭṭa (*fl.* 1649), the son of Śañkara Bhaṭṭa, the son of Nārāyaṇa Bhaṭṭa (b. 1513), the son of Rāmeśvara. The son of Bālā and Mahādeva was Divākara, who wrote the *Tithyarka* = *Tithyarkaprakāśa* at Kāśī in Saṃ. 1740 = A.D. 1683. Manuscripts:

Benares (1956) 11933. 148ff. Copied in Saṃ. 1740 or 1750 = A.D. 1683 or 1693.
Kunte B 91. 67ff. Copied in A.D. 1713. Property of Paṇḍita Jvālā Datta Prasāda of Lahore.
Benares (1956) 13725. 138ff. Copied in Saṃ. 1856 = A.D. 1799. With the *Anukramaṇikā* of Vaidyanātha.
CP, Kielhorn XIX 108. 63ff. Copied in Saṃ. 1877 = A.D. 1820. Property of Sadāśiva Dīkṣita of Sāgar.
Baroda 8431. 116ff. Copied in Śaka 1751 = A.D. 1829.
PUL I 243. 120ff. Copied in Śaka 1752 = A.D. 1830. With the *Anukramaṇikā* of Vaijanātha.
WHMRL G. 65. Ff. 1–20, 21/22, 23–43, 44/45, 46–65, 67, 67b–98, and 100–129. Copied in Saṃ. 1911 = A.D. 1854.
AS Bengal 2197 (G 1015). 68ff. Copied by Bālakṛṣṇa Bhaṭṭa on Wednesday 8 śuklapakṣa of Bhādrapada in Śaka 1831 = 22 September 1909.
Baroda 10858. 139ff.
Benares (1956) 11992. Ff. 1–2, 2b–79, 81–85, 85b–111, 111b–118, 118b–128, and 128b–136. Incomplete.
Benares (1956) 12286. 77ff. Incomplete.
Benares (1956) 12310. 41ff.
Benares (1956) 13498. Ff. 1–4, 7–37, and 40–102. Incomplete.
Calcutta Sanskrit College (Smṛti) 69. 110ff. Incomplete.
DC 7460. Ff. 2–8, 11–60, and 62–65. No author mentioned. From the Dīkṣit (B) Collection.
DC (Gorhe) App. 157. Property of Gañgādhara Rāmakṛṣṇa Dharmādhikārī of Puṇatāmbe, Ahmadnagar.
GJRI 3493/131. 144ff. No author mentioned.
GVS 813 (1789). Ff. 6–79 and 84. Incomplete.
Jaipur (II). 2 copies.
Kerala 6801 (7376). 2600 granthas.
Oudh (1879) VIII 18. 172pp. Property of Śyāma Lāla of Lucknow Zila.
PL, Buhler III E 124. 240ff. No author mentioned. Property of Kṛṣṇarāva Bhīmāśañkara of Vaḍodarā.
PUL I 244. 57ff. (ff. 52–53 missing). Incomplete.
SOI (List) 343.
VVRI 2476. 40ff. Incomplete.

The *Tithyarka* with the *Anukramaṇikā* of Vaijanātha was edited by Śrīkṛṣṇapanta Śāstrin, *AG* 8, Kāśī Saṃ. 1989 = A.D. 1932.

Verses 3–5 are:

śrīrāmeśvarasūrisūnur abhavan nārāyaṇākhyo mahān
yenākāry avimuktake suvidhinā viśveśvarasphāpanā/
tatputro vibudhādhipaḥ kṣititale śrīśañkaras tatsuto
jāto bhāskarapūjakaḥ pṛthuyaśāḥ śrīnīlakaṇṭho budhaḥ//
bhāradvājakule ᵓmale samabhavat śrībālakṛṣṇābhidhaḥ
sāhityāmṛtavārirāśir atulaḥ sarvadvijānāṃ guruḥ/
tatsūnuḥ prathamo mahāmaṇir iva prakhyātakīrtir guṇair
jāto nyāyanaye bṛhaspatisamo nāmnā mahādevakaḥ//
tatputreṇa divākareṇa viduṣā śrīnīlakaṇṭhaprabhor
dauhitreṇa budhaiḥ sudhārasasamāsvādyaḥ pareṣāṃ kṛte/

tithyarkaḥ kriyate praṇamya pitaraṃ bālāṃ tathā mātaraṃ
śrīkāntaṃ tapanaṃ śriyaṃ paśupatiṃ vācaṃ mahādevatām//

DIVĀNANDA MIŚRA

Author of a ṭīkā on the *Sarvārthacintāmaṇi* of Veṅkaṭeśa (*fl.* 1654). Manuscripts:

BORI 1014 of 1886/92. 36ff. Copied in Saṃ. 1934 = A.D. 1877.
PL, Buhler IV E 505. 54ff. (*Sārvārthacintāmaṇi* of Divānacanda). Property of Caturbhuja Bhaṭṭa of Khambhāliyāṃ.

DIVYATATTVA (*fl.* before 1000)

Author of a *Saṃhitā* mentioned by al-Bīrūnī (b. 973) in his *Fī taḥqīq mā li-ᵓl-Hind* (p. 121 ed. Hyderabad; vol. 1, p. 157 trans. Sachau).

DIVYASIṂHA MAHĀPĀTRA

A scion of the Vatsagotra, Divyasiṃha wrote a *Kālapradīpa* = *Kāladīpa* in Orissa, traditionally before the fourteenth century; it is cited by Gadādhara Rājaguru (*fl. ca.* 1725/1750). Manuscripts:

GOML Madras R 2999. 55ff. Grantha. Copied in 1919/20 from a manuscript belonging to Jugulakiśora Pāṇigrāhi of Parlakimedi.
AS Bengal 2777 (G 4085). 48ff. Uḍiya.
AS Bengal 2778 (G 5603 A). 33ff. Uḍiya. Incomplete.
AS Bengal 2779 (G 5588 B). 13ff. Uḍiya. Incomplete.
Bhubaneswar 17 (Dh. 41(B)). 42ff. Uḍiya. From Raṇapur, Puri.
Bhubaneswar 18 (Dh. 92(B)). Ff. 51–92. Uḍiya. From Raṇapur, Puri.
Bhubaneswar 19 (Dh. 129). 67ff. Uḍiya. Incomplete. From Bhubaneswar.
Bhubaneswar 20 (Dh. 108). 38ff. Uḍiya. From Parlakimindi, Ganjam.
Cuttack, Provincial Museum 37b. See NCC, vol. 4, p. 22.
CP, Kielhorn XIX 45. 46ff. Property of Nṛsiṃha Miśra of Sammalpur.
Śāstrī, Rep. (1895–1900), p. 15. *Divyasiṃhakārikā*, a verse abridgement by Divyasiṃha of his *Kāladīpa* and *Śrāddhadīpa*.
Śāstrī, Rep. (1901–1906), pp. 5–6. See NCC.
Viśvabhāratī 630. See NCC.

The *Kālapradīpa* was edited by Gopīnātha Kara, Cuttack 1914 (BM 14027. a. 1. (3)).

The first verse is:

praṇamya devaṃ śrīkṛṣṇaṃ bhavānīśaṅkarāv api/
tanyate kāladīpo ᵓyaṃ divyasiṃhena dhīmatā//

After the colophon is the verse:

śrīvatsagotrasamutpanno
divyasiṃhābhidhaḥ sudhīḥ/
kāladīpābhidhaṃ granthaṃ
kṛtavān kṛtināṃ mude//

DĪKṢITA MAṆIRĀMA (*fl. ca.* 1650/1700)

See Maṇirāma Dīkṣita (*fl. ca.* 1650/1700).

DĪKṢITA SĀṂVATSARA

Author of a ṭīkā on the *Samarasāra* of Rāmacandra (*fl.* 1447). Manuscripts:

BORI 202 of A 1883/84. 49ff.
GVS 2949 (3799). Ff. 3–30. Incomplete.
SOI 6119 = SOI (List) 433.

DĪNADAYĀLU PĀṬHAKA

The son of Vaṃśīdhara, the son of Vākpati of the Kauśikagotra, Dīnadayālu wrote a *Muhūrtabhairava*. Manuscripts:

Oudh V (1875) VIII 10. 116pp. Copied in A.D. 1850. Property of Gurusevaka of Faizabad Zillah.
Bombay U Desai 1410. 36ff. Incomplete (ends in verse 210).
Oudh XXI (1889) VIII 24. 110pp. Property of Raghuvara Prasāda of Gonda Zila.

Verse 4 is:

jātaḥ śrīkuśikaḥ kule sumatimān vidvān kavir vākpatir
devīdāsakulendupāṭhakavaro tatsūnuvaṃśīdharaḥ/
tatsūnuḥ prakaroti bhairavamuhūrtākhyaṃ priyaṃ sarvadā
nāmnā dīnadayālu vīkṣya bhaṇitaṃ pūrvaṃ munīnāṃ ca yat//

DĪNĀNĀTHA

The son of Kṛṣṇavilāsa of the Sandīpanagotra, Dīnānātha wrote a *Sarvasaṅgraha*. Manuscripts:

CP, Hiralal 6347. Property of Rāmprasād Tiwāri of Belkherā, Jubbulpore.
CP, Hiralal 6348. Property of Madanmohan of Raipur, Hoshangābād.
CP, Hiralal 6349. Property of Govindrām of Mālākherī, Hoshangābād.
CP, Kielhorn XXIII 169. 90ff. Property of Govindarāma Bhaḍajī of Sāgar.

The *Sarvasaṅgraha* was published with his own bhāṣāṭīkā by Baccū Jhā, Kalyāṇa-Mumbaī Saṃ. 1982, Śaka 1847 = A.D. 1925.

Verse 1 is:

gaṇeśavāggurūn natvā sandīpanakulodbhavaḥ/
dīnānāthaḥ subodhārthaṃ kurute sarvasaṅgraham//

The colophon begins: iti śrīkṛṣṇavilāsātmajadīnānāthaviracite.

DĪNĀNĀTHA JHĀ (*fl.* 1939/1951)

The son of Vaṃśīdhara, a Maithila Brāhmaṇa of the Kāśyapagotra, the pupil of Guṇānanda, and a resident of Baraunī, Muṅgera, Dīnānātha wrote the following works on jyotiṣa:

1. A *Pañcāṅgavijñāna*, published as *HSS* 104, Benares 1939; 2nd ed., Benares 1948; 4th ed., Vārāṇasī 1968.

2. An anvaya and Hindī ṭīkā, *Bhāvabodhinī*, on the *Jātakālaṅkāra* of Gaṇeśa (*fl.* 1613), completed on Monday 15 śuklapakṣa of Kārttika in Saṃ. 1998 = 3 November 1941, and edited by Kapileśvara Śāstrin, *KSS* 141, Benares 1950. At the end are the verses:

vihāre muṅgere prathitaguṇadhāmā mama purī
baraunī nāmnīyaṃ vilasati dvijāgryair gurujanaiḥ/
tadasyāṃ sañjāto laghumatir ahaṃ maithilakule
vinamro dīnānātha iti varavaṃśīdharajanuḥ//2//
guṇaśreṇī yasmin maṇigaṇa ivābhāti nitarāṃ
guṇānandaḥ prājño mama guruvaro jyotiṣi guruḥ/
avāpaṃ yatpādāmbujayugalapūjāyatamanā
ahaṃ prājñām ādyāṃ
sadayahṛdayānugrahalavaiḥ//3//

3. A Hindī ṭīkā, *Subodhinī*, on the *Dharācakra* of Lomaśa, which is adhyāya 24 of utthāna 13 of the *Lomaśasaṃhitā*; the example in the ṭīkā is for Friday 10 śuklapakṣa of Mārgaśīrṣa in Śaka 1863 = 28 November 1941. The *Subodhinī* was published in *HSS* 162, Benares 1944; 2nd ed., Vārāṇasī 1963. Verse 3 at the beginning is:

baraunīgrāmavāstavyo nirmalaḥ kāśyapodbhavaḥ/
śrīdīnānāthanāmāhaṃ bhāṣāṃ sodāhṛtiṃ bruve//

4. A Hindī ṭīkā, *Vimalā*, on the *Yoginījātaka*, published as *HSS* 145, Benares 1941.

5. A Hindī ṭīkā, *Bhāvabodhinī*, on the bhāvaphala from the *Bhṛgusaṃhitā*, published in *HSS* 163, Benares 1944; 2nd ed., Vārāṇasī 1963.

6. A Hindī ṭīkā, *Bhāvabodhinī*, on the *Kheṭakautuka* of Nabbāba Khānakhānā (1556/1627), published in *HSS* 166, Benares 1944; 2nd ed., Benares 1956.

7. A ṭippaṇī and pariśiṣṭa to the *Śiśubodha* of Kalādhara Śarman (*fl.* 1844), published in *HSS* 114, 2nd ed., Benares 1949.

8. A Hindī ṭīkā, *Bhāvabodhinī*, on the *Padmakośa* of Bhagavānadatta, published as *HSS* 210, Vārāṇasī 1951.

DUḤKHABHAÑJANA

Alleged author of the following works on jyotiṣa.

1. *Āryatulya*. Manuscript:

Oudh VIII (1876) VIII 3. 16pp. Copied in A.D. 1850. Property of Raghuvara Tivāri of Bārābānki Zila.

2. *Janmapaddhati*. Manuscript:

Oudh VIII (1876) VIII 11. 22pp. Property of Raghuvara Tivāri of Bārābānki Zila.

3. *Jātakasudhākara* = *Jātakayogasudhākara* = *Yogasudhākara*. Manuscripts:

Oudh VII (1875) VIII 3. 24pp. Copied in A.D. 1830. Property of Jānakīprasāda of Bārābānki Zila.
Oudh VIII (1876) VIII 30. 20pp. Copied in A.D. 1830. Property of Viśveśvara Bakṣa Tivāri of Bārābānki Zila.
Oudh VI (1875) VIII 6. 208pp. Property of Śivasahāya of Unao Zila.

4. *Muhūrtakalpākara*. Manuscript:

Oudh VIII (1876) VIII 25. 28pp. Property of Raghuvara Tivāri of Bārābānki Zila.

5. *Varṣapaddhati*. Manuscript:

Oudh VIII (1876) VIII 31. 20pp. Copied in A.D. 1830. Property of Raghunātha Upādhyāya of Bārābānki Zila.

6. *Sārasaṅgraha*. Manuscript:

Oudh VIII (1876) VIII 33. 14pp. Property of Raghunātha Upādhyāya of Bārābānki Zila.

DURGA (*fl.* before 1200)

Author of 7 verses giving bījas to the planetary parameters in the *Khaṇḍakhādyaka* (665) of Brahmagupta (b. 598); these are cited by Āmarāja (*fl. ca.* 1200) in his *Vāsanābhāṣya* (pp. 22–23). The first verse is:

śrīkhaṇḍakhādye karaṇe grahāṇāṃ
bījāni durgaḥ kurute yathā ca/
jñātvāntaraṃ kṣepabhavaṃ tathānyad
viśleṣajātaṃ grahayor viditvā//

DURGADEVA

Author of an *Aṅgavidyāparīkṣā*. Manuscript:

LDI 7534 (8223/1). Ff. 1–3v.

DURGADEVA

Author of a *Saṃvatsaraphala* or *Ṣaṣṭisaṃvatsaraphala*, sometimes said to be identical with or a part of the

Arghakāṇḍa of Durgadeva (*fl.* 1032), but see p. 5 of the ed. of the *Riṣṭasamuccaya*. Manuscripts:

GVS 2955 (869). 14ff. Copied in Saṃ. 1674 = A.D. 1617.
Poleman 4987 (Harvard 535). 14ff. Copied in Saṃ. 1687, Śaka 1552 = A.D. 1630.
LDI (LDC) 2394. 9ff. Copied in Saṃ. 1691 = A.D. 1634.
BORI 584 of 1895/1902. 14ff. Copied in Saṃ. 1703 = A.D. 1646.
LDI (LDC) 1556. 13ff. Copied in Saṃ. 1725 = A.D. 1668.
NPS 106 of 1941–43. Copied in Saṃ. 1759 = A.D. 1702. (*Sāṭhikā* in Hindī of Durgādevī ?). Property of the Nāgarīpracāriṇī Sabhā in Vārāṇasī.
LDI (LDC) 5502. 13ff. Copied in Saṃ. 1764 = A.D. 1707.
Baroda 3326. 22ff.
Baroda 9493. 43ff. Incomplete.
GVS 2942 (1631). 19ff.
LDI (DSC) 9726 = LDI (VDS) 1320 (9727). 6ff.
Leipzig 1122. 12ff. (f. 1 missing). Copied from a manuscript copied in Saṃ. 1403 = A.D. 1346.
PL, Buhler IV E 498. 10ff. Property of Śeṭha Bhīmaśī Māṇeka of Mumbaī.

DURGADEVA (*fl.* 1032)

A Digambara Jaina, the pupil of Saṃyamadeva, the pupil of Saṃyamasena, the pupil of Mādhavacandra, and a resident of Kumbhanagara (Kumbher near Bharatpur), Durgadeva wrote the following works on jyotiṣa in Śaurasenī Prākṛta:

1. *Riṣṭasamuccaya* in 261 verses, composed at the Śāntināthabhavana in Kumbhanagara on 11 śuklapakṣa of Śrāvaṇa in Saṃ. 1089 = *ca.* 21 July 1032, during the reign of one Lakṣmīnivāsa. Manuscripts:

Bombay, Ailaka Pannalalji Digambara Jaina Sarasvati Bhavana 1527/388. 10ff. Copied by Jagarāma, a Digambara Jaina, in Saṃ. 1981 = A.D. 1924. There are two other manuscripts in the same collection. See ed., p. 3, and NCC, vol. 4, p. 19.
Baroda 13190. 11ff. Photograph of a manuscript copied by Muni Samudra at Medinīpura and formerly belonging to Paṇḍita Sumaticandra Gaṇi. (*Kālajñāna*).
BORI 392 of 1879/80. 5ff.
Manuscript belonging to Paṇḍita Jugalkiśorajī Mukhtar. See ed., p. 3.

The *Riṣṭasamuccaya* was edited with a Saṃskṛta chāyā and an English translation by A. S. Gopani, *SJS* 21, Bombay 1945. Verses 258 and 260–261 are:

saṃjāo iha tassa cārucario nāṇaṃbudhoyā maī
sīso desajaī vibohaṇaparo ṇīsesabuddhāgamo/
nāmeṇaṃ siriduggaeva vidio vāgīsarāyaṇṇao
tenedaṃ raïyaṃ visuddhamaïṇā satthaṃ mahatthaṃ
  phuḍaṃ//
saṃvaccharaïgasahase volīṇe ṇavayasīi saṃjutte/
sāvaṇasukkeyārasi diahammi ya mūlarikkhaṃmi//
sirikuṃbhanayaraṇayae
  sirilacchinivāsanivaïrajjaṃmi/
sirisaṃtināhabhavaṇe muṇibhaviasammaüle ramme//

2. The *Arghakāṇḍa* in 149 verses. Manuscripts:

LDI 7384 (675). 10ff. Copied by Muni Vīrakalaśa at Pattana in Saṃ. 1566 = A.D. 1509.
Benares (1963) 34701. 20ff.
BORI 1 of 1898/99. 14ff.
GVS 2753 (3795). 20ff. Incomplete.
LDI 7382 (737). 12ff.
LDI 7383 (8223/2). Ff. 3v–5v.
LDI 7385 (7438/2). F. 3v. Incomplete (20 gāthās).
LDI 7386 (1801). 11ff. With a vṛtti.
Surat, Jainananda Pustakalaya at Gopipura 3. See Velankar, p. 15.

Verses 1–2 are:

namiūṇa vaḍḍhamāṇaṃ saṃyamadevaṃ
  nareṃdathuapāvaṃ/
vocchāmi agghakaṃḍaṃ bhaviyāṇa hiyaṃ
  payatteṇa//
viraguruparaṃparāe kamāgayā ettha
  sayalasasatthaṃ/
laddhūṇa maṇualoe niddiṭṭhaṃ duggaeveṇa//

DURGARṢI (DUṂGARṢI) LAKṢMĪDATTA MĀPĀR (*fl.* 1905/1914)

Author of a *Parvasiddhigrantha*, published at Amadāvāda in 1905 (BM 14053. b. 43), and of a *Grahabhavanapatha*, published at Ahmadabad in 1914 (IO 1. B. 16).

DURGASIṂHA

Author of a *Jyotiṣaratna*. Manuscript:

Osmania University B. 12/3. 19ff. Incomplete.

DURGASIṂHA

Author of a ṭīkā on the *Muhūrtacintāmaṇi* of Rāma (*fl.* 1600). Manuscript:

Osmania University B. 15/f.2. 17ff. Copied in A.D. 1837. Incomplete (vivāhaprakaraṇa).

DURGĀCARAṆA VIDYĀLAṄKĀRA

Author of an *Āyurdāyavinirṇaya*. Manuscript:

Benares (1963) 35501 = Benares (1906) 1558. Ff. 1–20, 3ff., ff. 1–8, 1f., 1f., ff.1–4, ff. 1–24, and ff. 1–10. Bengālī. With a yoginīdaśā, maṅgalācaraṇa, grahaṣaḍvarga, lagnasphuṭānayana, and grahāṇāṃ balābala.

DURGĀDATTA ŚARMAN (*fl.* 1963)

Author of a *Jyotiṣajagat* in Hindī, published at Dillī-Vārāṇasī-Paṭanā in 1963.

DURGĀDĀSA PUROHITA

Author of a *Bhaḍalīpurāṇa* in Marāṭhī. Manuscript:

LDI (LDC) 2158. 9ff. Copied in Saṃ. 1789 = A.D. 1732.

DURGĀDĀSA PRASĀDA

Author of an *Adhimāsaparīkṣā* published at Bombay (Mysore GOL B 3984).

DURGĀPRASĀDA

Author of a *Kṣetramiti* published at Kalyāṇapura (Mysore GOL B 3851) and at Lakno (Mysore GOL B 3873).

DURGĀPRASĀDA ŚARMAN

Author of a *Hāyanacandrodaya* published with a bhāṣāṭīkā at Bombay (Mysore GOL B 4029).

DURGĀPRASĀDA (*fl.* 1884)

Author of a Hindī bhāṣānuvāda of the *Bṛhatsaṃhitā* of Varāhamihira (*fl. ca.* 550), published at Lucknow in 1884 (IO 13. I. 6).

DURGĀPRASĀDA DVIVEDA (*fl.* 1891/1936)

The son of Haradevī and Sarayūprasāda, a resident of Paṇḍitapurī near Pilkhāva to the west of Ayodhyā, Durgāprasāda was patronized by the Mahārāja of Jayapura. He wrote on jyotiṣa the following works:

1. A Saṃskṛta and Hindī vyākhyā, *Vilāsī*, on the *Bījagaṇita* of Bhāskara (b. 1114), in Śaka 1813 = A.D. 1891. This was edited by Girijāprasāda Dviveda, 3rd ed., Lakṣmaṇapura 1941 (the preface is dated Jayapura Saṃ. 1973 = A.D. 1916). Verse 2 at the beginning is:

tātaśrīsarayūprasādacaraṇasvarvṛkṣasevāparo
mātṛśrīharadevyapārakaruṇāpīyūṣapūrṇāntaraḥ/
hṛtpadmabhramarāyamāṇagiriśo durgāprasādaḥ sudhīr
adhyetṛpratibhodgamāya kurute bījopari vyākṛtim//

2. The *Jaiminipadyāmṛta* with his own vṛtti, *Mūlakundalī*, composed in Jayapura in Śaka 1828 = A.D. 1906 and published at Bombay in 1925.

3. The *Pañcāṅgābhibhāṣaṇa*, published at Lucknow in [1918] (IO San. B. 814(m)).

4. A ṭīkā, *Upapattīnduśekhara*, on the gaṇitādhyāya of the *Siddhāntaśiromaṇi* of Bhāskara (b. 1114), edited by Girijāprasāda Dviveda, Ahmadābād 1936.

DURGĀRĀMA

Author of a *Grahaṇa āryā*. Manuscript:

Assam (1930) 18. See NCC, vol. 6, p. 247.

DURGĀŚAÑKARA

Author of a *Gautamajātaka*. Manuscript:

RORI Cat. II 5649. 9ff. Copied in Saṃ. 1884 = A.D. 1827. With the ṭīkā of Lakṣmīpati.

He is probably identical with Durgāśañkara (*fl. ca.* 1825/1850), the brother of Lakṣmīpati.

DURGĀŚAÑKARA

Alleged author of a ṭīkā on a *Mallāripaddhati*. Manuscript:

N-W P I (1874) 121. 15ff. Property of Jagannātha Jotishi of Benares.

DURGĀŚAÑKARA PĀṬHAKA (*fl. ca.* 1825/1850)

An Audīcya Brāhmaṇa, the son and pupil of Śivalāla Pāṭhaka, the brother of Lakṣmīpati, and a resident of Kāśī, Durgāśañkara cast a horoscope at the birth of Navanihāla Siṃha (1821/1840), for which he received a reward from Khaḍgasiṃha (*fl.* 1839/40), the successor to Raṇajit Siṃha (*fl.* 1799/1839) of Lāhora, and was associated with Lancelot Wilkinson (*fl.* 1834/1837), the agent of Sīhora, to whom he wrote on Thursday 2 śuklapakṣa of Śrāvaṇa in Saṃ. 1894 = 3 August 1837. He was later the astrologer of Viśvanātha, the Mahārāja of Rīvāṃ. His pupils included Lajjāśañkara Śarman and Hīrānanda Caturveda. See S. Dvivedin [1892] 119–120. He wrote a *Sarvasiddhāntatattvacūḍāmaṇi*. Manuscript:

BM 501 (Or. 5259). Ff. 1–4, 6, 8, 11–16, 18, 20, 22–96, 98–100, 102–112, 114–116, 118–124, 126–155, and 157–304. From Fortescue W. Porter.

He also wrote a *Sūryādigrahasādhanasiddhānta* that mentions "asmattatsarvasiddhāntīya." Manuscript:

RORI Cat. II 5653. 5ff. Copied by Vrajavāsī Sillū at Maṇikarnikātīra in Kāśī on Monday 8 kṛṣṇapakṣa of Āśvina in Saṃ. 1893 = 31 October 1836.

DURGĀŚAÑKARA UMĀŚAÑKARA ŚARMĀ MUḌEṬĪKARA (*fl.* 1909)

Author of a Gujarātī ṭīkā on the *Jātakālañkāra* of Gaṇeśa (*fl.* 1613), published at Bombay in Saṃ. 1966 = A.D. 1909 (IO 25. C. 38).

DURGĀSAHĀYA

Author of an *Abdaratna*. Manuscripts:

GJRI 890/2. 7ff. Copied in Saṃ. 1899 = A.D. 1842.
VVRI 4775. 9ff. Copied in Śaka 1785 = A.D. 1843.
Goṇḍal 3. 3ff. Copied in Saṃ. 1945 = A.D. 1888. Incomplete. No author mentioned.
Alwar 1709.
Benares (1963) 36488. 10ff. Incomplete. No author mentioned.
Kurukṣetra 27 (50052).
Radh. 33 and 43. No author mentioned. See NCC, vol. 1, rev. ed., p. 271.
Śāstrī, Not. 1911. 15. 7ff. Property of Paṇḍita Jayanārāyaṇa Vājapeyin of Patna.

The *Abdaratna* was published in Saṃ. 1918 = A.D. 1861; see Benares (1878) 137 and 138 = Benares (1869) XXXIII 5 and 6. 9ff.

DURGĀSAHĀYA

Author of a *Muhūrtasāgara*. Manuscript:

AS Bengal 2757 (G 10006). 28ff.

DURYODHANA

Author of a *Praśnacatuḥṣaṣṭi*. Manuscripts:

Kerala 10357 (L. 264). 475 granthas.
Kerala 10358 (T. 1112). 475 granthas.

DURYODHANA (*fl.* 1461)

The son of Vidyādhara, the son of Bhavaśarman of the Maudgalyagotra, a resident of Lavaṇīpurī, Duryodhana wrote a *Jñānapradīpacintāmaṇi* = *Praśnatantra*, which he completed on 8 kṛṣṇapakṣa of Bhādrapada in Saṃ. 1518 = *ca.* 29 August 1461. Manuscripts:

AS Bengal 7052 (G 5478) 25ff. Copied at Pāṭana on Sunday 13 kṛṣṇapakṣa of Jyeṣṭha in Saṃ. 1716, Śaka 1581 = *ca.* 5 June 1659 Julian.
RORI Cat. II 5567. 31ff. (f. 1 missing). Copied in Saṃ. 1838 = A.D. 1781.
SOI 11082.

The last 5 verses are:

x x x x x x x x x x x harijānanāmā
gauḍānvayeṣu lavaṇīpurīramyanivāsavāsī/
maudgalyagotras tu pavitravaṃśo
bhavaśarmanāmā śrutiyajñavettā//
tatputranāmākhyavidyādharākhyo
brahmajñadevārcanabhaktiraktaḥ/
prāsādavāpīkṛtadharmavidyo
yaśaḥprasiddho bahubhāgyayuktaḥ//
tasyaiva vaṃśe sutanur babhūva
jyotirvidāṃ vedavidāṃ cakāraḥ/
tayā sukṛtyā khalu jñānadīpaṃ
śrīduryodhanasya daivajñahetoḥ//
vedākṣaśatacatvāri racitaṃ jñānadīpakam/
aṣṭādaśāstrum (?) adhyāyaṃ
śrīduryodhanadhīkṛtam//
śrīvikramagate kāle nāgendutithivatsare/
nabhasyakṛṣṇam aṣṭamyām utpannaṃ
jñānadīpakam//

DURLABHA (*fl.* 932)

A resident of Multān, Durlabha wrote a *zīj* whose epoch is Śaka 854 = A.D. 932; this is known only from the *Fī taḥqīq mā li-ᵓl-Hind* of al-Bīrūnī (b. 973) (pp. 348 and 388 ed. Hyderabad; vol. 2, pp. 9–10 and 54 trans. Sachau).

DURLABHARĀJA (*fl.* 1160)

The son of Narasiṃha, the son of Rājapāla, the son of Jāhilla of the Prāgvāṭakula, a minister of finance of the Caulukya monarch Bhīma I (*fl. ca.* 1031/1065), Durlabharāja was made a mahattama by Kumārapāla (*fl. ca.* 1143/1172). He began the *Sāmudrikatilaka* in 800 āryās in *ca.* A.D. 1160; it was finished by his son, Jagaddeva (*fl. ca.* 1175). Manuscripts;

Anup 5273. 10ff. Copied in Saṃ. 1524 = A.D. 1467.
Udaipur 580. Copied in Saṃ. 1632 = A.D. 1575.
AS Bombay 401 = AS Bombay (Indraji) 90. 23ff. Copied in Saṃ. 1744 = A.D. 1687.
Kathmandu (1960) 467 (I 1195). 55ff. Nevārī. Copied in NS 808 = A.D. 1688. Incomplete.
BORI 568 of 1899/1915. 27ff. Copied in Saṃ. 1935 = A.D. 1878.
Baroda 13200. 45ff. Copied in April of A.D. 1928.
Anup 5272. 56ff.
Bombay U Desai 1505. 38ff.
Bombay U Desai 1506. 10ff. Incomplete (ends at 2, 7).
Bombay U Desai 1507. Ff. 11–13. Incomplete (continues Bombay U Desai 1506).
BORI 348 of 1879/80. 25ff.
BORI 569 of 1899/1915. 27ff.
BORI 190 of 1902/07. 13ff.
Poleman 5225 (Harvard 1110). 33ff. Incomplete.
Rajputana, p. 47. In Bikaner.
Rajputana, p. 54, Property of the State Library in Bikaner. Is this Anup 5272?

Verses 1–6 at the end are:

atrāsti ko ᵓpi vaṃśaḥ
prāgvāṭākhyas trilokavikhyātaḥ/
nṛpasaṃsadi vṛddhāyām
ālambanayaṣṭir abhavad yaḥ//
āsīt tatra vicitra-
śrīmajjāhillasaṃjñayā jātaḥ/
vyayakaraṇapadāmātyo
nṛpateḥ śrībhīmadevasya//
samajani tadaṅgajanmā
prathitaḥ śrīrājapāla iti nāmnā/
pratipakṣadvipasiṃhaḥ

śrīnarasiṃhaḥ sutas tasya//
śrīmān durlabharājas
tadapatyaṃ buddhidhāma sukavir abhūt/
yaṃ śrīkumārapālo
mahattamaṃ kṣitipatiḥ kṛtavān//
prakṣālayituṃ malam iva
vāṇī majjati vapurvidhāmbudhiṣu/
yasyāvikhyāsavasatī (?)
rājaturaṅgaśakunaprabandheṣu//
tenopajñātam idaṃ
puruṣastrīlakṣaṇaṃ tadanu kavinā/
tasyaiva sutena jagad-
devena samarthayāṃcakre//

DURVALI

Author of a *Tithinirṇaya*. Manuscript:

WHMRL E. 11. 2. Copied in Saṃ 1886 = A.D. 1829.

DULLAHA (*fl.* 1776)

At the request of Śiva, Dullaha wrote an udāharaṇa on the *Jātakapaddhati* of Śrīpati (*fl.* 1039/1056), which he completed on Thursday 5 śuklapakṣa of Āśvina in Śaka 1698 = 17 October 1776. Manuscripts:

Mithila 375. 35ff. Maithilī. Copied on Friday 1 kṛṣṇapakṣa of Āśvina in Śaka 1766 = 25 October 1844. Property of Babu Puruṣottama Jhā of Babhanagama, Supaul, Bhagalpur.
Mithila 375 A. 37ff. Maithilī. Incomplete. Property of Paṇḍita Dharmadatta Miśra of Babhanagama, Supaul, Bhagalpur.

The first 2 verses are:

bhāsvantam x x x praṇamyodāharaṇayojanam/
kriyate śīghrabodhāya dullahena mude mayā//
śāke vasvaṅkabhūte (°bhūpe) gatavati bhabhade cāśvine śuklapakṣe
pañcamyāṃ jīvavāre śubhadam iti mayā śrīśivānujñayā/
bāle bodyā vihīne paṭhati sati tadā racyate dullahena
śraipatyaṃ vai samantād vivaraṇam akhilaṃ kvailakhagrāmamadhye//

Dullaha also wrote a ṭippaṇī on the *Tājika* of Nīlakaṇṭha (*fl.* 1587). Manuscript:

Mithila 130. 5ff. Maithilī. Incomplete. Property of Babu Puruṣottama Jhā of Babhanagama, Supaul, Bhagalpur.

The first verse is:

śrīkṛṣṇacaraṇāmbhojaṃ natvā śrīdullaho mudā/
atha ślokānvayaṃ cakre nīlakaṇṭhyāṃ kvacit kvacit//

DEVA

Alleged author of a *Praśnasaṅgraha* or *Praśnamīta*. Manuscript:

ABSP 430. Ff. 2–14. Copied in Saṃ. 1863 = A.D. 1806. Incomplete.

DEVAKĪNANDANA

Author of a *Daivakīnandana*. Manuscript:

Śāstrī, Not. 1904. 107. 32ff. Bengālī. Copied by Rāmasundara Śarman. Property of Paṇḍita Rakṣākara Nyāyapañcānana of Dakṣiṇābhāga, Kālīgañja, Ḍhākā.

The second verse is:

ādau bhāskaram īśvaram x x x x sākṣāt surāṇāṃ varaṃ
viśveśvaraṃ (?) viśvagataṃ x x x x x x x x x x x/
nānāśāstram upāsya bhāskaramataṃ cālokya vārāhakaṃ
jyotiḥśāstram akalpayan navam idaṃ śrīdevakīnandanaḥ//

DEVAKĪNANDANA (*fl.* 1807/1838)

The son of Jīvānanda, the son of Lakṣmīdhara, a resident of Mallikā on the northern side of Mount Kūrma (Kumaon, U.P.), Devakīnandana wrote the following works on jyotiṣa.

1. A ṭīkā, *Ānandakanda*, on the *Kalpavallīpaddhati* of Viṭṭhala (*fl.* 1626), composed in Śaka 1729 = A.D. 1807. Manuscript:

Jammu and Kashmir 4002. 83ff.

Verses 3–6 are:

samīcīno dīnottamajanavilīnottamarataṃ
samāsīno ᵓhīnoditapadavurīṇo ᵓtra ca janaḥ/
vihīnodāsīno rasikajanapīno himagireḥ
samāsannāsīno madanasadanaṃ kūrmakudharaḥ//
ihāsīd visvāsī sunigamavidāṃ durmatibhidāṃ
śivāyāḥ sevāyāḥ paramavidhivijñānasunidhiḥ/
vidhijñānāṃ cūḍāmaṇisaraṇimāheyataraṇi-
praphulatpādābjaḥ pravarataralakṣmīsurabudhaḥ//
śeṣāśeṣārthavettā kaluṣitamanasāṃ cittabhettā ripūṇāṃ
madhye tāro yadīyaṅghrisarasiruha x m arcyo valānā mahāntaḥ/
yadgīrjyotirvivāde sakalavidhividāṃ nirvivādo hi vedo
jīvānandaḥ suto ᵓsyājani vimaladhiyāṃ sarvadānandamūrtiḥ//
teṣāṃ pādāmbhojalāṅghriprasādo
natvaivaitān devakīmātaraṃ ca/
satpaddhatyāṣ ṭippaṇaṃ kalpavalyāḥ
kurve horākovidānandakandam//

Verses 5–7 at the end are:

grāme tasmin mallikākhyāṃ dadhāne
kūrmākhyādrer uttare deśabhāge/
saṃsthe nānāśāstracarcāpravīṇa
āsīd daivajño hi lakṣmīdharākhyaḥ//
jyotiḥśāstre sarvagarvāpahārī
khyātaḥ pṛthvyāṃ sarvasiddhāntavettā/
śeṣoktīnāṃ cāpi sārasya vettā
jīvānandākhyo ᵓsya putro babhūva//
putras teṣāṃ goyamādrīnduśāke
tatpādābjādhyānasamprāptabodhaḥ/
granthaṃ horākovidānandakandaṃ
cakre pūrṇaṃ devakīnandanākhyaḥ//

2. The *Kṛpāpaddhati*, composed in Śaka 1736 = A.D. 1814. Manuscripts:

Alwar 1728.
Jammu and Kashimir 4000. 6ff.

3. The *Horāhaskara*, composed in Śaka 1760 = A.D. 1838. Manuscript:

Jammu and Kashmir 3987. 96ff.

DEVAKĪNANDANA (*fl.* 1882)

A resident of Haripura, Devakīnandana wrote an enormous *Muhūrtasindhu* = *Bṛhat Muhūrtasindhu* for Meharacandra in Śaka 1804 = A.D. 1882. This was published at Muṃbaī in 1885.

DEVAKĪNANDANA SIṂHA (*fl.* 1934)

Author of a *Jyotiṣaratnākara* in Hindī, of which the 1st khaṇḍa was published at Vārāṇasī in Saṃ. 1991 = A.D. 1934; 2nd ed., Vārāṇasī Saṃ. 2014 = A.D. 1957.

DEVAKĪRTI (*fl.* before 800)

A rājā often cited as an authority on jātaka—e.g., by Kalyāṇavarman (*fl. ca.* 800) in *Sārāvalī* 37,1; by Utpala (*fl.* 966/968) on *Bṛhajjātaka* 1, 19–20; 2,7; and 9,8 (see P. V. Kane [1948/49] 24); and by al-Bīrūnī (b. 973) in *Fī taḥqīq mā li-ᵓl-Hind* (p. 123 ed., Hyderabad; vol. 1, p. 158 trans. Sachau).

DEVAKṚṢṆA ŚARMAN (b. 9 November 1818)

The son of Rāmadhana Miśra, a Gauḍa Brāhmaṇa, Devakṛṣṇa studied jyotiṣa at the Kāśika Rājakīya Pāṭhaśālā under Lajjāśaṅkara. He taught jyotiṣa at Jambūnagara in Kāśmīra for nine years beginning in Śaka 1781 = A.D. 1859 at the request of Raṇavīra Siṃha, the Mahārāja of Kāśmīra from 1857. In 1868 he succeeded Nandarāma Śarman at the Kāśika Rājakīya Pāṭhaśālā. He died at Vārāṇasī in Śaka 1811 = A.D. 1889. Among his pupils was Sudhākara Dvivedin (*fl.* 1892/1907). See S. Dvivedin [1892] 125–126.

DEVACANDA

Author of a *Karmavipāka*, Manuscripts:

CP, Hiralal 697. Property of Rāmlāl of Dhūmā, Seonī.
CP, Hiralal 698. Property of Kārelāl of Śobhāpur, Chhindwārā.
CP, Hiralal 699. Property of Jagannātha Śukla of Hardā, Hoshangābād.
CP, Hiralal 700. Property of Ajodhyābhaṭ of Hardā, Hoshangābād.
CP, Hiralal 701. Property of Govindrām Bhaṭ of Hardā, Hoshangābād.

DEVADATTA

Author of a *Karapañcāṅga*. Manuscript:

IM Calcutta 1331. See NCC, vol. 3, p. 177.

DEVADATTA MIŚRA

Author of a *Laghusaṅgraha*. Manuscript:

Benares (1963) 37071 = Benares (1911–1912) 2077. 11ff. Copied in Saṃ. 1815 = A.D. 1758.

DEVADATTA (*fl.* 1662)

The son of Nāgeśa, the son of Govinda, the son of Keśava of the Bhāradvājagotra, and the uncle of Murāri (*fl.* 1665), Devadatta wrote a *Grahaprakāśa* in Śaka 1584 = A.D. 1662; see SATE 142–149. Manuscripts:

BM 474 D (Add. 26,448e). 11ff. See SATE 17.
BORI 149 of A 1883/84. 4ff. (*Grahalaghuprakāśa*).

Verses 1–3 are:

praṇamya nāgānanamantrapūrṇaṃ
guruṃ maṭāmbāṃ ravimukhyakheṭān/
sacchiṣyabodhārtham ahaṃ prakurve
grahaprakāśaṃ sulaghuprakāram//
dhyātvā viśveśvaraṃ devaṃ tathā siddhivināyakam/
smṛtvā gurupadāmbhojaṃ tato labdhvāvabodhakam//
jāto ᵓham agryeṇa manorathena
yasyānukampāmṛtavṛṣṭipṛṣṭaḥ/
pāraṃgataś cākhiladarśanānāṃ
nāgeśabhaṭṭam janakaṃ nato ᵓsmi//

Two further verses give his genealogy:

bhāradvājakule ᵓsya vipratilakaḥ śrotre paro naiṣṭiko mantrajñaḥ śrutiśāstravic ca kuśalaḥ śrīkeśavo devavit/
tatputro bhiṣajajñavedanipuṇaḥ smārtaparo daivavid
govindākhyatadātmajo ᵓticaturo nāgeśatatsūnunā//
śrīdevadattena kṛto hi samyak

siddhāntapakṣānugadṛṣṭigo varaḥ/
grahaprakāśo ᵓtilaghuprakāro
grāhyaḥ sudhībhiḥ pariśodhanīyaḥ//

Devadatta also wrote a ṭīkā on the *Grahaprakāśa*. Manuscript:

BM 474 E (Add. 26,448f). 2ff. See SATE 17.

The colophon begins: iti śrīmaddaivajñanāgeśātmajadevadattaviracitāyām.

DEVADATTA ŚĀSTRIN (*fl.* 1899)

Author of a *Ramalabhairava* = *Vijayacandra*, published at Kāśī in 1899 (BM 14053. cc. 40. (2)).

DEVADĀSA

The son of Nāmadeva, the son of Arjuna of the Gautamagotra and Mālavajñāti (?), Devadāsa wrote a *Devadāsaprakāśa*. Manuscripts:

AS Bengal 2681 (G 10620). 253ff. Copied on 4 kṛṣṇapakṣa of Vaiśākha in Saṃ. 1943 = *ca.* 21 May 1886. Formerly property of Bālamukunda.
AS Bengal 2682 (G 1433). 243ff.
Bikaner 816. 13ff. Incomplete (ends with malamāsanirṇaya).
Mitra, Not. 1832. 316ff. Maithilī. Property of the Rājā of Darbhāñgā.

The last verse is:

āsīd gautamagotrajo ᵓrjuna iti śrīmālavajñā⟨tiko⟩
vandyas tattanayo maharṣisadṛśaḥ śrīnāmadevaḥ kṛtī/
tasyopāsanakarmaṭhaḥ smṛticaṇaḥ śrīdevadāsaḥ sutas
tenākāri nibandharatnam akhilagranthārthasārapradam//

DEVADĀSA MIŚRA

Author of a *Tithinirṇaya*. Manuscript:

BORI 258 of 1887/91. 23ff. From Gujarāt.

DEVANANDIN

Author of a *Garbhaṣaḍāracakra*. Manuscript:

RJ 1642 (vol. 2, p. 270). 6ff. Property of Baḍā Terahapanthiyoṃ of Jayapura.

DEVANANDIN

Author of a *Svapnāvalī*. Manuscripts:

RJ 3133 (vol. 4, p. 295). 3ff. Copied on 13 śuklapakṣa of Bhādrapada in Saṃ. 1958 = *ca.* 25 September 1901.
RJ 3134 (vol. 4, p. 295). 3ff.

DEVANĀTHA ṬHAKURA TARKAPAÑCĀNANA

Author of a *Smṛtikaumudī* which deals, among other things, with tithis. Manuscripts:

Mithila I 437 C. 70ff. Copied in Saṃ. 1947 = A.D. 1890. Property of the Rāj Library, Darbhanga.
Darbhanga 134 (S 9). Ff. 31–33, 35, 67, and 70. Maithilī. Incomplete. No author mentioned.
Mithila I 437. 93ff. Maithilī. Property of Pandit Śrīkānt Jhā of Naḍuār, Jhañjhārpur, Darbhanga.
Mithila I 437 A. 166ff. Maithilī. Property of Pandit Dīnabandhu Jhā of Isahapur, Manīgāchī, Darbhanga.
Mathila I 473 B. 136ff. Maithilī. Property of Pandit MM. Rājināth Miśra of Saurāth, Madhubani, Darbhanga.
Mithila I 437 D. 168ff. Maithilī. Property of the Rāj Library, Darbhanga.

The colophon begins: iti tarkapañcānanamahopādhyāyadevanāthaṭhakurakṛtāyāṃ.

Probably a part of the *Smṛtikaumudī* is the *Kālakaumudī*. Manuscript:

Mithila I 60. 70ff. Maithilī. Incomplete. Property of Babu Chandradhārī Singh of Rauti Deaurhi, Madhubani, Darbhanga.

DEVABHADRA SŪRI (*fl. ca.* 1175)

The pupil of Śrīcandra Sūri (*fl. ca.* 1150), Devabhadra wrote a vṛtti on his guru's *Saṅgrahaṇīratna*. Manuscripts:

AS Bombay 1682. 45ff. Copied in Saṃ. 1482 = A.D. 1425. From Bhāu Dājī.
Paris BN (Senart) 275 (Sanscrit 1665). 51ff. Copied in A.D. 1427.
LDI 3095 (3633). 13ff. Copied by Mahaṃ Dāmāka of the Kāyasthajñāti in Saṃ. 1486 = A.D. 1429.
LDI 3096 (2263). 10ff. Copied by Vyāsa Padma, the son of Narbada of the Dīsāvālajñāti, at Vaṭapadra in Saṃ. 1488 = A.D. 1431.
LDI 3094 (3783). 23ff. Copied in Saṃ. 1504 = A.D. 1447.
BORI 815 of 1899/1915. 42ff. Copied in Saṃ. 1660 = A.D. 1603.
LDI 3087 (2619). 20ff. Copied in Saṃ. 1687 = A.D. 1630.
LDI (KS) 520 (10064). 52ff. Copied for Bharamādevī, the daughter of Maladhārī and the wife of Śā Rājasika, the son of Rupasī, the son of Śā Bhīmasī of the Śaṅkhavālagotra, in Saṃ. 1699(?) = A.D. 1642.
Agra, Vijayadharma Lakṣmī Jñānamandira 1294 and 1295. See Velankar, p. 410.
Ahmadabad, Bhandar of the Vimala Gaccha Upāśraya, Falusha's Pole 17 (26) and Haji Patel's Pole 34 (15 and 16), 35 (22, 24, 34, 38, and 43), and 37 (23), and 13 (14) with Udyotavimalagaṇi. See Velankar.

Ahmadabad, Dela Upāśraya Bhandar, ground floor 55 (2 to 10) and first floor 33 (3 and 4). See Velankar.
AS Bengal Jaina 7571.
Baroda 3008. 98ff.
Baroda, Haṃsavijayaji Maharaj at the Kantivijaya Bhandar 312 and 1373. See Velankar.
Berlin 1950 (or. fol. 742). 98ff. (ff. 21–32 missing).
Berlin (Jaina) 751 (or. fol 2419). 80ff.
Berlin (Jaina) 752 (or. fol 2673). 45ff.
BORI 106 of 1869/70. 72ff.
BORI 207 of 1873/74. 23ff. From Surat.
BORI 877 of 1892/95. 112ff.
BORI 850 of 1895/1902. 74ff.
BORI 1311 and 1312 of 1895/1902. See Velankar.
Cambay II 151. 360ff.
Chani, Bhandar of Kantivijayaji Maharaj 528. See Velankar.
Florence 653. 112ff.
Jaipur, Inner Bhandar of Harisāgaragaṇi 43 and Outer Bhandar 29. See Velankar.
Jesalmere 132(2). Ff. 187–275.
Jesalmere 260. 256ff.
Jesalmere, Bada Bhandar 14, 125, and 879. See Velankar.
Jesalmere, Bhandar of the Bhāṇṭhaki Kundi 71 and 275. See Velankar.
Jesalmere, Sambhavnath Temple 136. See Velankar.
Kaira, Bhandar of Sammatiratna Sūri 52 and 103. See Velankar.
LDI 3088 (5511). 64ff.
LDI (KS) 521 (10533). 9ff.
LDI (VDS) 491 (9831). 11ff.
Leningrad (1918) 188. 68ff.
Limdī 1233. See Velankar.
Mandvi, Anantanātha Mandira 17. See Velankar.
Mitra, Not. 2737. 112ff. Property of Rāya Dhanapat Siṃha, Bahādur, of Āliṃgañj.
Oxford 1367 (Sansk. d. 323) = Hultzsch 473. Ff. 1–3 and 10–26.
Oxford 1368 (Sansk. d. 324) = Hultzsch 474. 23ff.
Patan, Bhandar at the Agali Sheri 46 (21 and 22), 48 (25), 62 (18), 74 (35), 75 (25), and 77 (15). See Velankar.
Patan, Sangha Bhandar 21 (8) and 23 (78). See Velankar.
Patan, Sangha Bhandar, Limdi Pada 3 (27). See Velankar.
Patan, Vad. Pārśvanātha Pustaka Bhandar 5 (36). See Velankar.
Punjab 2714 and 2715. See Velankar.
Surat, Jainānanda Bhandar 115, 1516, and 2664. See Velankar.

The *Saṅgrahaṇīvṛtti* was edited by Muni Lalitavijaya, Bombay 1915 (BM 14101. d. 23 and IO 17. B. 40) and at Bhavnagar (see Velankar, p. 409). It ends:

śrīharṣapurīyagacchālaṅkāramaladhāriśrīmadabhayadevasūripaṭṭaratnaśrīhemacandrasūriśiṣyaśrīcandrasūricaraṇāmbujacañcarīkeṇa śrīmunicandrasūribhyo labdhapratiṣṭhena śrīdevabhadrasūriṇā viracitā.

Devabhadra is also alleged to have composed a vṛtti on the *Kṣetrasamāsa* of Jinabhadra (*fl.* 609), though this is probably the work of his pupil's pupil, Ānandasūri; see Velankar, p. 99.

DEVABHADRA PĀṬHAKA (*fl.* 1755)

The son of Bhāgīrathī and Balabhadra Pāṭhaka, the son of Gaṅgādhara Pāṭhaka, the son of Rāmacandra Pāṭhaka of the Nāgarajāti, and a pupil of Hariśaṅkara, Devabhadra wrote a vyākhyā on the *Nakṣatrasatrasūtra* of Baudhāyana, completed on Sunday 2 kṛṣṇapakṣa of Bhādrapada in Saṃ. 1812 = 21 September 1755. Manuscripts:

Mitra, Not. 4180. 45ff. Copied in Śaka 1753 = A.D. 1831. Property of AS Bengal.
CP, Kielhorn I B 36. 87ff. Copied in Śaka 1770 = A.D. 1848. Property of Bābā Śāstrī Bhāke of Chāndā.

The first 2 verses are:

śrīvighneśaṃ muniṃ baudhāyanaṃ kātyāyanaṃ gurum/
asmadvṛddhajanānān tu guruṃ ca hariśaṅkaram//
yāgakālavivektāraṃ gaṅgādharaṃ tu pāṭhakam/
pitaraṃ balabhadraṃ ca bhāgīrathīṃ tu mātaram//

The next to the last verse is:

netrenduvasume (?) varṣe ᵓsite nabhasi bhāskare/
pitṛbhe ca dvitīyāyāṃ sūtrabhāṣyam idaṃ kṛtam//

The colophon begins: iti śrīmanmahāyājñikanāgarajātīyapāṭhakaśrīrāmacandrasūnugaṅgādharapāṭhakavaṃśasambhūtapāṭhakaśrībalabhadrātmajadevabhadrakṛtau.

Devabhadra also wrote a *Vāravārdhuṣikasya Vārasaṅkhyāsaṃskāravidhi*. Manuscript:

Jammu and Kashmir 4683. 4ff. Copied in Saṃ. 1816 = A.D. 1759.

And he is apparently identical with the author of a *Grahayajñaprabodha*. Manuscript:

IM Calcutta 5139. See NCC, vol. 6, p. 256.

DEVARĀJA

The son of Varadārya or Varadarāja of the Atrigotra, Devarāja wrote a *Kuṭṭākāraśiromaṇi* explaining the algebra of Āryabhaṭa (b. 476). He also wrote a ṭīkā on this, the *Mahālakṣmīmuktāvalī*, in which he mentions Bhāskara (b. 1114). Manuscripts:

Mysore (1922) 4398. 35ff.
Mysore (1911 + 1922) B 596. 10ff.
Mysore (1911 + 1922) B 597. 52ff. With the ṭīkā.
Mysore (1922) B 975. 4ff. With the ṭīkā.
Tanjore D 11355 = Tanjore BL 11050. Ff. 6–59. Grantha. Incomplete. With the ṭīkā.

The *Kuṭṭākāraśiromaṇi* with the ṭīkā was edited by K. Seshacharya. *Maharaja's Sanskrit College Magazine* 5, 1929, 145 sqq. (see NCC, vol 4, p. 369), and from two of the Mysore manuscripts by B. D. Āpaṭe as *ASS* 125, Poona 1944. The first verse of the mūla is:

natvā ramādharaṇyau
varadāryasutena devarājena/
āryabhaṭācāryakṛtaḥ
kuṭṭākāraḥ prakāśyate spaṣṭam//

The colophon of the ṭīkā begins: ity atrikulābharaṇasya skandhatrayavedinaḥ siddhāntavallabha iti prasiddhāparanāmnaḥ śrīvaradarājācāryasya tanayena devarājena viracitāyāṃ.

DEVARĀJA = DEVARĀMA

Author of a *Muhūrtaparīkṣā* or *Muhūrtamuktāvalī*. Manuscripts:

PL, Buhler IV E 351. 4ff. (*Muhūrtaparīkṣā* of Devarāja). Property of Maṅgala Śaṅkara of Ahmadābād.
PL, Buhler IV E 358. 8ff. (*Muhūrtamuktāvalī* of Devarāma). Property of Lalubhāī Jośī of Ahmadābād.

DEVALA (*fl.* third or fourth century)

A well known authority on astrology and divination quoted often by Varāhamihira (*fl. ca.* 550), Utpala (*fl.* 966/968), and others; see P. V. Kane [1948/49] 6. In one place (*Bṛhatsaṃhitā* 86,1) Varāhamihira indicates that he was quoted by Ṛṣabha. There exists a *Kākaruta* in 32 verses ascribed to him (actually based on him). Manuscripts:

BORI 86 of 1892/95. 5ff. Copied in Saṃ. 1630 = A.D. 1573. Attribution to Devala from NCC, vol. 3, p. 296.
LDI 7458 (883). 4ff.
Udaipur, Sarasvati Bhandar 84, 78. See NCC.
WHMRL G. 20. g. Ff. 1–2.

Verse 1 is:

kākarutaṃ pravakṣyāmi devalena niveditam/
lābhālābhādikaṃ sarvaṃ yena jānanti mānavāḥ//

The colophon is: iti devalarṣikṛtakākarutam.

He is also alleged to be the author of a *Gomukhajananaśānti*. Manuscripts:

GOML Madras D 3289. 3pp. Telugu.
GOML Madras D 3292. 8pp. Nandināgarī.

DEVAŚĀLI MUNI

Author of a *Bhāvakārikā*. Manuscript:

Bombay U 501 B. F. 8.

DEVASŪRI

Author of a *Janmapradīpa*. Manuscripts:

BORI 1345 of 1884/87. 5ff. Copied in Saṃ. 1741 = A.D. 1684. From Gujarāt.
Ahmadabad, Dela Upāśraya Bhandar, first floor 24 (221 and 222). See Velankar, p. 129.

DEVASVĀMIN (*fl.* third, fourth, or fifth century)

An astrologer quoted by Varāhamihira (*fl. ca.* 550) in *Bṛhajjātaka* 7, 6–7 and by Utpala (*fl.* 966/968) *ad. loc.* See P. V. Kane [1948/49] 6.

DEVĀCĀRYA (*fl.* 689)

Author of a *Karaṇaratna* whose epoch is Śaka 611 = A.D. 689, and which is based on Āryabhaṭa (b. 476). Manuscripts:

Kerala 3045 (T. 559) = Kerala C 662 (C. 559). 24pp.
Mysore (1922) 4477. Ff. 46–54. No author mentioned.
Mysore (1911 + 1922) B 576. Ff. 156–168. No author mentioned.

The colophon begins: iti devācāryakṛtau.

DEVĀNANDA SŪRI (*fl.* 1398)

The pupil of Padmaprabha of the Pūrṇimā Gaccha, Devānanda wrote in Śaka 1320, Saṃ. 1455 = A.D. 1398 a *Kṣetrasamāsa* which is sometimes attributed to Candraprabha. Devānanda wrote his own vṛtti on this. Manuscripts:

LDI 2992 (1395/1). 10ff. Copied by Lalitasundara in Saṃ. 1536 = A.D. 1479. With a ṭippaṇī.
Ahmadabad, Dela Upāśraya Bhandar, first floor 33 (20). With his own vṛtti. See Velankar, p. 100.
Baroda, Hamsavijayaji Maharaj at the Kantivijaya Bhandar 1590. With his own vṛtti. See Velankar.
Baroda, Kantivijayaji 332. With his own vṛtti. See Velankar.
Chani, Bhandar of Kantivijayaji Maharaj 5. With his own vṛtti. See Velankar.
Chani, Bhandar of Kantivijayaji Maharaj 287. Ascribed to Candraprabha. See Velankar.
LDI 2991 (576). Ff. 32–52. With his own vṛtti.
Patan, New Sangha Bhandar, Paper 18 (5). With his own vṛtti. See Velankar.
Patan, Sangha Bhandar, Limdi Pada 5 (31). Ascribed to Candrapabha. See Velankar.
Poona, Fergusson College, Mandlik Library Suppl. 457. With his own vṛtti, See NCC vol., 5, p. 159.
Surat, Jainānanda Bhandar 472. Ascribed to Candraprabha. See Velankar.

DEVĪDATTA

Apparently the son of Muralīdhara and the pupil of Devīdāsa, Devīdatta wrote a *Jyotiṣakaustubha* and a ṭīkā on the same; we have only the 6th mayūkhoddīpana, on vāstu. Manuscript:

AS Bengal 7069 (G 6344). 43ff. Copied in Saṃ. 1841 = A.D. 1784.

The colophon begins: iti śrīdaivajñamuralīdharātmajagurudevīdāsaraghunāthacaraṇārcanasāvadhānadevīdattaviracitāyāṃ.

DEVĪDATTA JOŚĪ (*fl.* 1922)

Author of a *Sugamajyotiṣa* published with a Hindī translation at Allahabad in 1922 (IO San. B. 617), 2nd ed. Almora 1932 (NL Calcutta 180. Kb. 93. 9).

DEVĪDATTA (*fl.* 1885)

Author of a *Siṃhasthagurunirṇaya*, published at Chhapra in 1885 (BM).

DEVĪDAYĀLU (*fl.* 1906/1917)

Author of a pañcāṅga for Saṃ. 1963–1971 = A.D. 1906–1914, published at Lāhaura in 1906 (BM 14096. dd. 7) and of another, *Pañcāṅgadivākara*, for Saṃ. 1975 = A.D. 1918, published at Lāhaura in 1917 (BM 14055. ddd. 1. (1)).

DEVĪDAYĀLU BHĀRADVĀJA (*fl.* 1913)

Author of a pañcāṅga, *Tithipatrikā*, for Saṃ. 1970 = A.D. 1913, published at Amṛtasara in 1913 (BM 14096. b. 8. (3)).

DEVĪDĀSA

Alleged author of a ṭīkā on the *Tattvārthādhigama* of Umāsvāti (*fl.* first century); see Velankar, p. 156.

DEVĪDĀSA (*fl. ca.* 1600/1625)

The son of Lāla of the Bharadvājagotra, a resident of Kānyakubja, Devīdāsa was the uncle of Balabhadra (*fl.* 1655). The last mentions Devīdāsa's ṭīkās on the *Vyakta* or *Bījagaṇita* of Bhāskara (b. 1114) and on the *Śrīpatipaddhati* of Śrīpati (*fl.* 1039/1056) in his *Hāyanaratna:*

tasyātmajāḥ pañca babhūvur eṣāṃ
śrīdevidāsaḥ prathamo babhūva/
vyakte ca yaḥ śrīpatipaddhatau ca
ṭīkāṃ vyadhāc chiṣyagaṇasya tuṣṭyai//

DEVĪPRASĀDA ŚUKLA

Author of a *Yogadīpikā*. Manuscripts:

Oudh (1876–1878) VIII 2. 108pp. Copied in A.D. 1827. Ascribed to Devīdatte. Property of Mannālāla of Tirwā, Lucknow Zila.

Oudh IX (1877) VIII 8. 50pp. Copied in A.D. 1858. Property of Rāmadayāla of Lucknow.

Lucknow 520. D 37 Y (45707).

DEVĪSAHĀYA

Author of a *Muhūrtaracana*. Manuscript:

Lucknow 520. D 37 M (45547).

DEVĪSAHĀYA

The son of Kṛṣṇakaura, the son of Śobhārāma, the son of Mahādevapada, the son of Śivadattarāya of the Bharadvājagotra, Devīsahāya wrote a ṭīkā, *Līlāvatīvilāsa*, on the *Līlāvatī* of Bhāskara (b. 1114). Manuscripts:

AS Bengal 6918 (G 5503). 63ff. Copied in Saṃ. 1817 = A.D. 1760.

Jammu and Kashmir 2891. 64ff. Copied in Saṃ. 1908 = A.D. 1851.

Jammu and Kashmir 2837. 78ff.

N-W P I (1874) 35. 29ff. Property of Govinda Bhaṭṭa of Mirzapore.

VVRI 5745. 64ff.

At the end are the following verses:

śrīmadbharadvājamuneḥ kulābdhau
dvijādirājo ᵓjani puṇyakāyaḥ/
sa x uṇajātimaheśabhālā-
laṅkārabhūtaḥ śivadattarāyaḥ//
śrīmanmahādevapadāravinda-
nimagnacetāḥ sukṛtāmburāśiḥ/
tasmān mahādevapadābhidheyo
budhopameyaḥ suta udbabhūva//
āsīt tasya suto ᵓparo guṇigaṇagrāmāgraṇīr vādijij
jyotiḥśāstravicārasāranipuṇo bhūpālamālārcitaḥ/
śobhārāma iti prathām adhigato yatpādapaṅkte ruhadhyānānugrahavaibhavena nikhilāṃ vidyām avāpur janāḥ//
bhāskarād iva nāsatyau rāmāt kuśalavāv iva/
kṛṣṇaviṣṇū sahāyāntau śobhārāmāt sutāv ubhau//
. . . .

śrīkṛṣṇakauraḥ kila kīrtigauraḥ
śrīmatsyadevas tu tato babhūva/
yatpādapadmadvayasevanena
mādṛgjanaḥ sarvapumarthapātram//
devī sahāyī bhavati yasya sarveṣṭakarmasu/
śrīkṛṣṇakauratanayo babhūvānvarthanāmakaḥ//
tena devasahāyena yathāmatiṃ vinirmitaḥ/
līlāvatīvilāso ᵓyaṃ sanmodaṃ tanutāntaram//

DEVĪSIṂHA

Author of a *Siṃhasudhānidhi*. Manuscript:

Anup 5306. 284ff.

DEVENDRA

Author of a *Bhāvādhyāya*, which is perhaps a part of the *Jātakatilaka* of Devendrācārya. Manuscript:

RORI Cat. I 3225. 15ff. Copied by Jagannātha on Thursday 5 kṛṣṇapakṣa of Mārgaśīrṣa in Saṃ. 1892 = 10 December 1835.

The colophon begins: iti śrīdevendranāmākavikṛte.

DEVENDRĀCĀRYA

Author of a *Jātakatilaka*. Manuscript:

LDI (LDC) 969. 7ff. Copied in Saṃ. 1916 = A.D. 1859.

DAIVAJÑADĀSA = DAIVAJÑATĀNA

Author, at the request of Siṃha of the Matsya family, of an Āndhraṭīkā, *Jayacaryā*, on the *Narapatijayacaryā* of Ādityadeva. Manuscripts:

GOML Madras D 13938. Ff. 108–154. Telugu. Formerly the property of Velamūri Veñkambhaṭlugāri.
IO 6427 (Mackenzie III 97). Ff. 54–159. Telugu. From Colin Mackenzie.
IO 6428 (Mackenzie III 236b). 9ff. Telugu. Incomplete. From Colin Mackenzie.

The first 2 verses are:

śrīśāradāvighnavināyakaṃ ca
brahmāṇam īśaṃ grahasundaraṃ ca/
natvāndhrabhāṣāṃ prakaroti ṭīkāṃ
daivajñadāso jayacaryam ākhyam//
matsyānvayasamudbhūtaḥ siṃhākhyaḥ
siṃhavikramaḥ/
tasyādeśakṛtādeśo jayacaryākhyasañgrahe//

DAIVAJÑAVARA

Title of the author of a *Jyotiṣaratnākara*. Manuscripts:

Oppert II 1968. 88pp. Telugu. Property of Veñkaṭeśvarajosya of Siddhavaṭa, Kaḍapa.
Oppert II 2892. Property of Madirazu Bhagavanulu of Utukūru, Vissampeṭa, Kṛṣṇa.

DUALATACANDA GAṆI

Author of a *Muhūrtamuktāvalī*. Manuscript:

Benares (1963) 36384. 5ff. Copied in Saṃ. 1778 = A.D. 1721.

DUALATARĀMA

Author of a *Kāmadhenusāraṇī*. Manuscript:

LDI (LDC) 4824. 14ff. Copied in Saṃ. 1905 = A.D. 1848.

DYUMAṆI

Author of a *Grahasādhanopapatti;* this may be a part of the *Sūryasiddhānta*. Manuscript:

RORI Cat. III 12636. 56ff. Incomplete. With the *Vāsanā* of Nṛsiṃha.

DRAVYAVARDHANA (*fl. ca.* 500?)

Authority on śakuna cited by Varāhamihira (*fl. ca.* 550) in *Bṛhatsaṃhitā* 86, 2; see P. V. Kane [1948/49] 6. He is identified with an Aulikara mahārāja of Daśapura–Ujjayinī by V. V. Mirashi [1957]; see also D. C. Sircar [1959] and V. V. Mirashi [1959].

DRUPADA

Author of a *Tājakasāra*. Manuscript:

Jaipur (II). 33ff. Copied in Saṃ. 1691, Śaka 1556 = A.D. 1634.

DRUPADA MUNI

Author of a *Goprasūtilakṣaṇa*. Manuscript:

Jammu and Kashmir 851. 1f.

DVĀRAKĀNĀTHA YAJVAN

Author of a ṭīkā, *Śulbadīpikā*, on the *Śulbasūtra* of Baudhāyana, in which he refers to Āryabhaṭa (b. 476). This was edited by G. F. Thibaut [1874/77]; reprinted by Satya Prakash and Ram Swarup Sharman, New Delhi 1968.

DVIJARĀJA

(Title of the ?) author of a *Tithinirṇayasañgraha.* Manuscript:

Kerala 6799 (4958). 2250 granthas. Copied in Saṃ. 1890 = A.D. 1833.

Also ascribed to a Dvijarāja is an *Ahādikajanma*. Manuscript:

IM Calcutta 1436. See NCC, vol 1, rev. ed., p. 486.

DHANAÑJAYA

Dhanañjaya of the Vatsagotra wrote a *Jyotiścandrodaya* in more than 48 prakāśas in which the latest authority quoted is the *Rājamārtaṇḍa* of Bhojarāja (*fl. ca.* 1005/1056). Manuscripts:

GOML Madras R 4416. 236ff. Grantha. Copied in 1924/25 from a manuscript belonging to Maguṇirājaguru Mahāpātro of Kolasandhapuram, Aska, Ganjam. Incomplete (prakāśas 27–48).
CP, Hiralal 1866. Ascribed to Dhanurjaya. Property of Ārtodās Pāṭjoshī of Jagdalpur, Bastar.

GOML Madras R 3199. 196ff. Oriyā. Incomplete. Purchased in 1919/20 from Gopīnātha Tripāṭhi of Boyrani, Gañjam.

Near the beginning is the verse:

śrīmadvatsasuvaṃśapañkajavanaprodbodhacaṇḍa-
dyutir
daivajñapravaro dhanañjaya iti khyātas tu yo
bhūtale/
jyotiśśāstram idaṃ samīkṣya bahudhā so ᵓhaṃ tu
natvā gurūn
kurve jyautiṣacittakairavavanaprollāsacandroda-
yam//

DHANAPATI

Author of a *Jñānamuktāvalī*. Manuscripts:

RORI Cat. III 16812. 64ff. (ff. 1–2 missing) Copied by Kevalarāma on Wednesday 13 kṛṣṇapakṣa of of Bhādrapada in Saṃ. 1805 = A.D. 1748 (the date is irregular).
BORI 153 of A 1883/84. 37ff.

The colophon begins: iti śrīdhanapativiracitāyāṃ.

DHANARĀJA

The son of Mahātmā Vidyāvinoda, Dhanarāja wrote a ṭīkā in bhāṣā on the *Jātakārṇava* ascribed to Varāhamihira (*fl. ca.* 550). Manuscript:

Florence 276. 7ff.

The colophon begins: iti śrīvarāhamihirācāryakṛtajātakārṇavaṭīkāmahātmavidyāvinodatatputradhanarājakṛtaṭīkāyāṃ.

DHANARĀJA (*fl.* 1635)

The pupil of Bhojarāja (or Bhuvanarāja) Gaṇi, the pupil of Kalyāṇasāgara Sūri of the Añcala Gaccha, Dhanarāja completed a ṭīkā, *Dīpikā*, on the *Mahādevī* of Mahādeva (*fl.* 1316) at Padmāvatī in Mārwār, Rājasthān, on 8 śuklapakṣa of Jyeṣṭha in Saṃ. 1692 = *ca.* 13 May 1635 during the reign of Gajasiṃha Rāṣṭroḍa, the mahārāja of Mārwār from 1620 to 1638. Manuscripts:

Baroda 689. 32ff. Copied in Saṃ. 1719 = A.D. 1662.
BORI 340 of 1879/80. 38ff. Copied in Saṃ. 1722 = A.D. 1665.
LDI 7101 (7129). 48ff. Copied by Ṛṣi Buddhiśekhara Gaṇi, the pupil of Vācaka Bhāvaśekhara Gaṇi, for Ṛṣi Rājaśekhara Gaṇi of the Añcala Gaccha at Rājanagara in Saṃ. 1729 = A.D. 1672.
BORI 124 of 1899/1915. 29ff. Copied in Saṃ. 1733 = A.D. 1676.
BORI 497 of 1892/95. 25ff. Copied in Saṃ. 1734 = A.D. 1677.
PL, Buhler IV E 327. 59ff. Copied in Saṃ. 1752 = A.D. 1695. Property of Jīvanakuśala Gorajī of Bhuja.
LDI 7098 (5132). 35ff. Copied at Meḍatā on Sunday 14 śuklapakṣa of Āśvina in Saṃ. 1754 = 19 September 1697 Julian.
RORI Cat. III 15832. 75ff. Copied by Amara Bhaṭṭa Pālīvāla at Udayapura in Saṃ. 1760 = A.D. 1703 during the reign of Amarasiṃha II (1698/1710).
BORI 845 of 1887/91. 38ff. Copied in Saṃ. 1761 = A.D. 1704. From Gujarāt.
LDI 7100 (8877). 41ff. Copied by Paṇḍita Kuśalavijaya Gaṇi, the pupil of Paṇḍita Ratnavijaya Gaṇi, at Jesalamera in Saṃ. 1779 = A.D. 1722.
RORI Cat. III 13920. 32ff. Copied in Saṃ. 1794 = A.D. 1737.
LDI 7099 (7412). 29ff. Copied by Ṛṣi Goïnda, the pupil of Urajājī, at Pallikāpura on 1 śuklapakṣa of Phālguna in Saṃ. 1852, Śaka 1717 = *ca.* 8 March 1796.
Goṇḍal 255. 26ff. Copied on Thursday 7 kṛṣṇapakṣa of Phālguna in Saṃ. 1902 = 19 March 1846.
RORI Cat. III 11996(31). 33ff. Copied by Balabhadra at Yodhanagara in Saṃ. 1908 = A.D. 1851.
AS Bombay 254. 30ff. From Bhāu Dājī.
Baroda, Hamsavijayaji Maharaj in the Kantivijayaji Bhandar 487. See Velankar, p. 304.
Baroda, Library of Kantivijayaji 1241. See Velankar.
BORI 392 of 1880/81. 38ff.
RORI Cat. II 7136. 32ff. (ff. 1–8 missing). Incomplete.
RORI Cat. III 14037. 37ff.
SOI 9907. No author mentioned.

Verses 1–3 at the end are:

varṣe netranavāñgabhūparimite jyeṣṭhasya pakṣe site
ᵓṣṭamyāṃ sadguṇapṛkthamannarayute
padmāvatīpattane/
rājā hy utkaṭavairināgadamano
rāṣṭroḍavaṃśodbhavaḥ
śrīmān śrīgajasiṃhabhūpativaro ᵓsti śrīmaror
maṇḍale//
jaine śāsana evam añcalagaṇe satsajjanaiḥ saṃstute
kalyāṇodadhisūrayaḥ śubhakarā nandantu
bhūmaṇḍale/
tatsevākarabhojarājagaṇayo vidvadvarā vācakā
āsan sarvasudhīmanaḥkamalinīsambodhane
bhānavaḥ//
kheṭānāṃ hi purā kṛtā budhamahādevena yā sāraṇī
tasyā daivavidāṃ sukhārthajananīṃ vṛttiṃ varāṃ
vistarāṃ/
tacchiṣyo dhanarāja evam akarod dharṣeṇa
bahvādarair
bahvarthaiḥ sahitāṃ ca paṇḍitapadād āptaprasakter
guroḥ//

The colophon begins: ity āñcalikavācanācāryaśrībhuvanarājagaṇīndrāṇāṃ śiṣyapaṇḍitaśrīdhanarājakṛtā.

DHANEŚVARA

Author of a *Gaṇakapradīpa*. Manuscript:

Nagpur 490 (471). 33ff. Copied in Śaka 1562 = A.D. 1640. From Amaravati.

DHANEŚVARA

The son of Vīreśvara, the son of Someśvara of the Kuṣkaskulā (?), Dhaneśvara wrote a ṭīkā, *Līlāvatībhūṣaṇa*, on the *Līlāvatī* of Bhāskara (b. 1114), in which he mentions the *Amṛtakūpikā* of Sūrya (*fl.* 1541). Manuscripts:

AS Bombay 275. 95ff. Incomplete.
Baroda 3286. 188ff.
GVS 2750 (5428). Ff. 14–54. Incomplete.
VVRI 4617. 37ff. Incomplete.

At the end of the prakīrṇādhyāya is the verse:

mahyāḥ sannikaṭasthale parisarālaṅkārabhūte suvidvatkhyāte sujanāśrite dvijavaraḥ
kuśkaskulābhūṣaṇam/
yaḥ someśvara ity abhūc chubhamatir vīreśvaras tatsutas
tatsūnugrathite ᵓpy udāhṛtipathe ᵓgacchat prakīrṇakramaḥ//

The colophon begins: iti śrīmaddhaneśvaradaivajñaviracite.

DHANEŚVARA BHAṬṬA

Author of a (ṭīkā on the?) *Sūryasiddhānta*. Manuscript:

PL, Buhler IV E *443. 25ff. Copied in Saṃ. 1522 = A.D. 1465. Property of Harakharāma Śāstrī of Sihora.

DHANVANTARI

Author of a *Bṛhatkālajñāna*. Manuscript:

LDI 6658 (605). 26ff. Copied by Ṛṣi Rāmarṣi, the pupil of Viṣṇukumāra, in Ambikānagara in Saṃ. 1806 = A.D. 1749. With a *Bālāvabodha* in Old Gujarātī.

DHARAṆĪDHARA

The son of Viśvanātha, Dharaṇīdhara wrote a *Dharaṇīdharīpaddhati*. Manuscript:

RORI Cat. III 11029(5). 16ff. Copied in Saṃ. 1836 = A.D. 1779. With a ṭippaṇa.

DHARAṆĪDHARA

Author of a ṭīkā on the *Mādhavīya* of Mādhava (*fl. ca.* 1330/1385). Manuscript:

Oudh (1879) IX 8. 18pp. Copied in A.D. 1839. Property of Paṇḍit Śyām Lāl of Lucknow Zila.

DHARMAKHĀNA = DHARMARĀJA

A member of the Siṃhavaṃśa, Dharmakhāna wrote a *Jyotiḥsāra*. Manuscripts:

Śāstrī, Not. 1907. 111. 67ff. Bengālī. Copied in Śaka 1670 = A.D. 1748. Property of Paṇḍita Kṛṣṇadāsa Smṛtibhūṣaṇa of Dinajpur.
Benares (1963) 35481 = Benares (1905) 1425. 9ff. Incomplete. (*Jyautiṣasāra* of Dharmarāja).

The last verse is:

gandharvārṇavasindhunīrajamite saṃvatsare nirmale
māse mādhavasaṃjñake mṛdudhiyaḥ santoṣadaṃ jñānadam/
putrapremabharādimaṃ vyaracayad granthaṃ prayatnāt svayaṃ
rāṭhābhūṣaṇasiṃhavaṃśaprabhavaḥ śrīdharmakhānaḥ sudhīḥ//

I do not comprehend the chronogram in the first pāda.

DHARMAPĀṬHIN

A member of the Bhāradvājagotra, Dharmapāṭhin wrote a *Gaṇitādarśa* following the *Sūryasiddhānta* in 8 adhikāras:

1. madhya.
2. tithi.
3. grahasphuṭa.
4. dikcakrodayāsta.
5. upakaraṇa.
6. candragrahaṇa.
7. sūryagrahaṇa.
8. pariveṣa.

Manuscript:

GOML Madras R 3288. 11ff. Copied in 1920/21 from a manuscript belonging to Paṇḍita Gopīnāthānandaśarmagāru of Parlākimodi.

The first 2 verses are:

praṇamya rādhikākṛṣṇau tadājñāṃ śirasā vahan/
karomi gaṇitādarśaṃ sūryasiddhāntasaṃmatam//
bhāradvājasagotreṇa sudhiyā dharmapāṭhinā/
tanyate gaṇitādarśaḥ prītyai siddhāntasaṃvidām//

DHARMAMERU

Author of a stabaka in Old Gujarātī on the *Saṅgrahaṇīratna* of Śrīcandra (*fl. ca.* 1150). Manuscript:

LDI 3109 (60). 100ff. Copied in Saṃ. 1891 = A.D. 1834.

DHARMASĀGARA (*fl.* 1582)

Assistant to Hīravijaya Sūri of the Tapā Gaccha in writing a vṛtti on the *Jambūdvīpaprajñapti* in Saṃ. 1639 = A.D. 1582; see Velankar, p. 131.

DHARMĀDITYA

Author of a ṭīkā, *Bhāsvatītilaka*, on the *Bhāsvatī* of Śatānanda (*fl.* 1099). Manuscript:

Anup 4933. 9ff. Incomplete.

DHARMEŚVARA (*fl. ca.* 1600/1650)

The son of Rāmacandra, the son of Prabhākara, the son of Ratnākara, the son of Balabhadra, the son of Devadatta of the Vatsagotra, a Brāhmaṇa residing in Mālava, Dharmeśvara was a pupil of Śrīdhara and Nīlakaṇṭha. He wrote the following works on jyotiḥśāstra.

1. A ṭīkā, *Vāsanābhāṣya*, on the *Keśavapaddhati* of Keśava (*fl.* 1496/1507), composed for Bhāratha (or Bhātara) Sāhi. Manuscripts:

Oudh XIV (1881) VIII 30. 136pp. Copied in A.D. 1695. Property of Bhairavadatta of Unao Zila.
Mithila 92. 46ff. Maithilī. Copied in Śaka 1767, Sāl. San. 1253 = A.D. 1845. Property of Paṇḍita Umādatta Miśra of Salampur, Ghataho, Darbhanga.
Mithila 32. 37ff. Maithilī. Copied on Sunday 10 kṛṣṇapakṣa of Pauṣa in Śaka 1775 = 22 January 1854. Property of Paṇḍita Janārdana Miśra of Chanour, Manigāchī, Darbhanga.
Baroda 3133. 84ff. Copied in Saṃ. 1937 = A.D. 1880.
Kurukṣetra 189 (19557).
Mithila 32 A. 47ff. Maithilī. Property of Paṇḍita Sītārāma Pāṭhaka of Karṇpūr, Sukpur, Bhāgalpur.
PUL II 3414. 50ff. Incomplete.
SOI 8411. No author mentioned.
VVRI 2552. 17ff. Incomplete.
VVRI 3299. 15ff. Incomplete.

Verse 1 at the end is:

yo ᵓbhūd vatsakule prabhākarasamakhyāto dvijas tatsutaḥ
śrīrāmo gaṇakāgragaṇyagaṇitas tatsūnudharmeśvaraḥ/
śrīmadbhārathasāhirājamukuṭālaṅkārahāreṇa cājñaptaḥ keśavapaddhater vyaracayat sotpattiṭīkām imām//

2. A ṭīkā, *Anvayārthadīpikā*, on the *Camatkāracintāmaṇi* of Nārāyaṇa Bhaṭṭa. Manuscripts:

BORI 898 of 1886/92. 30ff. Copied in Saṃ. 1793 = A.D. 1736.
Benares (1963) 34757. 11ff. Copied in Saṃ. 1802 = A.D. 1745. Incomplete.
Oudh XX (1888) VIII 168. 12pp. Copied in A.D. 1764. Property of Paṇḍita Pratāpa Nārāyaṇa of Allahabad Zila.
RORI Cat. II 6629. 30ff. Copied by Vijayalāla in Saṃ. 1828 = A.D. 1771.
BORI 860 of 1891/95. 15ff. Copied in Śaka 1694 = A.D. 1772.
BORI 414 of 1895/98. 25ff. Copied in Saṃ. 1896 = A.D. 1839.
Goṇḍal 88. 25ff. Copied by Monajī Bhāī, the son of Rāvalamūla, on Saturday 13 kṛṣṇapakṣa of Māgha I in Saṃ. 1896, Śaka 1760 = 29 February 1840.
Oxford 1545 (Sansk. d 187) = Hultzsch 283b. 62ff. Copied in Saṃ. 1897 = A.D. 1840.
RORI Cat. III 10209. 12ff. Copied by Rāmadatta Jośī in Saṃ. 1900 = A.D. 1843. Incomplete (grahabhāvaphala).
RORI Cat. II 4668. 28ff. Copied by Keśavajī Jādavajī at Saradhāra in Saṃ. 1901 = A.D. 1844.
Baroda 3117. 44ff. Copied in Saṃ. 1906 = A.D. 1849.
Goṇḍal 94. 13ff. Copied by Kevala Dave at Bhuja on Friday 5 śuklapakṣa of Āṣāḍha I in Saṃ. 1911, Śaka 1776 = 30 June 1854.
RORI Cat. I 3130. 25ff. Copied by Umāśaṅkara at Kāśī in Saṃ. 1912 = A.D. 1855.
AS Bengal 7017 (G 2281) = Mitra, Not. 2666. 29ff. Copied by Mukundarāma in Saṃ. 1915 = A.D. 1858.
Benares (1963) 35358 = Benares (1903) 1294. 40ff. Copied in Saṃ. 1934 = A.D. 1877.
GOML Madras D 15785. 42ff. Copied on Sunday 7 śuklapakṣa of Bhādrapada in Saṃ. 1940 = 9 September 1883.
Goṇḍal 87. 29ff. Copied by Vāsudeva, the son of Mādhavajī Vyāsa, at Goṇḍala on Sunday 6 kṛṣṇapakṣa of Caitra in Saṃ. 1947 = A.D. 1891 (the date is irregular).
ABSP 449. 23ff. No author mentioned.
Alwar 1756.
Benares (1963) 34457. Ff. 1–24 and 27–38. Incomplete.
Benares (1963) 35817 = Benares (1913/1914) 2284. 37ff. Incomplete (grahabhāvaphala).
Benares (1963) 36499. 9ff. Incomplete.
Cocanada, Telugu Academy 2190. See NCC, vol. 6, p. 387.
IM Calcutta 3473, 3491, 3563, and 8134 (incomplete). See NCC.
Jaipur (II).
Jammu and Kashmir 4005. 19ff.
Kathmandu (1960) 101 (I 1199). 6ff. Incomplete.
Mithilā. See NCC.
Osmania University B. 82/7. 16ff.
Oudh XX (1888) VIII 110. 10pp. Property of Paṇḍita Pratāpa Nārāyaṇa of Allahabad Zila.
Rajputana, p. 57. From Alwar.
SOI 5981 = SOI (List) 362.

Viśvabhāratī 1532. See NCC.
VVRI 2380. 5ff. Incomplete.
WHMRL X. 57.

The *Anvayārthadīpikā* was published at Benares in 1856 (IO 362); at Kāśī in Saṃ. 1926 = A.D. 1869 (BM); at Benares in 1870 (IO 7. B. 40); at Delhi in Saṃ. 1929 = A.D. 1872 (BM and IO 1605); at Delhi in 1876 (IO 411); and at Calcutta in B.S. 1291 = A.D. 1883 (IO 395). The last verse is:

camatkāracintāmaṇeś cāruṭīkāṃ
cakārānvayārthaprabodhapradīpām/
sudaivajñadharmeśvaro mālavīyaḥ
pramodāya bhūdevavidvajjanānām//

3. A *Jātakapaddhati* in 6 adhyāyas. Manuscripts:

Anup 4622. 4ff. Copied in Saṃ. 1711 = A.D. 1654. Property of Maṇirāma Dīkṣita (*fl. ca.* 1675/1700).
Bombay U 494. 6ff.
Jammu and Kashmir 2885. 14ff.

The last verse is:

śrīmanmālavadeśajo dvijavaraḥ śrīdevadattātmabhūḥ
sarvajño balabhadra asya tanayo ratnākaro ᵓsyātmajaḥ/
yo ᵓbhūd vedanidhiḥ prabhākara iti śrīrāmacandro ᵓṅgabhūr
yasyāsyāpi sutaś cakāra matimān dharmeśvaraḥ paddhatim//

The first verse in the Kashmir manuscript is:

śrīdharaṃ nīlakaṇṭhaṃ ca natvā gurutaraṃ gurum/
tatprasādāt pravakṣyāmi jātake karmapaddhatim//

4. A *Muhūrtaśiromaṇi*. Manuscript:

Alwar 1910.

DHIYEŚVARA = DHEYEŚVARA

Author of a ṭīkā, *Budhavallabhā*, on the *Laghujātaka* of Varāhamihira (*fl. ca.* 550). Manuscripts:

Mithila 316. 31ff. Maithilī. Copied by Dharmadatta at Yokīgrāma on Sunday 8 kṛṣṇapakṣa of Āśvina in Śaka 1744, Sāl. San. 1230 = 6 October 1822. Property of Babu Candradeva Jhā of Mahinathapur, Jhanjharpur Bazar, Darbhanga.
PL, Buhler IV E 423. 35ff. Ascribed to Dheyeśvara. Property of Maṅgala Śaṅkara of Ahmadābād.

The first verse is:

praṇamya gaurīpatipādapaṅkajaṃ
sureśagandharvaṣaḍaṅghrisevitam/
karomi ṭīkāṃ budhavallabhāṃ imāṃ
dhiyeśvarākhyo nijayālpajāte//

DHĪRAJASIṂHA

Author of a *Gaṇitacandrikā* in Hindī. Manuscript:

NPS 30 A of 1906–08. Copied in Saṃ. 1899 = A.D. 1842. Property of Lālā Jānakīprasāda of Chatarapura.

DHĪRAVIJAYA

Author of a *Kāmadhenutithisāraṇi* in Gujarātī. Manuscript:

LDI (LDC) 1522. 4ff.

DHĪRĀNANDA KĀVYANIDHI (*fl.* 1891)

Author of a *Sāmudrika* published [NP] in 1891 (NL Calcutta 180. Kd. 89. 2). He also translated the *Bṛhatsaṃhitā* of Varāhamihira (*fl. ca.* 550) into Bengālī; this was edited by Pañcānana Tarkaratna, 2nd ed., Calcutta 1910 (BM 14055. d. 5. (2) and IO 22. D. 7).

DHĪREŚVARA

Author of a *Buddhipradīpa;* see R. Jha [A2. 1967]. Manuscripts:

Mithila 219. 12ff. Maithilī. Copied in Sāl. San. 1312 = A.D. 1904. Property of Paṇḍita Jayakṛṣṇa Jhā of Champa, Benipati, Darbhanga.
GJRI 3184/396. 4ff. Maithilī. Incomplete.

The first verse is given in a corrupt version in Mithila; I follow Jhā:

natvā hariṃ bhāskaraṃ bhāratīṃ ca
gaṇeśaṃ śivaṃ ceṣṭadevaṃ guruṃ ca/
sudhīreśvareṇa praṇītaṃ samastaṃ
samālocya śāstraṃ subuddhipradīpam//

DHṚTIKARA DVIVEDIN

Author of a *Daivajñavallabha*. Manuscripts:

Anup 4767. 56ff. Copied by Laghugovinda, the son of Bhaira, the son of Kālidāsa, at Tripurārigrāma in Saṃ. 1524 = A.D. 1467 during the rule of Kṛṣṇadāsa Mahāṭhakkura. Incomplete.
Jammu and Kashmir 2978. 159ff.
VVRI 2506. 152ff. Incomplete.

Verse 2 is:

bhāsvantaṃ praṇipatya bālagaṇakajñānārthasaṃsādhanaṃ
buddhvā gargavarāhalallavihitaṃ śāstraṃ tathānyaiḥ kṛtam/
vidvatpūrvapadaṃ kṛtī dhṛtikaro granthaṃ svamāñcalaṃ
saṅgṛhyātimahāphalaṃ tam aniśaṃ dhīrāḥ kurudhvaṃ kare//

The colophon begins: iti śrīdvivedīdhṛtikaraviracitāyāṃ.

DHAUÑKALASIṂHA (*fl.* 1748)

Author of a *Ramalapraśna* in Hindī. Manuscript:

NPS 50 of 1917–19. Copied Saṃ. 1918 = A.D. 1861. Property of the Sarasvatī Bhaṇḍāra at Lakṣmaṇakoṭa, Ayodhyā.

NAGNAJIT

An authority cited by Varāhamihira (*fl. ca.* 550) in *Bṛhatsaṃhitā* 57, 4 and 15, and by Utpala (*fl.* 966/968) on *Bṛhatsaṃhitā* 55, 31 and 57, 4. See P. V. Kane [1948/49] 13.

BEÑGALURU NAÑJUṆḌA ŚĀSTRIN (*fl.* 1912)

Author of a *Sarvajyotiṣaratna*, published at Billary in 1912 (IO 21. I. 21), and reprinted at Bellary in 1917 (IO 28. K. 2).

PAḺAṈIYAPPAṈ NAṬARĀCAṈ (b. 1932)

Author of a *Kalyāṇaṅkaḷ* in Tamil, published at Tiruppur in 1970.

KĪRANŪR NAṬARĀJAR (*fl.* 1665)

Author of a *Jātakālaṅkāra* in Tamil in Śaka 1587 = A.D. 1665. Manuscript:

GOML Madras (Tamil) R 80(a). Ff. 1–135. Tamil. Purchased in 1911/12 from Cawder Beg (= Kadir Baig) of Triplicane.

This has been published with his own commentary by V. K. Velu Nāyakar, Cenna 1964.

NANDAKUMĀRA DATTA (*fl.* 1857)

Author of a *Kākacaritra* published in his *Sarvajñānamañjarī*, which went through 17 editions between 1857 and 1898 (see IO, *Printed Books*, vol. 2, p. 1213, and vol. 4, p. 2393).

NANDAPAṆḌITA

The son of Devaśarman, Nandapaṇḍita wrote a *Jyotiḥśāstrasamuccaya*. Manuscripts:

AS Bengal 7054 (G 903) = Mitra, Not. 1762. 126ff.
Jammu and Kashmir 2873. 81ff.

At the end are 2 verses:

śrīnārāyaṇapādapaṅkajanatiprāptaprabodhodayo
vedān aṅgayutān sabhāratakhilān sāhityakāvyānvitān/
mīmāṃsaiśvarasaṅkhyabaudham akhilaṃ cārvākajainābhidhaṃ
granthaṃ yo vyavṛṇot sa paṇḍitakaviḥ śrīdevaśarmābhavat//
tasya śrīśitikaṇṭhabhaktinirato nandābhidho ᵓbhūt suto
vindā devapadāravindaśaraṇā yasya prasiddhā prasūḥ/
so ᵓyaṃ nātilaghuṃ na vistutataraṃ jyotirvidāṃ sevako
jyotiḥśāstrasamuccayaṃ vyaracayac chiṣyaughasamprārthitam//

NANDAPAṆḌITA

The son of Rāmapaṇḍita Dharmādhikāri, Nandapaṇḍita wrote a *Tattvamuktāvalī*. Manuscripts:

AS Bengal 2743 (G 5535). 28ff. Copied by Viśvanātha Kamaṭhāna at Kāśī in the kṛṣṇapakṣa of Phālguna in Saṃ. 1743 = February 1687.
AS Bengal 2744 (G 10003). 264ff. Copied by the Brāhmaṇa Khemarāma on Sunday 11 kṛṣṇapakṣa of Āṣāḍha in Saṃ. 1818 = 26 July 1761. With the ṭīkā, *Bālabhūṣā*, of Veṇīpaṇḍita.
Dharwar 698 (688). 246ff. Copied in Śaka 1705 = A.D. 1783.
Adyar Index 2376 = Adyar Cat. 34 K 22. 45ff.

The last verses are:

anantabhaṭṭahemādrikavivallabhamādhavaiḥ/
kṛtāḥ siddhāntasaritaḥ smṛtisindhau samāviśan//
tattvamuktāḥ samuddhṛtya smṛtisindhoḥ svayaṃ kṛtāt/
tattvamuktāvalīm etāṃ niramān nandapaṇḍitaḥ//

The colophon begins: iti śrīdharmādhikārirāmapaṇḍitātmajanandapaṇḍitakṛtā.

A part of his *Smṛtisindhu* is the *Kālanirṇayataraṅga* or *Kālanirṇayakautuka*. Manuscripts:

Anup 2655. 102ff.
Benares (1956) 13979. 116ff.
N-W P V (1880) Dharmaśāstra II 9. 106ff. Property of Ḍhuṇḍhirāja Śāstrī of Benares.

NANDARĀMA MIŚRA (*fl.* 1763/1778)

The son of Dīpacandra, Nandarāma wrote the following works on jyotiḥśāstra.

1. The *Grahaṇapaddhati* composed at Kāmyakavana in Saṃ. 1820 = A.D. 1763. Manuscripts:

RORI Cat. II 4104. 5ff. Copied in Saṃ. 1822 = A.D. 1765.
N-W P X (1886) A 6. 6ff. Property of Paṇḍita Gaṅgāsahāya of Alvara.
RORI Cat. II 4761. 6ff.

2. The *Svarapañcāśikā* composed at Kāmyakavana in Saṃ. 1822 = A.D. 1765. Manuscripts:

RORI Cat. II 4105. 3ff. Copied in Saṃ. 1832 = A.D. 1775.
RORI Cat. II 5318. 4ff. Copied by Haradeva Lālā in Saṃ. 1865 = A.D. 1808.

BORI 889 of 1884/87. 4ff. Copied in Saṃ. 1903, Śaka 1767 = A.D. 1845.
PUL II 4093. 4ff.
RORI Cat. II 5322. 7ff.
RORI Cat. III 15396(1). 5ff.
RORI Cat. III 18203. 4ff.

3. The *Goladarpaṇa*. Manuscripts:

Benares (1963) 35760. 17ff. Copied in Saṃ. 1824 = A.D. 1767.
Jodhpur 455. See NCC, vol. 6, p. 179.
SOI 2902 = SOI Cat. II: 1001–2902. 24ff. Copied in Śaka 1756 = A.D. 1834.

4. The *Praśnaratna* = *Keralīyapraśnaratna*, completed at Kāmyavana on 7 śuklapakṣa of Āśvina in Saṃ. 1824 = *ca.* 29 September 1767; he wrote his own ṭippaṇī. Manuscripts:

RORI Cat. III 11447. 27ff. Copied in Saṃ. 1828 = A.D. 1771. With his own ṭippaṇī.
RORI Cat. II 5338. 8ff. Copied in Saṃ. 1837 = A.D. 1780.
BORI 940 of 1886/92. 51ff. Copied in Saṃ. 1847 = A.D. 1790. With his own ṭippaṇī.
BORI 547 of 1899/1915. 15ff. Copied in Saṃ. 1847 = A.D. 1790.
RORI Cat. I. 2914. 35ff. (ff. 1–6 missing). Copied by Udayarāma at Savāī Jayapura in Saṃ. 1850 = A.D. 1793. With his own ṭippaṇī.
BORI 165 of A 1883/84. 48ff. Copied in Saṃ. 1875 = A.D. 1818. With his own ṭippaṇī.
Oudh VII (1875) VIII 12. 46pp. Copied in A.D. 1826. With his own ṭippaṇī. Property of Jānakīprasāda of Bārābānki Zila.
RORI Cat. I 3736. 25ff. Copied by Vinayacandra Muni at Subhaṭṭapura in Saṃ. 1887 = A.D. 1830. With his own ṭippaṇī.
BORI 425 of 1895/98. 11ff. Copied in Saṃ. 1895 = A.D. 1838.
Mithila 196. 38ff. Copied by Mayūra Daivajña at Pharakkābāda on Tuesday 3 śuklapakṣa of Kārttika in Śaka 1776 = 24 October 1854. With his own ṭippaṇī. Property of Paṇḍita Janārdana Miśra of Chanaur, Manigachi, Darbhanga.
RORI Cat. II 6377. 27ff. Copied in Saṃ. 1915 = A.D. 1858.
PUL II 3660. 26ff. Copied in Saṃ. 1932 = A.D. 1875.
Jammu and Kashmir 4115. 13ff. Copied in Saṃ. 1941 = A.D. 1884 from Alwar 1855.
Alwar 1855.
Alwar 1856. Within his own ṭīkā. 2 copies.
AS Bengal 7164 (G 7832). 4ff. With his own ṭippaṇī.
AS Bengal 7165 (G 4414). 43ff. Bengālī. With his own ṭippaṇī.
AS Bengal 7166 (G 7253). 19ff. (4ff. missing). With his own ṭippaṇī. Incomplete.
Bikaner 705. 22ff. With his own ṭīkā.
BORI 939 of 1886/92. 13ff.
CP, Kielhorn XXIII 85. 46ff. Property of Govindarāma Bhaḍajī of Sāgar.
Jaipur II. 26ff. With his own ṭippaṇī.
Mithila 196 A. 24ff. Maithilī. With his own ṭippaṇī. Property of Paṇḍita Lakṣmī Vallabha Jhā of Bhakharaini, Madhepur, Darbhanga.
Oudh XIV (1881) VIII 13. 18pp. Property of Govindaprasāda of Lucknow Zila.
RORI Cat. I 2562. 46ff. With his own ṭippaṇī.
RORI Cat. II 5635. 49ff. With his own ṭīkā.

The *Praśnaratna* was published with the Hindī ṭīkā, *Sundarī*, of Sundaralāla Śarman of Bombay in Saṃ. 1980 = A.D. 1923 (IO San. D. 942(a)); repr. at Bombay in Saṃ. 2010. Śaka 1875 = A.D. 1953. The last verses are:

āste yad vasudhāvibhūṣaṇamaṇau śrīmadvraje sadvraje
ramyaṃ kāmyavanaṃ trayīdhutamalās tasmin vasanti dvijāḥ/
śrīkṛṣṇāśrayadīpacandratanayo yo nandarāmābhidhas
teṣāṃ saṃskṛtavān prabandham amalaṃ satpraśnaratnāhvayam//
proktaṃ candronmīlanaṃ śuklavastrais
tac cāśuddhaṃ vijñanindyaṃ samantāt/
vācyaṃ tajjñaiḥ pakṣapātaṃ vihāyo-
tpātābhikhye ᵓsmin na teṣāṃ trapābhūt//
siddhāṣṭacandravarṣe ᵓśviyujaḥ sitapakṣasaptamyām/
pūrtim agāda grantho ᵓyaṃ śūnyābdhidvipramair vṛttaiḥ//

5. A ṭippaṇī on his *Praśnaratna*, completed on 11 śuklapakṣa of Bhādrapada in Saṃ. 1827 = *ca.* 30 August 1770. Manuscripts:

RORI Cat. III 11447. 27ff. Copied in Saṃ. 1828 = A.D. 1771.
BORI 940 of 1886/92. 51ff. Copied in Saṃ. 1847 = A.D. 1790.
RORI Cat. I 2914. 35ff. (ff. 1–6 missing). Copied by Udayarāma at Savāī Jayapura in Saṃ. 1850 = A.D. 1793.
BORI 165 of A 1883/84. 48ff. Copied in Saṃ. 1875 = A.D. 1818.
Oudh VII (1875) VIII 12. 46pp. Copied in A.D. 1826. Property of Jānakīprasāda of Bārābānki Zila.
RORI Cat. I 3736. 25ff. Copied by Vinayacandra Muni at Subhaṭṭapura in Saṃ. 1887 = A.D. 1830.
Mithila 196. 38ff. Copied by Mayūra Daivajña of Pharakkābāda on Tuesday 3 śuklapakṣa of Kārttika in Śaka 1776 = 24 October 1854. Property of Paṇḍita Janārdana Miśra of Chanaur, Manigachi, Darbhanga.
Alwar 1856. 2 copies.
AS Bengal 7164 (G 7832). 4ff.
AS Bengal 7165 (G 4414). 43ff. Bengālī.

AS Bengal 7165 (G 7253). 19ff. (4ff. missing). Incomplete.
Bikaner 705. 22ff.
Jaipur (II). 26ff.
Mithila 196 A. 24ff. Maithilī. Property of Paṇḍita Lakṣmī Vallabha Jhā of Bhakharaini, Madhepur, Darbhanga.
Mithila 196 B. 22ff. Maithilī. Property of Paṇḍita Mahīdhara Miśra of Lalabag, Darbhanga.
RORI Cat. I 2562. 46ff.
RORI Cat. II 5635. 49ff.

The last verse is:

saptadvyaṣṭenduvarṣasya bhādraśuklaśivātithau/
ṭippaṇīyaṃ mayā kḷptā saṅkṣiptārthaprakāśinī//

6 and 7. An *Iṣṭadarpaṇa* to which he wrote his own udāharaṇa. Manuscripts:

BORI 875 of 1886/92. 7ff. Copied in Saṃ. 1832 = A.D. 1775. (*Iṣṭadarpaṇodāharaṇa*).
RORI Cat. III 14947. 16ff. Copied by Rāmanārāyaṇa at Ajamera in Saṃ. 1912 = A.D. 1858. (*Iṣṭadarpaṇodāharaṇa*).
Jaipur (II). 56ff. (*Iṣṭadarpaṇodāharaṇa*).
N-W P I (1874) 13 = N-W P I (1874) 26. 8ff. (*Iṣṭadarpaṇodāharaṇa*). Property of Sāma Lāla of Benares. Is this PUL II 3274?
N-W P I (1874) 14 = N-W P I (1874) 27. 5ff. Property of Sāma Lāla of Benares. Is this PUL II 3275?
N-W P II (1877) B 48. 10ff. (*Iṣṭadarpaṇodāharaṇa*). Property of Vāgīśvarī Datta of Benares.
N-W P II (1878) A 4. 7ff. (*Iṣṭadarpaṇodāharaṇa*). Property of Mukundaji of Mathurā.
PUL II 3274. 8ff. With his own vyākhyā.
PUL II 3275. 5ff. With his own vyākhyā.

8. A *Saṅketacandrikā* = *Śataślokī*, written in Saṃ. 1834 = A.D. 1777. Manuscripts:

Jaipur (II). 6ff. Copied in Saṃ. 1834 = A.D. 1777.
Baroda 1164. 6ff. Copied in Saṃ. 1836 = A.D. 1779.
Leipzig 1078. 6ff. Copied in A.D. 1821.
Benares (1963) 36653 = Benares (1903) 1078. 7ff. Copied in Saṃ. 1937 = A.D. 1880.
Jammu and Kashmir 4138ga. 20ff. Copied in Saṃ. 1940 = A.D. 1883 from Alwar 1986.
Alwar 1986.
SOI 5982 = SOI (List) 363.

The last verse is:

śrutiguṇavasuśaśivarṣā-
kṣayanavamīpūrvadevagurau/
saṅketacandrikeyaṃ
vinirmitā nandarāmena//

9. A *Svarasāra* composed in Saṃ. 1835 = A.D. 1778. Manuscripts:

Jaipur (II). 8ff. Copied in Saṃ. 1835 = A.D. 1778.
RORI Cat. III 15084. 9ff. Copied in Saṃ. 1835 = A.D. 1778.
Kathmandu (1960) 217 (III 104). 7ff. Copied by Durgādatta on Friday 7 kṛṣṇapakṣa of intercalary Āṣāḍha in Saṃ. 1912 = 6 July 1855.
DC 7492. 10ff.
VVRI 1235. 4ff. Incomplete.

10. A *Patrikāgamanapraśnavicāra*. Manuscript:

Benares (1963) 36432 = Benares (1903) 1053. 7ff. No author mentioned in Benares (1963).

11. A *Yantrasāra*. Manuscripts:

BORI 851 of 1884/87. 37ff. Copied in Saṃ. 1859, Śaka 1724 = A.D. 1802. From Gujarāt.
BORI 504 of 1892/95. 21ff. Copied in Saṃ. 1887 = A.D. 1830.
Poleman 4723 (Columbia, Smith Indic 127). 27ff. Copied in Saṃ. 1891 = A.D. 1834.
Jammu and Kashmir 4942. 24ff.
RORI Cat. III 11340. 25ff. Incomplete.

12. A *Śrīkṛṣṇajanmapatra*. Manuscript:

Jammu and Kashmir 2895. 11ff. Incomplete.

13. A *Svaravicāra*. Manuscript.

RORI Cat. II 8413(8). 30ff. Incomplete.

NANDALĀLA (= NANDARĀMA) ŚARMAN (1804/1867)

A Sarayūpārīṇa Brāhmaṇa from Kaḍemānikapura, Prayāga, Nandarāma taught jyotiḥśāstra at the Kāśika Rājakīya Saṃskṛta Pāṭhaśālā in Vārāṇasī from Śaka 1757 = A.D. 1835 till his death in Śaka 1789 = A.D. 1867 at the age of sixty-three. See S. Dvivedin [1892] 125.

NANDIKEŚVARA

Author of an *Akṣarapraśna*. Manuscripts:

Dharwar (KRI) V 3 (2301). 66ff. Copied in Śaka 179 (1799 = A.D. 1877?).
Dharwar (KRI) V 4 (2535). A copy of Dharwar (KRI) 2301.

NANDIKEŚVARA

Author of a *Kālottara*. Manuscript:

Saṃskṛta Sāhitya Pariṣat, Calcutta II. F. 2. See NCC, vol. 4, p. 82.

NANDIKEŚVARA

Author of a *Jyotiḥsaṅgrahasāra*. Manuscript:

Mitra, Not. 1113. 6ff. Bengālī. Property of Vrajanātha Vidyāratna of Navadvīpa.

The first verse is:

dinanātham praṇamyādau nandikeśvaradhīmatā/
jyotiḥsaṅgrahasāro ᵓyam bhāṣayā likhyate mayā//

NANDIKEŚVARA (*fl. ca.* 1640)

The son of Mālajit Vedāṅgarāya (*fl.* 1643,) the son of Tigalābhaṭṭa, the son of Ratnabhaṭṭa, a resident of Śrīsthala in Gurgaradeśa, Nandikeśvara wrote a *Gaṇakamaṇḍana*. Manuscripts:

Benares (1963) 36507. Ff. 11–56 and 58–63. Copied in Saṃ. 1703 = A.D. 1646. Incomplete.
Jaipur (II). 29ff. Copied in Saṃ. 1703 = A.D. 1646.
Bombay U 402. 64ff. Copied on Sunday 2 śuklapakṣa of Śrāvaṇa in Saṃ. 1791 = 21 July 1734 Julian.
RORI Cat. II 5171. 52ff. (f. 46 missing). Copied in Saṃ. 1794 = A.D. 1737.
VVRI 2677. 17ff. Copied in Saṃ. 1809 = A.D. 1752.
PUL II 3322. 28ff. Copied in Saṃ. 1828 = A.D. 1771.
BORI 432 of A 1881/82. 19ff. Copied in Śaka 1705 = A.D. 1783.
BORI 886 of 1886/92. 19ff. Copied in Saṃ. 1843 = A.D. 1786.
VVRI 2633. 24ff. Copied in Saṃ. 1853 = A.D. 1796.
RORI Cat. II 9991. 30ff. (f. 1 missing). Copied by Moḍīrāma Brāhmaṇa at Sāhapurā in Saṃ. 1862 = A.D. 1805. Incomplete (gaṇitaprakaraṇa).
BORI 887 of 1886/92. 65ff. Copied in Saṃ. 1871 = A.D. 1814.
Goṇḍal 37. 26ff. Copied at Vāṃkanera in Baṃkapurī on Wednesday 1 śuklapakṣa of Śrāvaṇa in Saṃ. 1914 = 22 July 1857.
Alwar 1737.
AS Bengal 2745 (G 6343). Ff. 2–13, 16–20, 29–34, and 36–37.
Benares (1963) 35672 = Benares (1897–1901) 900. 3ff. Incomplete.
Benares (1963) 35682 = Benares (1905) 1513. Ff. 7–11 and 14–17. Incomplete.
Benares (1963) 36508. Ff. 2–27. Incomplete.
Benares (1963) 37266. Ff. 14–15. Incomplete.
BORI 530 of 1875/76. 62ff. From Dilhī.
BORI 466 of 1892/95. 44ff.
BORI 409 of 1895/98. 17ff.
IO 6337 (2743 E). 13ff. Incomplete (adhyāyas 1–2). From B. H. Hodgson.
Kathmandu (1960) 63 (III kha) = Kathmandu (1905) I 1412. 8ff. Incomplete (to puṣyārkapraśaṃsā).
VVRI 1050. 13ff. Incomplete.
VVRI 2501. 11ff. Incomplete.
VVRI 4720. 27ff. Incomplete.
WHMRL L. 26. e.

The last verses are:

śrīmadgurjaradeśe ᵓsti vipravṛndavibhūṣitam/
śrīsthalākhyaṃ puraṃ ramyaṃ
  puruhūtapuropamam//
tatrāsīj jyotiḥśāstrajño ratnabhaṭṭāhvayo dvijaḥ/
tajjaḥ śrītigalābhaṭṭaḥ sarvavidyāmahodadhiḥ//
tatputro mālajitsaṃjño vedavedāṅgapāragaḥ/
yena vedāṅgarāyeti prāptaṃ dillīśvarāt padam//
pitṛbhaktirataḥ prājñas tatsūnur nandikeśvaraḥ/
dvijaprītyai vyadhāt pūrvaṃ granthaṃ
  gaṇakamaṇḍanam//
jyotirnibandham akhilaṃ tu tathā muhūrta-
cintāmaṇiṃ gaṇakabhūṣaṇaratnamāle/
jyotirvidābharaṇasajjanavallabhākhyau
dṛṣṭvā trivikramaśatādi mayedam uktam//

NANDIN

An authority cited by Utpala (*fl.* 966/968) on *Bṛhatsaṃhitā* 8, 19; 35, 3; 52, 73; 85, 53; and 103, 60; see P. V. Kane [1948/49] 25. In one of these (on 52, 73) Nandin quotes Satya (*fl.* fourth century). He is probably the author of the *Nāndīyātrā* cited by Utpala on *Yogayātrā* 5, 19.

NANDISŪRI (*fl. ca.* 1747)

Author of a treatise on astronomy, *Kheṭatantra*, with tables; it refers to the 22nd 60-year cycle after Śaka 409, which is A.D. 1747/1806. Manuscripts:

GOML Madras D 13405. 42pp. Telugu. Incomplete (adhikāras 3 and 5–7).
GOML Madras D 13406. 20pp. Grantha and Telugu. Incomplete.
Kerala 4522 (2479 C). 125 granthas Telugu.
Kerala 4523 (2481 D). 135 granthas. Telugu.
Kerala 4524 (2519 Z 13). 18 granthas. Telugu. Incomplete.

The colophon begins: iti nandisūriviracite.

NANDĪŚVARA

Author of a *Dvīpavicāra* = *Yantramālā*, a description of Jambūdvīpa. Manuscript:

Mitra, Not. 2569. 10ff. Property of Bābu Rāmadāsa Sena of Bahrāmpur.

The colophon begins: iti nandīśvaravaradvīpavicāraḥ.

NABBĀBA KHĀNAKHĀNĀ (1556/1627)

See Khānakhānā (1556/1627).

NAYANASUKHA MIŚRA (*fl.* 1817)

Author of a *Prāṇakṛṣṇakriyāmbudhi* for Prāṇakṛṣṇa, a landowner near Calcutta, in 1817; this was published at Calcutta in 1818 (BM).

NAYANASUKHOPĀDHYĀYA (*fl.* 1730)

Author of the *Ukāra*, a translation of the *Ukarr Thāwadūsiyūs* or the Arabic version of the *Spherica* of Theodosius (*fl.* first century B.C.) made by Qusṭā ibn Lūqā al-Baᶜalbakī (d. 912), corrected by Thābit ibn Qurrā (834/901), and commented on by Naṣīr al-Dīn al-Ṭūsī (1201/1274). It is sometimes called *Kaṭara* (from quṭr, diameter). Manuscripts:

Baroda 8926. 44ff. Copied in Saṃ. 1787 = A.D. 1730. (*Kaṭara*).
Calcutta Sanskrit College 118. 54ff. Copied in Saṃ. 1787, Śaka 1652 = A.D. 1730.
Cambridge R. 15. 139b. Ff. 9–66. Copied in A.D. 1803.
Kerala 2329 (1506) = Congress, p. 33. 36ff. Copied in Saṃ. 1865 = A.D. 1808.
Baroda 9215(b). Ff. 160–221. (*Kaṭara*).
Baroda 11236. 117ff. (*Kaṭara*).
Benares (1963) 35762. Ff. 1–24, 31–122, 133–140, 21–23, 124–192, 1–82, and 1–56. With the *Siddhāntasāra* of Jagannātha.

The *Ukāra* begins in the Cambridge copy: atha ukārākhyo granthaḥ sāvajūsayusa(Theodosius)kṛto likhyate/··· idaṃ yūnāni(Greek)bhāṣātaḥ araba-(Arabic)bhāṣāyām abulaaccāsaahasasyā(Abū al-ᶜAbbās ibn Muᶜtaṣim)jñayā kustāvivirūkāvālvahvi (Qusṭā ibn Lūqā al-Baᶜalbakī) saṃjñena ··· grathitaṃ/idaṃ sāvitavinikusai(Thābit ibn Qurrā)saṃjñena śodhitam/narasīra(Naṣīr al-Dīn al-Ṭūsī)saṃjñena ṭīkā kṛtā/seyaṃ saṃskṛtaśabdair nayanasukhopādhyāyair nibadhyate//

NARACANDRA SŪRI (d. 1230)

The pupil of Devaprabha Sūri of the Harṣapurīya or Maladhāri Gaccha and a teacher of Vastupāla, the minister of Vīradhavala of Davalakha (*fl.* 1230/1231), a feudatory of the Caulukya mahārāja Bhīmadeva II (1178/1239), Naracandra died on 10 kṛṣṇapakṣa of Bhādrapada in Saṃ. 1287 = *ca.* 24 August 1230; see B. J. Sandesara [1953] 73–75. Among his works is a *Vastupālapraśasti*, ed. by Puṇyavijaya Sūri, *SJS* 5, Bombay 1961, pp. 21–23. On jyotiḥśāstra he wrote a *Jyotiṣasāra* = *Nāracandra* in 4 prakaraṇas, on which a ṭippaṇaka was written by Sāgaracandra Sūri. Manuscripts:

LDI 7016 (3523). 17ff. Copied by Maheśa Joṣi of the Moḍhajñāti in Saṃ. 1525 = A.D. 1468. With the ṭippaṇaka of Sāgaracandra. Incomplete (2 prakaraṇas).
LDI 6992 (1045). 6ff. Copied in Saṃ. 1529 = A.D. 1472.
Līṃbaḍī 1387 (548). 23ff. Copied in Saṃ. 1560 = A.D. 1503. With the ṭippaṇaka of Sāgaracandra.
Goṇḍal 179. 31ff. Copied in Saṃ. 1569 = A.D. 1512. With a ṭīkā.
BORI 536 of 1899/1915. 31ff. Copied in Saṃ. 1622 = A.D. 1565.
LDI (LDC) 4783. 13ff. Copied in Saṃ. 1627 = A.D. 1570. Incomplete (prakaraṇa I).
LDI (LDC) 1436. 23ff. Copied in Saṃ. 1649 = A.D. 1592.
RORI Cat. II 4352. 30ff. Copied at Koraṇṭānagara in Saṃ. 1651 = A.D. 1594. With the ṭippaṇaka of Sāgaracandra.
Baroda, Pra. Śrī. Kām. Vi. Sam. Śā. Saṃ. Copied by Udayasaubhāgya Muni, the pupil of Puṇyasaubhāgya, the pupil of Śaṅkarasaubhāgya, at Sāṅgāneranagara on Sunday 7 śuklapakṣa of Kārttika in Saṃ. 1653 = 17 October 1596. See Praśasti (1), p. 151.
PL, Buhler IV E 206. 5ff. Copied in Saṃ. 1654 = A.D. 1597. (*Nāracandrapaddhati*). Property of Śeṭha Bhīmaśī Māṇeka of Mumbaī. Buhler notes another copy.
RORI Cat. II 8333. 42ff. Copied by Vastā Matheṇa in Saṃ. 1663 = A.D. 1606. With a stabaka in Old Gujarātī.
RORI Cat. I 3008. 37ff. Copied by Syāmaliyā in Saṃ. 1664 = A.D. 1607. With the ṭippaṇaka of Sāgaracandra.
LDI 6998 (1628). 22ff. Copied by Harajī at Satyapura in Saṃ. 1669 = A.D. 1612.
LDI 6999 (7630). 9ff. Copied by Vācaka Guṇajī, the pupil of Lalitaprabha Sūri, at Nārolīdraṅga in Saṃ. 1675 = A.D. 1618
RORI Cat. III 17263. 36ff. (ff. 1–2 missing). Copied by Jinasoma Gaṇi, the pupil of Śrīsoma Gaṇi, at Nūtanapura in Saṃ. 1675 = A.D. 1618. With a ṭippaṇa.
LDI (LDC) 5019. 29ff. Copied in Saṃ. 1676 = A.D. 1619. Incomplete (prakaraṇas I–II).
RORI Cat. I 1997. 43ff. Copied by Vinayaprabha Sūri at Pattana in Saṃ. 1693 = A.D. 1636. With the ṭippaṇaka of Sāgaracandra.
BORI 929 of 1886/92. 89ff. Copied in Saṃ. 1698 = A.D. 1641. With the ṭippaṇaka of Sāgaracandra.
Goṇḍal 178. 17ff. Copied by Ṛṣi Nārāyaṇa at Sārakoṭa on Tuesday 3 (read 6) kṛṣṇapakṣa of Vaiśākha in Saṃ. 1702 = 6 May 1645 Julian.
Cāṇasmā, Ni. Vi. Jī. Ma. Pu. Copied by Padmasāgara Gaṇi of the Añcala Gaccha at Burahānapura in Saṃ. 1704 = A.D. 1647. See Praśasti (1), p. 216.
LDI 7013 (4157). 31ff. (ff. 1–5 missing). Copied by Mānaharṣa Muni, the pupil of Paṇḍita Merugaṇi, the pupil of Paṇḍita Dharmagaṇi, the pupil of Vācaka Samayakalaśa Gaṇi of the Bṛhatkaratara Gaccha, at Dahīravāsa in Saṃ. 1704 = A.D. 1647. Incomplete.
Florence 301. 22ff. Copied by Dharmaratna Sūri at Sirohīnagara in Saṃ. 1707 = A.D. 1650. With a ṭīkā.
RORI Cat. II 8392. 27ff. Copied by Sāṅgā Ṛṣi, the pupil of Mahimāsāgara of the Vijaya Gaccha, at

Gaṅgrāḍha in Saṃ. 1717 = A.D. 1660 during the reign of Auraṅgazeba (1658/1707). With a stabaka in Old Gujarātī.

RORI Cat. I 660. 28ff. Copied by Viśeśara Muni in Saṃ. 1724 = A.D. 1667. With a stabaka.

RORI Cat. III 16723. 23ff. (f. 4 missing). Copied by Haradāsa in Saṃ. 1728 = A.D. 1671.

RORI Cat. III 14591. 20ff. Copied at Jesalamera in Saṃ. 1734 = A.D. 1677. With a stabaka in Old Rājasthānī.

RORI Cat. III 13980. 33ff. (f. 32 missing). Copied by Hitasāgara Gaṇi, the pupil of Lābhasāgara, at Avantipārśvanātha in Tājapura in Saṃ. 1749 = A.D. 1692. With the ṭippaṇaka of Sāgaracandra.

IO 6345 (3315). 13ff. Copied by Paṇḍita Netasīha at Maulatrāṇa on Tuesday 4 śuklapakṣa of Vaiśākha in Saṃ. 1751 = 17 April 1694 Julian. With a ṭabā in Old Rājasthānī.

LDI 7009 (5371). 31ff. (f. 1 missing). Copied by Muni Dānavijaya for Muni Jasavijaya at Somesara in Saṃ. 1753 = A.D. 1696. With the ṭippaṇaka of Sāgaracandra. Incomplete.

RORI Cat. I 2577. 9ff. Copied by Nemaharṣa at Maroṭṭakaṭṭa in Saṃ. 1754 = A.D. 1697. With the ṭippaṇaka of Sāgaracandra.

LDI (LDC) — (between 5282 and 5299). 21ff. Copied in Saṃ. 1755 = A.D. 1698. With a ṭippaṇī.

RORI Cat. II 6821. 11ff. (f. 1 missing). Copied by Ratnā Paṇḍita, the pupil of Tiladhīra, at Jaitāraṇa in Saṃ. 1759 = A.D. 1702. Incomplete (prakaraṇa I).

RORI Cat. II 8339. 12ff. Copied by Padamasī, the pupil of Dayāvinaya, at Lūṇasara in Saṃ. 1760 = A.D. 1703.

RORI Cat. II 4747. 14ff. Copied at Kuṇḍāgrāma in Saṃ. 1761 = A.D. 1704. With a stabaka in Old Rājasthānī.

RORI Cat. I 3799. 31ff. Copied by Īśaradāsa at Sīrohī in Saṃ. 1762 = A.D. 1705. With a stabaka in Old Rājasthānī.

LDI 7012 (3108). 24ff. Copied by Ṛṣi Jayacanda at Jihānāvāda in Saṃ. 1764 = A.D. 1707 during the reign of Pātiśāha Ālamaśāha (= Aurangzib) (1658/1707).

LDI (LDC) 4693. 41ff. Copied in Saṃ. 1765 = A.D. 1708. With a stabaka.

Anup 4682. 28ff. Copied by Muni Śiva Ḍāṃbarajī at Vikramapura in Saṃ. 1766 = A.D. 1709. With a stabaka in Old Rājasthānī. Property of Lālacanda Mastrī.

LDI (KS) 1008 (11036) = LDI (KC) K/1008. 25ff. (f. 1 missing). Copied by Kṣamāsundara, the pupil of Jinasundara Sūri, in Saṃ. 1766 = A.D. 1709. Incomplete (ends in prakaraṇa II).

RORI Cat. III 17066. 39ff. Copied by Ḍuṅgaramalla, the pupil of Phatehadharma, in Saṃ. 1772 = A.D. 1715. With a stabaka in Old Rājasthānī.

LDI 7026 (8954). 33ff. Copied by Dīpavijaya Gaṇi, the pupil of Paṇḍita Labdhivijaya Gaṇi, the pupil of Vijayaprabha Sūri, for Muni Hīrajī at Belāgrāma in Saṃ. 1782 = A.D. 1725. With a ṭippaṇaka and a stabaka in Old Gujarātī.

Surat, Jainānanda Pustakālaya. Copied by Lavajī, the pupil of Vācaka Karmacandrajī, on Monday 11 kṛṣṇapakṣa of Kārttika in Saṃ. 1783 = 7 November 1726 Julian. See Praśasti (1), p. 302.

RORI Cat. II 4408(2). Ff. 7–14. Copied by Rājapāla Vairāgī, the pupil of Rūpa Ṛṣi, at Pālhanapura in Saṃ. 1786 = A.D. 1729.

RORI Cat. III 13827(9). 10ff. Copied by Saṃ. 1790 = A.D. 1733. Incomplete.

Goṇḍal 177. 27ff. Copied by Jeṭhā at Vaḍhavāṇa on Wednesday 10 śuklapakṣa of Mārgaśira in Saṃ. 1793 = 1 December 1736. With a ṭabā in Gujarātī.

RORI Cat. I 3783. 43ff. Copied in Saṃ. 1795 = A.D. 1738. With a ṭippaṇa in Old Rājasthānī.

RORI Cat. II 6776. 19ff. Copied in Saṃ. 1799 = A.D. 1742.

RORI Cat. III 10781. 51ff. Copied in Saṃ. 1802 = A.D. 1745. With a stabaka in Old Rājasthānī. Incomplete (prakaraṇa I).

RORI Cat. III 13761(16). Ff. 36–42. Copied in Saṃ. 1802 = A.D. 1745. Incomplete.

RORI Cat. III 13944. 44ff. Copied in Saṃ. 1804 = A.D. 1747. With a stabaka in Old Gujarātī.

LDI 6989 (7853). 9ff. Copied by Ṛṣi Devīcanda, the pupil of Paṇḍita Jesīṅghajī, at Bhalaḍārāvāḍa in Saṃ. 1805 = A.D. 1748.

LDI 7020 (8358). 13ff. Copied by Muni Puruṣottamavijaya, the pupil of Kastūravijaya, the pupil of Rucivijaya, the pupil of Paṇḍita Rūpavijaya, at Vijāpura for Kesaravijaya and Kapūravijaya in Saṃ. 1806 = A.D. 1749. With a *Bālāvabodha* in Old Gujarātī.

RORI Cat. III 11842. 13ff. Copied by Mānajī Mahātmā of Campāvatī at Savāī Jayapura in Saṃ. 1808 = A.D. 1751. Incomplete (prakaraṇa I).

RORI Cat. I 3770. 29ff. Copied by Mūlacanda Muni in Saṃ. 1809 = A.D. 1752. With the ṭippaṇaka of Sāgaracandra.

RORI Cat. II 7010. 27ff. Copied by Karmacaṇḍa Paṇḍita, the pupil of Guṇasundara Mahopādhyāya, in Saṃ. 1809 = A.D. 1752. Incomplete (prakaraṇa II).

RORI Cat. II 9477. 35ff. Copied by Raghucandra at Vikramapura in Saṃ. 1809 = A.D. 1752. With a stabaka in Old Rājasthānī.

RJ 3012 (vol. 4, p. 285). 26ff. Copied on 14 kṛṣṇapakṣa of Mārgaśīrṣa in Saṃ. 1810 = *ca.* 22 December 1753.

LDI 6997 (3593). 33ff. Copied by Udayadharma Muni at Jesalamera in Saṃ. 1812 = A.D. 1755. Incomplete (prakaraṇas I–II).

LDI 7021 (2067). 56ff. Copied by Paṇḍita Bhagavānasāgara, the pupil of Khuśālasāgara Gaṇi, the

pupil of Jayantasāgara Gaṇi, at Delavāḍā in Saṃ. 1817 = A.D. 1760. With a stabaka in Old Gujarātī. Incomplete (prakaraṇa I).
Florence 303. 26ff. Copied in Saṃ. 1819 = A.D. 1762. With a Hindī ṭippaṇa.
ABSP 7. 22ff. Copied in Saṃ. 1820 = A.D. 1763.
RORI Cat. I 3737. 20ff. Copied by Ratnacandra at Karmāvāsa in Saṃ. 1821 = A.D. 1764. With the ṭippaṇaka of Sāgaracandra. Incomplete (prakaraṇa I).
RORI Cat. III 15913(2). Ff. 4–36. Copied in Saṃ. 1822 = A.D. 1765.
RORI Cat. II 9577(2). Ff. 4–25. Copied by Rūpapurī Gusāīṃ, the pupil of Sugāla, at Jāṭa in Saṃ. 1823 = A.D. 1766.
LDI 7025 (1614). 44ff. Copied by Dayācandra Muni for Bhagavānadāsa, the pupil of Lalitavijaya, the pupil of Jñānavijaya, at Visalanagara in Saṃ. 1839 = A.D. 1782. With a stabaka in Old Gujarātī. Incomplete (prakaraṇas I–II).
RORI Cat. III 14040. 23ff. Copied by Labdhivijaya in Saṃ. 1839 = A.D. 1782. With a stabaka in Old Rājasthānī.
Līṃbaḍī 1388 bis (1051). 20ff. Copied in Saṃ. 1841 = A.D. 1784.
RORI Cat. III 16949. 43ff. Copied in Saṃ. 1845 = A.D. 1788. With a stabaka in Old Rājasthānī.
LDI (DSC) 9724. 7ff. Copied in Saṃ. 1846 = A.D. 1789.
AS Bombay 311. 14ff. Copied in Saṃ. 1851 = A.D. 1794. From Bhāu Dājī.
LDI (LDC) 5725. 42ff. Copied in Saṃ. 1861 = A.D. 1804.
RJ 3014 (vol. 4, p. 285). 37ff. Copied on 3 śuklapakṣa of Phālguna in Saṃ. 1864 = *ca.* 28 February 1808.
Poleman 5126 (Harvard 983). 11ff. Copied in Saṃ. 1873, Śaka 1738 = A.D. 1816.
LDI 6996 (7681). 30ff. Copied for Ṛṣi Indrabhāṇa, Ṛṣi Hukamacanda, and Ṛṣi Dayācanda of Vausivāla at Ānandapura in Saṃ. 1880 = A.D. 1823.
RORI (Jaipur) IV 136(1). 94ff. Copied in Saṃ. 1884 = A.D. 1827. (*Nāracandrasāraṇī*).
LDI (LDC) 4282. 15ff. Copied in Saṃ. 1888 = A.D. 1831.
RORI Cat. II 7666. 18ff. Copied by Amṛtavijaya at Daityāridurga in Saṃ. 1899 = A.D. 1842.
RORI Cat. III 17234(11). Ff. 61–98. Copied by Lakṣmīnārāyaṇa Dave at Sojata in Saṃ. 1905 = A.D. 1848.
LDI 7022 (7305). 43ff. Copied by Śivarāma Ṭhākora for Paṇḍita Ratnavijaya at Aṇahillapurapattana in Saṃ. 1906 = A.D. 1849. With a stabaka in Old Gujarātī.
LDI 6991 (192). 7ff. Copied by Ṛṣi Keśarīcanda of the Luñkāgaṇa at Vikramapura in Saṃ. 1950 = A.D. 1893.
RORI Cat. III 10801. 36ff. Copied in Saṃ. 1964 = A.D. 1907.
Agra, Vijayadharma Lakṣmī Jñānamandira 3071–3076 (3075 with the ṭippaṇaka of Sāgaracandra). See Velankar, p. 211.
Ahmabadad, Dela Upāśraya Bhandar, ground floor 67 (33, 34, and 35) and first floor 24 (165, 166, and 167). See Velankar.
Ahmadabad, Bhandar of the Vimala Gaccha Upāśraya, 19 (16) at Haja Patel's Pole and 8 (15) in possession of Uddyotavimalagaṇi. See Velankar.
Alwar 1822.
Anup 4683. 9ff. Property of Anūpasiṃha (*fl.* 1674/98).
Anup 4684. 9ff.
Anup 4685. 2ff.
Anup 4686. 14ff. Incomplete (prakaraṇa I).
AS Bengal 7136 (G 6896). 26ff.
AS Bengal 7137 (G 6625). 56ff. With a stabaka in Hindī.
Baroda 3385. 7ff.
Baroda 9485. 8ff. Incomplete (prakaraṇa I). With sāraṇī.
Baroda 11818. 6ff. Incomplete (to panotīphala).
Baroda, Library of Kantivijayaji 1115. With the ṭippaṇaka of Sāgaracandra. See Velankar.
Baroda, Hamsavijayaji Maharaj at the Kantivijaya Bhandar 925 (with the ṭippaṇaka of Sāgaracandra) and 990. See Velankar.
Benares (1963) 37015. Ff. 1–9 and 11–56 and 1f. With a ṭīkā. Incomplete.
Bikaner, Bada Upāśraya 1 (39, 50, and 66) and 3 (86). The last with the ṭippaṇaka of Sāgaracandra. See Velankar.
Bombay U 2406(66). No ff. given. Copied for Sivadharma, the pupil of Padmasundara Gaṇi. Incomplete (prakaraṇa II).
Bombay U 2406(69). No ff. given. Incomplete (prakaraṇa I).
BORI 51 of 1870/71. 13ff. Bought in Surat.
BORI 606 of 1884/86. 21ff.
BORI 538 of 1899/1915. 18ff.
BORI 753 of 1899/1915. 20ff.
Calcutta Sanskrit College 33. 29ff.
Cambay, Jñānavimalasūri Bhandar 133. With the ṭippaṇaka of Sāgaracandra. See Velankar.
Chani, Bhandar of Kantivijayaji Maharaj 311. With the ṭippaṇaka of Sāgaracandra. See Velankar.
CP, Hiralal 2519. Property of the Lokāgaccha Jain Mandir at Bālāpur, Akolā.
CP, Hiralal 2520. Property of Puttelāl Gaurisankar of Valgaon, Amraotī.
CP, Hiralal 2521. Property of Śivrām of Hoshangābād.
CP, Hiralal 2522. Property of Sāligrām of Hoshangābād.
CP, Hiralal 7273. Property of the Ḅalātkār Gaṇ Jain Mandir at Kārañjā, Akolā.

CP, Hiralal 7274. Property of the Sen Gaṇ Jain Mandir at Kārañjā, Akolā.
Florence 300. 5ff. Incomplete.
Florence 302. 10ff. Incomplete (prakaraṇa I).
IO 6346 (3384a). 16ff. With the ṭippaṇaka of Sāgaracandra. Incomplete (ends in prakaraṇa III). From H. T. Colebrooke.
Jaipur, Inner Bhandar of Harisāgaragaṇi 59. With the ṭippaṇaka of Sāgaracandra. See Velankar.
Jaipur, Outer Bhandar of Harisāgaragaṇi 45 (3c). With the ṭippaṇaka of Sāgaracandra. See Velankar.
Jesalmere, Sambhavnatha Temple 204. See Velankar.
Kaira, Bhandar of Sammatiratna Sūri 163. See Velankar.
LDI 6982 (6136). 54ff. Copied by Nayavijaya, the pupil of Punyavijaya Gaṇi, the pupil of Kanakavijaya Gaṇi, the pupil of Jayatilaka Sūri, at Lāsa.
LDI 6983 (2454). 27ff. (ff. 3, 4, 6, 8, and 9 missing). Incomplete.
LDI 6984 (8641/1). Ff. 1–18.
LDI 6985 (8839). 20ff. Incomplete.
LDI 6986 (7659). 17ff. Incomplete.
LDI 6987 (6440). 16ff. Incomplete.
LDI 6988 (191). 15ff.
LDI 6990 (7343/1). Ff. 1–9.
LDI 6993 (7300). 11ff. Incomplete.
LDI 6994 (8843). Ff. 6–20. Incomplete.
LDI 6995 (6724). 8ff. (f. 1 missing). Incomplete.
LDI 7000 (8936). 29ff. Incomplete (prakaraṇas I–II).
LDI 7001 (7023). 8ff. Incomplete (prakaraṇa I).
LDI 7002 (8948). 9ff. Copied by Bhāvaprabha Sūri of the Pūrṇimā Gaccha. Incomplete (prakaraṇa I).
LDI 7003 (6844). 20ff. Incomplete (prakaraṇa I).
LDI 7004 (6843). 19ff. (f. 1 missing). Incomplete (prakaraṇa I).
LDI 7005 (6685/4). Ff. 19–63. With a vṛtti.
LDI 7006 (1407). 15ff. With a vṛtti. Incomplete (prakaraṇa II).
LDI 7007 (3747). 23ff. Copied by Nayaharṣa, the pupil of Punyaharṣa, at Jālora. With the ṭippaṇaka of Sāgaracandra. Incomplete (ends in prakaraṇa II).
LDI 7008 (6569). 20ff. Copied by Kamalaharṣa. With the ṭippaṇaka of Sāgaracandra.
LDI 7010 (3577). Ff. 12–15. With the ṭippaṇaka of Sāgaracandra. Incomplete (prakaraṇa II).
LDI 7011 (4879). 31ff. With the ṭippaṇaka of Sāgaracandra. Incomplete (prakaraṇas I–II).
LDI 7014 (3133). 30ff. With the ṭippaṇaka of Sāgaracandra. Incomplete (prakaraṇas I–II).
LDI 7015 (1535). 6ff. With the ṭippaṇaka of Sāgaracandra. Incomplete (prakaraṇas I–II).
LDI 7017 (6192). 59ff. (ff. 1–8 missing). Copied for Lābhacanda Nanicanda. With the ṭippaṇaka of Sāgaracandra. Incomplete (prakaraṇa I).
LDI 7019 (1717). 9ff. (ff. 1–2 missing). With an avacūri.
LDI 7023 (7345). 10ff. With a stabaka in Old Gujarātī.
LDI 7024 (7420). 29ff. With a stabaka in Old Gujarātī.
LDI 7179 (3409/2). Ff. 1–2. Incomplete (lagnaghaṭikānayana). With a vyākhyā.
LDI (AKC) 1218. 44ff.
LDI (DJSC) 265. 14ff. With a stabaka.
LDI (KC) K/341. 21ff. With the ṭippaṇaka of Sāgaracandra. Incomplete.
LDI (KC) K/531. 22ff. With the ṭippaṇaka of Sāgaracandra.
LDI (KhC) 122 = LDI (VC) 122. 35ff.
LDI (KS) 1009 (10559). 22ff. With the ṭippaṇaka of Sāgaracandra.
LDI (KS) 1010 (10369). 24ff. With the ṭippaṇaka of Sāgaracandra. Incomplete (prakaraṇas I–II).
LDI (LDC) 4027. 13ff.
LDI (LDC) 4399/134. Ff. 223–240.
LDI (LDC) 4531. 1f. Incomplete (naṣṭajātaka). With an avacūri.
LDI (LDC) 4717. 32ff. With a stabaka.
LDI (LDC) 5280. 37ff. With a *Bālāvabodha.*
Līṃbaḍī 1388 (781). 9ff.
Mitra, Not. 2798. 32ff. Property of Rāya Dhanapat Siṃha, Bahādur, of Ājiṃgañj.
Paris BN 968 (Sans. Dév. 328) VIII.
Patan, Sangha Bhandar 56 (4) and 75 (103, 110, 120, and 139). 56 (4) and 75 (120) with the ṭippaṇaka of Sāgaracandra. See Velankar.
Patan, Vadi Pārśvanātha Pustaka Bhandar 17 (16) and 25 (15). With the ṭippaṇaka of Sāgaracandra. See Velankar.
PUL II 3589. Ff. 2–14. Incomplete.
RJ 1670 (vol. 2, p. 272). 19ff. Incomplete. Property of Baḍa Terahapanthiyoṃ of Jayapura.
RJ 3013 (vol. 4, p. 285). 17ff.
RORI Cat. I 587. 30ff. With a ṭippaṇa.
RORI Cat. I 605. 4ff. (*Sāroddhārajyotiṣa*).
RORI Cat. I 675. 15ff. With the ṭippaṇaka of Sāgaracandra. Incomplete (prakaraṇa I).
RORI Cat. I 1980. 25ff. Copied by Rājasundara Ṛṣi. With the avacūri of Rājasundara.
RORI Cat. I 3438. 29ff. (f. 1 missing). With the ṭippaṇaka of Sāgaracandra.
RORI Cat. I 3728. 32ff. With a stabaka in Old Rājasthānī. Incomplete (prakaraṇa I).
RORI Cat. II 5536. 10ff. (*Sāroddhāra*).
RORI Cat. II 5540. 18ff. (*Sāroddhāra*).
RORI Cat. II 6650. 26ff. With the ṭippaṇaka of Sāgaracandra. Incomplete (prakaraṇa I).
RORI Cat. II 8370. 46ff. Incomplete.
RORI Cat. II 8538. 29ff. Copied by Bhojarāja Ṛṣi at Riṇī.
RORI Cat. II 9505. 8ff.
RORI Cat. II 9774. 5ff. Incomplete.
RORI Cat. II 9797. 35ff. With a stabaka in Old Rājasthānī.

RORI Cat. III 10244. 27ff. With the ṭippaṇaka of Sāgaracandra. Incomplete (yantrakoddhāra to grahagocaraśuddhiyantra).
RORI Cat. III 11584(3). Ff. 64–82.
RORI Cat. III 13062. 17ff. (f. 1 missing). Incomplete (prakīrṇaprakaraṇa).
RORI Cat. III 14415(12). Ff. 30–72.
RORI Cat. III 14968. 66ff. With a ṭippaṇa. Incomplete.
RORI Cat. III 15489. 8ff. Incomplete (prakaraṇa I).
RORI Cat. III 16957. 11ff. (ff. 1–2 missing).
RORI Cat. III 17322. 10ff.
RORI (Jaipur) IV 67. 14ff.
SOI 537 = SOI Cat. I: 1418–537. No ff. given.
SOI 641 = SOI Cat. I: 1419–641. No ff. given. With a ṭabā.
SOI 3545 = SOI Cat. II: 1028–3545. 38ff.
Surat, Jainānanda Bhandar at Gopipura 1740 and 2920; *cf.* Praśasti(1), p. 302. See Velankar.
Surat, Jain Upāśraya Library and Cintāmaṇi Pārśvanātha Temple Library (the latter with the ṭippaṇaka of Sāgaracandra). See Velankar.
WHMRL G. 76. e. With the ṭippaṇaka of Sāgaracandra.
WHMRL G. 93. a. With the ṭippaṇaka of Sāgaracandra.
WHMRL Q. 23. g.

The *Nāracandra* was edited with a Gujarātī translation by Ratilāla Prāṇajīvanadāsa Sūḍīvāḷā, Surat 1913 (BM 14055. d. 19), and by Kṣamāvijaya Gaṇi at Bombay in 1938 (see Velankar). The first verse is:

śrīarhantaṃ jinaṃ natvā naracandreṇa dhīmatā/
sāram uddhriyate kiṃcij jyotiṣakṣīranīradheḥ//

NARACANDROPĀDHYĀYA (*fl.* 1167/1177)

The pupil of Siṃhasūri, the pupil of Uddyotanasūri of the Kāsadraha or Kāśahrada Gaccha, Naracandra (see B. J. Sandesara [1953] 74, fn. 1) wrote the following on jyotiḥśāstra.

1 and 2. A *Praśnaśata* and an avacūrṇi on it, both written in Saṃ. 1234 = A.D. 1177. Manuscripts:

BORI 1357 of 1884/87. 18ff. With his own avacūrṇi. Copied in Saṃ. 1572 = A.D. 1515.
RORI Cat. II 4900. 23ff. Copied in Saṃ. 1797 = A.D. 1740.
Agra, Vijayadharma Lakṣmī Jñānamandira 2270. See Velankar, p. 275.
Ahmadabad, Bhandar of the Vimala Gaccha Upāśraya 10 (18). With his own avacūrṇi. See Velankar.
Baroda 721. 6ff.
Baroda 3024. 10ff. (*Praśnaśatavṛttyuddhāra*).
BORI 388 of 1880/81. 5ff. With his own avacūrṇi.
Chani, Bhandar of Kantivijayaji Maharaj 239. With his own avacūrṇi. See Velankar.
Patan, Bhandar of the Agali Sheri 81 (86). With his own avacūrṇi. See Velankar.
Surat, Jainānanda Bhandar at Gopipura 828. With his own avacūrṇi. See Velankar.

3. A *Janmasamudra* = *Janmāmbhodhi* in 8 kallolas. Manuscripts:

Baroda 2799. 10ff. With a ṭīkā.
Baroda, Hamsavijayaji Maharaj at Kantivijaya Bhandar 273. With his own *Beḍāvṛtti*. See Velankar, p. 129.
Benares (1963) 36666. 11ff. (*Samudrajātaka*).
Chani, Bhandar of Kantivijayaji Maharaj 244. With his own *Beḍāvṛtti*. See Velankar.
LDI 6809 (3400). 7ff. With a ṭippaṇī.

The colophon begins: iti śrīkāśahradagacchīyaśrīsiṃhasūriśiṣyaśvetāmbaraśrīnaracandropādhyāyakṛte.

4. A ṭīkā, *Beḍāvṛtti*, on the *Janmasamudra*, completed at Campāvatī on Monday 14 śuklapakṣa of Phālguna in Saṃ. 1323 = 6 March 1167 during the reign of Kumārapāla (*ca.* 1143/1172). Manuscripts:

Anup 4601. 37ff. Copied by the son of Kamalākara Bhaṭṭa in Saṃ. 1707 = A.D. 1650.
BORI 277 of 1873/74. 83ff. Copied in Saṃ. 1707 = A.D. 1650. From Rānder.
Anup 4602. 32ff. Copied by Maṇirāma Dīkṣita at Śaivapura in Saṃ. 1711 = A.D. 1654.
Benares (1963) 34390. Ff. 1–65, 3ff., ff. 66–73, and 1f. Copied in Saṃ. 1798 = A.D. 1741.
GVS 2925 (861). 17ff. Copied on Saturday 3 śuklapakṣa of Mārgaśira in Saṃ. 1930 = 22 November 1873.
Anup 4603. 20ff.
Baroda, Hamsavijayaji Maharaj at Kantivijayaji Bhandar 273. See Velankar, p. 129.
Benares (1963) 36617. 81ff.
Chani, Bhandar of Kantivijayaji Maharaj 244. See Velankar.

The *Beḍājātaka* or *Beḍāvṛtti* was published with his own Hindī ṭīkā by Gopeśa Kumāra Ojhā in his *Triphalā*, Dillī-Vārāṇasī-Paṭanā 1971, pp. 159–271. The last two verses are:

śrīkāśahradagacchagucchataralaśrīdevacandrābdhiyuk-
śrīudyotanasūripaṭṭamukuṭaśrīsiṃhasūriprabhoḥ/
śiṣyaḥ śrīnaracandranāmavidito yo ᵓdhyāpako jñāpakaś
cakre janmasamudra eṣa sudhiyā tenārthagehaṃ jayī//
śrīmadvikramavatsare trinayanāghoṣe ᵓtra varṣe tapomāse śuddhacaturdaśīśaśidine campāvatīpaṭṭane/
caitye ᵓkāri kumārapālanṛpater vṛttiṃ ca kāśahradopādhyāyo naracandra indra⟨xx⟩paryāyarūpam imām//

5. A *Jñānacaturviṃśikā*. Manuscripts:

LDI 6889 (1753). 1f. Copied in Saṃ. 1708 = A.D. 1651. With an avacūrī.
LDI 6740 (3533). 1f. With an avacūrī.
LDI (SCC) Sag. 494/1. 1f. With an avacūrī.
Osmania University 125. 1f. With an avacūrī.
Patan, Bhandar at the Agali Sheri 80 (105). With an avacūrī. See Velankar, p. 147.

The *Jñānacaturviṃsī* with the avacūrī was edited by Āryendra Śarman, Hyderabad-Deccan 1956, on the basis of the Osmania University manuscript. Verse 24 is:

śrīkāsadrahagacchapo ᵓrbudagirinyastādināthaḥ purā
caikākī navamāsakalpavihṛtiḥ śrīsiṃhasūriprabhuḥ/
tannāmapratisābhidho gurur abhūd gotre ᵓsya
śiṣyaḥ śrutas
teneyaṃ caturārtham arthabahulā cakre
caturviṃśikā//

The colophon begins: iti śrīnaracandropādhyāyaracitā.

NARAPATI (*fl.* 1177)

The son of Naradeva of Dhārā in Mālava, Narapati wrote a *Narapatijayacaryā* = *Svarodaya;* according to the commentator, Harivaṃśa, on *Narapatijayacaryā* 1, 1–10, he wrote it at Aṇahilanagara during the reign of Ajayapāla, who was the Caulukya mahārāja from *ca.* 1174 to 1177:

vidyālaye mālavasaṃjñadeśe
dhārāpurīramyanivāsavāsī/
nānāgamajño nṛpalokapūjyo
budhaḥ prasiddho naradevanāmā//
svarabalaphalavettā dehatattveṣv abhijño
viditaśakunaśāstras tantramantrapravīṇaḥ/
kalitagaṇitasārāsāracūḍāmaṇijño
narapatir iti nāmnā tasya putro babhūva//
jñāne yaḥ sarvajño
nṛpagaṇapūjyaḥ sarasvatīsiddhiḥ/
tena kṛtaṃ śāstram idaṃ
pracuraguṇaṃ doṣarahitaṃ ca//
yo vetti śāstram etad
gurumukhakathitaṃ sadyuktisiddhaṃ ca/
vasati viśadā samagrā
karakamale tasya vijayaśrīḥ//
jitvā ripunṛpalakṣmīṃ
dadāti nijabhūpater na sandehaḥ/
etacchāstrajñabudhaś
caturvidhe caiva saṅgrāme//
śrīmaty aṇahilanagare
khyāte śrīajayapālanṛparājye/
śrīpatinarapatikavinā
racitam idaṃ tatrasaṃsthena.

See also Ādityadeva. The date of composition in some manuscripts is given as Tuesday 1 śuklapakṣa of Caitra in Saṃ. 1232 = 1 March 1177. There are commentaries by Narahari, Mahādeva, and Harivaṃśa. The manuscripts are:

Kathmandu (1960) 188 (I 1537). 67ff. Copied by Daivajña Nṛsiṃha in the Rājamaṇḍalī at Śrīpaśupatisthāna on Friday 4 śuklapakṣa of Bhādrapada in NS 400 = 30 August 1280 during the reign of Anantamalladeva (1274/1310).
Kathmandu (1960) 197 (I 1179). 77ff. Nevārī. Copied by Daivajña Gajarāja at Bhaktāpurī on Friday 12 śuklapakṣa of Bhādrapada in NS 522 = 14 July 1402. Incomplete.
BORI 33 of 1880/81. Ff. 1–68 and 1–43. There is noted on this the date Saṃ. 1471 = A.D. 1414.
PL, Buhler IV E 203. 127ff. Copied in Saṃ. 1487 = A.D. 1430. Property of Maṅgala Śaṅkara of Ahmadābād.
Baroda 6086. 73ff. Copied in Saṃ. 1510 = A.D. 1453.
RJ 3011 (vol. 4, p. 285). 148ff. (ff. 4–12 missing). Copied on 15 śuklapakṣa of Caitra in Saṃ. 1523 = *ca.* 30 March 1466. Incomplete.
LDI 7282 (531). 8ff. Copied in Saṃ. 1532 = A.D. 1475. Incomplete (sarvatobhadracakra).
PL, Buhler IV E 204. 54ff. Copied in Saṃ. 1572 = A.D. 1515. Property of Śivaśaṅkara Jośī of Ahmadābād. Buhler notes 5 other copies.
RORI Cat. II 8787. 60ff (f. 1 missing). Copied by Kṛṣṇa, the son of Goīyā Pāṭhaka of the Gauḍajñāti, a resident of Vaṭapadra, on Tuesday 7 śukla pakṣa of Bhādrapada in Saṃ. 1638 = 5 September 1581 Julian.
BORI 331 of 1882/83. 154ff. Copied in Saṃ. 1644 = A.D. 1587. From Gujarāt.
RORI Cat. III 11084(1). 78ff. Copied by Bhairavadāsa Vyāsa at Jodhapura in Saṃ. 1644 = A.D. 1587. Incomplete (begins with adhyāya 3).
Anup 4791. 44ff. Copied by Vidyāratna at Sarasvatīpattana in Saṃ. 1661 = A.D. 1604 during the reign of Dalapati.
LDI (LDC) 4016. 7ff. Copied in Saṃ. 1667 = A.D. 1610.
Kathmandu (1960) 196 (I 1172). 79ff. Copied in Śaka 1535 = A.D. 1613. Incomplete.
Anup 4795. 31ff. Copied by Narasapa in Śaka 1569 = A.D. 1647. Incomplete (cakroddhāra).
Leipzig 1159. 79ff. Copied in A.D. 1652. Incomplete (ends in bhūbalādhyāya).
Goṇḍal 171. Ff. 18–33. Copied in Saṃ. 1718, Śaka 1584 = A.D. 1661/62. With the ṭīkā of Narahari. Incomplete.
LDI 6791 (85). 121ff. Copied by Śivajī Ojhā, the son of Śīrāma, the son of Nārasiṃha, at Sthāmalānagara in Saṃ. 1725 = A.D. 1668.
RORI Cat. I 3444. 51ff. Copied by Vicārasāgara at Harṣapura in Saṃ. 1734 = A.D. 1677.
Jaipur (II). 106ff. Copied in Saṃ. 1736 = A.D. 1679.
Anup 4793. Ff. 90–107. Copied at Bījāpura in Saṃ. 1742 = A.D. 1685. Incomplete (pañcaratna).

BORI 437 of A 1881/82. 104ff. Copied in Saṃ. 1747 = A.D. 1690.
PUL II 3577. 71ff. (ff. 26, 30–33, and 56 missing). Copied in Saṃ. 1747 = A.D. 1690.
LDI (LDC) 1172. 50ff. Copied in Saṃ. 1750 = A.D. 1693.
Tanjore D 11467 = Tanjore BL 4205. 84ff. Copied on 4 kṛṣṇapakṣa of Phālguna in Śaka 1615 = *ca.* 4 March 1694. Incomplete.
RORI Cat. II 5940. 90ff. (f. 79 missing). Copied by Udayarāma, the son of Paramānanda, in Saṃ. 1758 = A.D. 1701.
RORI Cat. III 16787. 202ff. Copied by Rūparatna, the pupil of Sādhuratna, in Saṃ. 1766 = A.D. 1709.
ABSP 221. 84ff. Copied by Vaṃśīdhara on Monday 2 kṛṣṇapakṣa of Vaiśākha in Saṃ. 1781 = 27 April 1724 Julian.
RORI Cat. III 15318. 171ff. (ff. 21–24 missing). Copied in Saṃ. 1788 = A.D. 1731.
AS Bombay 382. 92ff. Copied by Vināyaka of Cittapūrṇanagara on 13 śuklapakṣa of Pauṣa in Śaka 1657 = *ca.* 15 December 1735. From Bhāu Dājī.
Bombay U Desai 1466. 105ff. Copied in Saṃ. 1808, Śaka 1673 = A.D. 1751.
Baroda 1386. 107ff. Copied in Saṃ. 1812 = A.D. 1755.
Goṇḍal 170. 23ff. Copied on Sunday 8 śuklapakṣa of Jyeṣṭha in Saṃ. 1820 = 19 June 1763. With the ṭīkā of Narahari.
Kathamandu (1960) 190 (II 330). 66ff. Copied by Śivadeva, the son of Rāma, in Śaka 1685 = A.D. 1763.
RJ 1669 (vol. 2, p. 272). 49ff. Copied in Saṃ. 1830 = A.D. 1773. Property of Baḍā Terahapanthiyoṃ of Jayapura.
RORI Cat. II 8530. 110ff. Copied by Nemavijaya at Dādhyāgrāma on Saturday 11 śuklapakṣa of Caitra in Saṃ. 1830 = 3 April 1773. With the ṭīkā of Narahari.
BORI 535 of 1875/76. 80ff. Copied in Saṃ. 1837 = A.D. 1780. From Dilhī.
ABSP 1112. 77ff. Copied by Nārāyaṇa Jyotirvid on Saturday 30 kṛṣṇapakṣa of Śrāvaṇa in Śaka 1703 = 18 August 1781.
Benares (1963) 37811. Ff. 1–16, 16b–40, and 40b–109. Copied in Saṃ. 1838 = A.D. 1781.
Benares (1963) 37846 = Benares (1878) 57 = Benares (1869) XIII 1. 52ff. Copied in Saṃ. 1849 = A.D. 1792.
RORI Cat. II 7836. 124ff. (ff. 1–3 missing). Copied in Saṃ. 1849 = A.D. 1792.
Calcutta Sanskrit College 56. 81ff. Copied in Saṃ. 1851 = A.D. 1794.
LDI (AKC) 733. 5ff. Copied in Saṃ. 1851 = A.D. 1794.
Benares (1963) 37773. Ff. 1, 3–56, and 58–59. Copied in Saṃ. 1852 = A.D. 1795. Incomplete.
IO 3116 (2701). 27ff. Copied by the son of Vaidyanātha on 10 śuklapakṣa of Bhādrapada in Saṃ. 1853 = 11 September 1796. With the *Jayaśrīvilāsa* of Gokulanātha. From Colin Mackenzie.
IO 3110 (745). Ff. 1–20 and 20b–58. Copied in A.D. 1799. Incomplete. From H. T. Colebrooke.
Benares (1963) 37847 = Benares (1903) 1195. Ff. 1–14 and 16–25. Copied in Saṃ. 1858 = A.D. 1801.
Oudh XIX (1887) XIX 2. 250pp. Copied in A.D. 1803.
Kerala 8080 (6948). 1400 granthas. Copied in Saṃ. 1865, Śaka 1730 = A.D. 1808.
IO 3111 (2297). 58ff. Copied in A.D. 1813 from IO 745. From Calcutta.
AS Bombay 381. 284ff. Copied in Śaka 1736 = A.D. 1814. From Bhāu Dājī.
Baroda 9287. 55ff. Copied in Saṃ. 1872 = A.D. 1815. Incomplete (svarodaya).
Mithila 431 G. 4ff. Maithilī. Copied in Śaka 1737 = A.D. 1815. Incomplete (svarodaya). Property of Paṇḍita Raghunātha Jhā of Sonakorthu, Manigachi, Darbhanga.
Benares (1963) 37907. 6ff. Copied in Saṃ. 1873 = A.D. 1816. Incomplete (ahibalacakra).
Oxford 1578 (Sansk. c. 107) = Hultzsch 297. Ff. 1–13, 16–24, and 26–29. Copied in Saṃ. 1873 = A.D. 1816. Incomplete.
Benares (1963) 37854. Ff. 1–51 and 1–140. Copied in Saṃ. 1878 = A.D. 1821. With the ṭīkā of Mahādeva.
RORI Cat. I 1747. 50ff. Copied in Saṃ. 1880 = A.D. 1823.
RORI Cat. III 11028. 204ff. (f. 1 missing). Copied in Saṃ. 1880 = A.D. 1823.
Benares (1963) 37784. Ff. 1–102 and 1–14. Copied in Saṃ. 1882 = A.D. 1825.
Ahmadnagar 299 (272/2). 108ff. Copied on 10 śuklapakṣa of Āṣāḍha in Śaka 1750 = *ca.* 21 June 1828.
Baroda 11151. 160ff. Copied in Śaka 1750 = A.D. 1828. With the *Jayalakṣmī* of Mahādeva.
BORI 927 of 1886/92. 319ff. Copied in Saṃ 1902 = A.D. 1845.
Mithila 431 D. 62ff. Copied by Ārttinātha on Monday 2 śuklapakṣa of Māgha in Śaka 1768 = 18 January 1847. Incomplete. Property of Paṇḍita Puṇyānanda Jhā of Chanaur, Manigachi, Darbhanga.
BORI 931 of 1886/92. 10ff. Copied in Saṃ. 1904 = A.D. 1847. Incomplete (pañcapakṣinirūpaṇa).
Goṇḍal 169. 13ff. Copied in Saṃ. 1906 = A.D. 1849. Incomplete (sarvatobhadracakra).
Benares (1963) 37826 = Benares (1905) 1498. Ff. 1–68 and 1–10. Copied in Saṃ. 1916 = A.D. 1859.
Jammu and Kashmir 2793. 55ff. Copied in Saṃ. 1918 = A.D. 1861. Incomplete.
Leningrad (1914) 311 (Ind. VI 24). 10ff. Copied in Saṃ. 1919 = A.D. 1862. With a *Jayalakṣmī*. Incomplete (sarvatobhadracakra).
Baroda 3169. 177ff. Copied in Saṃ. 1939 = A.D. 1882.
Kathmandu (1960) 189 (III 588). 63ff. Copied on 15 śuklapakṣa of Āṣāḍha in Saṃ. 1951 = *ca.* 17 July 1894.

GOML Madras R 2398. 60ff. Telugu. Copied in A.D. 1917/18 from a manuscript belonging to Jayanti Jogannagāru of Hamsavaram, Tuni, Godāvarī.
GOML Madras R 2472(h). Ff. 111–118. Copied in A.D. 1917/18 from a manuscript belonging to D. V. Vīrabhadra Somayājulugāru of Jegurupāḍu, Rajahmundry, Godāvarī.
GOML Madras R 2890. 66ff. Telugu. Copied in A.D. 1917/18 from a manuscript belonging to Jayanti Jogannagāru of Hamsavaram, Godāvarī. Incomplete (ends in adhyāya 4).
ABSP 51. 53ff. Copied by Paṇḍita Jñānodaya Muni during the rule of Jinaharṣa Sūri of the Kharatara Gaccha.
Adyar Index 3051–3052 =
  Adyar Cat. 28 A 43. 10ff. Telugu. Incomplete (adhyāyas 1–2).
  Adyar Cat. 28 A 44. 201ff. Telugu. With an Āndhraṭīkā. Incomplete.
  Adyar Cat. 28 A 46. 35ff. Grantha. Incomplete.
  Adyar Cat. 33 L 17. 214ff. Karṇāṭakī. Ascribed to Kālavarman Vidyādeva.
Alwar 1818. 2 copies.
Alwar 1990. Incomplete (saptanāḍikācakra).
Anup 4470. 1f. Incomplete (kākanīḍakālaśakuna).
Anup 4785. 22ff. Incomplete (cakroddhāra).
Anup 4786. 82ff.
Anup 4787. 384ff. Incomplete.
Anup 4788. 124ff. Incomplete. This is probably Bikaner 732. 125ff.
Anup 4789. 52ff. Copied by Haṃsarāja. Incomplete.
Anup 4790. 50ff. Incomplete.
Anup 4792. 24ff. Incomplete.
Anup 4794. 26ff. Incomplete.
Anup 4796. 43ff. Incomplete.
Anup 4797. 24ff. Incomplete.
Anup 4798. 22ff. Incomplete.
Anup 4799. 4ff. Incomplete (ekāśītipadasaṃvijñānapradīpaka). Property of Anūpasiṃha (*fl.* 1674/98).
AS Bombay 380. 85ff. Incomplete. From Bhāu Dājī.
AS Bombay 383. 44ff. From Bhāu Dājī.
AS Bombay 384. 43ff.
Baroda 1390. 112ff. With the *Jayalakṣmī* of Harivaṃśa.
Baroda 3170. 248ff.
Baroda 8042. 3ff. Incomplete (1 adhyāya from svarodaya).
Baroda 9354. 43ff. With the *Jayalakṣmī* of Mahādeva. Incomplete (dvādaśārādicakra).
Baroda 10271. *ca.* 40ff. Bengālī. Incomplete.
Baroda 11710. 206ff. With the ṭīkā of Narahari.
Baroda 12103(a). 39ff. Grantha. Incomplete (1 adhyāya).
Benares (1963) 37501. 92ff. Incomplete.
Benares (1963) 37560. Ff. 2–18 and 2ff. Incomplete (praśnasāra).
Benares (1963) 37708. 142ff. With the ṭīkā of Pratāpasiṃha. Incomplete.
Benares (1963) 37726. 18ff. Incomplete.
Benares (1963) 37777. 20ff. Incomplete.
Benares (1963) 37783. Ff. 1–25, 25b–38, and 38b–84, and 2ff. With the ṭīkā of Narahari. Incomplete.
Benares (1963) 37787. 22ff. With the *Jayalakṣmī* of Harivaṃśa. Incomplete.
Benares (1963) 37803. Ff. 1–15, 15b–29, and 41–104. Incomplete.
Benares (1963) 37804. 19ff. Incomplete (bhūbalādhyāya).
Benares (1963) 37805 = Benares (1897–1901) 217. Ff. 7–36. Incomplete.
Benares (1963) 37806 = Benares (1903) 1240. 175ff. Incomplete.
Benares (1963) 37807. Ff. 1–8 and 10–55. Incomplete.
Benares (1963) 37808. Ff. 1–25 and 25b–26. Incomplete.
Benares 1963) 37809. 12ff. Incomplete.
Benares (1963) 37810. Ff. 12–13. Incomplete.
Benares (1963) 37812. Ff. 8–13 and 31–37. Incomplete.
Benares (1963) 37813. Ff. 14–30 and 38–41. Incomplete.
Benares (1963) 37814. Ff. 1–122, 124–134, and 136–216. With the ṭīkā of Narahari. Incomplete.
Benares (1963) 37836. 50ff.
Benares (1963) 37837. Ff. 1–42 and 44–73.
Benares (1963) 37838. 6ff. Incomplete.
Benares (1963) 37845. 18ff. With the ṭīkā of Narahari. Incomplete.
Benares (1963) 37848. Ff. 1–7, 9–12, and 1–28. Incomplete.
Benares (1963) 37849. Ff. 1–64 and 66–79 and 2ff. Incomplete.
Benares (1963) 37850. 22ff. Incomplete.
Benares (1963) 37851 = Benares (1903) 1204. 53ff. Incomplete.
Benares (1963) 37852 = Benares (1878) 58 = Benares (1869) XIII 2. Ff. 1–22 and 47–61. Incomplete.
Benares (1963) 37853. Ff. 3–4, 6–21, 23–26, 31–62, 64–129, 131–153, 157–182, and 186. With a ṭīkā. Incomplete.
Benares (1963) 37855. 20ff. Incomplete.
Benares (1963) 37866. 7ff. Incomplete.
Benares (1963) 37919. Ff. 1, 3–4, and 6–9. Incomplete.
Berlin 1744 (or. fol. 571). 67ff.
Bikaner 691. 7ff. Incomplete (śakunaśāstra).
Bombay U 501 C. Incomplete (śanicakra).
Bombay U 534 A. Ff. 1–4. Incomplete (sarvatobhadracakra).
BORI 335 of 1879/80. 10ff. Incomplete (svarodayaprakaraṇa).
BORI 158 of A 1883/84. 93ff. With the *Jayalakṣmī* of Harivaṃśa.
BORI 159 of A 1883/84. 76ff. With the *Jayalakṣmī* of Harivaṃśa.
BORI 483 of 1892/95. 61ff.

BORI 484 of 1892/95. 16ff. Incomplete (sarvatobhadra).
BORI 524 of 1895/1902. 40ff.
BORI 322 of Vishrambag I. 159ff.
Calcutta Sanskrit College 57. 47ff.
Cambridge R. 15. 131. 73ff. Copied from a manuscript copied by Devaśarman in A.D. 1721.
Cambridge University 141 = Cambridge University Add. 2390.
CP, Hiralal 823 = CP, Hiralal 2411. Property of Dīnānāth of Singharī, Bilāspur.
CP, Hiralal 1598. (*Jyotiṣacakra*). Property of Gaurīśaṅkar of Gaṛhākoṭā, Saugor.
CP, Hiralal 2408. Property of Tukārām Pāṭhak of Yeodā, Amraotī.
CP, Hiralal 2409. Property of Gaṇeśbhaṭ Dakshiṇdās of Haṭṭā, Damoh.
CP, Hiralal 2410. Property of the Chaube family of Gaṛhā, Jubbulpore.
CP, Hiralal 2412, 2414, and 2415. Property of the Bhonsalā Rājās of Nāgpur. One of these is probably CP, Kielhorn XXIII 66. 147ff. With the ṭīkā of Narahari. Property of Jānojī Mahārāj of Nāgpur.
CP, Hiralal 2413. Property of Svāmī Rāmratan of Śobhāpur, Chhindwārā.
CP, Hiralal 6750. Property of Janārdan Śāstrī of Ganiyārī, Bilāspur.
CP, Hiralal 6751. Property of Rāmchandrarāv of Bilāspur.
CP, Hiralal 6752. Property of Wāsudev Kāle of Mulekheḍī, Bulḍānā.
DC 4074. 46ff.
DC 7908. 62ff.
GJRI 3163/375. 76ff. Incomplete.
GJRI 3164/376. Ff. 4–73. Incomplete.
GOML Madras D 15615. Ff. 49–51. Telugu. With the ṭīkā of Narahari. Incomplete (ahibalacakra).
GOML Madras D 17753. 8pp. Telugu. Incomplete (tatkālacandracakra).
GOML Madras R 7391. Ff. 56–76. Grantha. Incomplete. Purchased in 1939/40 from T. S. Krishna Aiyer of Triplicane, Madras.
IM Calcutta 1617. Incomplete (ahibalaprakaraṇa). See NCC, vol. 1, rev. ed., p. 486.
IO 3109 (2445). 77ff. From F. Buchanan.
IO 3112 (1984). 120ff. From Dr. John Taylor.
IO 3113 (744). 33ff. Bengālī. Incomplete. From H. T. Colebrooke.
IO 3114 (936). 98ff. Bengālī. From H. T. Colebrooke.
IO 3115 (1043). 44ff. Bengālī. From H. T. Colebrooke.
IO 6429 (Mackenzie III 96). 45ff. Telugu. With an Āndhraṭīkā. Incomplete (sarvatobhadracakra).
Jaipur (II). 43ff.
Jammu and Kashmir 2773ka. 99ff. Incomplete.
Jammu and Kashmir 3006. 176ff.
Jammu and Kashmir 3031. 24ff. Incomplete.
Jammu and Kashmir 3098. 18ff. Incomplete.
Jammu and Kashmir (2) 718. 93ff. With the ṭīkā of Viśveśvara.
Jammu and Kashmir (2) 719. 8ff. With the ṭīkā of Viśveśvara.
Kathmandu (1960) 191 (I 788). 5ff. Nevārī. Incomplete.
Kathmandu (1960) 192 (III 425). 12ff. Incomplete.
Kathmandu (1960) 193 (II 260). 62ff. Nevārī. Incomplete.
Kathmandu (1960) 194 (I 1209). 68ff. Incomplete.
Kathmandu (1960) 195 (III 195). 8ff. Incomplete.
Kathmandu (1960) 198 (I 1160). 76ff. Nevārī.
Kathmandu (1960) 200 (IV). 164ff. With the ṭīkā of Harivaṃśa.
Kathmandu (1960) 202 (I 1674). 58ff. Nevārī. With the ṭīkā of Harivaṃśa.
Kavīndrācārya 842.
Kerala 8081 (5198). 1200 granthas. Incomplete.
Kerala 8082 (6970). 600 granthas. Incomplete.
Kerala C 701 A (C 2535 A). 11ff. Grantha. Incomplete (kūrmacakra). Property of Vāsudeva Śarma of Vaṭṭapaḷḷi, Śucīndram.
Kurukṣetra 480 (50082).
Kurukṣetra 481 (50433).
LDI 6704 (572/2). Ff. 6–7. Incomplete (kūrmacakravicāra).
LDI 6922 (8890). 7ff. With a ṭīkā. Incomplete (trailokyadīpakacakra).
LDI 6972 (7092). 42ff. (ff. 9–25 and 39–41 missing). Incomplete.
LDI 6973 (6785). Ff. 5–57. Incomplete.
LDI 7161 (2210). 62ff. Incomplete (to adhyāya 5).
LDI 7457 (4060/1). Ff. 1–2. Incomplete (kākaruta).
LDI 7531 (4060/2). Ff. 2–4. Incomplete (śvānaruta).
LDI (KC) K/1062 = LDI (KS) 1006 (11090). 41ff. Incomplete.
Leipzig 1158. 35ff. Copied by Paṇḍita Udayavīra Gaṇi. Incomplete (ends in adhyāya 4).
Leipzig 1160. 94ff. (ff. 1–2 missing). With the *Jayalakṣmī* of Mahādeva. Incomplete (ends in bhūbalādhyāya).
Leumann 66. 25ff. Extracts from an Oxford manuscript.
Līṃbaḍī 1320 (1526). 4ff.
Lucknow ——— (46168).
Mithila 146. 66ff. Maithilī. Incomplete. Property of Paṇḍita Babujana Jhā of Sasipur, Jogiara, Darbhanga.
Mithila 431. 73ff. Maithilī. Property of the Raj Library at Baruary, Parsarma, Bhagalpur.
Mithila 431 A. 6ff. Maithilī. Incomplete. Property of Paṇḍita Umādatta Miśra of Salampur, Ghataho, Darbhanga.
Mithila 431 B. 80ff. Maithilī. Property of Paṇḍita Maṇīśvara Jhā of Lalganj, Jhanjharpur, Darbhanga.
Mithila 431 C. 51ff. Maithilī. Property of Paṇḍita Gaṅgādhara Jhā of Jonki, Deodha, Darbhanga.

Mithila 431 E. 42ff. Maithilī. Incomplete. Property of Paṇḍita Śaśinātha Miśra of Tarauni, Sakri, Darbhanga.
Mithila 431 F. 106ff. Maithilī. Incomplete. Property of Paṇḍita Vāsudeva Jhā of Sukpur, Bhagalpur.
Mitra, Not. 1093. 57ff. Bengālī. Incomplete. Property of Brahmavrata Sāmādhyāyī of Varddhamāna, Dhātrīgrāma.
Munich 366. Ff. 1–62, 74, and 161. Incomplete.
Munich 367. Ff. 1–2, 4–24, and 26–64. Incomplete.
Mysore 455 (494).
Mysore (1922) 132. 59ff. (sarvatobhadracakra).
Mysore (1922) 212. Ff. 57–97. (sarvatobhadracakra).
Mysore (1922) 810. 25ff. (sarvatobhadracakra).
Mysore (1922) 1723. 23ff. (sarvatobhadracakra).
Mysore (1922) 1798. Ff. 72–74. (sarvatobhadracakra).
Mysore (1911 + 1922) 2084. Ff. 107–140. (sarvatobhadracakra).
N-W P V (1880) A 10. 170ff. With a *Jayalakṣmī*. Property of Paṇḍita Mākhana Misra of Muttra.
Oudh XX (1888) VIII 157. 66pp. Property of Paṇḍita Pratāpa Nārāyaṇa of Allahabad Zila.
Oxford 1579 (Sansk. c. 108) = Hultzsch 298. 25ff.
Oxford 1596 (Walker 168). Ff. 122–136.
Oxford CS d. 759. 101ff.
Oxford CS e. 247. 36ff.
Poleman 5226 (U Penn 1831). 73ff.
Poleman 5227 (U Penn 1862). 2ff. Incomplete.
Poleman 5228 (Harvard 580). 11ff.
PUL II 3575. 77ff.
PUL II 3576. 57ff. Incomplete.
PUL II 3578. 34ff. Incomplete
PUL II 3579. 14ff. Incomplete (adhyāya 5).
Puri, Raghunandana Pustakālaya. See V. Raghavan in *JORMadras* 26, 1956–57, 76.
RORI Cat. I 3501. 160ff. (f. 74 missing).
RORI Cat. I 3800. 49ff.
RORI Cat. II 5555. 117ff. (f. 1 missing). With the *Jayalakṣmī* of Harivaṃśa. Incomplete.
RORI Cat. II 5772(4). Ff. 17–19. Incomplete (svarodayaprakaraṇa).
RORI Cat. II 5830. 126ff. (ff. 51 and 73 missing).
RORI Cat. II 6091. 43ff.
RORI Cat. II 6095. 2ff. Incomplete (koṣṭhacakra).
RORI Cat. II 6910. 91ff. (ff. 1–16, 41, 60–61, 63–64, 66, 68–70, 73–74, and 84–88 missing).
RORI Cat. II 7132. 80ff. (ff. 1–38 missing).
RORI Cat. II 7587. 97ff. (ff. 3–6 and 27 missing). With the *Jayalakṣmī* of Harivaṃśa. Incomplete.
RORI Cat. II 8217. 39ff. With the *Jayalakṣmī* of Harivaṃśa. Incomplete.
RORI Cat. II 9174. 87ff.
RORI Cat. II 9387. 72ff. Incomplete (to bhūbalanirūpaṇa).
RORI Cat. III 11141. 109ff. (f. 1 missing). Incomplete.
RORI Cat. III 11336. 66ff. (ff. 1–3, 5–8, 11–12, 21–22, 27, 29–30, 35–37, and 59 missing).
RORI Cat. III 11355. 37ff. (f. 1 missing). Incomplete.
RORI Cat. III 12243. 1f. Incomplete (sarvatobhadrayantra).
RORI Cat. III 15388. 3ff. Incomplete (svarodaya).
RORI Cat. III 15804. 18ff. Incomplete.
RORI Cat. III 15898. 44ff. Incomplete.
RORI Cat. III 16200. 106ff. Incomplete.
RORI Cat. III 17110. 4ff. Incomplete.
RORI Cat. III 17142(1). 11ff. With the ṭīkā of Lālacanda. Incomplete (svarodaya).
SOI 160 = SOI Cat. I: 1373–160. 3ff.
SOI 2699 = SOI Cat. II: 1025–2699. 100ff.
SOI 3483 = SOI Cat. II: 1026–3483. Ff. 1–16, 50, and 89–137.
SOI 3507 = SOI Cat. II: 1133–3507. 78ff. With the *Jayalakṣmī* of Harivaṃśa.
SOI 4444.
SOI 6030.
SOI 8393.
SOI 8574. With a Marāṭhī ṭīkā.
SOI 9595. With a Marāṭhī ṭīkā.
SOI 10109.
Tanjore D 11466 = Tanjore BL 4192. 69ff. Incomplete.
Tanjore D 11468 = Tanjore BL 4193. 73ff. Incomplete.
Tanjore D 11469 = Tanjore BL 10976. 73ff. Telugu. Incomplete.
Tanjore D 11470 = Tanjore BL 10977. 66ff. Grantha. Incomplete.
Tanjore D 11471 = Tanjore BL 10979. 45ff. Grantha. Incomplete.
Tanjore D 11472 = Tanjore BL 10978. 28ff. Telugu. Incomplete.
VVRI 2384. 27ff. Incomplete.
VVRI 2393. 25ff. With the *Jayalakṣmī* of Harivaṃśa. Incomplete.
VVRI 2394. 96ff. With the ṭīkā of Narahari. Incomplete.
VVRI 2395. 43ff. Incomplete.
VVRI 2495. 4ff. Incomplete (koṭayuddhanirṇayaprakaraṇa).
VVRI 2510. 165ff. With the *Jayalakṣmī* of Harivaṃśa. Incomplete.
VVRI 2610. 7ff. Incomplete.
VVRI 4533. 107ff. With the *Jayalakṣmī* of Mahādeva. Incomplete.
VVRI 4631. 68ff. With the *Jayalakṣmī* of Mahādeva. Incomplete.
WHMRL G. 29. a.

The *Narapatijayacaryā* has been published with the *Jayalakṣmī* of Harivaṃśa at Benares in 1882 (BM and NL Calcutta 180. Ka. 88. 3); at Lucknow in 1896 (NL Calcutta 180. Kb. 89. 5); at Meraṭha in [1900] (BM 14953. g. 15); at Meerut in 1902 (IO 2051); with the *Jayalakṣmī* of Harivaṃśa at Bombay in Saṃ. 1963 = A.D. 1906 (IO 18. H. 21 and NL Calcutta 180. Kb. 90. 11), reprinted at

Mumbaī in Saṃ. 1991, Śaka 1856 = A.D. 1934; and with his own *Subodhinī* by Gaṇeśadatta Pāṭhaka as *KSS* 205, Vārāṇasī 1971. The *Ahibalacakra* from it with the Hindī ṭīkā, *Śiśutoṣiṇī*, of Vindhyeśvarīprasāda Dvivedin was published as *VSG* 19, Banārasa 1955. (A part of ?) the *Narapatijayacaryā* is included in the *Bhṛgusaṃhitā* published at Meerut in 1920 (NL Calcutta 180. Ka. 92. 1–3). Verse 2 is:

vividhavibudhavandyāṃ bhāratīṃ vandamānaḥ
pracuracaturabhāvaṃ dātukāmāṃ janebhyaḥ/
narapatir iti loke khyātanāmābhidhāsye
narapatijayacaryānāmakaṃ śāstram etat//

The last verse in some manuscripts is:

vikramārkagate kāle pakṣāgnibhānuvatsare/
māse caitre site pakṣe pratipadbhaumavāsare//

NARASIṂHA

Author of a ṭīkā on a *Golīyarekhāgaṇita*. Manuscript:

RORI (Jaipur) IV 65. 45ff. Copied in Saṃ. 1935 = A.D. 1878.

NARASIṂHA = NṚSIṂHA

Author of a *Daivajñakaṇṭhābharaṇa; cf.* the *Daivajñaratna* of Narasiṃha and the *Daivajñabhūṣaṇa* of Nṛsiṃha (*fl.* between 1626 and 1685). Manuscripts:

Oudh XX (1888) VIII 78. 54pp. Copied in A.D. 1767. Ascribed to Nṛsiṃha. Property of Paṇḍita Pratāpa Nārāyaṇa of Allahabad Zila.
Oudh XX (1888) VIII 45. 36pp. Property of Paṇḍita Pratāpa Nārāyaṇa of Allahabad Zila.

NARASIṂHA

The son of Lakṣmaṇa of the Bhāradvājagotra and a resident of Vidurapura, Narasiṃha wrote a *Daivajñaratna; cf.* the *Daivajñakaṇṭhābharaṇa* of Narasiṃha and the *Daivajñabhūṣaṇa* of Nṛsiṃha (*fl.* between 1626 and 1685). Manuscript:

GOML Madras R 2596(c). Ff. 50–94. Telugu. Presented in 1917/18 by Vāśireḍḍi Candramaulīśvaraprasāda Bahadur, Zamindar of Muktyala, Kistna.

NARASIṂHA

Assumed author of a *Narasiṃhapaddhati*. Manuscript:

Goṇḍal 180. 10ff. Copied by Trīkamajī on Saturday 1 kṛṣṇapakṣa of Bhādrapada in Saṃ. 1788, Śaka 1653 = 4 September 1731 Julian. Incomplete.

NARASIṂHA

The son of Lakṣmaṇa of the Kāśyapagotra, Narasiṃha wrote a *Bhāvaphaladarśanadīpikā* in 12 prakaraṇas. Manuscript:

GOML Madras R 2343. 18ff. Telugu. Copied in A.D. 1916/17 from a manuscript belonging to Kavi Dakṣiṇāmūrtigāru of Masulipatam.

The colophon begins: iti śrīmatsakalaśāstrapravīṇasaṃskṛtāndhrabhāṣākavitādhurīṇakāśyapagotralakṣmaṇācāryatanayanarasiṃhācāryaviracita.

NARASIṂHA ŚĀSTRIN

Author of a *Jātakaśiromaṇi*. Manuscript:

Oppert II 1967. 16pp. Telugu. Property of Veṅkaṭeśvarajosya of Siddhavaṭa, Kaḍapa.

NARASIṂHA (*fl.* between 1807 and 1866 ?)

The son of Kāpurī (or Naupurī) Śiṅgaya, the son of Peddividvat of the Vādhūlagotra, Narasiṃha wrote a *Tithicakra* based on the *Tithicakra* of Mallikārjuna (*fl.* 1411 ?). In this he refers to the 23rd 60-year cycle after Śaka 409 = A.D. 487; this corresponds to A.D. 1807/1866. *Cf.* the *Kālacakra* of Nṛsiṃha. Manuscripts:

GOML Madras (Telugu) R 205(b). Ff. 39–41. Telugu. Incomplete. Restored in A.D. 1914/15.
GOML Madras R 2454(b). Ff. 12–16. Telugu. Copied in A.D. 1917/18 from a manuscript belonging to Jayanti Jogannagāru of Hamsavaram, Tuni, Godāvarī.
GOML Madras R 2454(c). Copied in A.D. 1917/18 from the same manuscript.

Verses 2–5 are:

tithicakraṃ yat praṇītaṃ mallikārjunasūriṇā/
kālena mahatā tasmin khilabhūte tadādarāt//
kāpurīśiṅgayāryasya nṛsiṃhena susūnunā/
etad eva sphuṭataraṃ kriyate saurasaṃmatam//
ṣaṣṭis trinayanaguṇitā
prabhavā gatābdasaṃmiśrā/
navagaganābdhisametaḥ
śakanṛpakālo bhaven nūnam//
śākaḥ surāgnicandraḥ
sauravyasto ᵓrkair māsayuk pṛthak/
trighnaḥ svāṣṭābdhine-
trāṃśayuktaḥ śaraśaśāṅkayuk//

NARAHARI

Author of a *Tithicakra*. Manuscript:

Wien (Univ.) 280.

NARAHARI

Author of a *Vāstucandrikā*. Manuscripts:

AS Bengal 2817 (G 3073). 10ff. Copied on Friday 14 śuklapakṣa of Pauṣa in Saṃ. 1934 = 18 January 1878.

AS Bengal 2818 (G 10229). 9ff. Copied on Thursday 6 śuklapakṣa of Phālguna in Saṃ. 1942 = 11 March 1886.

The colophon begins: iti śrīnaraharijyotiṣiviracitāyāṃ.

NARAHARI AYĀCITA

Author of a *Grahayajñaprayoga*. Manuscript:

Poona, Bhāratīya Itihāsa Saṃśodhaka Maṇḍala vi. 383/22. See NCC, vol. 6, p. 256.

NARAHARI ŚUKLA

Author of a *Jñānapradīpikā*. Manuscript:

Benares (1963) 37013 = Benares (1903) 1045. 135ff. Copied in Saṃ. 1871, Śaka 1736 = A.D. 1814.

NARAHARI SŪRI

The son of Nṛsiṃha Sūri, Narahari wrote a *Gaṇakavallabha*. Manuscripts:

GOML Madras R 2458(d). Ff. 12–15. Copied in A.D. 1917/18.
Baroda 7950. 150ff. Telugu. Ascribed to Nṛsiṃha Sūri.

NARAHARI (*fl. ca.* 1500)

The son and pupil of Narasiṃha, the son of Gaṇeśa, Narahari wrote a ṭīkā, *Vyākhyāplava*, on the *Narapatijayacaryā* of Narapati (*fl.* 1177) during the reign of Bhairavendra, who ruled Mithilā from *ca.* 1480 to *ca.* 1515. Manuscripts:

Anup. 4800. 96ff. Copied by Sāmaladāsa Sāṃgāuta at Āgarā in Saṃ. 1654 = A.D. 1597. Property of Mādhava Josī.
Mithila 432. 211ff. Maithilī. Copied in Lakṣmaṇa Saṃ. 501 = A.D. 1619 (?). Property of Paṇḍita Manamohana Jhā of Mangarauni, Madhubani, Darbhanga.
Goṇḍal 171. Ff. 18–33. Copied in Saṃ. 1718, Śaka 1584 = A.D. 1661/62. Incomplete.
Kathmandu (1960) 521 (I 211). 145ff. Nevārī. Copied in NS 810 = A.D. 1690. Incomplete.
Goṇḍal 170. 23ff. Copied on Sunday 8 śuklapakṣa of Jyeṣṭha in Saṃ. 1820 = 19 June 1763.
Mitra, Not. 2381. 130ff. Bengālī. Copied in Śaka 1693 = A.D. 1771. Property of Khaḍganātha Pāṭhaka of Bhaisdirāgrām, Jñodoyāḍā, Pūrṇiyā Zillā.
RORI Cat. II 8530. 110ff. Copied by Nemavijaya at Dadhyāgrāma on Saturday 11 śuklapakṣa of Caitra in Saṃ. 1830 = A.D. 1773 (the date is irregular).
Mithila 13. 8ff. Copied in Saṃ. 1942 = A.D. 1885. Incomplete (ahibalacakra). Property of Rāmacandra Jhā of Mahināthpur, Deodhā, Darbhanga.
Alwar 1820.
Baroda 11710. 206ff.
Benares (1963) 37724. Ff. 1–74 and 80–81. Incomplete.
Benares (1963) 37783. Ff. 1–25, 25b–38, and 38b–84, and 2ff. Incomplete.
Benares (1963) 37814. Ff. 1–122, 124–134, and 136–216. Incomplete.
Benares (1963) 37845. 18ff. Incomplete.
BORI 336 of 1879/80. 67ff.
Chamba 2. 480 pp.
CP, Hiralal 6753. Property of Pāṇḍuraṅg Joshi of Jāvalbutā, Bulḍānā.
CP, Hiralal 6754. Property of Ārtodās Pāṭ Joshi of Jagdalpur, Bastar.
CP, Kielhorn XXIII 66. 147ff. Property of Jānojī Mahārāj of Nāgpur.
GOML Madras D 15615. Ff. 49–51. Telugu. Incomplete (ahibalacakra).
IM Calcutta 1618 and 1619. Incomplete (ahibalacakra). See NCC, vol. 1, rev. ed., p. 486.
IO 6425 (Mackenzie II 43). 186 and 87ff. From Colin Mackenzie.
Jaipur II.
Jammu and Kashmir 3081. 3ff. Incomplete.
Kathmandu (1960) 522 (I 1194). 119ff. Incomplete.
Kathmandu (1960) 523 (I 1194). 43ff. Incomplete.
Leiden A 1.
Oudh XX (1888) VIII 159. 192pp. Property of Paṇḍita Pratāpa Nārāyaṇa of Allahabad Zila.
Oudh XX (1888) VIII 163. No ff. given. Property of Paṇḍita Pratāpa Nārāyaṇa of Allahabad Zila.
PUL II 3582. 108ff. Incomplete.
VVRI 2394. 96ff. Incomplete.

The first 2 verses are:

tātaṃ guruṃ ca vibudhaṃ narasiṃham īśaṃ
natvā tadīyadayayāvagatārthasārthaḥ/
etat svarodayasamudragatiprasiddhyai
vyākhyāplavaṃ naraharih prakaroty avaśyam//
śrībhairavendrapadapañkajasevanottha-
kīrtiḥ samastavibudhān asakṛt praṇamya/
yāce yadīha bhavati skhalanaṃ kadācit
tatrāvanaṃ kuruta vaṃśamahāśayatvāt//

At the end are the 3 verses:

śrīmāṇḍavaśaśineha vidite vaṃśe budhālañkṛte
khyāte śrotriyamaṇḍalīṣu mahati svācāracaryojjvale/
vedavyākaraṇāgamādinikaṣo naiyāyikaḥ satkavir
jyotiḥśāstravikāśanaikamihiro dhīro gaṇeśo ᵓbhavat//
tasyātmajo ᵓbhūn narasiṃhadhīro
nyāyāgamādyadbhutavidya ekaḥ/
vedasmṛtijyotiṣaśāstrasāra-
vyākhyānaśuddhaikamatir dvijendraḥ//
tasyātmajo naraharis tata eva buddhyā
vyākhyām imāṃ sakalaśiṣyajanānubandhāt/
naiyāyikaḥ samakarod viṣadārthasārthāṃ
nānāguṇeṣu kutukī mithilāvanīsthaḥ//

NAROTTAMA

Alleged author of a *Sarvasaṅgraha*. Manuscript:

ABSP 443. Ff. 14–19 and 22–46. Incomplete.

NARMADĀGIRI AVADHŪTA (*fl.* 1855/1856 ?)

Author of ṭīkās in Hindī on the following 2 texts.

1. A ṭīkā on the *Jātakālaṅkāra* of Gaṇeśa (*fl.* 1613). Manuscript:

AS Bengal 7041 (G 8431). 166ff. Copied on 13 śuklapakṣa of Phālguna in Saṃ. 1911 = 28 February 1855.

The colophon begins: iti śrīnarmadāgiriṇāvadhūtena viracitāyāṃ.

2. A ṭīkā, *Bālabodhinī*, on the *Ṣaṭpañcāśikā* of Pṛthuyaśas (*fl. ca.* 575). Manuscript:

AS Bengal 7366 (G 10025). Ff. 1–11 and ff. 1–37 (ff. 10 and 21 missing). Copied on Monday 10 kṛṣṇapakṣa of Caitra in Saṃ. 1912 = 28 April 1856.

The colophon begins: iti śrīnarmadāgiriṇā avadhūtena kṛtāyāṃ.

NALLAKOṆḌA KĀMĀBHAṬṬA

See Kāmābhaṭṭa.

NAVADVĪPA

Born into the family of Nityānanda, the cousin and follower of Caitanya (1485/1533), Navadvīpa wrote an *Adbhutasārasaṅgraha* based on the *Adbhutasāra* of Mahādeva Śarman. Manuscript:

AS Bengal 2580 (G 134) = Mitra, Not. 465. 20ff. Bengālī. Incomplete.

The first 2 verses are:

natvā mahāprabhuṃ kṛṣṇacaitanyākhyaṃ tadagrajam/
prabhuṃ nityānandarāmaṃ prabhum advaitam īśvaram//
nityānandavaṃśajena navadvīpena kenacit/
kriyate śāntisahitodbhutasārasya saṅgrahaḥ//

NAVANIDHIRĀMA (*fl.* 1907)

Author, with Lakṣmaṇadāsa, of a *Jātakasaṅgraha*, published with the Hindī ṭīkā of Kāśīrāma at Bombay in Saṃ. 1964 = A.D. 1907 (IO 21. I. 15).

NAVANĪTA NARTANA KAVI

Author of an *Ariṣṭanavanīta* in 6 paricchedas; there is a ṭīkā by Śrīdhara. Manuscripts:

Baroda 11367. 59ff. Copied in Śaka 1559 = A.D. 1637 from a manuscript copied in 14— (?). With the ṭīkā of Śrīdhara.
Kathmandu (1960) 22 (I 1306). 54ff. Copied on Wednesday 6 śuklapakṣa of Phālguna in NS 800 = 25 February 1680 Julian. With the ṭīkā of Śrīdhara.
Mithila 9 A. 10ff. Maithilī. Copied in Śaka 1664 = A.D. 1742. Property of Paṇḍita Śrīnandana Miśra of Kanhauli, Sakri, Darbhanga.
PL, Buhler IV E 6. 8ff. Copied in Saṃ. 1806 = A.D. 1749. Property of Harirāmaśāstrī of Aṅkaleśvara.
Mithila 9 D. 6ff. Maithilī. Copied in Śaka 1706 = A.D. 1784. Property of Paṇḍita Padmanābha Miśra of Lālabag, Darbhanga.
Mithila 9. 5ff. Maithilī. Copied by Śivanātha Śarman at Parihārapura on Monday 6 kṛṣṇapakṣa of Bhādrapada in Śaka 1712 = 27 September 1790. Property of Paṇḍita Mahīdhara Miśra of Lālabāg, Darbhanga.
Mithila 9 C. 8ff. Maithilī. Copied in Śaka 1718 = A.D. 1796. Property of Paṇḍita Janārdana Miśra of Chanaur, Manigāchi, Darbhanga.
Mithila 9 B. 6ff. Maithilī. Copied in Śaka 1739 = A.D. 1817. Property of Paṇḍita Dāmodara Jhā of Andhrāṭhārhī, Darbhanga.
Poleman 4723a (Columbia, Smith Indic 59). 7ff. Copied in Saṃ. 1874 = A.D. 1817. With the ṭīkā of Śrīdhara.
PUL II 3940. 10ff. Copied in Saṃ. 1895 = A.D. 1838. Incomplete (viṃśottarīdaśā). With the *Viṃśottarīdaśāpaddhati* of Nārāyaṇa Paṇḍita.
Benares (1963) 36690. 5ff. Copied in Śaka 1763 = A.D. 1841. Incomplete (viṃśottarīdaśānayanacakra).
Baroda 12626. 8ff. Copied in Saṃ. 1906 = A.D. 1849. Incomplete (viṃśottarīdaśānayanaprakāra). With the *Viṃśottarīdaśāpaddhati* of Nārāyaṇa Paṇḍita.
ABSP 1063. 10ff.
ABSP 1280. Ff. 2–4. Incomplete. No author mentioned.
Alwar 1711. With the ṭīkā of Śrīdhara.
Baroda 13365(c). Ff. 65–99. Nandināgarī. With the ṭīkā of Śrīdhara.
Baroda 13422(g). 7ff. Nandināgarī.
Benares (1963) 35008. 58ff. With the ṭīkā of Śrīdhara.
BORI 52 of B 1919/24. No ff. given.
BORI 53 of B 1919/24. No ff. given.
CP, Hiralal 2483. Property of Kanhaiyālāl Guru of Saugor.
GOML Madras D 13604. Ff. 27–39. Telugu.
GOML Madras D 13605. Ff. 64–71. Karṇāṭakī.
GOML Madras D 13606. Ff. 30–34. Telugu. Incomplete.
GOML Madras D 13607. Ff. 9–30. Grantha. With the ṭīkā of Śrīdhara. Incomplete (ends in pariccheda 6).
GOML Madras D 17374. 9pp. Telugu. Incomplete (ends in pariccheda 5).
Kathmandu (1960) 206 (I 619). 2ff. Incomplete.

Mithilā. See NCC, vol. 1, rev. ed., p. 370.
Mysore (1922) 1771. 8ff.
Mysore (1922) 4398. 36ff. With a *Laghugaṇita*.
Mysore (1911 + 1922) B 574. Ff. 59–66.
Mysore and Coorg 260. 1000 granthas. Incomplete. Property of Mahādeva Joyisa of Sringeri.
Mysore and Coorg 295. 1000 granthas. With the ṭīkā of Śrīdhara. Property of Nārāyaṇa Dīkṣita of Bommarasaiyana Agrahara.
N-W P IX (1885) A 5. 13ff. Property of Rājāji Jyautiṣi of Benares.
Oppert II 4468. Property of the Śaṅkarācāryasvāmimaṭha at Śṛṅgeri, Cikkamogulūr, Mysore.
PUL II 3255. 7ff.
Śṛṅgeri, Śaṅkara Nārāyaṇa Jyautiṣika 42. See NCC.
Tanjore D 11306 = Tanjore BL 10980e. Ff. 155–164. Grantha. Incomplete. (ends at VI 89).
Tanjore D 11307 = Tanjore BL 10982e. No ff. given. Grantha.
Tanjore D 11308 = Tanjore BL 12248e. No ff. given. Grantha.
Tanjore Supplement 1004. Incomplete. See NCC.

The *Navanītāriṣṭa* was published with a Telugu explanation by Vellāla Sītārāmayya at Madras in 1927 (IO San. B. 991(e)); there is also said to be an edition with a Karṇāṭaṭīkā published at Cāmarājanagara (Mysore GOL B 985). The first verse is:

śrīraṅgeśaṃ natvā
horāśāstrāmbudhīn samāsoḍhya/
navanītanartanakavir
ariṣṭanavanītam ājahne//

NAṢĪR AL-DĪN MUḤAMMAD AL-ṬŪSĪ (1201/1274)

The great polymath of thirteenth-century Iran, Naṣīr al-Dīn wrote voluminously in Arabic and Persian on the exact sciences. Of interest to us here is his *Risālat al-usṭurlāb* in Persian (see C. A. Storey, *Persian Literature*, vol. 2, pt. 1, London 1958, pp. 52–53), of which there is a Devanāgarī transliteration under the title: *Yantrarājaparīkṣā* of Nāsīruddīna Muhammada Tūsī. Manuscript:

Benares (1963) 34568. 62ff.

NĀGA DESIGA (*fl.* 1012)

Granted support for teaching mathematics, astronomy, and other subjects at Ummacige = Koṭavumacgi, Gadag, Dharwar, Mysore by Keśavayya, a mahāsāmantādhipati and mahādaṇḍanāyaka under the Cālukya monarch, Tribhuvanamalla Vikramādityadeva V (1008/1015), on Sunday 8 śuklapakṣa of Pauṣa in Śaka 934 = 23 November (?) 1012; see R. S. Panchamukhi [A3. 1929/30].

NĀGAJOŚĪ BHIÑGĀRAKARA

Also known as Kavināga, Nāgajośī wrote a *Buddhivilāsa*. Manuscripts:

Baroda 12384. 51ff. Copied in Śaka 1760 = A.D. 1838.
DC 7935. 34ff. Ascribed to Kavināga Daivajña.
Osmania University Ac/74/5. 33ff. Ascribed to Nāgeśa.

JUMANĀḶA NĀGAYYĀ MAHĀLIÑGAYYĀ (*fl.* 1910)

Author of a pañcāṅga in Saṃskṛta and Kannaḍa for Śaka 1833 = A.D. 1911, published at Jumanāḷa in 1910 (BM 14096. b. 27. (1)).

NĀGARA VĀCAKA (*fl.* first century A.D.)

See Umāsvāti Vācaka (*fl.* first century A.D.) and Velankar, p. 155.

NĀGAŚARMAN

Author of a karaṇa called *Gaṇakavallabha*. Manuscripts:

BORI 145 of A 1883/84. 17ff. Copied in Saṃ. 1485 = A.D. 1428.
RORI Cat. III 11247. 10ff. Copied by Gurudāsa in Saṃ. 1749 = A.D. 1692. With the *Padmalīlāvilāsinī* of Nārāyaṇa Paṇḍita.

The first verse is:

natvā gaṇendragirijāpatimādhavādīn
vidhyambikādinakarādinavagrahāṃś ca/
śrīnāgaśarmagaṇakaḥ sphuṭakheṭakarma
vakṣyāmy ahaṃ gaṇakavallabhanāmaśāstram//

NĀGĀCĀRYA

Author of a *Bṛhadramala*. Manuscript:

Nagpur 1270 (867). 29ff. Copied in Śaka 1724 = A.D. 1802. From Nasik.

NĀGEŚA = NĀGADEVA (*fl.* 1619)

The son of Śiva, the son of Khecara or Tukeśvara (Keśava ?) of the Gārgyagotra, Nāgeśa was the father of Śiva (*fl. ca.* 1650) and the teacher of Yādava (*fl.* 1663). He wrote the following works:

1. The *Grahaprabodha* in 36 verses, whose epoch is Śaka 1541 = A.D. 1619; there is an udāharaṇa and a sāriṇī by Yādava (*fl.* 1663). Manuscripts:

Baroda 3107. 4ff. Copied in Saṃ. 1842 = A.D. 1785.
PL, Buhler IV E 221. 9ff. Copied in Saṃ. 1854 = A.D. 1797. (*Parvaprabodha*). Property of Nānā Jośī of Nandurabāra.

AS Bombay 232. 32ff. Copied in Śaka 1735 = A.D. 1813. From Bhāu Dājī.
BORI 422 of A 1881/82. 6ff. Copied in Śaka 1799 = A.D. 1877. Ascribed to Śiva.
Benares (1963) 35648 = Benares (1903) 1226. 38ff. Copied in Saṃ. 1939 = A.D. 1882 (Saṃ. 1839 = A.D. 1782 in Benares (1903)). With the sāraṇī of Yādava.
Ānandāśrama 2618. With the udāharaṇa of Yādava.
Ānandāśrama 2619.
AS Bombay 233. 11ff. With the udāharaṇa of Yādava. From Bhāu Dājī.
Baroda 3108. Ff. 4–29. With the sāraṇī of Yādava.
Baroda 9435. 4ff. (parvādhikāra and 2 other adhikāras).
BM 464 (Add. 14,365c. A). Pp. 1–2. See SATE 9.
Bombay U 343. 5ff.
CP, Hiralal 1536. Ascribed to Śiva. Property of Gopāl Jayakrishṇa of Kuṭāsā, Akolā.
CP, Hiralal 1537. Ascribed to Śiva. Property of Śyāmrāj Rāmkrishṇa of Pāthroṭ, Amraotī.
CP, Hiralal 1538. Ascribed to Śiva. Property of the Balātkār Gaṇ Jain Mandir at Kārañjā, Akolā.
DC 369. Ff. 1 and 3–5.
PUL II 3350. 5ff.
SOI 7866.

The *Grahaprabodha* was published at Bombay (?) in 1833 (?) (BM Add. 14,357 III and Add. 14,365 II).

Verses 35–36 are:

āsīd gārgyakulaikabhūṣaṇamaṇir vidvajjanānandakṛt
śiṣyājñānatamonivāraṇaravir bhūmīpatiḥ pārthivaḥ/
jyotiḥśāstramahābhimānamahimaḥ
spaṣṭīkṛtabrahmadhīr
dhairyaudāryanidhis tu khecara (tukeśva⟨ra⟩) iti
khyāto mahīmaṇḍale//
tadātmajas taccaraṇaikabhaktis
tadvat prasiddhaḥ śivanāmadheyaḥ/
tadaṅgajo dṛggaṇitānusāraṃ
grahaprabodhaṃ vyatanoc ca nāgaḥ//

The colophon begins: iti śrīśivadaivajñātmajanāgeśadaivajñakṛtau.

2. A *Tithinirṇayatattva* = *Nirṇayatattva* in 102 verses, in which he mentions the *Nirṇayasindhu* of Kamalākara Bhaṭṭa (*fl.* 1612). Manuscripts:

Baroda 9299. 16ff. Copied in Śaka 1680 = A.D. 1758.
Bombay U Desai 207. 3ff. Copied in Saṃ. 1884 = A.D. 1827.
Bombay U 1049. 9ff. Copied by Rāmacandra Bhaṭṭa Reḍe on 13 kṛṣṇapakṣa of Śrāvaṇa in Śaka 1770 = *ca.* 26 August 1848.
AS Bengal 2791 (G 5860). 8ff. Copied by Viśvanātha Gāḍhava.
AS Bombay 313. 7ff. From Bhāu Dājī.
Baroda 13633 8ff.
Bombay U 1048. 6ff.
Bombay U 1050. 18ff. Copied by Sakhārāma Bhaṭṭa Bākra.
Mithila I 245. 6ff. Property of Paṇḍita Dāmodara Jhā of Sāhapur, Pandaul, Darbhanga.

The last verse is:

iti nirṇayasindhusārataḥ
pratimāsaprathito vinirṇayaḥ/
śivanandananāgadaivavit
tanoti nirṇayatattvasaṃjñakam//

The colophon begins: iti śrīśivajośīvitsutanāgadevaviracito.

3. A *Muhūrtadīpaka*. Manuscripts:

PL, Buhler IV E 347. 45ff. Copied in Saṃ. 1722 = A.D. 1665. Property of Aṇṇā Paṇḍita of Mulhera.
Benares (1963) 36466. 11ff. Incomplete. Ascribed to Śivasūnu.
Benares (1963) 36718. 7ff. Incomplete. Ascribed to Śivasūnu.
CP, Hiralal 4256. Property of Govindprasād Śāstrī of Jubbulpore.

NĀGOJI (NĀGEŚA) BHAṬṬA KĀLA
(*fl. ca.* 1700/1750)

The son of Satī and Śivabhaṭṭa, a Mahārāstrian Brāhmaṇa, Nāgoji was the pupil of Hari Dīkṣita, the son of Vīreśvara Dīkṣita, the son of Bhaṭṭoji Dīkṣita (*fl. ca.* 1600/1650), and was the protégé of Rāma, the lord of Śṛṅgavera. He was a prolific author in many fields, but especially in vyākaraṇa; see P. V. Kane [1930/62] vol. 1, pp. 453–456, and P. K. Gode [1955]. His works touching on jyotiḥśāstra are:

1 and 2. The *Madhyajātaka* on which he wrote a ṭīkā. Manuscript:

Bombay U Desai 1402. 77ff. Copied in Saṃ. 1789 = A.D. 1732. With the ṭīkā.

The colophon begins: iti śrīmatkālopanāmakanāgojibhaṭṭaviracite.

3. A *Tithinirṇaya* = *Tithinirṇayenduśekhara*. Manuscripts:

Kerala 6790 (1622). 1000 granthas. Copied in Saṃ. 1840 = A.D. 1783.
Kerala 6791 (5181). 1000 granthas. Copied in Saṃ. 1840 = A.D. 1783.
Benares (1956) 12623. 32ff. Copied in Saṃ. 1876 = A.D. 1819.
GVS 851 (3233). 6ff. Copied on Saturday 10 śuklapakṣa of Bhādrapada in Śaka (read Saṃ.) 1887 = 28 August 1830. (*Parvanirṇaya*).
Baroda 8343. 47ff. Copied in Śaka 1763 = A.D. 1841.
Benares (1956) 13024. 12ff.

Kerala 6792 (9701). 1000 granthas.
PL, Buhler III E 123. 22ff. Property of Sukheśvara Śāstrī of Ahmadābād. Buhler notes 2 other copies.

The *Tithinirṇaya* was edited by Viśvanātha Śāstrī in *CSS* 472, Benares 1940, pp. 51–103. The colophon begins: iti śrīmannāgojibhaṭṭaviracitas.

NĀTHA

See Śrīnātha.

NĀTHA

Author of a *Nāthapadyasaṅgraha*. Manuscript:

GJRI 962/74. 4ff.

NĀTHA (?) (*fl.* 1650)

The son of Murāri, Nātha (?) wrote a *Praśnamārga* in 32 adhyāyas in ME 825 = A.D. 1650. Manuscripts:

Adyar Index 3876 =
Adyar Cat. 29 G 30. 88ff. Grantha and Malayālam. Incomplete. No author mentioned.
Adyar Cat. 33 D 9. 202ff. Malayālam.
AS Bengal 7175 (G 6330). 162ff.
Alwar 1854.
Baroda 7873(c). 15ff. Grantha. Incomplete. No author mentioned.
Baroda 9840. 137ff. Grantha. No author mentioned.
Baroda 9843(e). 3ff. Grantha. Incomplete (adhyāya 23). No author mentioned.
Mysore (1911 + 1922) 2926. 103ff. No author mentioned.
N-W P VII (1882) 1. 187ff. No author mentioned. Property of Paṇḍita Lakṣmīnātha Śāstrī of Jeypore.
N-W P VII (1882) 2. 161ff. No author mentioned. Property of Paṇḍita Durgā Prasāda of Jeypore.
PL, Buhler IV E 260. 116ff. No author mentioned. Property of Caturbhuja Bhaṭṭa of Navānagara.
VVRI 266. 5ff. No author mentioned.
VVRI 271. 6ff. No author mentioned.

Adhyāyas 1–16 with the ṭippaṇī of Punnaśśeri Nampi Nīlakaṇṭha Śarman were published at Kalpathi-Palghat in 1926. At the end of the AS Bengal manuscript is the corrupt verse:

suto murārer lubdhodaye (?)
madhyavanānoyanāyanāthān (?)/
sa praśnamārgākhyam akārṣam etac
chāstraṃ sukhaṃ bodhayituḥ svaśiṣyān//

NĀTHADATTA

Author of a *Jyotirviveka*. Manuscript:

Calcutta Sanskrit College 32. 26ff.

NĀTHĀCĀRYA

Author of an *Adbhutasāgara*. Manuscript:

Jaipur (II).

NĀDADEVĀRYA

Jaina author of a Karṇāṭaṭīkā, *Bhāvārthaprakāśikā*, on adhyāyas 1–6 of the *Bṛhajjātaka* of Varāhamihira (*fl. ca.* 550). Manuscript:

GOML Madras R 406. 91ff. Karṇāṭakī. Purchased in 1911/12 from G. Śrīnivāsa Rao of Mysore.

The colophon begins: śrīmajjinendrabhaktinistandrajñānacandrikāsāndradaharakuharanādadevāryanim.

GURU NĀNĀK (1469/1530)

To Guru Nānāk, the founder of the Sikh religion, is ascribed a Hindī work on divination entitled *Prīcchā*. Manuscript:

BM (Hindī) 23 (Or. 2764). 47ff. Gurumukhī. From Rev. A. Fisher.

NĀNĀBHĀĪ

Author of a set of astronomical tables, the *Nānābhāīsāraṇī*. Manuscript:

Kathmandu (1960) 209 (I 1205). 73ff. (ff. 1–2 missing). Nevārī. Copied in Nep. Saṃ. 874 = A.D. 1754.

See also the *Nānābhāīsārasaṅgraha* of Nīlakaṇṭha.

NĀMADEVA

See Gaṇapati.

NĀMADEVA TUKĀRĀMA PĀVALE (*fl.* 1968)

The son of Tukārāma Nārāyaṇa Pāvale and a resident of Mahāḍa, Kulābā, Nāmadeva has written a number of books on astrology in Marāṭhī, among which is the *Vyāpāra mārtaṇḍa* published at Mahāḍa in 1968.

NĀMANĀRYA (*fl.* between 1687 and 1746)

The son of Śrīmūlasena (?) of the Kāśyapagotra, Nāmanārya wrote a *Gaṇitārṇava* in which he refers to the 60-year cycle beginning in Kali 4788 = A.D. 1687. Manuscripts:

GOML Madras R 7524. Ff. 114–127. Copied in 1940/41 from GOML Madras R 2602(e).
GOML Madras R 2602(a). Ff. 1–13. Telugu. Incomplete (ravicandrapadakāni). Presented in 1917/18 by U. Rāmayyagāru of Cintalapūḍi, Kistna.
GOML Madras R 2602(e). Ff. 58–73. Telugu. Presented with the above.
SOI 9464. No author mentioned.

NĀRACANDRA

Author of a *Bhuvanadīpaka*. Manuscripts:

CP, Hiralal 3697. Property of Viśbambharnāth of Ratanpur, Bilāspur.
CP, Hiralal 3698. Property of the Sen Gaṇ Jain Mandir at Kārañjā, Akolā.
CP, Hiralal 3699. Property of Vāsudev Golwalkar of Maṇḍlā.
PL, Buhler IV E 311. 23ff. Property of Bālakṛṣṇa Jośī of Ahmadābād.

NĀRADA

Author of an *Añgavidyā*. Manuscripts:

AS Bengal 7173 (G 5546) B. No ff. given. Copied by Devīdāsa near the Prahlādaghāṭa in Kāśī on Tuesday 8 kṛṣṇapakṣa of Jyeṣṭha in Saṃ. 1929 = 28 May (?) 1872.
RORI Cat. III 12708(3). 8ff.
RORI Cat. III 16704(4). F. 1v.

NĀRADA = GĀRGYA ṚṢI

Author of a *Kālacakra*. Manuscripts:

Tanjore D 11333 = Tanjore BL 11034. 24ff. Telugu.
Tanjore D 11334 = Tanjore BL 11035. 14ff. Telugu. Incomplete.
Tanjore D 11335 = Tanjore 15649. No ff. given. Telugu. Incomplete.
Tanjore D 11336 = Tanjore 15650. No ff. given. Incomplete.

NĀRADA

Author of a *Nāradapraśna* in 78 verses. Manuscripts:

Jammu and Kashmir 4092. 3ff. Copied in Saṃ. 1941 = A.D. 1884 (*Praśnanirṇaya*).
Adyar List = Adyar Index 3172 = Adyar Cat. 28 C 37. 10ff. Grantha.
Ānandāśrama 5008. (*Praśnanāradī*).
Bombay U 514. 41ff. (f.2 missing).
BORI 878 of 1887/91. 13ff. (*Laghupraśna*).
Mithila 150. 4ff. Maithilī. Property of the Citradhara Library at Tabhaka, Dalsingh Sarai, Darbhanga.
Mithila 200. 2ff. Maithilī. (*Praśnasañgraha*). Property of Paṇḍita Muktinātha Jhā of Baruari, Parsarma, Bhagalpur.
RORI Cat. III 10418. 1f. (*Laghupraśnanirṇaya*).
SOI 9590.

NĀRADA

An ṛṣi regarded as an authority on jyotiḥśāstra, appearing, for instance, as one of the interlocutors in the *Vṛddhagārgīsaṃhitā* and being cited by Varāhamihira (*fl. ca.* 550) in *Bṛhatsaṃhitā* 11,5 and 24,2; see P. V. Kane [1948/49] 13. He is the alleged author of a *Nāradasaṃhitā* on divination and muhūrtaśāstra, in 37 adhyāyas; this was extensively used by Viṣṇuśarman (*fl. ca.* 1370) in his *Muhūrtadīpikā*. Manuscripts:

Nagpur 961 (1243). 10ff Copied in Śaka 1480 = A.D. 1558. From Nasik.
Oudh VIII (1876) VIII 21. 124pp. Copied in A.D. 1640. Property of Nārāyaṇadatta of Bārābanki Zila.
Bombay U Desai 1383. 70ff. Copied in Saṃ. 1731 = A.D. 1674.
Benares (1963) 34732. Ff. 1–32 and 32b–40. Copied in Saṃ. 1787 = A.D. 1730.
RORI Cat II 9756. 70ff. Copied by Motīrāma at Nandagrāma in Saṃ. 1799 = A.D. 1742.
BORI 525 of 1895/1902. 56ff. Copied in Saṃ. 1823 = A.D. 1766.
CP, Kielhorn XXIII 67. 65ff. Copied in Śaka 1693 = A.D. 1771. Property of Sadāśivabhaṭṭa Ṭopale of Burhāṇpur.
Benares (1963) 36366 = Benares (1878) 124 = Benares (1869) XXVI 1. 23ff. Copied in Saṃ 1829 = A.D. 1772. Incomplete.
AS Bombay 312. 46ff. Copied in Śaka 1705 = A.D. 1783.
Berlin 862 (Chambers 469). 60ff. Copied in Saṃ. 1841 = A.D. 1784.
PL, Buhler IV E 207. 82ff. Copied in Saṃ. 1843 = A.D. 1786. Property of Nānā Jośī of Nandurabāra.
Kerala 8489 (2014). 1300 granthas. Copied in Saṃ. 1852 = A.D. 1795.
RORI Cat. I 2901 51ff. (f. 31 missing). Copied in Saṃ. 1857 = A.D. 1800.
DC 7903. 52ff. Copied in Śaka 1728 = A.D. 1806.
AS Bengal 2622 (G 2141) II. Ff. 3–4. Copied by Prahlādabhaṭṭa, the son of Gopāla, on Friday 5 śuklapakṣa of Pauṣa in Śaka 1733 = 20 December 1811. Incomplete (kākamaithunadarśanaśānti).
BORI 526 of 1895/1902. 60ff. Copied in Saṃ. 1880 = A.D. 1823.
Benares (1963) 36365 = Benares (1903) 1148. 62ff. Copied in Saṃ. 1882 = A.D. 1825.
Oudh XIII (1881) VIII 2. 290pp. Copied in A.D. 1825. Property of Keśavaprasāda of Unao Zila.
Calcutta Sanskrit College 58. 50ff. Copied in Saṃ. 1884 = A.D. 1827.
Bombay U Desai 1384. 49ff. Copied in Śaka 1771 = A.D. 1849.
Oxford CS d. 886(i). 87ff. Copied in Saṃ. 1925 = A.D. 1868.
Baroda 9211. 69ff. Copied in Śaka 1791 = A.D. 1869.
Benares (1963) 37063 = Benares (1878) 38 = Benares (1870–1880) 2. 69ff. Copied in Saṃ. 1932 = A.D. 1875 (date omitted in Benares (1870–1880) and (1963)).

Jammu and Kashmir 3998. 35ff. Copied in Saṃ. 1935 = A.D. 1878.
Baroda 1121. 54ff. Copied in Saṃ. 1947 = A.D. 1890.
Adyar Cat. 33 L 21. 95ff. Incomplete (*Nāradīya-jyotiṣa*).
Alwar 1823.
Ānandāśrama 2938.
Ānandāśrama 6660.
Anup 4813. 75ff. Copied by Gaṅgādhara on Sunday 14 śuklapakṣa of Phālguna in Saṃ. 1xx3.
AS Bengal 6973 (G 7830). 14ff. Incomplete (ends in adhyāya 5).
AS Bengal 6974 (G 10471). 19ff. Incomplete (ends in adhyāya 5).
Baroda 9234. 65ff.
Baroda 13310(a). 60ff. Grantha.
Baroda 13355(b). Ff. 180–203. Nandināgarī.
Benares (1963) 34754. Ff. 45–88. Incomplete.
Benares (1963) 37056 = Benares (1878) 113. 23ff. Incomplete.
Benares (1963) 37064. 35ff. Incomplete (to adhyāya 30).
BORI 160 of A 1883/84. 28ff.
CP, Hiralal 2527. Property of Shrīkrishṇa Pāṇḍuraṅg of Bālāpur, Akolā.
Florence 344. 1f. Incomplete (pallīvicāra).
GOML Madras D 3266. 2pp. (kākaviṣṭhāśānti).
GOML Madras D 3267. 3pp. (kākaviṣṭhāśānti).
GOML Madras D 3418. 16pp. Telugu (vāstuśānti).
GOML Madras D 13580. Ff. 1–107. Telugu. Incomplete.
GOML Madras D 13581. Ff. 8–18. Incomplete (adhyāya 25).
Jaipur (II). 56ff.
Kavīndrācārya 814.
Kerala 8490 (5168). 1200 granthas. Incomplete.
Kerala 8491 (14241 N). 800 granthas. Incomplete.
Kotah 250. 58pp.
Līṃbaḍī 1389 (1599). 2ff. Incomplete (mṛtyuyogādivicāra).
Mysore 452 (473).
Mysore 466 (1267).
Mysore (1922) 1535. 200ff.
Mysore (1922) 1799. 51ff.
Mysore (1911 + 1922) B 574. 15ff. With a Telugu ṭīkā. Incomplete (3 adhyāyas).
Mysore and Coorg 289 (*Jyotiṣakāṇḍanāradīya*). Property of Gopāla Śāstrī of Kadaba.
Mysore and Coorg 297. 1500 granthas. Property of the Śṛṅgeri Maṭha at Śṛṅgeri.
Mysore and Coorg 320. 1000 granthas. (*Laghunāradīya*). Property of Mahādeva Joyisa of Śṛṅgeri.
N-W P IX (1885) A 1. 82ff. Property of Dvarikādatta Vyāsa of Benares.
Oppert I 6952. (*Bṛhannāradīya*). Property of Puligaḍḍa Aruṇācalaśāstrī of Kottapeṭa, Vijayanagaram, Vizagapatam.
PUL II 3590. 56ff.
RORI Cat. III 15469. 53ff.
Sastri, Rep. (1893–94). 170pp. Telugu. Incomplete.
SOI 9557.
SOI 9930.
Tanjore D 18170 = Tanjore BL 9457. 59ff. Telugu. Incomplete.
VVRI 2527. 36ff. Incomplete.
VVRI 3805. 40ff. Malayālam.

The *Nāradasaṃhitā* was published at Vārāṇasī in 1905 (BM 14053. ccc. 56. (1) and NL, Calcutta 180. Kc. 90. 13 and 180. Kc. 90. 21); ed. by Rasikamohana Caṭṭopādhyāya, 2nd ed., Calcutta BS 1321 = A.D. 1915 (BM 14055. c. 1. (4) and IO San. D. 44); and with the Hindī ṭīkā, *Saralā*, of Vasatirāma Śarman at Bambaī in Saṃ. 1994, Śaka 1859 = A.D. 1937, reprinted at Bambaī in 1957. Verses 2–3 name the 18 authorities on jyotiḥśāstra:

brahmācāryo vasiṣṭho ᵓtrir manuḥ paulastyaromaśau/
marīcir aṅgirā vyāso nāradaḥ śaunako bhṛguḥ//
cyavano yavano gargaḥ kaśyapaś ca parāśaraḥ/
aṣṭādaśaite gambhīrā jyotiḥśāstrapravarttakāḥ//

NĀRADA

Alleged author of a *Nāradasiddhānta*. Manuscript:

Kavīndrācārya 862. With the vyākhyā of Kālidāsa.

NĀRADA

Supposed author of a *Nāradīyasaṅgrahasāra*. Manuscript:

IO 6404 (Mackenzie III. 235b). 4ff. Nandināgarī. Incomplete (jātalakṣaṇa). From Colin Mackenzie.

NĀRADA

Author of a *Pañcāśadakṣaraphala*. Manuscripts:

Benares (1963) 37440. 8ff.
Benares (1963) 37556. 4ff. (*Pañcadaśākṣara*).

NĀRADA

Alleged author of a *Mayūracitraka*. Manuscripts:

RORI Cat. I 25. 17ff. Copied by Raṅganātha at Gokula in Saṃ. 1848 = A.D. 1791.
RORI Cat. I 2907. 22ff. (f. 12 missing). Copied by Udayarāma in Saṃ. 1850 = A.D. 1793.
Leipzig 1115. 11ff. (ff. 5–6 missing). Copied in A.D. 1802.
Oxford CS c. 315(v). 19ff. Copied for Ṭhākuradāsa, the son of Puṣkara, and for Harasena, Nandakiśora, Yugalakiśora, and Devakīnandana on Sunday 4 śuklapakṣa of Āṣāḍha śuddha in Saṃ. 1866, Śaka 1731 = 16 July 1809.
RORI Cat. III 11016. 11ff. Copied by Vṛddha Ṛṣi in Saṃ. 1868 = A.D. 1811.

BORI 961 of 1886/92. 15ff. Copied in Saṃ. 1877 = A.D. 1820.
AS Bengal 6967 (G 6349). 21ff. Copied by the son of Śivarāja on Friday 14 kṛṣṇapakṣa of Māgha in Saṃ. 1883 = 24 February 1827.
Benares (1963) 37163. 15ff. Copied in Saṃ. 1889 = A.D. 1832.
RORI Cat. III 12038. 55ff. Copied by Jeṣṭhyeśvara Śarman at Jodhapura in Saṃ. 1893 = A.D. 1836.
BORI 544 of 1875/76. 22ff. Copied in Saṃ. 1894 = A.D. 1837. From Dilhī.
SOI 3393 = SOI Cat. II: 1064–3394(*sic*). 10ff. Copied in Saṃ. 1895 = A.D. 1838.
RORI Cat. II 6643. 20ff. Copied by Bhavānīdāsa Miśra at Jayanagara in Saṃ. 1896 = A.D. 1839.
Mithila 256. 18ff. Maithilī. Copied by Tulasīdatta Śarman on Saturday 11 kṛṣṇapakṣa of Bhādrapada in Saṃ. 1887 (read 1897), Śaka 1762 = A.D. 1840 (the date is irregular for both Saṃ. 1887 and 1897). Property of Paṇḍita Dharmadatta Miśra of Babhangama, Supaul, Bhagalpur.
RORI Cat. II 4284. 19ff. Copied by Sīrapāṇi Sūri in Saṃ. 1897 = A.D. 1840.
BORI 962 of 1886/92. 19ff. Copied in Saṃ. 1898 = A.D. 1841.
Bombay U 526. 16ff. Copied by Yajñeśvara Dīkṣita Sānye on 1 kṛṣṇapakṣa of Āśvina in Śaka 1772 = *ca.* 22 October 1850.
Goṇḍal 253. 24ff. Copied by Bāla Bhaṭṭa Gālanekara on Monday 12 śuklapakṣa of Phālguna in Saṃ. 1920 = 21 March 1864.
Benares (1963) 34913. 23ff. Copied in Saṃ. 1922 = A.D. 1865.
CP, Hiralal 3886 and 3887. Property of Janārdan Śāstrī of Ganiyārī, Bilāspur.
CP, Hiralal 3888. Property of Govind Sundar Śāstrī of Piñjaḍ, Akolā.
CP, Hiralal 3889. Ascribed to Varadācārya. Property of Śāligrām of Hoshangābād.
Kurukṣetra 759 (19913).
Leipzig 1116. 26ff. (f. 1 missing).
Poleman 5231 (U Penn 1816). 25ff.
RORI Cat. I 2899. 19ff. Incomplete.
RORI Cat. II 6222. 10ff.
RORI Cat. III 17200. 36ff.
VVRI 1680. 16ff.

NĀRADA

Author of a *Mātṛkāśakunāvalī*. Manuscript:

SOI 3315 = SOI Cat. II: 1067–3315. 3ff. Copied in Saṃ. 1908 = A.D. 1851.

NĀRADA

Author of a *Sāmudrika*. Manuscripts:

Adyar Index 6975 =
Adyar Cat. 19 E 56. 50ff. Malayālam. Incomplete.
Adyar Cat. 19 E 57. 26ff. Grantha.
Adyar Cat. 19 E 59. 11ff. Grantha. Incomplete (adhyāya 1).
Adyar Cat. 19 E 60. 10ff. Grantha.
Adyar Cat. 19 E 61. 3ff. Grantha. Incomplete (32 verses).
Adyar Cat. 33 B 4. 316ff. (*sic* !). Oriyā.
Leipzig 1173. 5ff.
Pattan, Saṅghavī Pāḍā 116 (13). Ff. 75–82. See Pattan, p. 81.

NĀRADA

Author of a *Svapnādhyāya*, alleged to be a part of the *Nāradīyasaṃhitā*. Manuscripts:

Tanjore D 11478 = Tanjore 13898. 6ff.
Tanjore D 11479 = Tanjore BL 935a. No ff. given. Grantha. Incomplete.
Tanjore D 11480 = Tanjore 15662. No ff. given. Incomplete.
Tanjore D 11481 = Tanjore 13896. No ff. given. Incomplete.
Tanjore D 11482 = Tanjore 15663. No ff. given. Incomplete.
Tanjore D 11483 = Tanjore BL 4322. 5ff. Incomplete.
Tanjore D 11484 = Tanjore 15664. No ff. given. Incomplete.
Tanjore D 11485 = Tanjore 15665. No ff. given. Incomplete.

NĀRĀYAṆA

Author of an *Uparāgakriyākrama* in 5 adhyāyas:

1. paryantaviṣaya.
2. somagrahaṇādāya.
3. sūryagrahaṇādāya.
4. ādāyaviṣaya.
5. vyatīpātagrahamauḍhyāvabodha.

Manuscripts:

Kerala 2519 (C. 2116 B) = Kerala C 656 B. 22ff. Malayālam. Copied in ME 1023 = A.D. 1848. Formerly property of S. Vāsudevan Mūs of Maṅgalappaḷḷi Illam, Tiruvalla.
GOML Madras D 13396. Ff. 1–8. Telugu. Copied on Tuesday 13 kṛṣṇapakṣa of Vaikāśi in a śuklasaṃvatsara = A.D. 1869 (?).
GOML Madras D 14020. Ff. 1–21. Telugu. With an Āndhraṭīkā. Incomplete (ends in adhyāya 3).
Kerala 2516 (CM 531 A) = Kerala C 655 A. 11ff. Malayālam. Incomplete. Formerly property of Śrīdharan Parameśvaran Mūttatu of Vaikom.
Kerala 2517 (8324 F). 225 granthas. Malayālam.
Kerala 2518 (8376 A). 225 granthas. Malayālam.
Kerala 2520 (3651 I). 130 granthas. Malayālam. Incomplete.
Lucknow 520. N 24 U (45769).
Viśvabhāratī 1389. Incomplete (adhyāyas 1–3). See NCC, vol. 2, p. 371.

The last verse is:

gurupādāmbujadhyānaśuddhāntaḥkaraṇena vai/
nārāyaṇena racita uparāgakriyākramaḥ//

NĀRĀYAṆA

Author of a ṭīkā, *Karmapradīpikā*, on the *Līlāvatī* of Bhāskara (b. 1114); this is sometimes ascribed to Mādhava. Manuscripts:

GOML Madras R 3497. 59ff. (ff. 1–2 and 57–59 are blank). Grantha. Copied in ME 1025 = A.D. 1850. Ascribed to Mādhava. Presented in 1920/21 by Lakṣmīnārāyaṇa Ayyar Avargal of Nāraṇammanapuram, Tinnevelly.
Adyar List = Adyar Index 5456 = Adyar Cat. 40 C 20. 94ff. Malayālam.
Baroda 6354. 53ff. (f. 44 missing). Grantha. Incomplete. No author mentioned.
GOML Madras D 13484. Ff. 60–119.
Kerala—(770 A). See NCC, vol. 3, p. 199.
Lucknow 520. N 24 K (46042).
PUL II 3917. 103ff. Malayālam.
PUL II 3918. 20ff. Grantha. Incomplete.

The last verse is:

etan nārāyaṇākhyena racitaṃ karmadīpakam/
santiṣṭhatu paraṃ loke namāmy āryabhaṭaṃ sadā//

NĀRĀYAṆA

Author of a *Kāladīpikā*. Manuscript:

Adyar Index 1237 = Adyar Cat. 34 G 23. 56ff. Malayālam.

NĀRĀYAṆA

Author of a *Dharmapravṛtti* which deals, among other matters, with tithis. Manuscripts:

Baroda 12427. 6ff. and 324ff. Copied in Saṃ. 1664 = A.D. 1607.
Osmania University 865. 122ff. Nandināgarī. Copied in A.D. 1636.
Anup 2430. 72ff. Copied by Nārāyaṇa, the son of Ananta, in Śaka 1627 = A.D. 1705.
Baroda 10544. 135ff. Copied in Saṃ. 1773 = A.D. 1716.
Baroda 13659. 139ff. Copied in Śaka 1662 = A.D. 1740.
Osmania University B/3/14. 105ff. Copied in A.D. 1751. Incomplete.
IO 1562 (1343). 103ff. Copied in A.D. 1799. From H. T. Colebrooke.
Baroda 12797. 17ff. Copied in Saṃ. 1904 = A.D. 1847. Incomplete (āhnika).
Anup 2429. 86ff.
Anup 2431. 113ff.
Anup 2432. 109 and 3ff.
Anup 2433. 13ff. Incomplete.
Baroda 171. 164ff.
Baroda 1032. 26ff. Incomplete (to vivāha).
Baroda 1033. 72ff. (ff. 21–22 and 37–40 missing). Incomplete.
Baroda 8020. 82ff. Incomplete (to vivāha).
Baroda 8033. 9ff. Incomplete (aśaucanirṇaya).
Baroda 8556. 11ff. Incomplete (dānavidhi).
Baroda 10306(b). 35ff. Telugu. Incomplete (to vivāha).
Baroda 12841. 52ff. Incomplete (śayanavidhi to aśaucaprakaraṇa).
Baroda 13398(a). 120ff. Nandināgarī. Incomplete.
Baroda 13441. Ff. 219–259. Nandināgarī. Incomplete.
IO 1560 (2172). 157ff. From Gaikawar.
IO 1561 (2063). 197ff. From Gaikawar.
IO 1563 (1663). 144ff. Incomplete. From H. T. Colebrooke.
Jammu and Kashmir 2680. 13ff. Incomplete.
Jammu and Kashmir 2695. 97ff. Incomplete.
Mithila I 236. 4ff. Maithilī and Devanāgarī. Incomplete. Property of the Rāj Library at Darbhanga.
Osmania University 67/12. 2ff. Incomplete.
Osmania University 996/A. 13ff. Telugu. Incomplete.
Osmania University 1179. 36ff. Telugu. Incomplete.
Osmania University A 620. 62ff. Telugu. Incomplete.
Osmania University B. 140/21/a. 12ff. Incomplete.
Oudh XX (1888) IX 112. 318pp. Property of Paṇḍita Pratāpa Nārāyaṇa of Allahabad Zila.

The *Dharmapravṛtti* was published with a Telugu tātparya at Madras in 1895 (IO 22. BB. 39). Verse 2 is:

dharmapravṛttiḥ śriyate kukalau nārāyaṇena tu/
viduṣāṃ karmaniṣṭhānāṃ
saṃmatidharmavardhanam//

NĀRĀYAṆA

Author of a *Praśnaprakāśa*. Manuscript:

PL, Buhler IV E 256. 49ff. From Khambhāliyām.

NĀRĀYAṆA

Author of a *Muhūrtadīpaka*. Manuscripts:

Kerala 13797 (1055 B). 400 granthas. Malayālam.
Kerala 13798 (5835 B). 400 granthas. Malayālam.

NĀRĀYAṆA

The son of Vāvadeva, Nārāyaṇa wrote a *Muhūrtaratnāvalī*. Manuscript:

Kathmandu (1960) 331 (I 1207). 51ff. Incomplete.

The last verse is:

śrīvāvadevāgnimatas tanūja-
nārāyaṇāgnyāhitanirmitāyām/
muhūrtaratnāvalisaṃjñakāyāṃ
yātrābhidho ᵓyaṃ stabakaḥ samāptaḥ//

NĀRĀYAṆA

Author of a *Lakṣmīnārāyaṇavilāsa.* Manuscript:

Osmania University 627/b. 35ff. Telugu. Incomplete (adhyāyas 1–2).

VĀMORI NĀRĀYAṆA

Author of a *Sabhākaumudī.* Manuscripts:

Jammu and Kashmir 2981. 42ff. Copied in Śaka 1778 = A.D. 1856. Incomplete.
Anup 5239. Ff. 2–113. Incomplete.
CP, Hiralal 6280. Property of Śrīnivāsrāv of Ratanpur, Bilāspur.
Kavīndrācārya 836. No author mentioned.
Tanjore D 11634 = Tanjore BL 4191. 140ff. Telugu. Incomplete.

Verse 1 is:

praṇamyendirāṃ jyotiṣābdheḥ sakāśāt
samādāya sāraṃ tathā dharmaśāstrāt/
sabhākaumudī tanyate samyatuṣṭyai
vidhijñena vāmorinārāyaṇena//

NĀRĀYAṆA

Author of a *Sphuṭadarpaṇa.* Manuscript:

CP, Kielhorn XXIII 184. 42ff. Property of Maṇinandapaṇḍita of Sammalpur.

NĀRĀYAṆA

Author of a *Horāpradīpa* in 20 adhyāyas, said in the Tanjore catalog to be a ṭīkā on the *Bṛhajjātaka* of Varāhamihira (*fl. ca.* 550). Manuscripts:

GOML Madras R 2394. No ff. given. Telugu. Copied in 1917/18 from a manuscript belonging to Elamañci Varāhanarasiṃha Śāstrī of Puṭṭakoṇḍa near Bikkavolu, Godāvarī.
Tanjore D 11673 = Tanjore BL 10981(b). Ff. 19–65. Telugu. With an Āndhraṭīkā.

The colophon begins: iti nārāyaṇācāryakṛtau.

NĀRĀYAṆA JYOTIRVIT

Author of a *Kālasāra; cf.* the *Kāladīpikā* of Nārāyaṇa. Manuscript:

Baroda 10921. Ff. 1–97 and 7ff. Copied in Śaka 1717 = A.D. 1795.

GAJAPATI NĀRĀYAṆA DEVA

A resident of Parlakimedi, Nārāyaṇa wrote an *Āyurdāyakaumudī.* Manuscript:

Cuttack 140. See NCC, vol. 2, p. 151.

NĀRĀYAṆA DHARMĀDHIKĀRI

Author of a *Lakṣaṇakāṇḍa.* Manuscript:

VVRI 1747. 39ff. Incomplete.

NĀRĀYAṆA PAṆḌITA

Author of a karaṇa entitled *Padmalīlāvilāsinī.* Manuscripts:

RORI Cat. III 11247. 10ff. Copied by Gurudāsa in Saṃ. 1749 = A.D. 1692. With the *Gaṇakavallabha* of Nāgaśarman.
BORI 162 of A 1883/84. 6ff. Copied in Śaka 1747 = A.D. 1825.
RORI Cat. III 11334. 9ff. Incomplete (to candraśṛṅgottarādhikāra).

NĀRĀYAṆA PAṆḌITA

Author of a *Viṃśottarīdaśāpaddhati* = *Nārāyaṇīpaddhati* in 265 verses, based on the *Gaurījātaka.* Manuscripts:

Bombay U Desai 1432. 16ff. Copied in Saṃ. 1667 = A.D. 1610.
PUL II 3940. 10ff. Copied in Saṃ. 1895 = A.D. 1838. Ascribed to Nārāyaṇa Bhaṭṭa, and connected with the *Navanītajātaka.*
Baroda 12626. 8ff. Copied in Saṃ. 1906 = A.D. 1849. With a *Daśācakroddhāra* and the viṃśottarīdaśānayanaprakāra of the *Navanītajātaka.*
Alwar 1825.

Verse 265 is:

śrīnārāyaṇapaṇḍitaprakaṭitāsau paddhatiḥ saddhitā
buddher vṛddhisamṛddhisiddhijananī
mugdhaprabodhapradā/
gaurījātakajātayuktijanitā satsaṃpradāyāgatā
vyaktāvyaktabahuprakārakaraṇā jāgarti martyeṣu ca//

NĀRĀYAṆA BHAṬṬA

Author of a *Grahayajñakalpavallī.* Manuscript:

Poona, Fergusson College, Mandlik Library, p. 78. See NCC, vol. 6, p. 255.

NĀRĀYAṆA BHAṬṬA

Author of a *Camatkāracintāmaṇi* in 114 verses, similar to that of Rājarṣi Bhaṭṭa; there is a commentary by Dharmeśvara (*fl. ca.* 1600/1650). Manuscripts:

Oudh XX (1888) VIII 65. 16pp. Copied in A.D. 1596. Property of Paṇḍita Pratāpa Nārāyaṇa of Allahabad Zila.
Anup 4578. 4ff. Copied by Narasiṃha at Vikramapura in Saṃ. 1698 = A.D. 1641.
Anup 4576. 5ff. Copied by Matiharṣa at Āsopā in Saṃ. 1703 = A.D. 1646.
Baroda 3375. 3ff. Copied in Saṃ. 1714 = A.D. 1657.
Baroda 9434. 16ff. Copied in Śaka 1604 = A.D. 1682.
RORI Cat. I 1787. 15ff. Copied by Premajī at Pattana in Saṃ. 1742 = A.D. 1685. With the Old Rājasthānī stabaka of Vekara Dvija.
RORI Cat. III 15326. 8ff. Copied by Jñānasāgara, the pupil of Lābhodaya, at Vairāṭanagara in Saṃ. 1756 = A.D. 1699.
Goṇḍal 89. 16ff. Copied at Kandanapura on 2 kṛṣṇapakṣa of Vaiśākha in Saṃ. 1762 = *ca.* 28 April 1705. With a Gujarātī ṭīkā.
Anup 4579. 7ff. Copied by Jīvana in Saṃ. 1767 = A.D. 1710.
RORI Cat. I 3269. 14ff. Copied by Pramodavijaya in Saṃ. 1768 = A.D. 1711. With an Old Rājasthānī stabaka.
RORI Cat. I 655. 12ff. Copied by Trikama Ṛṣi, the pupil of Govindajī, in Saṃ. 1789 = A.D. 1732. With the Old Rājasthānī stabaka of Rājarṣi.
BORI 898 of 1886/92. 30ff. Copied in Saṃ. 1793 = A.D. 1736. With the *Anvayārthadīpikā* of Dharmeśvara.
Benares (1963) 34756. 7ff. Copied in Saṃ. 1802 = A.D. 1745.
Benares (1963) 34757. 11ff. Copied in Saṃ. 1802 = A.D. 1745. With the *Anvayārthadīpikā* of Dharmeśvara. Incomplete.
RORI Cat. III 12898. 11ff. Copied by Jayavijaya Gaṇi at Māṇḍavī in Saṃ. 1802 = A.D. 1745. With a *Bālabodhinī* in Old Rājasthānī.
RORI Cat. I 611(1). 11ff. Copied in Saṃ. 1808 = A.D. 1751. With an Old Rājasthānī artha.
RORI Cat I 611(2). Ff. 12–13. Copied in Saṃ. 1808 = A.D. 1751. (dvādaśabhāvavicāra).
RORI Cat. I 3797. 12ff. Copied by Kuśala Harṣa in Saṃ. 1810 = A.D. 1753. With an Old Rājasthānī stabaka.
RORI Cat. III 11198. 11ff. Copied by Parasarāma Jośī at Jayanagara in Saṃ. 1816 = A.D. 1759.
AS Bengal 7015 (G 7764). 6ff. Copied in Saṃ. 1817 = A.D. 1760.
Leipzig 1099. 10ff. Copied by Lakṣmīrāma at Jayapura in A.D. 1763.
Oudh XX (1888) VIII 168. 12pp. Copied in A.D. 1764. With the *Anvayārthadīpikā* of Dharmeśvara. Property of Paṇḍita Pratāpa Nārāyaṇa of Allahabad Zila.
RORI Cat. III 17047. 15ff. Copied by Śivakīrti Gaṇi, the pupil of Lakṣmīkīrti, at Bhojāvāriṇī in Saṃ. 1822 = A.D. 1765. With an Old Rājasthānī stabaka.
Goṇḍal 92. 9ff. Copied by Jayakṛṣṇa Dīkṣita Bhaṭṭa on Sunday 1 śuklapakṣa of Śrāvaṇa in Saṃ. 1824 = 26 July 1767.
RORI Cat. II 6629. 30ff. Copied by Vijayalāla in Saṃ. 1828 = A.D. 1771. With the *Anvayārthadīpikā* of Dharmeśvara.
Leipzig 1097. 14ff. Copied by Dayāśaṅkara, a pupil of Upādhyāya Ṭaṅka Viṣṇurāma, in A.D. 1780.
VVRI 2373. 12ff. Copied in Saṃ. 1845 = A.D. 1788.
Benares (1963) 34620. 10ff. Copied in Saṃ. 1847 = A.D. 1790.
RORI Cat. III 17923. 14ff. Copied in Saṃ. 1850 = A.D. 1793. With an Old Rājasthānī stabaka.
Benares (1963) 36502. 17ff. Copied in Saṃ. 1859, Śaka 1724 = A.D. 1802.
RORI Cat. I 3768. 9ff. Copied at Bagaḍīdurga in Saṃ. 1860 = A.D. 1803.
Goṇḍal 91. 10ff. Copied in Saṃ. 1861 = A.D. 1804.
PUL II 3383. 7ff. Copied in Saṃ. 1861 = A.D. 1804.
RJ 450 (vol. 3, p. 245). 7ff. Copied at Jayapura on 4 śuklapakṣa of Jyeṣṭha in Saṃ. 1866 = *ca.* 17 May 1809 during the reign of Jagatasiṃha (1803/1818). Property of Ṭholiyoṃ of Jayapura.
VVRI 2389. 6ff. Copied in Saṃ. 1866 = A.D. 1809. Incomplete.
Udaipur 534. Copied in Saṃ. 1871 = A.D. 1814.
RORI Cat. I 3798. 30ff. Copied at Devalī in Saṃ. 1877 = A.D. 1820. With an Old Rājasthānī artha.
Mithila 64 B. 7ff. Maithilī. Copied in Śaka 1744 = A.D. 1822. Property of Paṇḍita Dāmodara Jhā, Andhratharhi, Darbhanga.
RORI Cat. III 10915. 6ff. Copied by Jorāvarasāgara, the pupil of Hīrasāgara, at Jodhapura in Saṃ. 1880 = A.D. 1823.
RORI Cat. I 3171. 44ff. Copied in Saṃ. 1884 = A.D. 1827. With an Old Rājasthānī artha.
Mithila 64 C. 20ff. Maithilī. Copied in Sāl. San. 1237 = *ca.* A.D. 1829. Property of Paṇḍita Phuddī Jhā of Awama, Jhanjharpur, Darbhanga.
AS Bengal 7019 (G 7791). 11ff. Copied in Saṃ. 1889 = A.D. 1832.
Leipzig 1030. 96ff. Copied in A.D. 1832. With the *Jātakābharaṇa* of Ḍhuṇḍhirāja (*fl. ca.* 1525).
Leipzig 1096. 10ff. Copied in A.D. 1834.
BORI 414 of 1895/98. 25ff. Copied in Saṃ. 1896 = A.D. 1839. With the *Anvayārthadīpikā* of Dharmeśvara.
Goṇḍal 90. 4ff. Copied on Saturday 2 kṛṣṇapakṣa of Mārgaśīrṣa in Saṃ. 1896 = 21 December 1839.
Goṇḍal 88. 25ff. Copied by Monajī Bhāī, the son of Rāvalamūla, on Saturday 13 kṛṣṇapakṣa of Māgha I in Saṃ. 1896, Śaka 1760 = 29 February 1840. With the *Anvayārthadīpikā* of Dharmeśvara.
Oxford 1545 (Sansk. d. 187) = Hultzsch 283a. 62ff. Copied in Saṃ. 1897 = A.D. 1840. With the *Anvayārthadīpikā* of Dharmeśvara.
Mithila 64. 8ff. Maithilī. Copied on Monday 3 śuklapakṣa of Bhādrapada in Śaka 1765, Sāl. San.

1251, Saṃ. 1899, Lakṣ. Saṃ. 744 = 28 August 1843. Property of Babu Ṭhīṭhara Jhā of Babhanagāmā, Supaul, Bhagalpur.
RORI Cat. III 10209. 12ff. Copied by Rāmadatta Jośī in Saṃ. 1900 = A.D. 1843. With the *Anvayārthadīpikā* of Dharmeśvara.
RORI Cat. II 4668. 28ff. Copied by Keśavajī Jādavajī at Saradhāra in Saṃ. 1901 = A.D. 1844. With the *Anvayārthadīpikā* of Dharmeśvara.
Baroda 3117. 44ff. Copied in Saṃ. 1906 = A.D. 1849. With the *Anvayārthadīpikā* of Dharmeśvara.
PL, Buhler IV E 95. 44ff. Copied in Saṃ. 1906 = A.D. 1849. (*Camatkāracintāmaṇiṭīkā*). Property of Uttamarāma Jośī of Ahmadābād.
RORI Cat. III 13067. 15ff. Copied by Kuṃvarajī, the son of Vastā Purohita, at Rāvaṇapura in Saṃ. 1907 = A.D. 1850.
Vangiya Sahitya Parishat 656. 7ff. Bengālī. Copied in Śaka 1773 = A.D. 1851.
Jammu and Kashmir 1189. 9ff. Copied in Saṃ. 1910 = A.D. 1853.
Goṇḍal 94. 13ff. Copied by Kevala Dave at Bhuja on Friday 5 śuklapakṣa of Āṣāḍha I in Saṃ. 1911, Śaka 1776 = 30 June 1854. With the *Anvayārthadīpikā* of Dharmeśvara.
RORI Cat. III 16341. 21ff. Copied in Saṃ. 1911 = A.D. 1854. With an Old Rājasthānī stabaka.
RORI Cat. I 3130. 25ff. Copied by Umāśaṅkara at Kāśī in Saṃ. 1912 = A.D. 1855. With the *Anvayārthadīpikā* of Dharmeśvara.
AS Bengal 7017 (2281) = Mitra, Not. 2666. 29ff. Copied by Mukundarāma in Saṃ. 1915 = A.D. 1858. With the *Anvayārthadīpikā* of Dharmeśvara.
Mithila 64 A. 12ff. Maithilī. Copied in Śaka 1784 = A.D. 1862. Property of Paṇḍita Śrīnandana Miśra of Kanhauli, Sakri, Darbhanga.
RORI Cat. I 3226. 16ff. Copied by Jayaśaṅkara Vyāsa in Saṃ. 1919 = A.D. 1862. With an Old Gujarātī stabaka.
Goṇḍal 93. 12ff. Copied by Murāri Bhaṭṭa, the son of Jagannātha Bhaṭṭa, on 11 śuklapakṣa of Kārttika I in Saṃ. 1921, Śaka 1786 = *ca.* 10 November 1864.
Benares (1963) 35358. 40ff. Copied in Saṃ. 1934 = A.D. 1877. With the *Anvayārthadīpikā* of Dharmeśvara.
GOML Madras D 15785. 42ff. Copied on Sunday 7 śuklapakṣa of Bhādrapada in Saṃ. 1940 = 9 September 1883. With the *Anvayārthadīpikā* of Dharmeśvara.
Goṇḍal 87. 29ff. Copied by Vāsudeva, the son of Mādhavajī Vyāsa, at Goṇḍala on Sunday 6 kṛṣṇapakṣa of Caitra in Saṃ. 1947 = A.D. 1891 (the date is irregular). With the *Anvayārthadīpikā* of Dharmeśvara.
ABSP 449. 23ff. With the *Anvayārthadīpikā* of Dharmeśvara.
Allahabad, Municipal Museum 172. With a ṭīkā. See NCC, vol. 6, p. 386.
Alwar 1756. With the *Anvayārthadīpikā* of Dharmeśvara.
Anup 4570. 16ff. Formerly property of Kauṇḍina Bhaṭṭa Poṭa.
Anup 4571. 13ff.
Anup 4572. 12ff.
Anup 4573. 11ff. Incomplete. Formerly property of Maṇirāma Dīkṣita (*fl. ca.* 1675/1700).
Anup 4574. 7ff. Formerly property of Anūpasiṃha (*fl.* 1674/1698).
Anup 4575. 6ff. Incomplete.
Anup 4577. 25ff. With other jyautiṣa material.
AS Bengal 7002 (G 6421) III. Ff. 13–20.
AS Bengal 7016 (G 7826). 13ff.
AS Bengal 7049 (G 4336). 16ff. Copied at Dadhicyapura.
AS Bengal 7122 (G 7925) IV. 14ff.
Benares (1963) 34457. Ff. 1–24 and 27–38. With the *Anvayārthadīpikā* of Dharmeśvara. Incomplete.
Benares (1963) 34792 = Benares (1878) 179. Ff. 2–7. Incomplete.
Benares (1963) 34793. 12ff. Incomplete.
Benares (1963) 35817. 37ff. With the *Anvayārthadīpikā* of Dharmeśvara.
Benares (1963) 36499. 9ff. With the *Anvayārthadīpikā* of Dharmeśvara. Incomplete.
CP, Kielhorn XXIII 28. 25ff. Property of Javāhara Śāstrī of Chāndā.
Florence 282. 13ff.
Florence 283. 19ff.
IM Calcutta 3537, 8129 (incomplete), 8134 (with the *Anvayārthadīpikā* of Dharmeśvara; incomplete), 8203 (incomplete), and 9596. See NCC.
IO 6403 (Bühler 267). 11ff. Incomplete. From G. Bühler.
Jaipur (II).
Jammu and Kashmir 2832. 7ff. (strījātaka).
Jammu and Kashmir 2918. 4ff.
Jodhpur 466 and 467. Each with a bhāṣāṭīkā. See NCC.
Kathmandu (1960) 96 (I 1188). 14ff. Nevārī.
Kathmandu (1960) 97 (I 1412). 9ff. Incomplete.
Kathmandu (1960) 98 (I 1112). 7ff. Incomplete.
Kathmandu (1960) 99 (III 425). 5ff.
Leipzig 1098. 10ff.
Mithila 64 D. 9ff. Maithilī. Property of Paṇḍita Rudrānanda Jhā of Parsarma, Bhagalpur.
Osmania University 121/10/b. 8ff. Incomplete.
Osmania University B. 82/7. 16ff. With the *Anvayārthadīpikā* of Dharmeśvara.
Oudh XX (1888) VIII 110. 10pp. With the *Anvayārthadīpikā* of Dharmeśvara. Property of Paṇḍita Pratāpa Nārāyaṇa of Allahabad Zila.
Oxford CS g. 14. 21ff.
PrSB 965 (or. oct. 758; now at Marburg). 10ff.
RJ 1649 (vol. 2, p. 270). 5ff. Property of Baḍā Terahapanthiyoṃ of Jayapura.
Rajputana, p. 57. From Alwar.

RORI Cat. I 663. 13ff.
RORI Cat. I 3128. 12ff. With a ṭīkā.
RORI Cat. I 3168. 19ff. With a ṭīkā.
RORI Cat. III 11196. 17ff.
RORI Cat. III 13107. 7ff.
RORI Cat. III 15329. 5ff. Copied by Gopālacanda at Vairāṭa.
RORI Cat. III 16082. 9ff. (strījātaka).
RORI Cat. III 16946. 5ff.
RORI Cat. III 17151. 8ff. With an Old Rājasthānī stabaka.
RORI (Jaipur) I 501 = Vidyābhūṣaṇa 501. 5ff. Incomplete.
RORI (Jaipur) IV 66. 6ff.
SOI 5981 = SOI (List) 362. With the *Anvayārthadīpikā* of Dharmeśvara.
Udaipur, Nathdvārā 186, 6 (incomplete) and 7 (with a ṭīkā). See NCC.
Viśvabhāratī 171 and 1532 (with the *Anvayārthadīpikā* of Dharmeśvara). See NCC.
VVRI 1895. 8ff. Incomplete.
VVRI 2380. 5ff. With the *Anvayārthadīpikā* of Dharmeśvara. Incomplete.
VVRI 2559. 7ff.
VVRI 2560. 6ff.
VVRI 2561. 2ff. Incomplete.
VVRI 2563. 7ff. Incomplete.
VVRI 2566. 15ff.
WHMRL D. 114. b.
WHMRL X. 57. With the *Anvayārthadīpikā* of Dharmeśvara.

The *Camatkāracintāmaṇi* has been published with the *Anvayārthadīpikā* of Dharmeśvara at Benares in 1856 (IO 362); with the same ṭīkā at Kāśī in Saṃ. 1926 = A.D. 1869 (BM); with a Marāṭhī anuvāda, 2nd ed., Poona 1869 (IO 399); with the *Anvayārthadīpikā* of Dharmeśvara at [Benares] in 1870 (IO 7. B. 40); with the same ṭīkā at Delhi in 1872 (BM and IO 1605); with the same ṭīkā at Delhi in 1876 (IO 411); with the same ṭīkā, edited by Rasikamohana Caṭṭopādhyāya, Calcutta BS 1291 = A.D. 1883 (IO 395 and NL Calcutta 180. Kb. 88. 1(1)); with the Sinhalese translation of H. D. Fernando Tambi-Appu Gurunnānse at [Colombo] in 1891 (BM 14053. cc. 63. (1)); with the Bengālī translation of Rāmagopāla Jyotirvinoda, edited by Kṣatranātha Jyotiratna, Calcutta 1895 (NL Calcutta 180. Kc. 90. 8(3)); with the Marāṭhī bhāṣāntara of Mahādeva Bhāskara Goḍabole, Poona 1915 (IO San. D. 605(b)); with the Hindī anuvāda of Madanamohana Pāṭhaka, Benares 1916 (IO San. B. 162(b)), reprinted Bombay 1919 (IO San. B. 948(b)) and Benares 1924 (IO San. B. 935(a)); with the same Hindī anuvāda and the Bengālī translation of Surendranātha Bhaṭṭācārya, 2nd ed., Calcutta 1936 (NL Calcutta 180. Kc. 93. 17); with the Hindī ṭīkā, *Bhāvaprabodhinī*, of Gaṇapatideva Śāstrin as *HSS* 45, Banārasa 1935, 2nd ed. Banārasa 1948, and 3rd ed. Vārāṇasī 1963; and edited with his own anvaya by Gaṇeśadatta Pāṭhaka, Benares 1966. There is an English translation by Kṛṣṇanātha Raghunāthajī, published at Bombay in 1894 (BM 14053. b. 31. (1) and IO 1258).

Verse 1 is;

lasatpītapaṭṭāmbaraṃ kṛṣṇacandraṃ
mudā rādhayāliṅgitaṃ vidyuteva/
ghanaṃ sampraṇamyātra nārāyaṇākhyaś
camatkāracintāmaṇiṃ sampravakṣye//

The last verse is:

camatkāracintāmaṇau yat khagānāṃ
phalaṃ kīrtitaṃ bhaṭṭanārāyaṇena/
paṭhed yo dvijas tasya rājñāṃ samakṣe
pravaktuṃ na cānye samarthā bhaveyuḥ//

See Nārāyaṇa Bhaṭṭa (*fl.* 1758).

NĀRĀYAṆA BHAṬṬA

Author of a *Tithinirṇaya* = *Tithivākyanirṇaya; cf.* the *Tithinirṇayaratnamālā* of Nārāyaṇa Svāmin. Manuscripts:

DC (Gorhe) App. 156. 14ff. Copied in Śaka 1969 = A.D. 1777. Property of Gaṅgādhara Rāmakṛṣṇa Dharmādhikārī of Puṇatāmbe, Ahmadnagar.
Benares (1956) 12125. 22ff.
DC (Gorhe) App. 155. 16ff. Property of Gaṅgādhara Rāmakṛṣṇa Dharmādhikārī of Puṇatāmbe, Ahmadnagar.
Tanjore D 18591 = Tanjore BL 150. 35ff. Incomplete.
Tanjore D 18592 = Tanjore BL 196 25ff.
Tanjore D 18593 = Tanjore BL 12323. 34ff.
Tanjore D 18594 = Tanjore 16340. 2ff. Incomplete.

NĀRĀYAṆA BHAṬṬA

Author of a *Pañcapakṣiśakunāvalī*. Manuscript:

LDI 7485 (2746). 5ff. Copied by Joṣī Mughārāma in Saṃ. 1796 = A.D. 1739.

This may be identical with the *Nārāyaṇī śakunavantī*. Manuscript:

SOI 6549.

NĀRĀYAṆA BHAṬṬA

Author of a ṭīkā on the *Bhuvanadīpaka* of Padmaprabhu Sūri, edited by Rasikamohana Caṭṭopādhyāya, Calcutta 1884 (IO 395).

NĀRĀYAṆA BHAṬṬA

Author of a *Santānapradīpa*. Manuscript:

Jammu and Kashmir 3015. 97ff. Incomplete.

NĀRĀYAṆA BHAṬṬA

Author of a ṭippaṇa on a *Sannipātakalikā*, presumably that of Auvunikurṇaka. Manuscript:

N-W P II (1877) B 6. 14ff. Property of Rāma Prasāda of Benares.

NĀRĀYAṆA ŚARMAN

Author of a *Kālacakravivaraṇa* in 90 verses describing the construction of an astronomical instrument, the samayasūcakayantra or kālayantra; he himself wrote a ṭīkā on this. Manuscripts:

IO 6310 (Mackenzie 11 47c). 9ff. From Colin Mackenzie.
IO 6311 (Mackenzie II 47d). Ff. 10–31. (*Kālacakravivaraṇaṭīkā*). From Colin Mackenzie.

The colophon begins: iti nārāyaṇaśarmaviracitaṃ.

NĀRĀYAṆA ŚARMAN CAKRAVARTIN

Author of a *Śāntitattvāmṛta*. Manuscripts:

Mitra, Not. 2477. 61ff. Bengālī. Copied in BS 1217 = *ca.* A.D. 1810. Property of Rājā Rājendranārāyaṇa Deva, Bahādur, of Calcutta. Is this Mitra, Not. 536?
IO 1760 (917). 80ff. Bengālī. From H. T. Colebrooke.
Mitra, Not. 536. 61ff. Bengālī. Property of Rājā Rādhākānta Deva, Bahādur, of Calcutta.

Verse 1 is:

natvā gopīkāntaṃ
matvā ca vividhamunivākyāni/
śrīnārāyaṇaśarmā
śāntikatattvāmṛtaṃ tanute//

The colophon begins: iti śrīnārāyaṇacakravartikṛtaṃ.

NĀRĀYAṆA SŪRI

Author of a vivṛti on the *Vṛttaśataka* of Maheśvara (*fl. ca.* 1100/1150). Manuscript:

PUL II 3955. 34ff. (ff. 10 and 28 missing).

NĀRĀYAṆA SŪRI

Author of a *Sāmudrikasāra*. Manuscripts:

Benares (1963) 34822. 27ff. Incomplete.
Benares (1963) 34885. Ff. 2 and 2b–14. Incomplete.

NĀRĀYAṆA SVĀMIN

The pupil of Sarvajña, Nārāyaṇa wrote a *Tithinirṇayaratnamālā; cf.* the *Tithinirṇaya* of Nārāyaṇa Bhaṭṭa. Manuscript:

AS Bengal 2789 (G 10728 B). 71ff.

The colophon begins: iti śrīsarvajñaśiṣyabhagavannārāyaṇasvāmiracitā.

NĀRĀYAṆA PAṆḌITA (*fl.* 1356)

The son of Nṛsiṃha or Narasiṃha, Nārāyaṇa wrote the following works on mathematics.

1. The *Bījagaṇitāvataṃsa* on algebra. Manuscripts:

Benares (1963) 35579 = Benares (1878) 94. 13ff. Incomplete.
Jaipur (II). 87ff.

The beginning of the *Bījagaṇitāvataṃsa* was edited from the Benares manuscript by K. S. Shukla [A3. 1969/70]; see also S. Dvivedin [1892] 85–86; B. Datta [1931c] and [1933]; and R. Garver [1932]. The colophon begins: iti sakalakalānidhinarasiṃhanandanagaṇitavidyācaturānananārāyaṇapaṇḍitaviracite.

2. The *Gaṇitakaumudī* on mathematics in 14 vyavahāras, completed on Thursday 2 kṛṣṇapakṣa of Kārttika in Śaka 1278 = 10 November 1356. The vyavahāras in the edition are:

I prakīrṇakavyavahāra.
(II) miśravyavahāra.
(III) śreḍhīvyavahāra.
(IV) kṣetravyavahāra.
(V) khātavyavahāra.
(VI) citivyavahāra.
IX (sic) kuṭṭakavyavahāra.
X vargaprakṛti.
XI bhāgādāna.
XII rūpādyaṃśāvatāra.
XIII aṅkapāśa.
XIV bhadragaṇita.

Manuscripts:

Cambridge R. 15. 140 41ff. Bengālī. Copied in A.D. 1791. Incomplete (vyavahāras 13 and 14).
IO 2883 (596 B). 37ff. Copied on Thursday 4 kṛṣṇapakṣa of Māgha in Saṃ. 1848, Śaka 1712 = 9 February 1792. Incomplete (vyavahāras 13 and 14).
Baroda 3097. 31ff. Copied in Śaka 1813 = A.D. 1891. Incomplete (vyavahāra 13 ?).
Anup 4490. 49ff. Incomplete.
Baroda 3096. 28ff. Incomplete (vyavahāra 13).
Benares (1963) 35668. 29ff. Incomplete (vyavahāra 13).
Calcutta Sanskrit College 71(1). Pp. 1–39.
LDI (LDC) 4071. 37ff.
PL, Buhler IV E 47. 32ff. Property of Uttamarāma Jośī of Ahmadābād.
Poona, Fergusson College, Mandlik Library, Suppl. 495. Incomplete (prastarādigaṇita of vyavahāra 13). See NCC, vol. 1, rev. ed., p. 60.
RORI Cat. II 4720. 37ff.

The *Gaṇitakaumudī*—at least the available portions thereof—was edited from a manuscript which had belonged to his father, Sudhākara Dvivedin, by Padmākara Dvivedin, *PWSBT* 57, 2 pts., Benares 1936–1942; see also P. Dvivedin [1925]; B. Datta and A. N. Singh [1935/38] passim; and, on vyavahāra 14, S. Cammann [1968/69] 274 sqq. The last 5 verses are:

āsīt saujanyadugdhāmbudhir avanisuraśreṇimukhyo jagatyāṃ
prakhyaḥ śrīkaṇṭhapādadvayanihitamanāḥ śāradāyā nivāsaḥ/
śrautasmārtārthavettā sakalaguṇanidhiḥ śilpavidyāpragalbhaḥ
śāstre śastre ca tarke pracuratarag atiḥ śrīnṛsiṃho nṛsiṃhaḥ//
tatsūnur asti gaṇitārṇavakarṇadhāraḥ
śrīśāradāpracuralabdhavaraprasādaḥ/
nārāyaṇaḥ pṛthuyaśā gaṇitasya pāṭīṃ
śrīkaumudīm iti mude guṇināṃ pracakre//
yāvat sapta kulācalāḥ kṣititale yāvac catuḥ sāgarā
yāvat sūryamukhā grahāś ca gagane yāvad dhruvas tārakāḥ/
stheyāt tāvad iyaṃ sadoditavatī śrīkaumudī kaumudo-
pūrasvacchayaśaḥpravāhasubhagā nārāyaṇendoḥ stutā//
nārāyaṇānanasudhākaramaṇḍalotthāṃ
ca turyasūktiracanāmṛtabinduvṛndām/
prītyaiva sajjanacakoragaṇāḥ pibantu
śrīkaumudīm uditahṛtkumudaḥ sadaitām//
gajanagaravimitaśāke
durmukhavarṣe ca bāhule māsi/
dhātṛtithau kṛṣṇadale
gurau samāptigataṃ gaṇitam//

JAGADGURU NĀRĀYAṆA BHAṬṬA (b. 1513)

The son of Rāmeśvara Bhaṭṭa of Pratiṣṭhāna and the grandfather of Kamalākara Bhaṭṭa (*fl.* 1612), Nārāyaṇa was born in Caitra of Śaka 1435 = 6 March–4 April 1513, and became a leading paṇḍita in Benares. He wrote a ṭīkā on the *Vṛttaratnākara* in A.D. 1545. See P. V. Kane [1930/62] vol. 1, pp. 419–421. Among his numerous works are a vivaraṇa on the *Kālanirṇaya* of Mādhava. Manuscripts:

Calcutta Sanskrit College (Smṛti) 49. 12ff. Copied by Govardhana Dīkṣita Tripāṭhin on Wednesday 14 kṛṣṇapakṣa of Āśvina in Saṃ. 1692 = 28 October 1635.
Baroda 12025. 12ff. Copied in Saṃ. 1699 = A.D. 1642.
Baroda 4039. 13ff. Copied in Saṃ. 1729 = A.D. 1672.
Calcutta Sanskrit College (Smṛti) 50. 99ff. Copied on 11 śuklapakṣa of Phālguna in Saṃ. 1866 = *ca.* 15 March 1810.
Baroda 8351. 15ff. Copied in Śaka 1759 = A.D. 1837.
Anup 1667 = Bikaner 861 A. 11ff.
Anup 1668 = Bikaner 861 B. 19ff.
Baroda 9034. Ff. 2–14. Incomplete.
Bikaner 861 C. 17ff.
Bikaner 861 D. 13ff.
GOML Madras R 2853. 13ff. Incomplete. Presented in 1918/19 by Śukla Kṛṣṇāji of Gujarātipeṭa, Vizagapatam.
Oppert I 3713. Property of Marutvāṅguḍi Svāmiśāstrī of Kumbhaghoṇam.
Oppert I 3768. Property of the Śaṅkarācāryamaṭha at Kumbhaghoṇam.
Oppert II 6233. Property of Narasiṃhācārya of Kumbhaghoṇam.
Oppert II 7314. (*Tithinirṇaya*). Property of Vaidyanāthaśāstrī of Nalluceri, Tanjore.
Poleman 2919 (U Penn 289). 17ff.

Verse 1 is:

sūrirāmeśvarasyādyaḥ sūnur nārāyaṇaḥ kṛtī/
kṛtavān mādhavācāryasaṅgrahaślokanirṇayam//

Nārāyaṇa is also the author of a *Prayogaratna*, published at Bombay in 1915. A part of this is the *Navagrahamakha*. Manuscript:

PUL I 308. 20ff.

NĀRĀYAṆA (*fl.* 1525 or 1559)

The son of Rāma, Nārāyaṇa wrote a *Grahaṇalikhanānukrama* = *Amṛtakumbha*, apparently in Saṃ. 1582 = A.D. 1525 though some manuscripts give the date Saṃ. 1616, Śaka 1481 = A.D. 1559. Manuscripts:

PL, Buhler IV E 5. 40ff. Copied in Saṃ. 1683 = A.D. 1626. (*Amṛtakumbha*). Property of Lāḍhorāvala of Khambhāliyāṃ. Buhler notes another copy.
RORI Cat. I 645. 37ff. Copied by Tulasivyāsa, the son of Śivarāma, in Saṃ. 1806 = A.D. 1749, apparently from a manuscript copied in Saṃ. 1616 = A.D. 1559.
Goṇḍal 46. 16ff. Copied at Kākikāgrāma in Saṃ. 1860 = A.D. 1803 from a manuscript copied in Saṃ. 1616 = A.D. 1559.
Goṇḍal 4. 33ff. Copied on Monday 6 kṛṣṇapakṣa of Phālguna in Saṃ. 1879 = 3 March 1823.
Baroda 2373. 24ff. Copied in Saṃ. 1884 = A.D. 1827. (*Amṛtakumbha*). Ascribed to Vārāyaṇa.
Baroda 10289. 18ff. Copied in Saṃ. 1892 = A.D. 1835.
Goṇḍal 5. 14ff. Copied on Friday 10 śuklapakṣa of Māgha in Saṃ. 1909, Śaka 1772 = 18 February 1853. Incomplete.
Ānandāśrama 2112.
AS Bombay 231 = AS Bombay (Indraji) 84. 13ff. Incomplete.
BORI 150 of A 1883/84. 17ff.
Rajputana, p. 47. (*Amṛtakumbha*). From Bikaner.

NĀRĀYAṆA (*fl.* 1571/1572)

The son of Ananta Agnihotrin, the son of Hari, the son of Kṛṣṇa, the son of Ananta of the Kauśikagotra, a Vājasaneyin Mādhyandinīya Brāhmaṇa residing at Sāsamaṇūra, Nārāyaṇa, the father of Gaṅgādhara (*fl.* 1586), wrote the following works at Ṭāpara to the north of Devagiri; see S. Dvivedin [1892] 78–79.

1. The *Muhūrtamārtaṇḍa* in Śaka 1493 = A.D. 1571. There are 11 prakaraṇas:

1. tyājya.
2. nakṣatra.
3. saṃskāra.
4. vivāha.
5. agnyādhāna.
6. gṛha.
7. yātrā.
8. miśra.
9. anadhyāya.
10. gocara.
11. saṅkrānti.

Cf. the abridgment by Nīlakaṇṭha (*fl.* 1680). Manuscripts:

DC 3303. 40ff. Copied in Saṃ. 1659 = A.D. 1602. From the Dīkṣit (A) Collection.

Anup 4989. 18ff. Copied in Śaka 1556 = A.D. 1634. Formerly property of the Jyotiṣarāja = Vīrasiṃha (b. 1617).

Berlin 2230 (or. fol. 1491). Ff. 3–157. Copied on 4 kṛṣṇapakṣa of Śrāvaṇa in Saṃ. 1703 = *ca.* 20 July 1646. With his own *Mārtaṇḍavallabhā*.

Nagpur 1625 (1470). 37ff. Copied in Śaka 1574 = A.D. 1652. From Nasik.

DC 4071. 38ff. Copied in Śaka 1576 = A.D. 1654. From the Dīkṣit (A) Collection.

LDI (LDC) 1291. 61ff. Copied in Saṃ. 1715 = A.D. 1658. With his own *Mārtaṇḍavallabhā*.

VVRI 4604. 63ff. Copied in Saṃ. 1732 = A.D. 1675. With a ṭīkā. Incomplete.

LDI 7126 (5649). 9ff. Copied in Saṃ. 1739 = A.D. 1682.

LDI 7125 (5305). 17ff. Copied by Bhīmajī, the pupil of Mahimāprabha Sūri, in Saṃ. (read Śaka) 1620 = A.D. 1698.

PL, Buhler IV E 354. 116ff. Copied in Saṃ. 1755 = A.D. 1698. With his own *Mārtaṇḍavallabhā*. Property of Bālambhaṭṭa of Surata. Buhler notes 6 other copies.

GVS 2894 (3120). 120ff. Copied on Sunday 3 kṛṣṇapakṣa of Āśvina in Saṃ. 1759 = 27 September 1702 Julian. With his own *Mārtaṇḍavallabhā*.

Poleman 4996 (U Penn 1876). 28ff. Copied in Saṃ. 1776, Śaka 1641 = A.D. 1719.

Benares (1963) 35812 = Benares (1913–1914) 2362. 26ff. Copied in Saṃ. 1777 = A.D. 1720. With a ṭippaṇa.

Benares (1963) 34322. 26ff. Copied in Saṃ. 1782, Śaka 1647 = A.D. 1725.

GVS 2895 (4153). 82ff. Copied on Thursday 13 kṛṣṇapakṣa of Vaiśākha I in Saṃ. 1784 = 6 April 1727 (?). With his own *Mārtaṇḍavallabhā*.

Calcutta Sanskrit College 102. Ff. 2–23. Copied in Saṃ. 1785, Śaka 1650 = A.D. 1728.

Baroda 3244. 27ff. (f. 1 missing). Copied in Saṃ. 1786 = A.D. 1729.

Jaipur (II). 15ff. Copied in Saṃ. 1787 = A.D. 1730.

RORI Cat. II 4671. 82ff. Copied by Nṛsiṃhadeva Agravāla at Jayapura in Saṃ. 1789 = A.D. 1732. With his own *Mārtaṇḍavallabhā*.

RORI Cat. II 9384. 25ff. Copied in Saṃ. 1791 = A.D. 1734.

Benares (1963) 35377. 79ff. Copied in Saṃ. 1799, Śaka 1664 = A.D. 1742. With his own *Mārtaṇḍavallabhā*.

Benares (1963) 36566. Ff. 14–16 and 18–42. Copied in Saṃ. 1803 = A.D. 1746. Incomplete.

DC 9422. 116ff. Copied in Śaka 1674 = A.D. 1752. With his own *Mārtaṇḍavallabhā*.

BORI 717 of 1883/84. 20ff. Copied in Śaka 1677 = A.D. 1755. From Mahārāṣṭra.

Kathmandu (1960) 327 (I 1173). 27ff. Copied in NS 875 = A.D. 1755. Incomplete.

DC 6115. 35ff. Copied in Śaka 1679 = A.D. 1757. From the Kesarī Marāṭhā Collection.

Oxford CS c. 315(i). 12ff. Copied on Sunday 6 (read 9) śuklapakṣa of Mārgaśīrṣa in Saṃ. 1814 = 18 December 1757.

Benares (1963) 36247. Ff. 4–10 and 12–20. Copied in Saṃ. 1818 = A.D. 1761. Incomplete.

BORI 118 of A 1879/80. 158ff. Copied in Śaka 1683 = A.D. 1761. With his own *Mārtaṇḍavallabhā*.

GVS 2893 (3021). 34ff. Copied on 3 śuklapakṣa of Phālguna in Saṃ. 1824 = *ca.* 19 February 1768. With a ṭīkā. Incomplete. No author mentioned.

Benares (1963) 36190 = Benares (1913–1914) 2361. 17ff. Copied in Saṃ. 1827, Śaka 1692 = A.D. 1770.

PL, Buhler IV E 353. 22ff. Copied in Saṃ. 1830 = A.D. 1773. Property of Uttamarāma Jośī of Ahmadābād. Buhler notes 12 other copies.

RORI Cat. III 16079(1). 4ff. Copied by Raghuvaraprasāda at Kāśī in Saṃ. 1831 = A.D. 1774. With the *Bālavivekinī* of Vitta. Incomplete.

Bombay U 444. 22ff. Copied by Yajñeśvara Sānye in Śaka 1702 = A.D. 1780.

RORI Cat. II 4732. 26ff. Copied in Saṃ. 1837 = A.D. 1780.

RORI Cat. II 6130. 56ff. Copied by Bālacanda at Gvāliyara in Saṃ. 1837 = A.D. 1780. With his own *Mārtaṇḍavallabhā*.

GVS 2892 (5267). Ff. 9–34. Copied on Tuesday 12 kṛṣṇapakṣa of Māgha in Saṃ. 1837 = 20 February 1781.

BORI 426 of A 1881/82. 129ff. Copied in Śaka 1703 = A.D. 1781. With his own *Mārtaṇḍavallabhā*.

GVS 2891 (3825). Ff. 1–3, 15–19, and 23. Copied on Monday 15 śuklapakṣa of Āśvina in Saṃ. 1841, Śaka 1707 = 17 October 1785. Incomplete.
Benares (1963) 35378. Ff. 1 and 3–9. Copied in Saṃ. 1845 = A.D. 1788. Incomplete.
AS Bombay 321. 192ff. Copied in Saṃ. 1849 = A.D. 1792. With his own *Mārtaṇḍavallabhā*.
Calcutta Sanskrit College 103. 33ff. Copied in Saṃ. 1849 = A.D. 1792.
RORI Cat. II 8886. 100ff. Copied by Tulasīrāma in Saṃ. 1849 = A.D. 1792. With his own *Mārtaṇḍavallabhā*.
Nagpur 1627 (1554). 16ff. Copied in Śaka 1715 = A.D. 1793. From Nasik.
IO 3023 (2528e). 21ff. Copied in A.D. 1795. From Gaikawar.
Ahmadnagar 310 (290/15). 175ff. Copied on 6 śuklapakṣa of Mārgaśīrṣa in Śaka 1720 = *ca.* 12 December 1798. Ascribed to Gaṇeśa.
AS Bombay 320. 17ff. Copied in Saṃ. 1855 = A.D. 1798. From Bhāu Dājī.
RORI Cat. II 5246. 27ff. Copied in Saṃ. 1855 = A.D. 1798. Incomplete (ends in prakaraṇa 11).
Goṇḍal 295. 101ff. Copied by Jagannātha at Saradhāra on Saturday 9 śuklapakṣa of Vaiśākha in Saṃ. 1856 = 11 May 1799. With his own *Mārtaṇḍavallabhā*.
Nagpur 1633 (1772). 134ff. Copied in Śaka 1722 = A.D. 1800. With his own *Mārtaṇḍavallabhā*. From Nagpur.
Poleman 4993 (Columbia, Smith Indic 162). 19ff. Copied in Śaka 1722 = A.D.1800.
Osmania University B. 9/19. 77ff. Copied in A.D. 1801. With his own *Mārtaṇḍavallabhā*.
RORI Cat. III 13906. 44ff. Copied by Bakhtāvaramalla Ṛṣi at Nāgaura in Saṃ. 1858 = A.D. 1801. With his own *Mārtaṇḍavallabhā*.
Benares (1963) 36137. 21ff. Copied in Saṃ. 1859 = A.D. 1802.
Benares (1963) 34575. Ff. 1–7 and 10–19. Copied in Saṃ. 1860 = A.D. 1803. Incomplete.
RORI Cat. II 7048. 27ff. Copied by Dayāśaṅkara Vyāsa in Saṃ. 1862 = A.D. 1805.
Leipzig 1075. 18ff. Copied in A.D. 1806.
BORI 176 of A 1883/84. 104ff. Copied in Saṃ. 1866 = A.D. 1809. With his own *Mārtaṇḍavallabhā*.
Benares (1963) 36573. 25ff. Copied in Saṃ. 1869 = A.D. 1812.
Goṇḍal 296. 109ff. Copied in Saṃ. 1871 = A.D. 1814. With his own *Mārtaṇḍavallabhā*.
SOI 2293 = SOI Cat. I: 1375–2293. Ff. 3–26. Copied in Saṃ. 1872 = A.D. 1815.
Benares (1963) 36574. 16ff. Copied in Saṃ. 1873 = A.D. 1816.
Baroda 5754. 190ff. Copied in Saṃ. 1874 = A.D. 1817. With his own *Mārtaṇḍavallabhā*.
Bombay U 445. 195ff. Copied by Jagadīśa Dharmādhikāri of Nasirābāda on 11 kṛṣṇapakṣa of Śrāvaṇa in Saṃ. 1874, Śaka 1739 = *ca.* 7 August 1817. With his own *Mārtaṇḍavallabhā*.
Benares (1963) 36249. 23ff. Copied in Saṃ. 1876 = A.D. 1819.
Bombay U 446. Ff. 31–142. Copied by Bābadeva Śarman, the son of Rāmacandra, the son of Govindabhaṭṭa Ṭhākura, on Sunday 9 śuklapakṣa of Māgha in Śaka 1741 = 22 January 1820. With his own *Mārtaṇḍavallabhā*. Incomplete.
Benares (1963) 36571. 20ff. Copied in Śaka 1744 = A.D. 1822.
Benares (1963) 37222. 177ff. Copied in Saṃ. 1879, Śaka 1744 = A.D. 1822. With his own *Mārtaṇḍavallabhā*.
Benares (1963) 36065. Ff. 1–15 and 18–52. Copied in Saṃ. 1882 = A.D. 1825. Incomplete.
Benares (1963) 37046 = Benares (1903) 1312. Ff. 1–57, 57b–94, and 1–3. Copied in Saṃ. 1882, Śaka 1747 = A.D. 1825. With his own *Mārtaṇḍavallabhā* and a *Candrasāraṇī*.
Baroda 3246. 27ff. Copied in Saṃ. 1883 = A.D. 1826.
Goṇḍal 298. 17ff. Copied by Dhelā, the son of Āmbā Vyāsa, on Saturday 12 śuklapakṣa of Āṣāḍha in Saṃ. 1883 = 15 July 1826.
Benares (1963) 36570. 22ff. Copied in Saṃ. 1884 = A.D. 1827.
Benares (1963) 35381. 23ff. Copied in Saṃ. 1885 = A.D. 1828.
SOI 2366 = SOI Cat. I: 1376–2366. 27ff. Copied in Saṃ. 1885 = A.D. 1828.
SOI 2372 = SOI Cat. I: 1377–2372. 155ff. Copied in Śaka 1750 = A.D. 1828. With his own *Mārtaṇḍavallabhā*.
Katrak 639. 45ff. Copied in Saṃ. 1885, Śaka 1751 = A.D. 1829. No author mentioned.
BORI 889 of 1891/95. 21ff. Copied in Śaka 1752 = A.D. 1830.
Oxford CS c. 316(iv). 23ff. Copied by a son of Gopāla for himself and his brothers, Choṭīlāla and Pannālāla, on Monday 4 kṛṣṇapakṣa of Phālguna in Saṃ. 1887, Śaka 1752 = 28 February 1831.
Leipzig 1071. 144ff. Copied in A.D. 1832. With his own *Mārtaṇḍavallabhā*.
Poleman 4991 (Columbia, Smith Indic 61). 32ff. Copied in Saṃ. 1889 = A.D. 1832.
Oxford CS c. 315(viii). 17ff. Copied on 14 śuklapakṣa of Bhādrapada in Saṃ. 1890 = *ca.* 28 August 1833.
RORI Cat. II 4887. 23ff. Copied in Saṃ. 1890 = A.D. 1833.
LDI (LDC) 638. 20ff. Copied in Saṃ. 1893 = A.D. 1836.
Osmania University 137/5 A. 65ff. Copied in A.D. 1837. With his own *Mārtaṇḍavallabhā*.
Benares (1963) 37219. 22ff. Copied in Saṃ. 1895 = A.D. 1838.
Oudh XIII (1881) VIII 10. 200pp. Copied in A.D. 1840. With his own *Mārtaṇḍavallabhā*. Property of Dīna Dayāla of Rae Bareli Zila.

SOI 3222 = SOI Cat. II: 1083–3222. 12ff. Copied in Śaka 1762 = A.D. 1840.
DC 129. 18ff. Copied in Saṃ. 1898 = A.D. 1841.
RORI Cat. II 4664. 20ff. Copied by Līlādhara, the son of Puruṣottama, in Saṃ. 1900 = A.D. 1843.
Poleman 4995 (U Penn 1819). 24ff. Copied in Śaka 1766 = A.D. 1844.
Benares (1963) 36332. 24ff. Copied in Saṃ. 1903 = A.D. 1846. Incomplete (ends in prakaraṇa 11).
LDI (LDC) 2200. 164ff. Copied in Saṃ. 1903 = A.D. 1846. With his own *Mārtaṇḍavallabhā*.
RORI Cat. II 6765. 35ff. Copied by Moṭī in Saṃ. 1903 = A.D. 1846.
VVRI 4763. 15ff. Copied in Saṃ. 1903 = A.D. 1846.
Kathmandu (1960) 326 (III 441). 35ff. Copied on 9 kṛṣṇapakṣa of Śrāvaṇa in Saṃ. 1905 = *ca.* 23 August 1848.
RORI Cat. II 9037. 57ff. Copied in Saṃ. 1905 = A.D. 1848. With an anvaya in Old Rājasthānī.
Benares (1963) 37223. 118ff. Copied in Śaka 1771 = A.D. 1849. With his own *Mārtaṇḍavallabhā*.
Benares (1963) 36248. 18ff. Copied in Saṃ. 1909 = A.D. 1852.
GJRI 3199/411. Ff. 1–9 and 11–23. Copied in Saṃ. 1909 = A.D 1852. Incomplete.
Goṇḍal 299. 22ff. Copied on Tuesday 1 kṛṣṇapakṣa of Bhādrapada II in Saṃ. 1909 = 28 September 1852.
Ahmadnagar 309 (223/7). 125ff. Copied on 10 śuklapakṣa of Pauṣa in Śaka 1775 = *ca.* 8 January 1854. With a ṭīkā.
Calcutta University 954. 17ff. Copied in Saṃ. 1911 = A.D. 1854.
Goṇḍal 294. 159ff. Copied by Mayāśaṅkara Hari Śukla at Vāṅkānera on Thursday 4 śuklapakṣa of Māgha in Saṃ. 1913 = 29 January 1857. With his own *Mārtaṇḍavallabhā*.
Benares (1963) 36143. 37ff. Copied in Saṃ. 1914 = A.D. 1857.
PUL II 3814. 26ff. Copied in Saṃ. 1914 = A.D. 1857.
Goṇḍal 297. 25ff. Copied in Saṃ. 1915 = A.D. 1858.
RORI Cat. II 9204. 222ff. Copied in Saṃ. 1915 = A.D. 1858. With his own *Mārtaṇḍavallabhā*.
PUL II 3817. 43ff. Copied in Saṃ. 1920 = A.D. 1863.
Benares (1963) 36825. Ff. 19–37. Copied in Śaka 1787 = A.D. 1865. With a ṭīkā. Incomplete.
RORI Cat. III 13112. 26ff. Copied by Icchārāma Purohita at Rādhanapura in Saṃ. 1922 = A.D. 1865.
Goṇḍal 300. 115ff. Copied in Saṃ. 1925 = A.D. 1868. 2 copies. Incomplete.
PUL II 3815. 29ff. Copied in Saṃ. 1929 = A.D. 1872.
RORI Cat. III 11297. 35ff. (f. 1 missing). Copied in Saṃ. 1937 = A.D. 1880.
Nagpur 1626 (1249). 96ff. Copied in Śaka 1803 = A.D. 1881. From Nasik.
LDI 7127 (496). 123ff. (ff. 4–25 missing). Copied in Saṃ. 1957 = A.D. 1900. With his own *Mārtaṇḍavallabhā*. Incomplete.
Adyar Index 4783 =
Adyar Cat. 8 D 74. 64ff.
Adyar Cat. 11 C 38. 36ff. (ff. 1–7 missing). Incomplete (ends in prakaraṇa 4).
AHRS 74. No author mentioned.
AHRS 220. No author mentioned.
Alwar 1907. 3 copies. With his own *Mārtaṇḍavallabhā*.
Ānandāśrama 786. With a ṭīkā.
Ānandāśrama 1830.
Ānandāśrama 2463.
Ānandāśrama 2464. With a ṭīkā.
Ānandāśrama 3548.
Ānandāśrama 3843.
Ānandāśrama 5006.
Ānandāśrama 6855. With a ṭīkā.
Ānandāśrama 7387.
Ānandāśrama 8107. With a ṭīkā.
Anup 4988. 27ff. This is probably Bikaner 684. 26ff.
AS Bengal 2699 (G 10614). 60ff. With his own *Mārtaṇḍavallabhā*.
AS Bombay 319. 21ff. From Bhāu Dājī.
Baroda 1185. 128ff. With his own *Mārtaṇḍavallabhā*.
Baroda 1541. 62ff. With his own *Mārtaṇḍavallabhā*. Incomplete.
Baroda 1542. 195ff. Said to have been copied in Saṃ. 1798 = A.D. 1741 and in Śaka 1770 = A.D. 1848. With his own *Mārtaṇḍavallabhā*.
Baroda 1675. 134ff. With his own *Mārtaṇḍavallabhā*. Incomplete (through vāstuprayoga).
Baroda 2484. 54ff. With his own *Mārtaṇḍavallabhā*.
Baroda 3230. 82ff. With his own *Mārtaṇḍavallabhā*.
Baroda 3245. 21ff.
Baroda 13629. 130ff. With his own *Mārtaṇḍavallabhā*.
Benares (1963) 34562. Ff. 14 and 23. Incomplete.
Benares (1963) 34574. Ff. 15 and 22 and 1f. Incomplete. No author mentioned.
Benares (1963) 35054. 18ff. No author mentioned.
Benares (1963) 35144. F. 5. Incomplete. No author mentioned.
Benares (1963) 35379. Ff. 3–4. Incomplete.
Benares (1963) 35380 = Benares (1878) 146 = Benares (1869) XXXIV 2. 5ff. Incomplete.
Benares (1963) 35477. Ff. 2–55. With his own *Mārtaṇḍavallabhā*. Incomplete. This is probably Benares (1897–1901) 614. 53ff.
Benares (1963) 35683. 38ff. Incomplete. This may be Benares (1878) 6. 37ff. and Benares (1869) II 1. 36ff. Copied in Saṃ. 1850 = A.D. 1793.
Benares 1963) 36041. Ff. 2–3, 6–7, 9–11, and 38–52. With his own *Mārtaṇḍavallabhā* (called *Rājavallabhā*). Incomplete.
Benares (1963) 36136. 86ff. With a ṭīkā. Incomplete.
Benares (1963) 36139. 14ff. Incomplete.
Benares (1963) 36140. 33ff.
Benares (1963) 36142. Ff. 1–21 and 21b–99. With his own *Mārtaṇḍavallabhā*.
Benares (1963) 36329. 31ff. With a ṭīkā. Incomplete.
Benares (1963) 36565. Ff. 1 and 3–5. Incomplete.
Benares (1963) 36567. Ff. 1–9. Incomplete.

Benares (1963) 36568. Ff. 24–30. Incomplete.
Benares (1963) 36569. 16ff. With a ṭīkā. Incomplete.
Benares (1963) 36572. 16ff. Incomplete.
Benares (1963) 37218. 26ff. Incomplete.
Benares (1963) 37220. Ff. 257–267. With his own *Mārtaṇḍavallabhā*. Incomplete.
Benares (1963) 37221. 209ff. With his own *Mārtaṇḍavallabhā*.
Benares (1963) 37275. 1f. Incomplete.
Berlin 879 (Chambers 324). 145ff. With his own *Mārtaṇḍavallabhā*.
BM 490 (Add. 14,360a). 6ff. From Major T. B. Jervis.
BM 491 (Add. 14,364a). 83pp. With his own *Mārtaṇḍavallabhā*. From Major T. B. Jervis.
Bombay U Desai 1412. 4ff. Incomplete (to I 24).
BORI 967 of 1886/92. 11ff.
BORI 888 of 1891/95. 17ff. Incomplete.
BORI 552 of 1895/1902. 102ff. With his own *Mārtaṇḍavallabhā*.
BORI 189 of Vishrambag I. 23ff. Many ff. missing. No author mentioned.
Cambridge University Add. 2512 = Cambridge University 261.
Cambridge University Add. 2544 = Cambridge University 293.
CP, Hiralal 4263. Property of Govind Joshi of Jubbulpore.
CP, Hiralal 4264. Property of Munnālāl of Jubbulpore.
CP, Hiralal 4265. Property of Govindbhaṭṭ of Jubbulpore.
CP, Hiralal 4266. Property of Govindprasād Śāstrī of Jubbulpore.
CP, Hiralal 4267. Property of the Balātkār Gaṇ Jain Mandir at Kārañjā, Akolā.
CP, Hiralal 4268. Property of Śrīnivāsarāv of Ratanpur, Bilāspur.
CP, Hiralal 4269. Property of Ghanśyām Wāmanbhaṭṭ of Mangrulpīr, Akolā.
CP, Hiralal 4270. Property of Bālkrishṇa Śeṇḍe of Gourjhāmar, Saugor.
CP, Hiralal 4271. Property of Bāpu Kavimaṇḍan of Bāsim, Akolā.
CP, Hiralal 4272. Property of Paraśurām Anant of Bāsim, Akolā.
CP, Hiralal 4273. Property of Śrikrishṇa Manohar of Bāsim, Akolā.
CP, Hiralal 4274. Property of Bhagvān Hari of Bāsim, Akolā.
CP, Hiralal 4275. Property of Rāmchandra Bābāji of Akoṭ, Akolā.
CP, Hiralal 4276. Property of Rāgho Viśvanāth Śāstrī of Murtizāpur, Akolā.
CP, Hiralal 4277. Property of Bājirāv Śāstrī of Murtizāpur, Akolā.
CP, Hiralal 4278. Property of Tukārām Govind Pāthak of Yeodā, Amraotī.
CP, Hiralal 4279. Property of Krishṇarāv Pāthak of Śendurjanā, Amraotī.
CP, Hiralal 4280. Property of Vāsudev Mahādev Tāre of Pāthroṭ, Amraotī.
CP, Hiralal 4281. Property of Hari Nīlkaṇṭh Joshi of Valgaon, Amraotī.
CP, Hiralal 4282. Property of Janārdan Māruti of Kholāpur, Amraotī.
CP, Hiralal 4283. Property of Gopāl Nārāyaṇ of Bhātkulī, Amraotī.
CP, Hiralal 4284 and 4285. Property of the Bhonsalā Rājās of Nāgpur.
CP, Hiralal 4286. Property of Nārāyaṇ Purāni of Hardā, Hoshangābād.
CP, Hiralal 4287. Property of Govindarām Bhaṭṭ of Hardā, Hoshangābād.
CP, Hiralal 4288. Property of Keśavrāv of Khurai, Saugor.
CP, Hiralal 4289. Property of Pāṇḍu Tānā Bhaṭṭ of Dewalgaon Rājā, Bulḍānā.
CP, Hiralal 4292. With a ṭīkā. Property of Mādhavrāv of Damoh.
CP, Hiralal 4293. With a ṭīkā. Property of the Bhonsalā Rājās of Nāgpur.
CP, Hiralal 4294. With a ṭīkā. Property of Bājirāv Śāstrī of Murtizāpur, Akolā.
CP, Hiralal 4295. With a ṭīkā. Property of Govind Śāstrī of Mañgalā, Bilāspur.
CP, Ḳielhorn XXIII 118. 26ff. Property of Javāhara Śāstrī of Chāndā.
DC 231. Ff. 4–39. No author mentioned. From the Dīkṣit (A) Collection.
DC 394. Ff. 2–3. No author mentioned.
DC 395. Ff. 2–28. No author mentioned.
DC 8682. Ff. 1–57, 66–68, and 91–98. With his own *Mārtaṇḍavallabhā*. From the Shrotriya Collection.
DC 8756. 18ff. From the Shrotriya Collection.
DC (Gorhe) App. 286 and 287. Property of Śañkara Bālakṛṣṇa Lumpāṭhakī of Puṇatāmbe, Ahmadnagar.
DC (Gorhe) App. 288 and 289. Property of Gañgādhara Rāmakṛṣṇa Dharmādhikārī of Puṇatāmbe, Ahmadnagar.
GJRI 1054/166. 146ff. Maithilī. With his own *Mārtaṇḍavallabhā*.
GJRI 3200/412. 20ff. Maithilī.
GOML Madras R 6954. Ff. 3–27. Telugu. With his own *Mārtaṇḍavallabhā*. Incomplete (ends in prakaraṇa 3). Purchased in 1938/39 from C. V. Rajagopalan of Komaleswaranpet, Madras.
GVS 2889 (1771). 5ff. Incomplete. No author mentioned.
GVS 2890 (2396). 27ff. Incomplete.
GVS — (857). Ff. 8–117. No author mentioned.
GVS — (3797) 9ff. No author mentioned.
GVS — (3853). Ff. 1, 8–16, 18–38, and 63. No author mentioned.
GVS — (3862 A). Ff. 12–19.

GVS — (3871). Ff. 10–13 and 21–22. No author mentioned.
GVS — (4196). 5ff. No author mentioned.
GVS — (5698). Ff. 17–20. No author mentioned.
IO 3024 (2684a). 15ff. Telugu. From Colin Mackenzie.
IO 3025 (2460). 121ff. With his own *Mārtaṇḍavallabhā*. From Gaikawar.
Jaipur (II). 2ff.
Jaipur (II). 68ff. With his own *Mārtaṇḍavallabhā*.
Jammu and Kashmir 841. 17ff.
Kavīndrācārya 807. With a ṭīkā. No author mentioned.
Kerala 13886 (9511 A). 400 granthas. Grantha. No author mentioned.
Kerala 13887 (13980 B). 100 granthas. With a ṭīkā. Incomplete. No author mentioned.
Kerala 13888 (1514). 3000 granthas. With his own *Mārtaṇḍavallabhā*.
Kerala 13889 (2022). 2900 granthas. With his own *Mārtaṇḍavallabhā*. Incomplete.
Kerala 13890 (10218). 2750 granthas. With his own *Mārtaṇḍavallabhā*. Incomplete.
Kerala 13891 (14240 P). 550 granthas. With his own *Mārtaṇḍavallabhā*. Incomplete.
Kotah 268. 46pp. No author mentioned.
Kurukṣetra 819 (50115).
LDI 7124 (2512). 30ff. (ff. 1–13 missing). Incomplete.
LDI (LDC) 3389/2. Ff. 13–28.
Leipzig 1072. 19ff. With his own *Mārtaṇḍavallabhā*. Incomplete (to III 14).
Leipzig 1073. 33ff. With his own *Mārtaṇḍavallabhā*. Incomplete (to VI 16).
Leipzig 1074. 15ff.
Leningrad (1914) 298 (Ind. II 95). Ff. 1–8 and 10.
Mysore (1922) 1766. 89ff. With his own *Mārtaṇḍavallabhā*.
Mysore (1922) C 590. Ff. 10–60. With his own *Mārtaṇḍavallabhā*.
Nagpur 1621 (888). 30ff. With his own *Mārtaṇḍavallabhā*. From Nasik.
Nagpur 1628 (2414). Ff. 5–28. No author mentioned. From Nagpur.
Nagpur 1629 (2542). 37ff. From Nagpur.
N-W P I (1874) 91. 30ff. Property of Trilochana Jotishi of Benares.
Oppert I 6637. Property of Durbha Rāmaśāstrulu of Maḍḍi near Padmanābha, Vizagapatam.
Oppert II 209. Property of the Jāghīrdār of Āraṇi, North Arcot.
Oppert II 478. Property of Subrahmaṇyadīkṣitar of Cidambaram, South Arcot.
Oppert II 3020. Property of Śiṣṭla Sākṣayya of Vissampeṭa, Kṛṣṇa.
Osmania University 137/2. 63ff. With his own *Mārtaṇḍavallabhā*. Incomplete.
Osmania University B. 9/9. 23ff. Incomplete (prakaraṇas 1–7).
Osmania University B. 9/20. 28ff. With his own *Mārtaṇḍavallabhā*. Incomplete (prakaraṇas 1–4).
Oudh XXII (1890) VIII 11. 200pp. With his own *Mārtaṇḍavallabhā*. Property of Kedāranātha of Āgrā Zila.
Oxford 787 (Walker 210b). Ff. 79–103.
Paris BN 212 H (Sans. dév. 311). F. 102. With a ṭīkā. Incomplete. Acquired May 1842.
Poleman 4992 (Columbia, Smith Indic 89). Ff. 1–4, 7–10, 12–16, and 22.
Poleman 4994 (U Penn 689). 23ff.
Poleman 4997 (U Penn 1787). 52ff. With a ṭīkā.
Poleman 4999 (U Penn 1789). 3ff. Incomplete (pallīsaratha).
PUL II 3816. 15ff.
RORI Cat. I 616. 31ff. Copied by Ratnacanda at Māṇḍavī.
RORI Cat. I 3220. 25ff.
RORI Cat. II 4709. 78ff. With his own *Mārtaṇḍavallabhā*. Incomplete.
RORI Cat. II 4729. 121ff. With his own *Mārtaṇḍavallabhā*.
RORI Cat II 4758(1). Ff. 1–9.
RORI Cat. II 5525. 47ff. (f. 26 missing). With his own *Mārtaṇḍavallabhā*. Incomplete (prakaraṇa 4).
RORI Cat. II 9136. 98ff. With his own *Mārtaṇḍavallabhā*. Incomplete.
RORI Cat. III 11029(1). 12ff. Incomplete.
RORI Cat. III 11851. 20ff. Incomplete (ends in prakaraṇa 11).
RORI Cat. III 15355. 42ff. Copied by Śivalāla.
RORI Cat. III 16182. 21ff.
SOI 2374 = SOI Cat. I: 1378–2374. 128ff. With a ṭīkā.
SOI 4985 = SOI (List) 47.
SOI 5022.
SOI 5604.
SOI 5605 = SOI (List) 224. With a ṭīkā.
SOI 9565.
SOI 9905. With his own *Mārtaṇḍavallabhā*.
Tanjore D 11565 = Tanjore BL 4306. 9ff. Incomplete.
Tanjore D 11566 = Tanjore BL 4305. 37ff. Incomplete.
Tanjore D 11567 = Tanjore BL 11009. 42ff. Grantha. Incomplete.
Tanjore D 11568 = Tanjore TS 1007. No ff. given. Incomplete.
Tanjore D 11569 = Tanjore BL 4308. 6ff. Incomplete.
Tanjore D 11570 = Tanjore 15682. No ff. given. Incomplete.
VVRI 1209. 114ff. With a ṭīkā.
VVRI 6720. 70ff. With a ṭīkā. Incomplete.
WHMRL M. 2. f. No author mentioned.
WHMRL M. 3. c.
WHMRL M. 8. a.

The *Muhūrtamārtaṇḍa* has been often published:

at an unknown place in Saṃ. 1893 = A.D. 1836 (SOI Cat. II: 1084–3858);
at Benares in 1854 (IO 216 & 353);
with the *Mārtaṇḍavallabhā* at Mumbaī in Śaka 1783 = A.D. 1861 (BM and IO 24. D. 11 & 24);
with the *Mārtaṇḍavallabhā* at Puṇyagrāma in Śaka 1787 = A.D. 1865 (BM);
with the *Mārtaṇḍavallabhā* at Kāśī in Saṃ. 1926 = A.D. 1869 (BM);
with the *Mārtaṇḍavallabhā*, edited by Rāmacandra Śāstrī, Madras 1871 (BM);
with the *Mārtaṇḍavallabhā* at Lucknow in 1879 (BM);
with the *Mārtaṇḍavallabhā*, edited by Rāvajī Śrīdhara Gondhalekara, Mumbaī Śaka 1816 = A.D. 1894;
with the *Mārtaṇḍavallabhā* and a Marāṭhī translation by Viṣṇu Vāsudeva Śāstrin at Poona in 1897 (IO 1390), reprinted at Bombay in 1907 (NL Calcutta 180. Kb. 90), 2nd ed. [Bombay] 1917 (IO 13. K. 28);
with the Āndhraṭīkā of Nori Guruliṅga Śāstrī at Madras in 1901 (BM 14053. ccc. 38 and IO 1913);
with the *Mārtaṇḍavallabhā*, edited by Maṇirāma Śāstrī, Mumbaī Śaka 1826 = A.D. 1904;
with the Gujarātī translation of Girijāśaṅkara Chaganalāla Vyāsa at Tintoi, Ahmadabad in 1916 (BM 14055. d. 30 and IO San. C. 271);
with the Gurajātī translation of Someśvara Dvārakādāsa at Bombay in 1921 (IO San. D. 714);
with the *Mārtaṇḍavallabhā* and a Hindī ṭīkā, *Sudhā*, of Rāmateja Pāṇḍya, edited by Sītārāma Pāṇḍya, Benares 1938;
with the Saṃskṛta and Hindī ṭīkā, *Mārtaṇḍaprakāśikā*, of Kapileśvara Śāstrin as *KSS* 145, Benares 1947.

Verses 1–3 at the end are:

śrīmatkauśikapāvano haripadadvandvārpitātmā haris
tajjo ᵓnanta ilāsurārcitaguṇo nārāyaṇas tatsutaḥ/
khyātaṃ devagireḥ śivālayam udak tasmād udak ṭāpara-
grāmas tadvasatir muhūrtabhavanaṃ mārtaṇḍam akrākarot//
yaḥ ṣaṣṭyā yutaśatavṛttabaddham enaṃ
mārtaṇḍaṃ paṭhati naraḥ sa viśvapūjyaḥ/
bahvāyuḥsukhadhanaputramitrabhṛtyān
samprāpnoty avikaladhīś ca tīrthasiddhim//
tryaṅkendrapramite varṣe śālivāhanajanmataḥ/
kṛtas tapasi mārtaṇḍo ᵓyam alaṃ jayatūdgataḥ//

2. A *Laghumuhūrtamārtaṇḍa*, also composed in Saṃ. 1628 = A.D. 1571; this may be identical with the *Muhūrtamārtaṇḍa*. Manuscripts:

Osmania University 121/14. 17ff. Copied in A.D. 1834.
DC 7020. 14ff. Incomplete. No author mentioned.
Osmania University 121/13. 14ff. Incomplete.
RORI Cat. II 5772(3). Ff. 3–17.
SOI 162 = SOI Cat. I: 1374–162. 30ff.
SOI 2567 = SOI Cat. II: 1082–2567. 14ff.

3. A ṭīkā, the *Mārtaṇḍavallabhā*, on his own *Muhūrtamārtaṇḍa*, composed in Śaka 1494 = A.D. 1572; it is sometimes ascribed to his father, Ananta. See also Nīlakaṇṭha. Manuscripts:

Poleman 4998 (Harvard 391). Ff. 32–49, 51–88, and 90–97. Copied in Saṃ. 1699, Śaka 1564 = A.D. 1642. Incomplete.
Berlin 2230 (or. fol. 1491). Ff. 3–157. Copied on 4 kṛṣṇapakṣa of Śrāvaṇa in Saṃ. 1703 = *ca.* 20 July 1646.
LDI (LDC) 1291. 61ff. Copied in Saṃ. 1715 = A.D. 1658.
PL, Buhler IV E 355. 104ff. Copied in Saṃ. 1717 = A.D. 1660. Property of Bālakṛṣṇa Jośī of Ahmadābād. Buhler notes 2 other copies.
LDI 7128 (8934). 54ff. Copied in Saṃ. 1739 = A.D. 1682.
AS Bengal 2700 (G 8709). 90ff. Copied for Jñānānanda Guru in Saṃ. 1741 = A.D. 1684 from a manuscript copied by Devavandya Ghasṛṇeśa for Mādhava in Śaka 1557 = A.D. 1635.
Benares (1963) 36251. Ff. 1–17 and 17b–72. Copied in Saṃ. 1746 = A.D. 1689. Incomplete.
PL, Buhler IV E 354. 116ff. Copied in Saṃ. 1755 = A.D. 1698. Property of Bālambhaṭṭa of Surata. Buhler notes 6 other copies.
GVS 2894 (3120). 120ff. Copied on Sunday 3 kṛṣṇapakṣa of Āśvina in Saṃ. 1759 = 27 September 1702 Julian.
GVS 2895 (4153). 82ff. Copied on Thursday 13 kṛṣṇapakṣa of Vaiśākha I in Saṃ. 1784 = 6 April 1727 (?).
RORI Cat. II 4671. 82ff. Copied by Nṛsiṃhadeva Agravāla at Jayapura in Saṃ. 1789 = A.D. 1732.
Benares (1963) 35377. 79ff. Copied in Saṃ. 1799, Śaka 1664 = A.D. 1742.
DC 9422. 116ff. Copied in Śaka 1674 = A.D. 1752.
Benares (1963) 36564 = Benares (1878) 7 = Benares (1869) II 2. Ff. 1–57 and 59–136. Copied in Saṃ. 1813 = A.D. 1756.
BORI 118 of A 1879/80. 158ff. Copied in Śaka 1683 = A.D. 1761.
Benares (1963) 36252. 127ff. Copied in Saṃ. 1819 = A.D. 1762.
Benares (1963) 35218. 81ff. Copied in Saṃ. 1833 = A.D. 1776.
RORI Cat. II 6130. 56ff. Copied by Bālacanda at Gvāliyara in Saṃ. 1837 = A.D. 1780.
BORI 426 of A 1881/82. 129ff. Copied in Śaka 1703 = A.D. 1781.
Benares (1963) 34549. Ff. 3–25, 45–100, and 102–137. Copied in Saṃ. 1846 = A.D. 1789. Incomplete.
Mitra, Not. 1737. 120ff. Copied in Saṃ. 1847 = A.D. 1790. Property of Paṇḍita Kālīcaraṇa Upādhyāya of Lālagolā, Murshidābād Zilā.

AS Bombay 321. 192ff. Copied in Saṃ. 1849 = A.D. 1792.
RORI Cat. II 8886. 100ff. Copied by Tulasīrāma in Saṃ. 1849 = A.D. 1792.
Goṇḍal 295. 101ff. Copied by Jagannātha at Saradhāra on Saturday 9 śuklapakṣa of Vaiśākha in Saṃ. 1856 = 11 May 1799.
Nagpur 1633 (1772). 134ff. Copied in Śaka 1722 = A.D. 1800. From Nagpur.
Osmania University B. 9/19. 77ff. Copied in A.D. 1801.
RORI Cat. III 13906. 44ff. Copied by Bakhtāvaramalla Ṛṣi at Nāgaura in Saṃ. 1858 = A.D. 1801.
BORI 432 of 1895/98. 129ff. Copied in Saṃ. 1861 = A.D. 1804.
BORI 176 of A 1883/84. 104ff. Copied in Saṃ. 1866 = A.D. 1809.
BORI 502 of 1892/95. 55ff. Copied in Saṃ. 1866 = A.D. 1809. Ascribed to Ananta.
Goṇḍal 296. 109ff. Copied in Saṃ. 1871 = A.D. 1814.
Baroda 5754. 190ff. Copied in Saṃ. 1874 = A.D. 1817.
Bombay U 445. 195ff. Copied by Jagadīśa Dharmādhikāri of Nasirābāda on 11 kṛṣṇapakṣa of Śrāvaṇa in Saṃ. 1874, Śaka 1739 = *ca.* 7 August 1817.
LDI (LDC) 3193. 132ff. Copied in Saṃ. 1874 = A.D. 1817.
Bombay U 446. Ff. 31–142. Copied by Bābadeva Śarman, the son of Rāmacandra, the son of Govindabhaṭṭa Ṭhākura, on Sunday 9 śuklapakṣa of Māgha in Śaka 1741 = 22 January 1820. Incomplete.
Oudh XII (1880) VIII 6. 216pp. Copied in A.D. 1821. No author mentioned. Property of Jagannātha of Gauri, Unao Zila.
Benares (1963) 37222. 177ff. Copied in Saṃ. 1879, Śaka 1744 = A.D. 1822.
Benares (1963) 37046 = Benares (1903) 1312. Ff. 1–57, 57b–94, and 1–3. Copied in Saṃ. 1882, Śaka 1747 = A.D. 1825. With a *Candrasāraṇī*.
SOI 2372 = SOI Cat. I: 1377–2372. 155ff. Copied in Śaka 1750 = A.D. 1828.
LDI (LDC) 3192. 18ff. Copied in Saṃ. 1889 = A.D. 1832.
Leipzig 1071. 144ff. Copied in A.D. 1832.
Osmania University 137/5 A. 65ff. Copied in A.D. 1837.
Jammu and Kashmir 843. 80ff. Copied in Saṃ. 1895 = A.D. 1838.
Oudh XIII (1881) VIII 10. 200pp. Copied in A.D. 1840. Property of Dīna Dayāla of Rae Bareli Zila.
LDI (LDC) 2200. 164ff. Copied in Saṃ. 1903 = A.D. 1846.
Benares (1963) 37223. 118ff. Copied in Śaka 1771 = A.D. 1849.
Goṇḍal 294. 159ff. Copied by Mayāśaṅkara Hari Śukla at Vāṅkānera on Thursday 4 śuklapakṣa of Māgha in Saṃ. 1913 = 29 January 1857.
RORI Cat. II 9204. 222ff. Copied in Saṃ. 1915 = A.D. 1858.
Jammu and Kashmir 2937. 33ff. Copied in Saṃ. 1917 = A.D. 1860.
LDI 7127 (496). 123ff. (ff. 4–25 missing). Copied in Saṃ. 1957 = A.D. 1900. Incomplete.
Adyar Cat. 11 D 111. 232ff.
Adyar Cat. 34 J 72. 210ff. (f. 1 missing).
Alwar 1907. 3 copies.
Anup 4990. 112ff.
Anup 4991. 118ff. Incomplete.
AS Bengal 2699 (G 10614). 60ff.
Baroda 1185. 128ff.
Baroda 1541. 62ff. Incomplete.
Baroda 1542. 195ff. Said to have been copied in Saṃ. 1798 = A.D. 1741 and in Śaka 1770 = A.D. 1848.
Baroda 1675. 134ff. Incomplete (through vāstuprayoga).
Baroda 2484. 54ff.
Baroda 3230. 82ff.
Baroda 9487. 112ff.
Baroda 13629. 130ff.
Benares (1963) 35217. 45ff. Incomplete.
Benares (1963) 35315. 111ff. Incomplete. No author mentioned.
Benares (1963) 35477. Ff. 2–55. Incomplete. This is probably Benares (1897–1901) 614. 53ff.
Benares (1963) 36041. Ff. 2–3, 6–7, 9–11, and 38–52. (*Rājavallabhā*). Incomplete.
Benares (1963) 36138. 29ff. Incomplete.
Benares (1963) 36141. Ff. 2–104. Incomplete.
Benares (1963) 36142. Ff. 1–21 and 21b–99.
Benares (1963) 36250. Ff. 1–15 and 17–37. Incomplete.
Benares (1963) 37220. Ff. 257–267. Incomplete.
Benares (1963) 37221. 209ff.
Berlin 879 (Chambers 324). 145ff.
BM 491 (Add. 14,364a). 83pp. From Major T. B. Jervis.
BORI 718 of 1883/84. 60ff.
BORI 552 of 1895/1902. 102ff.
Calcutta Sanskrit College 104. Ff. 1–18 and 79–139.
Calcutta Sanskrit College 105. Ff. 4–78.
DC 2318. Ff. 28–38, 38b–230, and 5ff. No author mentioned.
DC 8677. Ff. 1–49 and 51–83. From the Shrotriya Collection.
DC 8682. Ff. 1–57, 66–68, and 91–98. From the Shrotriya Collection.
DC 8773. Ff. 103–108, 113–124, and 131–182. From the Shrotriya Collection.
GJRI 1054/166. 146ff. Maithilī.
GOML Madras R 6954. Ff. 3–27. Telugu. Incomplete (ends in prakaraṇa 3). Purchased in 1938/39 from C. V. Rajagopalan of Komaleswaranpet, Madras.
IO 3025 (2460). 121ff. From Gaikawar.

Jaipur (II). 68ff.
Kathmandu (1960) 328 (I 1201). 137ff. Nevārī. Incomplete
Kathmandu (1960) 329 (III 441). 156ff. Incomplete.
Kerala 13888 (1514). 3000 granthas.
Kerala 13889 (2022). 2900 granthas. Incomplete.
Kerala 13890 (10218). 2750 granthas. Incomplete.
Kerala 13891 (14240 P). 550 granthas. Incomplete.
Kurukṣetra 821 (19683).
Leipzig 1072. 19ff. Incomplete (to III 14).
Leipzig 1073. 33ff. Incomplete (to VI 16).
Mysore (1922) 1766. 89ff.
Mysore (1922) C 590. Ff. 10–60.
Nagpur 1621 (888). 30ff. From Nasik.
N-W P I (1874) 1. 290ff. Said to have been copied in Saṃ. 1490 = A.D. 1433. Property of Trilochana Jotishi of Benares.
N-W P II (1877) B 5. 11ff. Incomplete (gocara). Property of Rāma Prasāda of Benares.
N-W P II (1877) B 14. 16ff. Property of Bholā Datta of Benares.
N-W P II (1877) B 26. 3ff. Incomplete (gṛha). Property of Rāma Prasāda of Benares.
N-W P II (1877) B 71. 18ff. Incomplete (vivāha). Property of Bholā Datta of Benares.
N-W P II (1877) B 86. 50ff. Property of Vāgīśvarī Datta of Benares.
N-W P II (1877) B 94. 10ff. Incomplete (yātrā). Property of Rāma Prasāda of Benares.
N-W P II (1877) B 99. 5ff. Incomplete (gṛha). Property of Rāma Prasāda of Benares.
N-W P II (1878) B 20. 112ff. Property of Mākhanji of Mathurā.
Osmania University 137/2. 63ff. Incomplete.
Osmania University B. 9/20. 28ff. Incomplete (prakaraṇas 1–4).
Oudh XVIII (1885) VIII 1. 212pp. Property of Nandarāma of Gonda Zila.
Oudh XXII (1890) VIII 11. 200pp. Property of Kedāranātha of Āgrā Zila.
Oxford CS d. 763(i). 149ff.
PUL II 3818. 142ff. Incomplete (to prakaraṇa 4).
PUL II 3819. 41ff. Incomplete.
PUL II 3820. 48ff. Incomplete (to prakaraṇa 4).
Rajputana, p. 7. Ascribed to Ananta. From Ujjain.
RORI Cat. II 4709. 78ff. Incomplete.
RORI Cat. II 4729. 121ff.
RORI Cat. II 5525. 47ff. (f. 26 missing). Incomplete (prakaraṇa 4).
RORI Cat. II 9136. 98ff. Incomplete.
SOI 9905.
Tanjore D 11571 = Tanjore BL 4807. 24ff. Incomplete (prakaraṇa 8). No author mentioned.

The editions of the *Mārtaṇḍavallabhā* have been listed above with those of the *Muhūrtamārtaṇḍa*. Verses 1–3 at the end are:

āsīt sāsamaṇūranāmanagare śrīkauśikasyānvaye
ᵓnanto vājasaneyipūjyacaraṇo mādhyandinīyāgraṇīḥ/
kṛṣṇas tattanayaḥ śrutismṛtividām agre sarejyo haris
tatputraḥ śrutivit tadātmajavaro ᵓnanto ᵓgnihotrī guruḥ//
tatputras tadanugrahāttadhiṣaṇo nārāyaṇaṣ ṭāpara-
grāme śiṣyagaṇecchayā nijakṛtagranthasya ṭīkāṃ sphuṭām/
cakre ᵓsyāṃ kṛpayā paropakṛtaye śodhyaṃ duruktaṃ budhair
mādṛkṣasya vilokya dhārṣṭyam api te kupyanti no sajjanāḥ//
sukhanidhipuruṣārthakṣmāsamābhiḥ samābhiḥ
parimitaśakakāle jātamārtaṇḍaṭīkām/
likhati paṭhati vipraḥ so ᵓtra bhūyād dharitryāṃ
sukhanidhipuruṣārthakṣmāsamo vā kṣamāvān//

NĀRĀYAṆA (*fl. ca.* 1635/1678)

Cintāmaṇi of the Devarātagotra, a resident of Dadhigrāma on the Payoṣṇī in Vidarbha, had 5 sons, of whom the oldest was Rāma; Rāma had 2 sons by Videhaputrī, Trimalla and Gopirāja; Trimalla's son was Ballāla; Ballāla, who married Goji, had 5 sons, of whom 3 were Rāma, Kṛṣṇa (*fl. ca.* 1600/1625), and Govinda; Govinda's son was Nārāyaṇa, who, like his uncle, worked in Kāśī, studying under Muniśvara Viśvarūpa (b. 1603). See S. Dvivedin [1892] 85 and S. B. Dikshit [1896] 284. He wrote the following three commentaries.

1. An udāhṛti on the *Grahalāghava* of Gaṇeśa (b. 1507). Manuscripts:

AS Bengal 6859 (G 4292). 44ff. (f. 1 missing). Copied on Thursday 30 (read 13) kṛṣṇapakṣa of Āṣāḍha in Saṃ. 1692, Śaka 1558 = 2 July 1635.
IM Calcutta 9306. See NCC, vol 6, p. 260.
PL, Buhler IV E 75. 33ff. Copied in Saṃ. 1904 = A.D. 1847. (*Grahalāghava*). Property of Motīrāma of Dhrāṅgadhrā.
LDI 6730 (1178). 84ff. Copied for Uttamarṣi in Saṃ. 1917 = A.D. 1860.
Benares (1963) 37200. Ff. 1–59 and 1f. Incomplete.
WHMRL D. 72.

The last verse is:

govindadaivajñasutena kāśyām
udāhṛtiḥ khecaralāghavasya/
nārāyaṇenālpamanīṣituṣṭyai
kṛtā dadhigrāmanivāsinā hi//

2. A ṭīkā, sometimes entitled *Jātakakaustubha*, on the *Jātakapaddhati* of Keśava (*fl.* 1496/1507), composed in Śaka 1600 = A.D. 1678. Manuscripts:

Benares (1963) 35065. 38ff. Copied in Saṃ. 1769 = A.D. 1712. Alleged to be accompanied by the vyākhyā of Govinda.
VVRI 2553. 59ff. Copied in Saṃ. 1828 = A.D. 1771. Incomplete.

SOI 3339 = SOI Cat. II: 989–3339. 6ff. Copied in Saṃ. 1842, Śaka 1707 = A.D. 1785.
Anup 4620. 26ff. Copied by Kuṃjā in Saṃ. (read Śaka) 1710 = A.D. 1788 (?). Formerly property of Hariścaraṇa.
Benares (1963) 36219. Ff. 47–56, 55b–56b, and 59–60. Copied in Saṃ. 1852 = A.D. 1795. Incomplete.
Osmania University B. VIII/9. 137ff. Copied in A.D. 1809.
Anup 4619. 44ff. Incomplete.
Benares (1963) 36105. 46ff. Incomplete.
Bombay U Desai 1359. Ff. 37–86. Incomplete (begins with verse 14).
Mysore (1911 + 1922) 2301. Ff. 133–202. (*Jātakakaustubha*).
Oudh XX (1888) VIII 130. 128pp. Property of Paṇḍita Pratāpa Nārāyaṇa of Allahabad Zila.
Oudh XXI (1889) VIII 3. 128pp. Property of Paṇḍita Vindhyeśvarī Prasāda of Gonda Zila.
VVRI 2653. 89ff. Incomplete.

Verses 1–5 and 8–9 at the end are:

abhūd dvijāgryo dadhiśabdapūrva-
grāme ⟨payoṣṇī⟩vikaṭe ᵓtiramye/
cintāmaṇir daivavidambujārkaḥ
śrīdevarātānvayaratnabhūtaḥ//
pañcābhavan tattanayā guṇāḍhyāḥ
pārthā ivaiṣāṃ prathamo hi rāmaḥ/
videhaputryāṃ tanayāv abhūtāṃ
rāmāt trimallābhidhagopirājau//
trimallasūnur gaṇakābjasūryo
ballālasaṃjñaḥ śivabhaktiyuktaḥ/
pañcātmajās tajjanitā hi teṣāṃ
jyeṣṭhas tu rāmo varajaḥ sa kṛṣṇaḥ//
yenākāri suvāsanaṃ suruciraṃ vyākhyānam arkodite
bīje śrīpatijātakasya vivṛtiḥ sodāhṛtir nirmalā/
jyotiḥśāstramahārṇavasya culukenāgastyavat
prāśanaṃ
prāptā yāvanasārvabhaumavaśato bhūtis tathā
gauravam//
govindasaṃjño gaṇako variṣṭhaḥ
kṛṣṇānujas tattanayas tv akārṣīt/
nārāyaṇaḥ keśavajātakādhva-
vyākhyāṃ saduddeśavicāraramyām//
pakṣonasarvayuk svarganighno bhāntraḥ
śakonmitiḥ/ (?)
rudronasarvayuk (?) khābhrarasacandre
samāyutāḥ//
tādṛkśakādau govindasūnunā nirmitāmalā/
vyākhyoddeśavicārāḍhyā keśavīyajanuḥpathaḥ//

3. A ṭīkā on the *Varṣapaddhati* or *Tājikapaddhati* of Keśava (*fl.* 1496/1507); probably a mistake for the *Jātakapaddhatiṭīkā*. Manuscript:

Oudh V (1875) VIII 15. 60pp. Copied in A.D. 1864. Property of Śrīkṛṣṇa of Ayodhyā.

NĀRĀYAṆA VANDYAGHAṬĪYA (*fl.* 1681)

A resident of Khanākula Kṛṣṇanagara in the Hugli District of Bengal, Nārāyaṇa composed a *Smṛtisarvasva* or *Smṛtitattva* which follows Raghunandana (*fl.* 1520/1570). In it he mentions Śaka 1603 = A.D. 1681. Manuscripts:

AS Bengal 2097 (G 3959). 134ff. Bengālī. Copied by Utsavānandadeva Śarman. Property of Kṛṣṇadeva Śarman on 23 Śrāvaṇa of Śaka 1740, Sāl. San. 1225 = *ca.* 23 August 1818. On another leaf is recorded the birth of the first son of Śrīvaṃśi Caṭṭopādhyāya at Daśadaṇḍa on Tuesday 29 Āṣāḍha of Śaka 1742 = 8 August 1820.
Śāstrī, Not. 1900. 417. 194ff. Bengālī. Copied in Śaka 1754 = A.D. 1832. Property of Paṇḍita Rāmānuja Bhaṭṭācārya of Viṣṇupura, Vākuḍā.
AS Bengal 2098 (G 5020). 102ff. Bengālī.
IO 1487 (1196). 132ff. Bengālī. From H. T. Colebrooke.

The first verse is:

śrīrāmaṃ jagatām īśaṃ praṇamya tasya tuṣṭaye/
tanoti smṛtisarvasvaṃ śrīmannārāyaṇaḥ sudhīḥ//

The colophon begins: iti vandyaghaṭīyaśrīnārāyaṇadevaśarmaṇā.

NĀRĀYAṆA SĀMUDRIKA (*fl. ca.* 1725)

The son of Mādhava Śrīgāṃvakara (or Śrīgrāmakara) (*fl. ca.* 1700) of the Kaśyapagotra and the younger brother of Dādābhāi (*fl.* 1719), Nārāyaṇa, a Cittapāvana Brāhmaṇa, composed: a *Horāsārasudhānidhi;* a vṛtti, *Daivajñasantoṣiṇī,* on the *Manuṣyajātaka* of Samarasiṃha (*fl.* 1274); a *Gaṇakapriyā;* a *Svarasāgara;* and a *Tājikasārasudhānidhi.* These are listed in the last verse of his *Tājikasārasudhānidhi:*

horāsārasudhānidhir viracitaḥ pūrvaṃ mayā jātake
vyākhyā vai narajātakasya racitā daivajñasantoṣiṇī/
praśne vai gaṇakapriyā nigaditā yuddhādisiddhyai
tathā
prokto hi svarasāgaras tad anu tārtīye sudhāyā
nidhiḥ//

See S. B. Dikshit [1896] 292 and S. L. Katre [1942b]. The following of these 5 works survive.

1. The *Horāsārasudhānidhi.* Manuscripts:

Ānandāśrama 1339. (*Jātakasudhānidhi*). (?)
AS Bengal 7375 (G 10404). 17ff. Incomplete (ends in dīrghāyuryoga).
Benares (1963) 34378. 22ff. Incomplete (pañcamabhāvavicāra).
Benares (1963) 34379. 91ff. Incomplete.
Benares (1963) 34380. 73ff. Incomplete (ends with caturthabhāva).

Benares (1963) 34528. 104ff.
Benares (1963) 35829 = Benares (1905) 1435. Ff. 94–121. Incomplete (rājayoga to ariṣṭa).
Benares (1963) 36863. 29ff. Incomplete. No author mentioned.
Bombay U Desai 1454. 24ff. Incomplete (strījātaka).
Bombay U Desai 1458. 249ff., 252ff., 16ff., and 2ff.
Jammu and Kashmir 3077. 72ff. Incomplete.

Verses 2–3 are:

śrīgrāmodbhavakaśyapānvayanidhiḥ sāmudrikajñaḥ sudhīḥ
śrīmān mādhavasaṃjñako dvijavaro vighneśasevārataḥ/
dādābhāīti tasmād ajani guṇagaṇaiḥ pūjitas tattvavettā
siddhāntānāṃ ca kartā munijanaviditas tatsamaḥ ko ᵓpi nānyaḥ//
tasmāl labdhavarānujo ᵓlpamatimān nārāyaṇo ᵓhaṃ bruve
horāsārasudhānidhiṃ gaṇitavittuṣṭyai camatkārikam/
śrīgargādimatād viśiṣṭam akhilaṃ saṅgṛhya sāraṃ paraṃ
yasya jñānabalena mokṣapadavīṃ prāpnoti niḥsaṃśayam//

2. The ṭīkā, *Daivajñasantoṣiṇī* or *Karmaprakāsikāvṛtti*, on the *Manuṣyajātaka* of Samarasiṃha (*fl.* 1274). Manuscripts:

VVRI 2592. 33ff. Copied in Saṃ. 1835 = A.D. 1778. Incomplete.
Bombay U 419. 65ff. (f. 28 missing). Copied in Saṃ. 1870 = A.D. 1813.
PUL II 3593. 18ff. Copied in Saṃ. 1918 = A.D. 1861. Incomplete (niṣekādhyāya).
AS Bengal 6989 (G 267) = Mitra, Not. 1524. 32ff. Copied in Saṃ. 1931 = A.D. 1874. Incomplete (ends in adhyāya 20). No author mentioned.
VVRI 4612. 35ff. Copied in Saṃ. 1954 = A.D. 1897.
Baroda 13972. 37ff. Copied in A.D. 1940.
Alwar 1894.
AS Bengal 6990 (G 5514) 67ff. No author mentioned.
Baroda 11308. Ff. 7–23. Incomplete.
Baroda 12300. 98ff.
Benares (1963) 34914. 42ff.
Benares (1963) 37065. 72ff. Incomplete (ends in adhyāya 20).
Bombay U 420. 58ff. Incomplete (ends in adhyāya 15).
Bombay U Desai 1349. Ff. 1–41 (also numbered ff. 31–71). No author mentioned.
BORI 844 of 1887/91. 54ff.
Jammu and Kashmir 828. 56ff.
Jammu and Kashmir 3068. 31ff. Incomplete.
Jammu and Kashmir 3995. 41ff.
Poleman 4989 (U Penn 1842). 10ff. Incomplete.
Poleman 4990 (U Penn 1844). 24ff. Incomplete.
Rajputana, p. 30. From Jaisalmer.
SOI 6063.
SOI 9521.
WHMRL Q. 23. j.
WHMRL Q. 23. o.

The colophon begins: iti sāmudrikopanāmakanārāyaṇakṛta.

3. The *Svarasāgara*. Manuscript:

Bombay U Desai 1512. 167ff.

The last 4 verses are identical (save presumably for the very last) with the last 4 verses of the *Tājikasārasudhānidhi*.

4. The *Tājikasārasudhānidhi* in a gaṇitatantra (3 adhyāyas) and a varṣatantra (5 adhyāyas). Manuscripts:

Bombay U Desai 1374 and 1375. Ff. 1–42 (gaṇitatantra) and ff. 43–110 (varṣatantra). Copied in Saṃ. 1786 = A.D. 1729.
Oudh XII (1880) VIII 4. 214pp. Copied in A.D. 1812. Property of Jagannātha of Gauri, Unao Zila.
CP, Kielhorn XXIII 55. 77ff. Copied in Śaka 1738 = A.D. 1816. Property of Lakṣmaṇa Śāstri of Śāgar.
SOI 6040 = SOI (List) 394. Copied in Saṃ. 1873 = A.D. 1816. Is this identical with the previous manuscript?
AS Bengal 7114 (G 2930). 2ff. Copied by Jayakṛṣṇa Miśra on Saturday 1 śuklapakṣa of Caitra in Saṃ. 1880 = 12 April 1823 (?). Incomplete (adhyāya 3).
Oxford 784 (Wilson 428). 79ff. Copied in A.D. 1831.
VVRI 2350. 23ff. Copied in Saṃ. 1900 = A.D. 1843.
RORI (Jaipur) IV 68. 81ff. Copied in Saṃ. 1914 = A.D. 1857.
Benares (1963) 37036. 144ff. Copied in Saṃ. 1928 = A.D. 1871.
Alwar 1800 (*Tājikadivākara*).
Alwar 1808.
AS Bengal 7113 (G 10206). 42ff. Incomplete.
Benares (1963) 35435. Ff. 1–2, 1f., and ff. 3–23. Incomplete.
Benares (1963) 37035. 11ff. Incomplete.
Benares (1963) 37187. 21ff. Incomplete.
BORI 521 of 1895/1902. 122ff.
CP, Hiralal 1991. Property of Rāmchandra Bābāji of Akoṭ, Akolā.
Mithila 136. 79ff. Maithilī. Incomplete (ends in adhyāya 6). Property of Paṇḍita Dharmadatta Miśra of Babhangama, Supaul, Bhagalpur.
N-W P I (1874) 78. 56ff. Property of Jagannātha Śukla of Benares.
N-W P I (1874) 81. 110ff. Property of Jagannātha Jotishi of Benares.
N-W P II (1877) A 10. 18ff. Property of Chaṇḍī Datta of Benares.
Oudh (1877–1878) VIII 18. 128pp. Property of Kṛṣṇa Datta of Sitapur Zila.

RORI Cat. I 2932. 49ff. Incomplete.
RORI Cat. III 16811. 201ff. (ff. 1–6 missing).
SOI 9581.

Verses 23–25 at the end are:

āsīt kāruṇyavārāṃ nidhir avanipatiprakhyamūrdhā maṇīnāṃ
vṛndair nīrājitāṅghrir jalanidhivacasāṃ pālane yaḥ sarasvān/
siddhāntānāṃ śaraṇyaḥ phaṇipatilapitaṃ yasya kaṇṭhe vyaraṃsīd
bhūdevo mādhavākhyaḥ paśupatinagare śrīśapādābjasevī//
tasmāj jātau sutau dvau pravaramatiyutau sarvavidyānidhānau
jyeṣṭho dādākhyabhaṭṭaḥ sakalavasumatīmaṇḍalakhyātanāmā/
jyotiḥśāstre vivasvān atulaguṇagaṇaḥ sūryasiddhāntaṭīkāṃ
yo ᵓkārṣīd adbhutārthāṃ smaraharacaraṇāsaktacetā nitāntam//
putras tasmāt kaniṣṭho jalanidhivacasāṃ mārmikaḥ satyavādī
śrīmān nārāyaṇākhyo niravadhikaruṇāpūrṇacittāntarātmā/
śiṣyaiḥ saṃprārthito ᵓhaṃ pravaranijadhiyā tājakagrantham enam
ākārṣaṃ tena tuṣṭo gajavadanavibhūr bhāratīprāṇanāthaḥ//

NĀRĀYAṆA BHAṬṬA (*fl.* 1758)

Author of a ṭīkā in Old Rājasthānī, the *Camatkāracintāmaṇi*, on the *Jātakasāra*. Nārāyaṇa copied RORI 6391 himself according to the catalog. Manuscripts:

RORI Cat. II 6391. 11ff. Copied at Kṛṣṇagaḍha on Friday 6 (read 9) Phālguna in Saṃ. 1814 = 17 March 1758.
RORI Cat. II 6393. 17ff. Copied on Tuesday 1 śuklapakṣa of Bhādrapada in Saṃ. 1827, Śaka 1692 = 21 August 1770.

The last verse is that of the *Camatkāracintāmaṇi* of Nārāyaṇa Bhaṭṭa.

camatkāracintāmaṇau yat khagānāṃ
phalaṃ kīrtitaṃ bhaṭṭanārāyaṇena/
paṭhed yo dvijas tasya rājñāṃ sabhāyāṃ
samakṣaṃ pravaktuṃ na cānye samarthāḥ//

The colophon begins: iti śrījātakasāragranthe vidvannārāyaṇakṛtacamatkāracintāmaṇibhāṣāṭīkā.

ĀRYAN NĀRĀYAṆAN MŪSSATU (1842/1902)

A resident of Vayaskara near Kottayam in Kerala and a member of the Plāntol family of Āyurvedic physicians, Nārāyaṇan wrote a *Nakṣatravṛttāvali* in 27 verses simultaneously giving lunar positions and praising Viśākham Tirunāḷ, Mahārāja of Travancore from 1880 to 1885. See K. K. Raja [1958] 268 and S. V. Iyer [A3. 1971] 32–33.

NĀRĀYAṆA BHAṬṬA (*fl.* 1893)

A Kāśmīrī Paṇḍita and the brother of Paṇḍita Sahaja Bhaṭṭa, Nārāyaṇa wrote a pañcāṅga for Laukika Saṃvat 4969 = A.D. 1893, the *Nakṣatrapattrikā Kāśmīrikī*. Manuscript:

IIL Oxford Stein 307. 14ff. Śāradā.

NĀRĀYAṆA (*fl.* 1905)

Author of a Hindī version, *Subodhinī*, of the *Daivajñavallabhā* of Śrīpati (?), published at Bambaī in Saṃ. 1962 = A.D. 1905 (BM 14053. ccc. 53), reprinted Bombay 1915–1916 (IO 22. E. 2).

NĀRĀYAṆA CINTĀMAṆI PURANDARE VASAĪKAR (*fl.* 1913/1914)

Author of a pañcāṅga in Saṃskṛta and Marāṭhī for Śaka 1835 = A.D. 1913, published at Bombay in 1913 (BM 14096. a. 3. (2)), and of another for Śaka 1836 = A.D. 1914, published at Bombay in 1914 (BM 14096. a. 3. (3)).

NĀRĀYAṆACANDRA BHAṬṬĀCĀRYA JYOTIRBHŪṢAṆA (*fl.* 1897)

Author of a *Horāvijñānarahasya* = *Jyotiṣakalpavṛkṣa*, published with a Bengālī translation, Calcutta 1897 (NL Calcutta 180. Kb. 89. 2), 2nd ed., Calcutta 1912 (IO 26. F. 38 and NL Calcutta 180. Kb. 90. 2).

NĀRĀYAṆADĀSA

Author of a *Jñānasvarodaya*. Manuscripts:

CP, Hiralal 1849. Property of Śivaśaṅkarlāl of Murgākherā, Narsinghpur.
CP, Hiralal 1850. Property of Rāmnārāyaṇ of Mohāsā, Hoshangābād.

NĀRĀYAṆADĀSA SIDDHA GOSVĀMIN (*fl. ca.* 1525?)

The son of Nayajā and Brahmadāsa, the pupil of Harijī Śarman, and probably a follower of Caitanya (1486/1533), Nārāyaṇadāsa, a Kāyastha, wrote an astrological work variously called *Praśnavaiṣṇava*, *Praśnārṇavaplava*, and *Vaiṣṇavaśāstra*, in 15 adhyāyas. Manuscripts:

Benares (1963) 36765 = Benares (1878) 49 = Benares (1869) XI 2. Ff. 1–8 and 12–42. Copied in Saṃ. 1600 = A.D. 1543. Incomplete.
LDI 7228 (7303). 16ff. (f. 1 missing). Copied by Dāmodara at Jvālāpura on Saturday 2 śuklapakṣa of Mārgaśīrṣa in Saṃ. 1662 = 2 November 1605 Julian. Incomplete.

ABSP 58. 64ff. (f. 33 missing). Copied on 13 śuklapakṣa of Āṣāḍha in Saṃ. 1670 = *ca.* 19 June 1613.
RORI Cat. III 10107. 52ff. Copied by Maheśa Jośī, the son of Śyodāsa, in Saṃ. 1687 = A.D. 1630.
Tanjore D 11516 = Tanjore BL 4313. 56ff. Copied by Śrotriya at the Madanadahanageha in Śaka 1563 = A.D. 1641.
VVRI 1695. 61ff. Copied in Saṃ. 1703 = A.D. 1646.
Bombay U 521. 77ff. Copied by Rāma on 8 kṛṣṇapakṣa of Bhādrapada in Saṃ. 1715 (but 1765 in Devanāgarī) = *ca.* 9 September 1658.
RORI Cat. III 16908. 27ff. Copied in Saṃ. 1715 = A.D. 1658.
LDI (LDC) 696. 55ff. Copied in Saṃ. 1717 = A.D. 1660.
Anup 4889. 49ff. Copied by Mathena Jośī in Saṃ. 1719 = A.D. 1662. Formerly property of Anūpasiṃha (*fl.* 1674/1698).
Benares (1963) 36763 = Benares (1878) 48 = Benares (1869) XI 1. 69ff. Copied in Saṃ. 1748 = A.D. 1691.
Anup 4887. 61ff. Copied by Yati Khetasīha at Gaḍha Ādhivanī in Saṃ. 1750 = A.D. 1693 during the reign of Anūpasiṃha (1674/1698).
Benares (1963) 37208. Ff. 1–68 and 10ff. Copied in Saṃ. 1756 = A.D. 1699, Saṃ. 1757 = A.D. 1700, and Śaka 1522 = A.D. 1600 (read Śaka 1622 = A.D. 1700). With the *Samarasāra* of Rāmacandra and a bhāvaphala from a *Hillāja*.
Bombay, Kielhorn 11. 60ff. Copied in A.D. 1714.
PUL II 3666. Ff. 8–34. Copied in Saṃ. 1773 = A.D. 1716.
Benares (1869) XXXV 3. 10ff. Copied in Saṃ. 1785 = A.D. 1728.
Goṇḍal 199. 61ff. Copied in Saṃ. 1798 = A.D. 1741.
Oxford 1555 (Sansk. d. 208) = Hultzsch 331. 73ff. Copied in Saṃ. 1799 = A.D. 1742.
RORI Cat. I 1156. 23ff. Copied by Mayārāma at Jayapura in Saṃ. 1804 = A.D. 1747.
Mithila 192. 22ff. Maithilī. Copied by Manabodha Śarman at Kāśī on Sunday 15 śuklapakṣa of Śrāvaṇa in Śaka 1672 = 5 August 1750 Julian. (*Praśnabhairava*). Property of Paṇḍita Sādhu Jhā of Yamathari, Jhanjharpur, Darbhanga.
Benares (1963) 34414. 33ff. Copied in Saṃ. 1825 (read 1828), Śaka 1693 = A.D. 1771.
RORI Cat. I 3044. 32ff. Copied by Jānakīdāsa of Kalyāṇapurī in Saṃ. 1833 = A.D. 1776.
Benares (1963) 36764. 61ff. Copied in Saṃ. 1834, Śaka 1699 = A.D. 1777.
BORI 941 of 1886/92. 31ff. Copied in Saṃ. 1835 = A.D. 1778.
RORI Cat. III 11029(4). 28ff. Copied in Saṃ. 1836 = A.D. 1779. Incomplete.
AS Bengal 7170 (G 9472). Copied by Sukhānanda on 30 Kārttika in Saṃ. 1838, Śaka 1708 (read 1703) = *ca.* 15 November 1781.
GJRI 991/103. Ff. 1–46, 49–52, 55–57, 60–62, and 65–67. Copied in Saṃ. 1841 = A.D. 1784. Incomplete.
Goṇḍal 200. 62ff. Copied on Saturday 2 kṛṣṇapakṣa of Phālguna in Saṃ. 1848 = 10 March 1792.
Goṇḍal 201. 47ff. Copied at Doṣpura on 3 kṛṣṇapakṣa of Vaiśākha in Saṃ. 1855 = *ca.* 3 May 1798.
CP, Kielhorn XXIII 86. 71ff. Copied in Saṃ. 1856 = A.D. 1799. Property of Govindarāma Bhaḍajī of Sāgar.
BORI 348 of 1880/81. 52ff. Copied in Saṃ. 1860 = A.D. 1803.
BORI 213A of 1883/84. No ff. given. Copied in Saṃ. 1879 = A.D. 1822. From Gujarāt.
Oudh XX (1888) VIII 19. 94pp. Copied in A.D. 1822. Property of Paṇḍita Pratāpa Nārāyaṇa of Allahabad Zila.
Benares (1963) 35827 = Benares (1903) 1288. 34ff. Copied in Saṃ. 1885 = A.D. 1828.
RORI Cat. I 3704. 25ff. Copied by Gumāna Muni in Saṃ. 1886 = A.D. 1829.
Benares (1963) 34413. 47ff. Copied in Saṃ. 1888, Śaka 1753 = A.D. 1831.
WHMRL G. 104. a. Ff. 1–19, 21, and 21b–23. Copied by Śraddha Arṣi, the pupil of Rāmakṛṣṇa Ṛṣi, the pupil of Pūjya Rāmasuṣa Ṛṣi, at Ṣarḍanagara on Thursday 5 kṛṣṇapakṣa of Vaiśākha in Saṃ. 1896 = 2 May 1839.
RORI Cat. II 4719. 47ff. (ff. 13–15 missing). Copied by Kevalacanda Gokulajī at Bañkāpurī in Saṃ. 1915 = A.D. 1858.
Bombay U Desai 1483. 59ff. Copied in Śaka 1786 = A.D. 1864.
Benares (1963) 34932. 52ff. Copied in Saṃ. 1923 = A.D. 1866.
AS Bengal 7173 (G 5546) A. 34ff. Copied by Devīdāsa near the Prahlādaghāṭa in Kāśī on Tuesday 8 kṛṣṇapakṣa of Jyeṣṭha in Saṃ. 1929 = 28 May (?) 1872.
Kathmandu (1960) 418 (III 576). 41ff. Copied on 7 śuklapakṣa of Vaiśākha in Saṃ. 1957 = *ca.* 5 May 1900.
ABSP 79. 55ff.
Alwar 1859.
Ānandāśrama 2306.
Ānandāśrama 3545.
Ānandāśrama 7911.
Anup 4890 = Bikaner 706. 39ff.
AS Bengal 7171 (G 55) = Mitra, Not. 784. 24ff. Incomplete.
AS Bengal 7172 (G 1281). 54ff. Bought by Yajñeśvara Bhaṭṭa Someśvara; formerly property of the son of Kṛṣṇa.
Baroda 3195. 35ff.
Baroda 9626. 26ff. Incomplete.
Benares (1963) 34415. Ff. 1–3 and 5–40. Incomplete.
Benares (1963) 34945. 59ff.
Benares (1963) 34532. 54ff.
Benares (1963) 36413. Ff. 2–13. Incomplete.

Benares (1963) 36414. 45ff.
Berlin 880 (Chambers 582). 34ff.
Bharatpur S 16.
Bombay U 522. 14ff. Incomplete (ends at IV 10).
BORI 442 of A 1881/82. 56ff.
BORI 340 of 1882/83. 21ff. (f. 1 missing). From Gujarāt.
BORI 402 of 1884/86. 25ff. (f. 1 missing).
BORI 831 of 1884/87. 30ff. From Gujarāt.
BORI 903 of 1891/95. 31ff.
Cambridge University 159 = Cambridge University Add. 2408. No author mentioned.
CP, Hiralal 3186. (*Praśnavinoda*). Property of Kārelāl of Śobhāpur, Chhindwārā.
CP, Hiralal 3187. Property of Śāligrām of Hoshangābād.
CP, Hiralal 3188. Property of Vāsudev Golwalkar of Maṇḍlā.
CP, Hiralal 3189. Property of Śivrām of Hoshangābād.
CP, Hiralal 5448. No author mentioned. Property of Rāmnāth of Jubbulpore.
CP, Kielhorn XXIII 90. 11ff. Property of Javāhara Śāstrī of Chāndā.
DC 171. 74ff. No author mentioned.
GJRI 992/104. 47ff. Incomplete.
GOML Madras D 13976. Ff. 64–81. Telugu. Incomplete (ends in adhyāya 10).
GOML Madras D 13977. 12pp. Telugu. Incomplete (ends in adhyāya 15).
IO 6358 (Mackenzie III 85) = Mackenzie 41. 38ff. Telugu. From Colin Mackenzie.
Jaipur (II). 60ff.
Jammu and Kashmir 2765kha. 48ff.
Jammu and Kashmir 2880. 36ff.
Kotah 296. 54pp.
Kurukṣetra 653 (19718).
Kurukṣetra 654 (19993).
LDI (LDC) 1352. 70ff.
LDI (LDC) 3329/114. Ff. 91–115.
Mysore (1955) 5167. 62ff. Grantha.
Nagpur 1213 (1251). 4ff. (*Praśnabhairava* of Brahmadāsa). From Nasik.
N-W P II (1877) A 14. 30ff. Property of Chaṇḍī Datta of Benares.
Oppert II 1984. 68pp. Telugu. Ascribed to Brahmadāsa. Property of Veṅkaṭeśvarajosya of Siddhavaṭa, Kaḍapa.
Oppert II 4742. No author mentioned. Property of the Śaṅkarācāryasvāmimaṭha of Śṛṅgeri, Cikkamogulūr, Mysore.
Osmania University B. 11/14. 30ff.
Osmania University B. 95/18/a. 13ff. Incomplete (adhyāyas 8–9).
Oudh XIV (1881) VIII 1. 24pp. Property of Govindaprasāda of Lucknow Zila.
Oudh XIV (1881) VIII 4. 15pp. (*Praśnavinoda*). Property of Govindaprasāda of Lucknow Zila.
Oudh XX (1888) VIII 46 = Oudh XX (1888) VIII 95. 98pp. Property of Paṇḍita Pratāpa Nārāyaṇa of Allahabad Zila.
Oxford 786 (Walker 157b). Ff. 69–120.
Oxford 1554 (Sansk. d. 196) = Hultzsch 304. 91ff.
Oxford CS d. 780(ii). 30ff.
Paris BN (Senart) 247 (Sanskrit 1709). 8ff. Incomplete (utpātaphala).
PL, Buhler IV E 174. 58ff. (*Tājakavaiṣṇava* of Siddha). Property of Bālakṛṣṇa Jośī of Ahmadābād.
PL, Buhler IV E 264. 127ff. Property of Khuśāla Bhaṭṭa of Ahmadābād, Buhler notes another copy.
Poleman 4988 (Harvard 509). 44ff.
PUL II 3664. 53ff.
PUL II 3665. 36ff.
PUL II 3667. 29ff. Incomplete (ends in adhyāya 13).
PUL II 3668. 12ff. Incomplete.
RORI Cat. I 3116. 39ff.
RORI Cat. I 3160. 74ff.
RORI Cat. II 5269. 34ff.
RORI Cat. II 5534. 34ff. Incomplete.
RORI Cat. II 6432. 17ff.
RORI Cat. II 7050. 27ff.
RORI Cat. II 9822. 9ff. Incomplete (to IX 9).
RORI Cat. III 14347(1). 83ff.
RORI (Jaipur) IV 83. 20ff. Incomplete (ends in adhyāya 8).
SOI 1 = SOI Cat. I: 1344–1. 16ff.
SOI 4273.
SOI 4434.
SOI 8391.
SOI 9943.
Tanjore D 11517 = Tanjore BL 4314. 36ff.
WHMRL I. 148.
WHMRL M. 21. f.
WHMRL V. 69.

The *Praśnavaiṣṇava* was published at Kāśī in Saṃ. 1926 = A.D. 1869 (BM); edited by Nārāyaṇa Śāstrin as *CSS* 2, Kāśī Saṃ. 1953 = A.D. 1896 (NL Calcutta 180. Kc. 89. 9); and edited with a Marāṭhī translation by Gaṇeśa Śāstrī Deśiṃgakara Jyotiṣī, Belgaum 1925 (IO San. B. 1285). Verses 1–2 are:

nārāyaṇaṃ paramapūruṣam ādidevaṃ
jyotirmayaṃ śubhakaraṃ ca carācareśam/
śāntaṃ praṇamya śirasā dvijapuṅgavānāṃ
praśnārṇavaplavam ahaṃ prakaromi śāstram//
śrībrahmadāsanayajātanayaḥ suvidvān
śrīmān gusāṃyinṛpatir yadunāthabhaktaḥ/
vārāhatājikamukundamataṃ samīkṣya
nārāyaṇaḥ paramaśāstram idaṃ cakāra//

The last verse is:

kāyasthavaṃśāmbunidheḥ pṛthivyāṃ
śrībrahmadāsaḥ śaśalāñchano ᵓbhūt/
tāreva devīnayajā ca tābhyāṃ
nārāyaṇo jño haribhakta āsīt//

This is followed by 6 verses of his guru, introduced by:
harijīśarmā nārāyaṇadāsāyāśirvādam imaṃ dadau —

śrībrahmadāsanayajātanayena yena
govindapādasarasīruhaṣaṭpadena/
praśnārṇavaplavam idaṃ racitaṃ hi śāstraṃ
nārāyaṇo °stu bhuvane sukhabhāk sadaiva//
eke kāvyavilāsamātranipuṇāḥ śāstrān abhijñāḥ pare
vidvāṃso na kavitvamātraracaṇāḥ kartuṃ paraṃ jānate/
vidvattā kavitā ca yatra na ca te sadviṣṇubhaktāśrayāḥ
kāyasthaḥ kavibhaktapaṇḍitapatir jīyāt sa nārāyaṇaḥ//
hṛtpañkaje yasya sadaiva viṣṇur
vāgdevatā yadvadane vibhāti/
govindamantrālapanena siddho
nārāyaṇaḥ so °stu śriyā sametaḥ//
śrīmadgusāṃyinṛpates tv ajire sadaiva
lakṣmīr vinodayatu bhaktapriyāgragasya/
yenākhilāgamavidāṃ dvijapuñgavānāṃ
jñānāya śāstram amalaṃ gaditaṃ pṛthivyām//
jayati jagati viṣnor bhaktadhuryaḥ kriyāvān
praśamitabhavabhītir brahmadāsātmajo °yam/
nikhilakaluṣachetrī mohadātrī ca yasya
sphurati hariharoktiḥ siddhanārāyaṇasya//
śrīsiddhanārāyaṇadāsagranthaṃ
ye vaidyakaṃ vāpy atha jyotiṣaṃ vā/
vilokayiṣyanti nṛpāṃ gaṇe te
pūjyāḥ bhaviṣyanti harir jagāda//

harijīvadattam āśīrvādaślokaṣaṭkam idaṃ siddhaye °stu.

PAMUJĀHA NĀRĀYAṆAPPĀ

Author of a *Jyotiṣaśāstra*. Manuscript:

Osmania University 1089. 25ff. Telugu. Incomplete.

NĀRĀYAṆAPRASĀDA MUKUNDARĀMA (SĪTĀRĀMA) (*fl.* 1904)

Author of a *Lagnajātaka*, published with a Hindī ṭīkā at Bareilly in 1904 (IO San. B. 840(d)), and reprinted at Bombay in Saṃ. 1973 = A.D. 1916 (IO San. B. 153(c)).

PAṆḌITA NĀRĀYAṆAPRASĀDA (*fl.* 1911/1916)

Author of a Hindī ṭīkā on the *Lagnacandrikā* of Kāśīnātha, published at Bombay in Saṃ. 1973 = A.D. 1916 (IO 12. L. 40), and of a Hindī translation of the *Bhāvakutūhala* of Jīvanātha Jhā (*fl. ca.* 1846/1900), edited by Gajanāna Śarman, Bombay Saṃ. 1968 = A.D. 1911 (BM 14053. dd. 19).

NĀRĀYAṆAPRASĀDA MIŚRA (*fl.* 1912/1915)

Author of a *Saṃvatsarīpaddhati*, published with his own Hindī ṭīkā at Bombay in 1912 (IO 22. H. 12); of a *Yoginīśaṭaka*, published with his own Hindī ṭīkā at Bombay in 1913 (IO San. C. 156(i)); and of a *Camatkārajyotiṣa*, published with his own Hindī ṭīkā at Bombay in 1915 (IO San. C. 102).

NĀRMADA = NARMADĀDEVA (*fl. ca.* 1375)

The father of Padmanābha (*fl. ca.* 1400) and the grandfather of Dāmodara (*fl.* 1417), Nārmada (see S. B. Dikshit [1896] 255) wrote a *Nabhogasiddhi* following the Brahmapakṣa. Manuscript:

Benares (1963) 35098. 58ff. Copied in Saṃ. 1613 = A.D. 1556. With sāraṇī.

NĀHNIDATTA

The pupil of Śrīpati, Nāhnidatta wrote a short astrological work in 25 verses called variously *Pañcaviṃśatikā*, *Vyavahāraśāstra*, *Bālavivekinī*, *Bālabodhinī*, and *Bālabodhadīpikā;* there is a ṭīkā, *Vyavahāraprakāśikā*, by Ḍhuṇḍhirāja. Manuscripts:

Anup 5183. 2ff. Copied by Āsakaraṇa at Meḍatā in Saṃ. 1651 = A.D. 1594. Ascribed to Mātṛdatta.
Bombay U Desai 1390. Ff. 10–24. Copied in Saṃ. 1658 = A.D. 1601. With the *Vyavahāraprakāśikā* of Ḍhuṇḍhirāja.
Anup 4897. 9ff. Copied by Sārasvata Haricaraṇu Makaranda at Mathurā in Saṃ. 1707 = A.D. 1650. Ascribed to Śīlāhnidatta.
RORI Cat. III 15488. 15ff. Copied by Śyāma Upādhyāya in Saṃ. 1743 = A.D. 1686. With a ṭīkā.
AS Bengal 7191 (G 7001) I. No ff. given. Copied by Rājanandana on 14 śuklapakṣa of Āṣāḍha II in Saṃ. 1782 = *ca.* 11 July 1725 Julian.
BORI 947 of 1886/92. 4ff. Copied in Saṃ. 1788 = A.D. 1731. No author mentioned.
Bombay U Desai 1389. 32ff. Copied in Saṃ. 1818 = A.D. 1761. With the *Vyavahāraprakāśikā* of Ḍhuṇḍhirāja.
BORI 70 of A 1882/83. 10ff. Copied in Saṃ. 1820 = A.D. 1763. (*Jyotirnirṇaya* of Nāndidatta, the pupil of Śrīpati).
Baroda 9776. 4ff. Copied in Saṃ. 1821 (?) = A.D. 1764 (?). Ascribed to Mātkidatta.
Jaipur (II). 4ff. Copied in Saṃ. 1823 = A.D. 1766. Ascribed to Vahnidatta.
Benares (1963) 34729. 4ff. Copied in Saṃ. 1835 = A.D. 1778. No author mentioned.
PUL II 3676. Ff. 80–84. Copied in Saṃ. 1840 = A.D. 1783.
Baroda 3388. 5ff. Copied in Saṃ. 1844 = A.D. 1787. Ascribed to Mātkidatta.
Calcutta Sanskrit College 66. 7ff. Copied in Saṃ. 1845 = A.D. 1788. With a ṭīkā. (*Bālavivekinī* of Śrīpati with the ṭīkā of Iśānadatta).
Benares (1963) 34709. 6ff. Copied in Saṃ. 1855 = A.D. 1798. No author mentioned.

PL, Buhler IV E 278. 5ff. Copied in Saṃ. 1859 = A.D. 1802. Ascribed to Kīdatta. Property of Khuśāla Bhaṭṭa of Ahmadābād.
BORI 426 of 1895/98. 4ff. Copied in Śaka 1735 = A.D. 1813. Ascribed to Lāhnidatta.
Bendall. Copied in A.D. 1823. With a ṭīkā. From Benares.
RORI Cat. III 14885. 3ff. Copied in Saṃ. 1885 = A.D. 1828.
Mithila 151. 5ff. Maithilī. Copied at Daḍibhaṅgāgrāma on Sunday 7 śuklapakṣa of Kārttika in Śaka 1765, Sāl. San. 1251 = 29 October 1843. Property of Paṇḍita Mahīdhara Miśra of Lalbag, Darbhanga.
Goṇḍal 214. 4ff. Copied by Vanamālī in Saṃ. 1901 = A.D. 1844. No author mentioned.
Mithila 156. 8ff. Maithilī. Copied by Bābū Lāla, the son of Vacakaniśarman, in Śaka 1766, Sāl. San. 1253 = A.D. 1844, from a manuscript copied on Monday 9 kṛṣṇapakṣa of Caitra in Śaka 1681 = 7 April 1760. With a ṭīkā. Property of Babu Mārkaṇḍeya Miśra of Babhangama, Supaul, Bhagalpur.
Benares (1963) 35540 = Benares (1897–1901) 609. 4ff. and 1f. Copied in Saṃ. 1908, Śaka 1772 = A.D. 1850. No author mentioned.
Poleman 5145 (U Penn 1796). 5ff. Copied by Dāmodara Sahasrabuddhe on Thursday 1 kṛṣṇapakṣa of Mārgaśīrṣa in Saṃ. 1930, Śaka 17⟨9⟩5 = 4 December 1873.
Anup 4900. 3ff. Ascribed to Lalladatta. This is probably Bikaner 624. 32ff.
Baroda 3200. 2ff. Ascribed to Śrītāhnidatta.
Benares (1963) 34627. 11ff. With a ṭīkā. No author mentioned.
Benares (1963) 35530 = Benares (1897–1901) 551. 8ff. With a ṭīkā. Incomplete. No author mentioned.
Benares (1963) 36037. Ff. 2–21. With a ṭīkā. Incomplete. No author mentioned.
Benares (1963) 36082. Ff. 1–2 and 4–11. With a ṭīkā. No author mentioned.
Benares (1963) 36089. 6ff. No author mentioned. Probably identical with Benares (1903) 1280. 8ff. No author mentioned.
Benares (1963) 36857. 4ff.
Benares (1963) 37082. 5ff. Bengālī. Incomplete.
Benares (1963) 37096. 11ff. With a ṭīkā. Incomplete. No author mentioned.
Bombay U Desai 1388. 4ff. With a ṭīkā. Incomplete (verses 1–7).
Bombay U Desai 1391. 12ff. With a ṭīkā.
BORI 151 of A 1883/84. 12ff. With a ṭīkā.
Calcutta Sanskrit College 67. 4ff. Ascribed to Prīṇāhnidatta.
GJRI 970/82. Ff. 3–5. Maithilī. Incomplete.
GJRI 3167/379. 7ff.
GJRI 3168/380. Ff. 1–6 and 8. Incomplete.
GJRI 3169/381. 5ff.
GJRI 3170/382. 3ff. Maithilī.
GJRI 3171/383. 6ff. Maithilī.
GJRI 3261/473. 8ff. Maithilī.
GVS 2851 (643). Ff. 1–2 and 4–6. Incomplete. Ascribed to Lāhnidatta.
Jaipur (II). 5ff.
Mithila 151 A. 13ff. Maithilī. Property of Paṇḍita Dāmodara Jhā of Andhratharhi, Darbhanga.
Mithila 156 A. 5ff. Maithilī. With a ṭīkā. Property of Paṇḍita Pañcānana Jhā of Sukpur, Bhagalpur.
Mithila 156 B. 6ff. Maithilī. Property of Paṇḍita Suvaṃśa Lāla Jhā of Pachagacchia, Bhagalpur.
Mithila 156 C. 4ff. Maithilī. Incomplete. Property of Paṇḍita Arjuna Thākur of Parsarma, Bhagalpur.
Mithila 156 D. 6ff. Maithilī. Property of Paṇḍita Jīvanātha Jhā of Sukpur, Bhagalpur.
Mithila 210. 2ff. Maithilī. Property of Paṇḍita Śrīnandana Miśra of Kanhauli, Sakri, Darbhanga.
Oxford CS b. 98(v). 10ff. Bengālī. No author mentioned.
PL, Buhler IV E 277. 5ff. Ascribed to Mahidatta. Property of Śivaśaṅkara Jośī of Ahmadābād.
RORI Cat. I 1988. 4ff. (f. 1 missing).
RORI Cat. II 4865. 7ff. Ascribed to Lāhnīdatta.
WHMRL B. 21. o. Ascribed to Śrīpati.
WHMRL M. 12. d.
WHMRL O. 31. d.
WHMRL Y. 44.

The *Pañcaviṃśatikā* was edited by Muralīdhara Jhā, Benares 1902 (BM 14053. a. 11. (2)); published with Hindī and Maithilī translations at Darabhaṅgā in Saṃ. 1967 = A.D. 1910 (BM 14055. d. 6); with the Hindī ṭīkā of Baccū Śarman at Darabhaṅgā in [1911] (BM 14053. b. 38. (4)); at Darbhanga in [1924] (IO San. B. 844(d)); and edited by Rāmateja Pāṇḍeya, Kāśī [ND]. Verse 25 is:

bulānāṃ śubhakarmaśastasamayajñānaṃ kilaikaikataḥ
ślokād astv iti saṅkalayya manasaḥ ślokāṃś caturviṃśatim/
pūrvācāryakṛtān vilokya bahuśo jyotirnibandhān bahūṃś
cakre śrīpatipādapadmamadhupaḥ śrīnāhnidatto dvijaḥ//

NIḤŚAṄKU (*fl. ca.* 500)

A disciple of Āryabhaṭa (b. 476) cited by Bhāskara (*fl.* 629) in his *Āryabhaṭīyabhāṣya* on Kālakriyā 10; see P. C. Sengupta, *The Khaṇḍakhādyaka*, Calcutta 1934, p. xix.

NIKAṢĀRĀKṢASĪ

Author of a *Praśnasāra* = *Keralī*. Manuscripts:

AS Bengal 7179 (G 7900) A. 5ff. Copied by Śivasahāya Brāhmaṇa on Saturday 8 kṛṣṇapakṣa of Bhādrapada in Saṃ. 1937 = 25 September 1880.

Benares (1963) 37575. 3ff.
Benares (1963) 37576. Ff. 1–2, 5–13, and 15–16, and 2ff. Incomplete.

The colophon begins: iti śrīnikaṣārākṣasīracitā.

NIJĀNANDA

Author of a ṭīkā, *Bījālavāla*, on the *Bījagaṇita* of Bhāskara (b. 1114). Manuscript:

Baroda 3202. 253ff. Copied in Saṃ. 1935 = A.D. 1878.

NIJĀNANDA

Author of a vivṛti, *Subodhinī*, on the *Nīlakaṇṭhī* of Nīlakaṇṭha (*fl.* 1572/1587). Manuscript:

Benares (1963) 37182. 29ff. Copied in Śaka 1787 = A.D. 1865. Incomplete (ends with dvādaśabhāva).

NITYAPRAKĀŚA BHAṬṬĀRAKA

Author of a vivṛti, *Visamākṣara*, on the *Bṛhajjātaka* of Varāhamihira (*fl. ca.* 550). Manuscripts:

Kerala 10892 (L. 548A). 2000 granthas. Malayālam. Copied in ME 881 = *ca.* A.D. 1705. Incomplete.
Kerala 10888 (527B). 3600 granthas. Malayālam.
Kerala 10889 (4261). 3600 granthas. Malayālam.
Kerala 10890 (8927). 3500 granthas. Malayālam. Incomplete.
Kerala 10891 (8976). 3500 granthas. Malayālam. Incomplete.
Kerala 10893 (C. 2117). 3550 granthas. Malayālam. Incomplete.
Kerala 10952 (C. 136). 3000 granthas. Malayālam. Incomplete.
Kerala 10953 (T. 90). 3000 granthas. Incomplete.
Kerala 10954 (5797). 5000 granthas. Malayālam. With a Keralabhāṣā. Incomplete.

NITYĀNANDA

Author of an *Iṣṭakālaśodhana* = *Iṣṭakālasādhana*. Manuscripts:

Goṇḍal 10. 6ff. Copied on Saturday 9 kṛṣṇapakṣa of Māgha in Saṃ. 1782 = 12 February 1726. No author mentioned.
Nagpur 215 (2047). 5ff. Copied in Śaka 1738 = A.D. 1816. From Nagpur. No author mentioned.
SOI 3357 = SOI Cat. II: 980–3357. 3ff. Copied in Saṃ. 1932 = A.D. 1875.
CP, Hiralal 475. Property of Viśvambharnāth of Ratanpur, Bilāspur.
IM Calcutta 1102. See NCC, vol. 2, p. 260.
Jaipur (II). 2 manuscripts.
N-W P I (1874) 67. 4ff. (*Niṣekavicāra*). Property of Pūrṇānanda Jotishi of Benares.
N-W P I (1874) 110. 8ff. Property of Jagannātha Jotishi of Benares.
SOI 7853. Incomplete (niṣekādhika). No author mentioned.

NITYĀNANDA

Author of a *Saṃvatsaravinirṇaya*. Manuscript:

Benares (1963) 35019. 15ff. Copied in Saṃ. 1822 = A.D. 1765.

NITYĀNANDA (*fl.* 1628/1639)

The son of Devadatta, the son of Nārāyaṇa, the son of Lakṣmaṇa, the son of Icchā Ḍulīnahaṭṭa, a Gauḍa Brāhmaṇa of the Mudgalagotra and a resident of Indrapurī (= Delhi), Nityānanda composed two astronomical treatises under Shāh Jahān (1628/1658); see M. M. Patkar [1938b] 172–173.

1. The *Siddhāntasindhu*, composed for Vāsafa Khān (Āsaf Khān; d. 1641), minister of Shāh Jahān, in Saṃ. 1685 = A.D. 1628. Manuscripts:

Alwar 2014.
Anup 5332. 28ff. Incomplete (khaṇḍa 3).
Anup 5333. 14ff. Incomplete (khaṇḍa 3).
Jaipur (II). 441ff.
Jaipur (II). 536ff.
Jaipur (II). 442ff.
Rajputana, p. 57. From Alwar.
SOI 9410. Incomplete (kāṇḍa 2). No author mentioned.

2. The *Siddhāntarāja*, following the sāyana system (i.e., using a tropical instead of a sidereal length of the year), composed in Saṃ. 1696 = A.D. 1639. This work contains the following chapters:

I gaṇitādhyāya

1. mīmāṃsā.
2. madhyama.
3. spaṣṭa.
4. tripraśna.
5. candragrahaṇa.
6. sūryagrahaṇa.
7. śṛṅgonnati.
8. bhagrahayuti.
9. bhagrahāṇām unnatāṃśādisādhana.

II golādhyāya.

1. bhuvanakośa.
2. golabandha.
3. yantra.

See S. Dvivedin [1892] 101–102 and S. B. Dikshit [1896] 289–290. Manuscripts:

AS Bombay 264. 8ff. Copied by Sukhānanda, the son of Vāhālajī, in Saṃ. 1725, Śaka 1590 = A.D. 1668. Incomplete (yantrādhyāya). From Bhāu Dājī.

Benares (1963) 35741. 84ff. Copied in Saṃ. 1804 = A.D. 1747.
Benares (1963) 37079. = Benares (1878) 68 = Benares (1870–1880) 9. 85ff. Copied in Saṃ. 1895 = A.D. 1838 (in Benares (1878) said to have been copied in Saṃ. 1936 = A.D. 1879).
Benares (1963) 34466. Ff. 1–36 and 39ff. Copied in Saṃ. 1915 = A.D. 1858. Incomplete.
Alwar 2005.
Alwar (1884), pp. 110–111. Incomplete (golādhyāya).
BORI 206 of A 1883/84. 47ff.
N-W P V (1880) B 22. 21ff. No author mentioned. Property of Pandits Rāmacandra and Udayānanda of Ulwar.
SOI 9366.
WHMRL V. 36.

The last verses are:

śrutismṛtivyākaraṇapravīṇair
viśiṣṭaśiṣṭācaraṇaikadakṣaiḥ/
śrīmatkurukṣetrasamīpasaṃsthā
dvijottamair indrapurī vibhāti//
tasyāṃ vasan gauḍakulaprasūto
ᵓnuśāsanenāpi ḍulīnahaṭṭaḥ/
icchābhidho mudgalagotrajanmā
babhūva pūrvaṃ satapā manīṣī//
tasyātmajaḥ śrautavidhiprayukto
vedāntaśāstrādikapāragāmī/
tapodhano jyautiṣaśāstradakṣo
vicakṣaṇo lakṣmaṇasaṃjñako ᵓbhūt//
tadīyaputras tapasā variṣṭhaḥ
sadā gariṣṭho dvijamaṇḍalīṣu/
sadā sadācārayuto manīṣī
nārāyaṇo dharmaparāyaṇo ᵓbhūt//
tasyātmajo jyautiṣaśāstradakṣo
vidyāvinodeṣu vilagnacetāḥ/
jitendriyaḥ satyatapaḥsametaḥ
śrīdevadatto ᵓsti narendramānyaḥ//
nityānandas tasya putro dvijānām
ājñākārī sūryalabdhaprasādaḥ/
ṣaḍgobhūpair vikramārkasya śāke
yāte cakre sarvasiddhāntarājam//

3. A *Śāhajahāṃgaṇita*. Manuscript:

Anup 5191. 12ff.

NITYĀNANDA PANTA PARVATĪYA (*fl.* 1932)

Author, with Gopāla Śāstrin Nene (*fl.* 1932/1936), of a *Varṣakṛtyadīpaka*, completed on Thursday 5 śuklapakṣa of Māgha in Saṃ. 1988 = 10 February 1932 and published as *KSS* 96 at Benares in 1932, reprinted Vārāṇasī 1967.

NIDHINĀTHA

Author of a *Praśnavibodhanī* in 49 verses. Manuscript:

Mithila 199. 2ff. Maithilī. Property of Babu Ṭhiṭhara Jhā of Babhangama, Supaul, Bhagalpur.

Verses 1–2 are:

gajānanaṃ namaskṛtya dineśaṃ girijāpatim/
kriyate nidhināthana manaḥpraśnavibodhanī//
śrīkṛṣṇacaraṇaṃ natvā gurvīśapadapaṅkajam/
tāṃ vakti śrīnidhiś caiva keralādyanusaṃmatām/

NIVĀRAṆACANDRA CAUDHURĪ (*fl.* 1906)

Author of a *Bṛhajjyotiṣasiddhānta*, published at Calcutta in [1906] (IO 20. F. 38), 4th ed., Calcutta 1942 (NL Calcutta 180. Kc. 94. 6).

NĪRĀJANAGIRI

Author of a *Grahaphala*. Manuscript:

CP, Kielhorn XXIII 24. 34ff. Property of Gaḍīpanta Paṭalavāra of Chāndā.

NĪLAKAṆṬHA

Author of a *Grahalāghavasāraṇī*, based on the *Grahalāghava* (1520) of Gaṇeśa (b. 1507). Manuscripts:

PL, Buhler IV E 508. 95ff. Copied in Saṃ. 1905 = A.D. 1848. (*Sāraṇīkoṣṭaka*). Property of Prāṇaśaṅkara of Mulī.
Benares (1963) 34362. 17ff.
Jaipur (II).
LDI (LDC) 5230. 10ff. With an udāharaṇa in Gujarātī.
Udaipur 524.

NĪLAKAṆṬHA

The son of Rāghava of the family of Harihara Kavi Paṇḍitādhirāja, Nīlakaṇṭha wrote a *Jātakapaddhati* = *Janmapaddhati* in 59 verses. Manuscripts:

Mithila 74. 7ff. Maithilī. Copied by Ratan Śarman for Trilokanātha Śarman on Monday 8 śuklapakṣa of Śrāvaṇa I in Śaka 1739 = 21 July 1817. Property of Babu Puruṣottama Jhā of Babhangama, Supaul, Bhagalpur.
Mithila 89 A. 8ff. Maithilī. Copied in Śaka 1750 = A.D. 1828. Property of Paṇḍita Mahīdhara Miśra of Lalbag, Darbhanga.
Mithila 89. 9ff. Maithilī. Copied by Vacanū Śarman on Tuesday 9 śuklapakṣa of Śrāvaṇa in Śaka 1763 = 27 July 1841. Property of Paṇḍita Mahīdhara Miśra of Lalbag, Darbhanga.
GJRI 925/37. 26ff. Maithilī. With a ṭīkā. Incomplete.
GJRI 964/76. 8ff. Maithilī. Incomplete.

Verse 59 is:

haricaraṇaparā sadā vilakṣyā
dvijakulabhūṣaṇarāghavāt suto yaḥ/
hariharakavipaṇḍitādhirāja-
vaṃśaja imāṃ sa cakāra nīlakaṇṭhaḥ//

NĪLAKAṆṬHA

Author of a ṭīkā on a *Jyotiṣatantragrantha*. Manuscript:

GVS 2814 (3751). Ff. 9–10 and 13–14. Incomplete.

NĪLAKAṆṬHA

Author of a *Tithiratnamālā*. Manuscript:

PL, Buhler IV E 187. 20ff. Property of Mūlajī Jīvana Paṇḍyā of Sīhora.

NĪLAKAṆṬHA

Author of a *Tithyādikṛtya*. Manuscripts:

Kurukṣetra 404 (19886).
Kurukṣetra 405 (19887).

NĪLAKAṆṬHA

Author of a *Nānābhāīsārasaṅgraha;* see Nānābhāī. Manuscript:

Kathmandu (1960) 210 (I 1208). 10ff. Nevārī. With a Nepālī bhāṣya.

NĪLAKAṆṬHA

Assumed author of a *Nīlakaṇṭhagaṇita;* this may be the *Gaṇitasaukhya* of the *Ṭoḍarānanda*. Manuscript:

GVS 2840 (1645). 7ff. Copied on Saturday 15 (read 5) śuklapakṣa of Pauṣa in Saṃ. 1681, Śaka 1546 = 4 December 1624 Julian. Incomplete. No author mentioned.

NĪLAKAṆṬHA

Author of a *Bālabhodikā*. Manuscript:

Benares (1963) 34492. Ff. 1–102, 104–119, and 121. Incomplete.

NĪLAKAṆṬHA

Alleged author of a ṭīkā on sections of the *Muhūrtamārtaṇḍa* of Nārāyaṇa (*fl.* 1571/1572); they are probably in fact parts of the latter's *Mārtaṇḍavallabhā*. Manuscripts:

N-W P II (1877) B 22. 25ff. Incomplete (gocara). Property of Rāma Prasāda of Benares.
N-W P II (1877) B 25. 5ff. Incomplete (gṛhapraveśa). Property of Rāma Prasāda of Benares.
N-W P II (1877) B 77. 43ff. Incomplete (nakṣatra). Property of Bholā Datta of Benares.
N-W P II (1877) B 78. 38ff. Incomplete (vivāha). Property of Bholā Datta of Benares.
N-W P II (1877) B 97. 34ff. Incomplete (gṛha). Property of Rāma Prasāda of Benares.

NĪLAKAṆṬHA

Alleged author of a *Muhūrtamuktāvalī* published with a bhāṣāṭīkā at Bombay, [ND] (Mysore GOL B 1645).

GĀRGYAKERALA NĪLAKAṆṬHA SOMAYĀJIN (b. *ca.* 14 June 1444)

The son of Jātavedas, a Nampūtiri Brāhmaṇa of the Gārgyagotra, Nīlakaṇṭha was born in the Keḷallūr illam (Keralasadgrāma), said to be the present Eṭamana illam, in Tṛ-k-kaṇṭiyūr (Kuṇḍapura) near Tirur, Kerala. He studied under Dāmodara, the son of Parameśvara (*ca.* 1380/1460), at Ālattūr (Aśvatthagrāma), Kerala. See K. V. Sarma [1956/57a]; K. K. Raja [1963] 143–152; and K. M. Marur and C. T. Rajagopal [1944]. His works include the following.

1. The *Golasāra* in 3 paricchedas containing 56 verses. Manuscripts:

Chalakkudi, Cochin, Rama Varma Maru Thampuran. Incomplete (to III 24). See edition, p. ix.
GOML Madras R 5151(a). Ff. 1–3. Grantha. Copied in 1925/26 from a manuscript belonging to Tippan Nambūdirippāḍ of Ponnurkottamana, Perumbavur, Travancore. Incomplete (II–III).
IO 6301 (Burnell 17e). 24ff. Malayālam. From A. C. Burnell.
Kerala 5065 (5867 B). 70 granthas. Malayālam. Incomplete (II–III).
Kerala 5066 (8358 E). 70 granthas. Malayālam.
Kerala 5067 (C. 1024 E) = Kerala C 633 E. 37ff. Malayālam. Formerly property of the Rājā of Cirakkal.
Kerala 5068 (T. 846 A). 70 granthas. Copied from Kerala 5067.
Kerala 5069 (C. 1869 B) = Kerala C 636 B. 15ff. Malayālam. Incomplete. (to III 24). Formerly property of Brahmadattan Nambūri of Kūḍallūr.

The *Golasāra* was edited from all these manuscripts and translated by K. V. Sarma as *VIS* 47, Hoshiarpur 1970. The colophon begins: iti gārgyakeralanīlakaṇṭhaviracite.

2–3. The *Siddhāntadarpaṇa* in 32 verses, on which Nīlakaṇṭha wrote his own ṭīkā. Manuscripts:

Kerala —— (475 D). Copied in A.D. 1551. See edition, p. 11.
Kerala C 633 F (C. 1024 F). 4ff. Malayālam. Copied in A.D. 1551. Formerly property of the Rājā of Cirakkal.
GOML Madras R 5151(b). Ff. 3–5. Grantha. Copied in 1925/26 from a manuscript belonging to Tippan Nambūdirippāḍ of Ponnurkottamana, Perumbavur, Travancore. Said to be a copy of Kerala 5867 C in edition.

IO 6302 (Burnell 17f). 3ff. Malayālam. From A. C. Burnell.
Kerala —— (5867 C). Copied by Śaṅkara. See edition, p. 11.
Kerala —— (8358 I). See edition, p. 12.
Kerala C 636 C (C. 1869 C). 4ff. Incomplete (begins with verse 13). Formerly property of Brahmadattan Nambūri of Kūḍallūr.
Trivandrum Palace Library 975. With his own ṭīkā. Incomplete (verses 2–7 and 17–27). See edition, p. 12.

The *Siddhāntadarpaṇa* with 2 derivative works was edited from all the manuscripts except that in Madras by K. V. Sarma as *ALS* P 30, Madras [1956]. Verse 18 gives the date Kali 4545 = A.D. 1444; the commentary states that: svajanmakālajñāpanārthaṃ caivam uktam/ tadahargaṇaś ca tyajāmyajñatāṃ tarkair iti. The ahargaṇa 1,660,181 corresponds to *ca.* 14 June 1444. The colophon begins: iti gārgyakeralasadgrāmanīlakaṇṭhaviracitaṃ.

4–5. The *Candracchāyāgaṇita* in 31 verses, on which Nīlakaṇṭha wrote his own ṭīkā. Manuscripts:

GOML Madras R 5185(b). Ff. 26–35. Grantha. Copied in 1925/26 from a manuscript belonging to Tippan Nambūdirippāḍ of Ponnūrkottamana, Perumbāvūr, Travancore. With his own vyākhyāna.
AHRS 23. No author mentioned.
Kerala 5348 (L. 1167 D). 30 granthas. Malayālam. Incomplete.
Kerala 5349 (475 I). 50 granthas. Malayālam. Incomplete.
Kerala 5350 (5862 B). 275 granthas. Malayālam. With his own vyākhyā. Incomplete.
Kerala 5688 (931 D). 45 granthas. Malayālam. With a Keralabhāṣā. No author mentioned.

The first verse of the vyākhyā is:

janmasthitihṛtayaḥ syur
jagato yasmāt praṇamya tad brahma/
candracchāyāgaṇitaṃ
kartrā vyākhyāyate ᵓsya gārgyeṇa//

6. The *Tantrasaṅgraha* = *Kriyākalāpa*, composed in A.D. 1501 in 8 adhyāyas; there are commentaries by Śaṅkara Vāriyar (*fl.* 1556), the *Laghuvivṛti*, and by a Nampūtiri of Tṛpparaṅgoṭṭu (Śvetāraṇya). Manuscripts:

Kerala 6457 (660 B). 500 granthas. Malayālam. Copied in ME 770 = *ca.* A.D. 1594.
Kerala 6476 (697). 5000 granthas. Malayālam. Copied in ME 920 = *ca.* A.D. 1744. With a Keralabhāṣā.
Kerala 6462 (C. 224 C) = Kerala C 641 C. 10ff. Malayālam. Copied in ME 928 = *ca.* A.D. 1752. Formerly property of Valia Rājā Avl. of Eḍapaḷḷy.
GOML Madras R 3205. 94ff. Copied in 1920/21 from a manuscript belonging to the Rājā of Cirakkal. With the *Laghuvivṛti* of Śaṅkara.
Adyar Index 2427 = Adyar Cat. 34 I 3. 84ff. Grantha.
Baroda 1884(b). 19ff. Malayālam.
GOML Madras D 13426. Ff. 105–128.
GOML Madras R 3877(i). Ff. 89–101. Grantha (ends in adhyāya 3). Purchased in 1921/22 from Aṇṇāsvāmi Śāstrigal of Bhavani, Coimbatore.
GOML Madras R 6122(a). Ff. 77–82. Grantha. With a ṭīkā. Incomplete (I 2 to II 34). Purchased in 1937/38 from E. Śrīnivāsācāryar of Srīperumbūdūr, Chingleput.
Kerala 6456 (475 E). 500 granthas. Malayālam.
Kerala 6458 (831 A). 500 granthas. Malayālam.
Kerala 6459 (5612 C). 500 granthas. Malayālam.
Kerala 6460 (8324 G). 500 granthas. Malayālam.
Kerala 6461 (10629 D). 500 granthas. Malayālam.
Kerala 6463 (C. 1079 E). 500 granthas. Malayālam.
Kerala 6464 (C. 2371 B) = Kerala C 730 B. 50ff. Malayālam. Formerly property of Tuppan Tuppan Nambūri of Punnarkoṭṭu, Manakal.
Kerala 6465 (10835). 325 granthas. Malayālam. Incomplete.
Kerala 6466 (8351). 2500 granthas. Malayālam. With the *Laghuvivṛti* of Śaṅkara.
Kerala 6467 (8905). 2500 granthas. Malayālam. With the *Laghuvivṛti* of Śaṅkara.
Kerala 6468 (10643). 2500 granthas. Malayālam. With the *Laghuvivṛti* of Śaṅkara.
Kerala 6469 (C. 524) = Kerala C 694. 106ff. Malayālam. With the *Laghuvivṛti* of Śaṅkara. Formerly property of Nārāyaṇan Rāman Piṣāroḍi of Kidaṅgūr.
Kerala 6470 (C. 809 A) = Kerala C 697 A. 132ff. Malayālam. With the *Laghuvivṛti* of Śaṅkara. Formerly property of Kṛṣṇa Vāriar of Vaḍaketteruvu Tiruvārpu.
Kerala 6471 (C. 994) = Kerala C 696. 153ff. Malayālam. With the *Laghuvivṛti* of Śaṅkara. Formerly property of Tuppan Tuppan Nambūri of Punnarkoṭṭu, Manakkal.
Kerala 6472 (C. 1079 D). 2500 granthas. Malayālam. With the *Laghuvivṛti* of Śaṅkara.
Kerala 6473 (L. 944 A). 2500 granthas. Malayālam. With the *Laghuvivṛti* of Śaṅkara.
Kerala 6474 (T. 94) = Kerala C 968. 206pp. With the *Laghuvivṛti* of Śaṅkara.
Kerala 6475 (C. 134 A) = Kerala C 695 A. 70ff. Malayālam. With the *Laghuvivṛti* of Śaṅkara. Incomplete. Formerly property of Kṛṣṇan Keśavan of Periṇād, Quilon.
Kerala 6477 (8358 D). 1400 granthas. Malayālam. With a Keralabhāṣā. Incomplete.
Lucknow 510. N 61 T (45736).
Lucknow 520. N 62 T (45755).
Lucknow 520. N 62 Tv (45868–45869). With a vyākhyā.

PUL II 3493. 25ff. Malayālam.
RAS (Whish) 134 (Whish 134). 129ff. Malayālam. With the *Laghuvivṛti* of Śaṅkara.

The *Tantrasaṅgraha* with the *Laghuvivṛti* of Śaṅkara was edited from 10 of the Kerala manuscripts (8351, 8905, 10643, C. 134 A, C. 524, C. 809 A, C. 994, C. 1079 D, L. 944 A, and T. 94) by Suranad Kunjan Pillai as *TSS* 188, Trivandrum 1958.

7. A ṭīkā, *Bhāṣya*, on the *Āryabhaṭīya* (499) of Āryabhaṭa (b. 476), omitting the Daśagītikā, composed for Netranārāyaṇa, the head of the Nampūtiri Brāhmaṇas; in the commentary on Kālakriyā 12–15 he states that he had observed a total eclipse of the Sun on 6 March 1467 (Oppolzer 6358) and an annular eclipse at Anantakṣetra on 28 July 1501 (not in Oppolzer). In it he quotes his own *Tantrasaṅgraha* and *Siddhāntadarpaṇa*. Manuscripts:

Kerala 1843 (C. 996) = Kerala C 646. 116ff. Malayālam. Copied in ME 877 = *ca.* A.D. 1701. Formerly property of Nārāyaṇan Rāma Piṣāroḍi.
GOML Madras R 5261. 142ff. Copied in 1926/27 from a manuscript belonging to Nārāyaṇa Nambūdirippāḍ of Pūmalli-mana, Chalasseri, Malabar. Incomplete (Kālakriyā and Golapāda).
Baroda 9886(b). 80ff. Malayālam. Incomplete (Golapāda).
Baroda 9887. 176ff. Malayālam. Incomplete (ends in Golapāda).
Kerala 1837 (5848). 950 granthas. Malayālam. Incomplete.
Kerala 1838 (C. 157) = Kerala C 647. 56ff. Malayālam. Incomplete. (Golapāda 25–50). Formerly property of Eṇṇakāṭṭu Koṭṭāram.
Kerala 1839 (C. 1415 A) = Kerala C 649 A. 94ff. Malayālam. Incomplete. Formerly property of Ravi Varma Valia Koil Tampurān of Kiḷimānur Palace.
Kerala 1841 (L. 1347 A). 3000 granthas. Malayālam. Incomplete.
Kerala 1842 (C. 527 A) = Kerala C 645 A. 123ff. Malayālam. Incomplete (to Golapāda 42). Formerly property of Nārāyaṇan Rāma Piṣāroḍi.
Kerala 1844 (C. 1875) = Kerala C 648. 80ff. Malayālam. Incomplete. Formerly property of Nārāyaṇabhaṭṭatiri of Tiruvalla.
Kerala 1845 (C. 2160 E) = Kerala C 637 E. 112ff. Malayālam. Incomplete. Formerly property of Vaittiyapa Pillai of Mānnār.
Kerala 1846 (C. 2160 F) = Kerala C 637 F. 21ff. Malayālam. Incomplete. Formerly property of Vaittiyapa Pillai of Mānnār.
Kerala 1847 (T. 296). 2700 granthas. Incomplete.
Trivandrum Palace Library 870. See NCC, vol. 2, p. 172.

The *Āryabhaṭīyabhāṣya* was edited by K. Sāmbaśiva Śāstrī as *TSS* 101 (Gaṇitapāda), Trivandrum 1930; as *TSS* 110 (Kālakriyā), Trivandrum 1931; and, by Suranad Kunjan Pillai, as *TSS* 185 (Golapāda), Trivandrum 1957. The colophon to the Gaṇitapāda begins:

iti śrīkuṇḍagrāmajena gārgyagotreṇāśvalāyanena bhāṭṭena keralasadgrāmagṛhasthena śrīśvetāraṇyanāthaparameśvarakaruṇādhikaraṇabhūtavigraheṇa jātavedaḥputreṇa śaṅkarāgrajena jātavedomātulena dṛggaṇirmāpakaparameśvaraputraśrīdāmodarāttajyotiṣām ayanena ravita āttavedāntaśāstreṇa subrahmaṇyasahṛdayena nīlakaṇṭhena somasutā viracitavividhagaṇitagranthena.

8–9. Nīlakaṇṭha refers to his lost *Grahaṇanirṇaya* and *Sundararājapraśnottara* in his *Āryabhaṭīyabhāṣya*; in the second of these he responds to questions posed by Sundararāja (*fl. ca.* 1475), the author of the ṭīkā, *Laghudīpikā*, on the *Vākyakaraṇa*.

10. A *Grahaṇagrantha* in which he refers to his own *Āryabhaṭīyabhāṣya*. Manuscript:

Trivandrum Palace Library 975. Incomplete. See *Golasāra*, pp. xx–xxi, and NCC, vol. 6, p. 248.

NĪLAKAṆṬHA (*fl.* 1569/1587)

The son of Padmāmbā and Ananta (*fl. ca.* 1575), the son of Cintāmaṇi of the Gārgyagotra, a resident of Dharmapura on the Godāvarī in Vidarbha, Nīlakaṇṭha was the father of Govinda (b. 2 October 1569), the brother of Rāma (*fl.* 1600/1601), and the pupil of Śiva. See S. Dvivedin [1892] 68–69; S. B. Dikshit [1896] 275; and M. M. Patkar [1938b] 167. One of the leading astrologers at Kāśī in the late 16th century, Nīlakaṇṭha was asked to compose the jyotiṣa sections of the *Ṭoḍarānanda* compiled for Ṭoḍaramalla (*fl.* 1565/1589); see P. V. Kane [1930/62] vol. 1, pp. 421–423, and *Ṭoḍarānanda*, vol. 1, ed. P. L. Vaidya, *Gaṅga OS* 5, Bikaner 1948. These include the following.

1. The *Jyotiḥsaukhya* contains a *Saṃhitāskandha*, composed in Śaka 1494 = A.D. 1572. The work includes the following vilāsas:

1. śākhāvatāra.
2. daivajñapraśaṃsā.
3. arkacāra.
4. candracāra.
5. bhaumacāra.
6. budhacāra.
7. gurucāra.
8. śukracāra.
9. śanaiścaracāra.
10. rāhucāra.

11. ketucāra.
12. agastyasaptarṣidhruvādicāra.
13. kūrmavibhāga.
14. nakṣatravyūha.
15. grahabhakti.
16. grahayuddhasamāgama.
17. grahaśṛṅgāṭakādiyoga.
18. garbhalakṣaṇa.
19. vātacakra.
20. sadyovarṣa.
21. sandhyālakṣaṇa.
22. digdāhalakṣaṇa.
23. bhūkampalakṣaṇa.
24. ulkālakṣaṇa.
25. pariveṣalakṣaṇa.
26. indracāpalakṣaṇa.
27. gandharvanagaralakṣaṇa.
28. pratisūryalakṣaṇa.
29. rajas.
30. nirghātalakṣaṇa.
31. arghakāṇḍa.
32. vibudhavaikṛta.
33. agnivaikṛta.
34. vṛkṣavaikṛta.
35. sasyādivaikṛta.
36. vṛṣṭivaikṛta.
37. nadyādivaikṛta.
38. prasavavaikṛta.
39. paśupakṣivaikṛta.

Manuscripts:

BORI 317 of 1882/83. 171ff. Copied by Nṛsiṃha Jośī at the Viśveśvarasaṃnidhi in Kāśī on 3 kṛṣṇapakṣa of Phālguna in Saṃ. 1672 = *ca.* 24 February 1616. Bought by Ratneśvara, the son of Divākara Paṇḍyā, for Vīreśvara in Saṃ. 1788 = A.D. 1731.
Anup 2383. 148ff. Copied in Saṃ. 1702 = A.D. 1645.
Jaipur (II). 118ff. Copied in Saṃ. 1728 = A.D. 1671.
BORI 915 of 1886/92. 124ff. Copied in Saṃ. 1846 = A.D. 1789.
DC 7914. 135ff. Copied by Tryambaka, the son of Ḍhuṇḍhirāja, at Śirṣlapura in Śaka 1734 = A.D. 1812. Incomplete (grahācāravividhotpāta).
Kathmandu (1960) 158 (III 474). 123ff. Copied by Kṛṣṇalāla Śarman on Tuesday 12 śuklapakṣa of Śrāvaṇa in Saṃ. 1903, Śaka 1768 = 4 August 1846.
Alwar 1795.
Anup 2384. 154ff.
Anup 2385. 80ff. Incomplete.
Anup 5236. 2ff. Incomplete (saptarṣicāra).
Baroda 11021. 188ff.
Kotah 251. 377pp.
Kotah 252. 135pp.
PUL II 3240. 4ff. Incomplete (agastyasaptarṣidhruvādicāra).
PUL II 3545. 16ff.

Verses 3–4 are:

yatkīrtyā vijito bhujaṅgamapatiḥ pātālam adhyāsta yad-
vāṇīnaipuṇamādhurīṣu vijito vācāṃ patiḥ svargataḥ/
dagdhārivrajayatpratāpaśikhino dhūmasya lekhāṃ vyadhād
indau lakṣmaṇaṭoḍaro vijayate sāmrājyalakṣmīṃ śritaḥ//
tadājñayā jyautiṣasaukhyam etad
ārabhyate paṇḍitasaukhyahetoḥ/
śrīnīlakaṇṭhena guruprasādāt
samūlasiddhāntanibaddhayukti//

The second section of the *Jyotiṣasaukhya* is the *Gaṇitasaukhya*, composed in Śaka 1494, Kali 4673 = A.D. 1572. Manuscripts:

Anup 2379. 28ff.
Anup 2380. 35ff. With an udāharaṇa.

The third section of the *Jyotiḥsaukhya* is the *Horāsaukhya*, which includes the following adhyāyas:

1. rāśiprabheda.
2. khacarayonibalābala.
3. miśraka.
4. kāraka.
5. viyonicintā.
6. niṣeka.
7. sūtika.
8. ariṣṭa.
9. ariṣṭabhaṅga.
10. āyurdāya.
11. antardaśā.
12. aṣṭavarga.
13. karmājīva.
14. rājayoga.
15. rājayogabhaṅga.
16. nābhasayoga.
17. candrayogasūryayoga.
18. dvigrahayoga.
19. trigrahayoga.
20. pravrajyā.
21. nakṣatraguṇa.
22. rāśiphala.
23. bhāgaphala.
24. bhāvavicāra.
25. aniṣṭayoga.
26. strījātaka.
27. niryāṇavicāra.
28. naṣṭajātaka.
29. dreṣkāṇarūpa.

Manuscripts:

Kathmandu (1960) 162 (I 1167). 102ff. Copied by Devadatta, the son of Purohita Rāma, on Friday 30 Vaiśākha in Saṃ. 1639 = 10 May 1583.
Anup 2382. 51ff.

CP, Hiralal 6917. Property of Govindbhaṭṭ of Jubbulpore.
Jaipur (II). 16ff. Incomplete (naṣṭajātaka).

Another section of the *Jyotiṣasaukhya* is the *Yātrāsaukhya*. Manuscript:

Anup 2378. 119ff.

2. The *Vivāhasaukhya*, which includes the following sections:

1. vivāha.
2. caturthikā.
3. dvirāgamanavadhūpraveśau.
4. rājābhiṣeka.

Manuscripts:

Kathmandu (1960) 161 (I 1203). 86ff. Copied by Harajī, the son of Purohita Rāma, on Sunday 6 kṛṣṇapakṣa of Śrāvaṇa in Saṃ. 1640 = 16 August 1584.
BORI 868 of 1884/87. 89ff. Copied in Saṃ. 1646 = A.D. 1589.
BORI 916 of 1886/92. 53ff. Copied in Saṃ. 1802 = A.D. 1745.
Leningrad (1914) 306 (Ind. V 96). 68ff. Copied in Saṃ. 1873 = A.D. 1816.
Anup 2386. Ff. 3–49. Incomplete.
BORI 869 of 1884/87. 63ff.
Kotah 255. 95pp.

3. The *Vāstusaukhya*. Manuscripts:

Benares (1963) 37213. Ff. 2–30. Copied in Saṃ. 1734 = A.D. 1677.
Leningrad (1914) 305 (Ind. V 95). 28ff. Copied on 7 śuklapakṣa of Vaiśākha in Saṃ. 1873 = *ca.* 3 May 1816.
AS Bengal 2813 (G 3068). 52ff. Copied in Saṃ. 1884 = A.D. 1827.
Benares (1963) 36529 = Benares (1903) 1115. 44ff. Copied in Saṃ. 1886 = A.D. 1829.
Kathmandu (1960) 159 (III 474). 25ff. Copied by Kṛṣṇalāla Śarman on Sunday 13 śuklapakṣa of Jyeṣṭha in Saṃ. 1903, Śaka 1768 = 7 June 1846.
Anup 2381. 29ff.
AS Bengal 2814 (G 5524). 18ff.
DC 7912. 34ff.
N-W P VIII (1884) 7. 27ff. Property of Pandit Kripaldatta of Benares.
N-W P IX (1885) B 1. 35ff. Property of Pandit Śyāmā Caraṇa of Benares.

Verses 2–3 are:

yatpādāmbujamādhvīkaṃ saṃsevyāvāptasanmatiḥ/
jayaty eṣa guruḥ sākṣād ananto bhaktavatsalaḥ//
govindapādakamaladvandvayojitamānasaḥ/
gṛhasaukhyaṃ nīlakaṇṭho brūte śrīṭoḍarājñayā//

Probably a part of this saukhya is the gṛhapraveśaprakaraṇa. Manuscript:

Kathmandu (1960) 160 (III 474). 12ff. Copied by Kṛṣṇalāla Śarman in Nepāladeśa on Monday 6 śuklapakṣa of Āśvina in Śaka 1767 = 6 October 1845.

4. The *Vyavahārasaukhya* (apparently different from that discussed by P. V. Kane [1930/62] vol. 1, p. 421, using BORI 366 of 1875/76; I assume Anup 2371 and 2372 contain the same text). Manuscripts:

AS Bengal 7117 (G 5530). 70ff. Copied on 13 kṛṣṇapakṣa of Mārgaśīrṣa in Saṃ. 1699 = *ca.* 8 December 1642. Formerly property of Paṇḍita Raghunātha Mālavīya.
Anup 2373. 59ff.
Kotah 253. 59pp.

Verse 2 is:

herambam ambām api ca praṇamya
govindapādāmbujaniṣṭhacetāḥ/
jagaddhitāya vyavahārasaukhyaṃ
brūte sphuṭaṃ ṭoḍaramallabhūpaḥ//

5. The *Saṃskārasaukhya*. Manuscripts:

Anup 2362. Ff. 1–30, 30b–32, and 35–37. Copied by Narasiṃghadāsa Kāyastha in Saṃ. 1737 = A.D. 1680.
Anup 2363 = Bikaner 748. 59ff.
Benares (1963) 35969. 16ff. Incomplete.
Kotah 254. 47pp.

6. The *Samayasaukhya*. Manuscripts:

Alwar 1525.
Anup 2375 = Bikaner 1033. 57ff. Formerly property of Kavīndrācārya Sarasvatī (*fl. ca.* 1600/1675).

There are also a number of manuscripts which the catalogs call only *Ṭoḍarānanda*.

PL, Buhler III E 104. 420ff. Copied in Saṃ. 1631 = A.D. 1574. Property of Rāmabhaṭṭa Agnihotrī of Ahmadābād.
Benares (1963) 35134. 35ff. Copied in Saṃ. 1855 = A.D. 1798.
Baroda 13964. 73ff. Copied in Saṃ. 1996 = A.D. 1939.
Adyar Index 7632.
Ānandāśrama 5088.
CP, Hiralal 1893. Property of Kanhaiyālāl Guru of Saugor.
CP, Kielhorn XXIII 45. 52ff. Property of Lakṣmaṇa Śāstri of Sāgar.
DC 7909. 140ff.
DC 7910. 116ff.
DC 7911. 76ff.
Jammu and Kashmir 2572. 104ff. Incomplete.
Kotah 256. 41pp.

Lucknow 610. T 40 T (45766).
Oudh (1879) VIII 1.112 pp. Property of Śyāma Lāla of Lucknow Zila.
PUL II 4003. 3ff. Incomplete (saṃvatsarānayanotpatti).
VVRI 1474. 72ff. Incomplete.

Nīlakaṇṭha's most popular work was the *Tājikanīlakaṇṭhī*, frequently called simply *Nīlakaṇṭhī*, based on Islamic astrology; there is a ṭīkā, *Rasālā* (1622), by his son Govinda (b. 2 October 1569); another, *Udāhṛti* (1629), by Viśvanātha (*fl.* 1612/1630); a third, *Śiśubodhinī*, by Mādhava (*fl.* 1633); and finally the *Śrīphalavardhinī* of Harṣadhara. The *Nīlakaṇṭhī* consists of 2 tantras, to which a 3rd, Nīlakaṇṭha's *Praśnakaumudī*, is often added:

I saṃjñātantra, sometimes called *Jātakapaddhati* or *Tājikapaddhati;* composed in Saṃ. 1644 = A.D. 1587.

1. grahādhyāya.
2. ṣoḍaśayoga.
3. sahama.
4. pātyāyinīdaśā.

II varṣatrantra or samātantra, completed on 8 śuklapakṣa of Āśvina in Śaka 1509 = *ca.* 29 September 1587.

1. varṣaphala.
2. muthahāphala.
3. ariṣṭa.
4. bhāvavicāra.
5. daśāphala.
6. māsadinaphala.

III praśnatantra, which contains quotations from various works; it is sometimes called *Praśnakaumudī* or *Jyotiṣkaumudī.*

1. grahasvarūpa and bhāvavicāra.
2. bhāvapraśna.
3. viśeṣapraśna.
4. prakīrṇaka.

Manuscripts:

Anup 4709. 23ff. Copied by Padmākara in Śaka 1557 = A.D. 1635.
AS Bengal 7327 (G 1842) = Mitra, Not. 2451. 6ff. Copied by Vaidyanātha, the son of Mahādeva Bhaṭṭa, the son of Rāghavasūri Bhaṭṭa, on 8 śuklapakṣa of Kārttika in Saṃ. 1721 = *ca.* 16 October 1664 Julian. Incomplete (saṃjñātantra).
Anup 4711. 16ff. Copied by Acyuta in Saṃ. 1724 = A.D. 1667. Incomplete (saṃjñātantra missing).
Anup 4707. Ff. 2–29. Copied by Lālajī Bhaṭṭa, the son of Gokula Bhaṭṭa of the Udīcyajñāti, in Saṃ. 1727 = A.D. 1670.
Anup 4712. 11ff. Copied by Lālajī Bhaṭṭa of the Udīcyajñāti in Saṃ. 1727 = A.D. 1670. Incomplete (saṃjñātantra).
Baroda 3174. 11ff. Copied in Saṃ. 1728 = A.D. 1671. With the udāhṛti of Viśvanātha. Incomplete.
LDI (LDC) 2665. 46ff. Copied in Saṃ. 1729 = A.D. 1672. Incomplete (*Tājikapaddhati*).
Nagpur 1017 (1215). 62ff. Copied in Śaka 1609 = A.D. 1687. With a vivṛti. From Nasik.
Anup 4708 = Bikaner 694. 36ff. Copied by Mīṇṭhāka, the son of Gopīnātha of the Nāgajñāti, at Rāyacū in Saṃ. 1746 = A.D. 1689. Formerly property of Anūpasiṃha (*fl.* 1674/1698).
Benares (1963) 37185. 87ff. Copied in Saṃ. 1746 = A.D. 1689. With the ṭīkā of Viśvanātha.
RORI Cat. II 4706. 31ff. (ff. 27–29 missing). Copied by Raghuvatsa in Saṃ. 1752 = A.D. 1695. With the ṭīkā of Viśvanātha. Incomplete (varṣatantra).
Oudh XX (1888) VIII 27. 46pp. Copied in A.D. 1697. Property of Paṇḍita Pratāpa Nārāyaṇa of Allahabad Zila.
PL, Buhler IV E 249. 34ff. Copied in Saṃ. 1756 = A.D. 1699. Incomplete (*Praśnakaumudī*). Property of Bālakṛṣṇa Jośī of Ahmadābād.
AS Bengal 7288 (G 250). 33ff. Copied by Ḍhuṇḍhirāja Tāṭaka on Thursday 10 kṛṣṇapakṣa of Āśvina of Saṃ. 1767, Śaka 1632 = 5 October 1710 Julian from a manuscript copied by Gaṅgādhara of the Audīcyajñāti at Kāśī on Saturday 10 kṛṣṇapakṣa of Mārgaśīrṣa in Śaka 1582 = 15 December 1660 Julian. Incomplete (varṣatantra).
Jaipur (II). 39ff. Copied in Saṃ. 1769 = A.D. 1712. Incomplete (varṣatantra).
Benares (1963) 34502. 20ff. Copied in Saṃ. 1777 = A.D. 1720. Incomplete (varṣatantra).
Benares (1963) 36602. 7ff. Copied in Saṃ. 1781 = A.D. 1724.
Bombay U Desai 1372. 54ff. Copied in Saṃ. 1786 = A.D. 1729. Incomplete (varṣatantra).
Bombay, Kielhorn 15. 72ff. Copied in A.D. 1731. Incomplete (varṣatantra). Ascribed to Divākara.
Jaipur (II). 45ff. Copied in Saṃ. 1792 = A.D. 1735. Incomplete (varṣatantra).
BORI 478 of 1892/95. 332ff. Copied in Saṃ. 1800 = A.D. 1743. With the *Rasālā* of Govinda.
Mithila 132. 12ff. Maithilī. Copied by Manabodha Śarman in Śaka 1665 = A.D. 1743. Incomplete (*Tājikapaddhati*). Property of Paṇḍita Baccā Jhā of Hanuman Nagar, Lohat, Darbhanga.
Oudh XX (1888) VIII 26. 22pp. Copied in A.D. 1745. Property of Paṇḍita Pratāpa Nārāyaṇa of Allahabad Zila.
Poleman 5003 (Columbia, Smith Indic 132). 44ff. Copied in Śaka 1667 = A.D. 1745. Incomplete (varṣatantra).
Benares (1963) 34500. 19ff. Copied in Saṃ. 1810 = A.D. 1753. Incomplete (varṣatantra).
Goṇḍal 442. 12ff. Copied in Saṃ. 1813 = A.D. 1756. Incomplete (saṃjñātantra).
LDI (LDC) 3389/5. Ff. 74–97. Copied in Saṃ. 1814 = A.D. 1757. Incomplete (varṣatantra).

AS Bombay 263. 41ff. Copied in Śaka 1682 = A.D. 1760. With the udāhṛti of Viśvanātha. Incomplete (saṃjñātantra). From Bhāu Dājī.

BORI 865 of 1891/95. 32ff. Copied in Saṃ. 1817 = A.D. 1760. Incomplete (*Jyotiṣkaumudī*).

IO 3048 (2521a). 38ff. Copied by Nāthurāma at Savāī Jaipura on Thursday 3 śuklapakṣa of Caitra in Saṃ. 1817, Śaka 1682 = 20 March 1760. From Gaikawar.

RORI Cat. II 8355. 19ff. Copied by Khemā Jatī at Daśapura in Saṃ. 1818 = A.D. 1761. With the ṭīkā of Viśvanātha.

Benares (1963) 36768 = Benares (1878) 182. 24ff. Copied in Saṃ. 1820 = A.D. 1763. Incomplete (*Jyautiṣakaumudī*).

IO 3047 (2692) 27ff. Copied by Darabāri Brāhmaṇa on 2 śuklapakṣa of Pauṣa in Saṃ. 1822, Śaka 1688 = *ca.* 13 December 1765. Formerly property of Khageśvara, the son of Viśvambhara of the Vatsagotra, a Brāhmaṇa from Kānyakubja. From Colin Mackenzie.

RORI Cat. II 6064. 101ff. Copied by Kāśīnātha Daivajña in Saṃ. 1822 = A.D. 1765. With the ṭīkā of Viśvanātha.

RORI Cat. II 5128. 26ff. Copied by Jñānasundara of the Upakeśa Gaccha at Gūrjarapura in Saṃ. 1825 = A.D. 1768.

Leipzig 1132. 16ff. Copied by Rādhākṛṣṇa in A.D. 1771. Incomplete (praśnatantra).

RORI Cat. III 15382. 105ff. Copied by Maujīrāma Dīkṣita at Vairāṭa in Saṃ. 1828 = A.D. 1771. With the ṭīkā of Viśvanātha. Incomplete (saṃjñātantra).

RORI Cat. II 6311. 30ff. Copied by Manapūra in Saṃ. 1832 = A.D. 1775. Incomplete (varṣatantra and saṃjñātantra).

Poleman 5000 (U Penn 690). 23ff. Copied in Śaka 1699 = A.D 1777.

Poleman 5004 (Columbia, Smith Indic 160, pt. 1). 24ff. Copied in Saṃ. 1834 = A.D. 1777. Incomplete (varṣatantra).

PUL II 3505. Ff. 3–37. Copied in Saṃ. 1834 = A.D. 1777. Incomplete (varṣatantra).

RORI Cat. II 4738. 34ff. Copied by Cimanalāla Brāhmaṇa at Jayapura in Saṃ. 1837 = A.D. 1780.

BORI 880 of 1887/91. 70ff. Copied in Saṃ. 1838 = A.D. 1781. With the ṭīkā of Viśvanātha. Incomplete (saṃjñātantra).

CP, Kielhorn XXIII 39. 21ff. Copied in Śaka 1704 = A.D. 1782. Incomplete (*Jyautiṣakaumudī*). Property of Jānojī Mahārāja of Nāgpur.

AS Bengal 7090 (G 4381). 47ff. Copied by Rājanārāyaṇadeva Śarman in Śaka 1705 = A.D. 1783.

BORI 440 of 1895/98. 63ff. Copied in Saṃ. 1840 = A.D. 1783. Incomplete (varṣatantra).

RORI Cat. I 2916. 46ff. (ff. 4 and 8–10 missing). Copied in Saṃ. 1840 = A.D. 1783. Incomplete.

Benares (1963) 34402. 23ff. Copied in Saṃ. 1841, Śaka 1706 = A.D. 1784. Incomplete (*Praśnakaumudī*).

Calcutta Sanskrit College 48. 20ff. Copied in Saṃ. 1841, Śaka 1897 (read 1707!) = A.D. 1784. Incomplete (varṣatantra).

Mithila 129 G. 43ff. Maithilī. Copied in Śaka 1706 = A.D. 1784. Property of Paṇḍita Dharmadatta Miśra of Babhangama, Supaul, Bhagalpur.

RORI Cat. III 11846. 27ff. Copied by Gopīnātha, the son of Rāmeśvara, in Saṃ. 1841 = A.D. 1784. With the ṭīkā of Viśvanātha. Incomplete (varṣatantra).

RORI Cat. III 11852. 73ff. Copied in Saṃ. 1841 = A.D. 1784. With the ṭīkā of Viśvanātha. Incomplete (saṃjñātantra).

RORI Cat. II 9634. 19ff. Copied in Saṃ. 1843 = A.D. 1786. Incomplete (varṣatantra).

BORI 920 of 1886/92. 141ff. Copied in Saṃ. 1845 = A.D. 1788. With the ṭīkā of Viśvanātha.

Goṇḍal 137. 36ff. Copied in Saṃ. 1846 = A.D. 1789. With the ṭīkā of Viśvanātha. Incomplete (varṣatantra and saṃjñātantra).

Oudh III (1873) VIII 7. 54pp. Copied in A.D. 1789. Incomplete (*Jyotiḥkaumudī*). Property of Paṇḍita Chhoṭe Lāla of Oonao Zillah.

RORI Cat. III 16442. 15ff. Copied in Saṃ. 1846 = A.D. 1789. Incomplete (ṣoḍaśayoga).

Benares (1963) 34841. 19ff. Copied in Saṃ. 1848 = A.D. 1791.

Benares (1963) 35193. 15ff. Copied in Saṃ. 1848 = A.D. 1791. Incomplete (saṃjñātantra).

Benares (1963) 35189. 20ff. Copied in Saṃ. 1850 = A.D. 1793. Incomplete (*Praśnakaumudī*).

Benares (1963) 35192. 15ff. Copied in Śaka 1716 = A.D. 1794. Incomplete (saṃjñātantra).

GJRI 951/63. 7ff. Copied in Saṃ. 1851 = A.D. 1794. Incomplete.

LDI (LDC) 3202. 15ff. Copied in Saṃ. 1851 = A.D. 1794. Incomplete (*Jātakapaddhati*).

BM 495 (Or 5249). 20ff. Copied by Paṇḍita Vasatasundara at Vikramapura on Saturday 8 śuklapakṣa of Vaiśākha in Saṃ. 1853 = 14 May 1796. Incomplete (saṃjñātantra). From H. Jacobi.

IO 3050 (1546c). 47ff. Copied by Jinadatta Ṛṣi on Wednesday 10 śuklapakṣa of Phālguna in Saṃ. 1853 = 8 March 1797. With the udāhṛti of Viśvanātha. Incomplete (saṃjñātantra). From H. T. Colebrooke.

Benares (1963) 35823 = Benares (1897–1901) 645. 26ff. Copied in Saṃ. 1854 = A.D. 1797. Incomplete (*Jyautiṣakaumudī*).

RORI Cat. III 13985. 19ff. Copied by Bhaktāvaramala Muni, the pupil of Sarūpacanda, at Haridurga in Saṃ. 1854 = A.D. 1797. Incomplete (ends in varṣatantra).

Poleman 5009 (U Penn 670). 15ff. Copied in Saṃ. 1855, Śaka 1720 = A.D. 1798. Incomplete (saṃjñātantra).

RORI Cat. II 8362. 58ff. (ff. 1–2 missing). Copied in Saṃ. 1855 = A.D. 1798. With the ṭīkā of Viśvanātha.

IO 3055 (1122b). 19ff. Copied by Mūlacanda in Saṃ. 1856 = A.D. 1799. Incomplete (*Praśnakaumudī*). From H. T. Colebrooke.

IO 3056 (2305). Ff. 1–14 and 16–18. Copied from IO 1122b in A.D. 1799. Incomplete (*Praśnakaumudī*). From Calcutta.

IO 3045 (1122a). 29ff. Copied in A.D. 1800. Incomplete (saṃjñātantra and varṣatantra). From H. T. Colebrooke.

IO 3046 (2306). 22ff. Copied from IO 1122a in A.D. 1800. Incomplete (saṃjñātantra and varṣatantra). From Calcutta.

RORI Cat. II 8221. 16ff. Copied by Bhavānīrāma Jatī in Saṃ. 1857 = A.D. 1800. With the ṭīkā of Viśvanātha.

Benares (1963) 36338. Ff. 1–34, 36–96, and 96b–102. Copied in Śaka 1723 = A.D. 1801. With the ṭīkā of Viśvanātha. Incomplete.

DC 2331. 42ff. Copied in Śaka 1723 = A.D. 1801.

DC 7436. 28ff. Copied in Śaka 1723 = A.D. 1801. Incomplete (varṣatantra).

PUL II 3502. 21ff. Copied in Saṃ. 1858 = A.D. 1801. Incomplete (saṃjñātantra).

Calcutta Sanskrit College 51. 24ff. Copied in Saṃ. 1859 = A.D. 1802. Incomplete (saṃjñātantra).

BORI 155 of A 1883/84. 28ff. Copied in Saṃ. 1860 = A.D. 1803. With the *Rasālā* of Govinda. Incomplete (saṃjñātantra).

DC (Gorhe) 78. 63ff. Copied by Rāghava, the son of Āppājī Khāṇḍekara, at Cikaṇagrāma on Thursday 3 kṛṣṇapakṣa of Mārgaśīrṣa in Śaka 1725 = 1 December 1803. With the ṭīkā of Viśvanātha. Incomplete (saṃjñātantra).

RORI Cat. II 5648. 18ff. Copied by Śivabagasa at Vārāṇasī in Saṃ. 1860 = A.D. 1803. With the ṭīkā of Mādhava.

DC (Gorhe) 79. 28ff. Copied by Rāghava, the son of Āpājī Khāṇḍekara, on Thursday 13 kṛṣṇapakṣa of Jyeṣṭha in Śaka 1726 = 5 July 1804. Incomplete (varṣatantra).

AS Bengal 7101 (G 7811 A). 26ff. Copied on 8 śuklapakṣa of Vaiśākha in Saṃ. 1862 = *ca.* 5 May 1805.

LDI (LDC) 3389/4. Ff. 62–73. Copied in Saṃ. 1862 = A.D. 1805. Incomplete (saṃjñātantra).

RORI Cat. II 5473. 38ff. Copied in Saṃ. 1862 = A.D. 1805. With the *Rasālā* of Govinda. Incomplete (saṃjñātantra).

Goṇḍal 138. 107ff. Copied by Mādhavajī Vyāsa at Goṇḍala in Saṃ. 1864 = A.D. 1807 during the reign of Indradevājī Jāḍejā (1799/1810). With a ṭīkā.

Poleman 5005 (U Penn 712). Ff. 2–57. Copied in Saṃ. 1864 = A.D. 1807. Incomplete (varṣatantra).

Osmania University B. 46/6. 36ff. Copied in A.D. 1809. Incomplete (saṃjñātantra and varṣatantra).

RORI Cat. II 5804. 25ff. Copied in Saṃ. 1866 = A.D. 1809. Incomplete (saṃjñātantra).

Oxford 1552 (Sansk. d. 192) = Hultzsch 291. 22ff. Copied in Saṃ. 1869, Śaka 1734 = A.D. 1812. Incomplete (praśnatantra).

Mithila 129. 32ff. Maithilī. Copied by Ṣarṅgapāṇi Śarman on Wednesday amāvāsyā of the kṛṣṇapakṣa of Vaiśākha in Śaka 1736 = 4 May 1814. Property of Babu Ṭhīṭhara Jhā of Babhangama, Supaul, Bhagalpur.

Mithila 129 C. 35ff. Maithilī. Copied in Śaka 1737 = A.D. 1815. Property of Paṇḍita Dāmodara Jhā of Andhraṭharhi, Darbhanga.

RORI Cat. II 6445. 47ff. Copied by Ratanavijaya at Rūpanagara in Saṃ. 1872 = A.D. 1815 during the reign of Kalyāṇasiṃha. Incomplete (varṣatantra and saṃjñātantra).

SOI 2589 = SOI Cat. II: 1032–2589. 71ff. Copied in Saṃ. 1873, Śaka 1739 = A.D. 1816/17. With the udāharaṇa of Viśvanātha. Incomplete (ends in the sahamādhyāya).

Benares (1963) 37179. 20ff. Copied in Saṃ. 1874 = A.D. 1817. Incomplete (varṣatantra).

GJRI 1085/197. Ff. 41–74. Copied in Saṃ. 1874 = A.D. 1817. Incomplete (varṣatantra, ending at aniṣṭādhyāya 74).

Poleman 5007 (U Penn 703). 41ff. Copied in Saṃ. 1876 = A.D. 1819. Incomplete (varṣatantra).

Benares (1963) 37226. Ff. 16–71. Copied in Saṃ. 1878, Śaka 1743 = A.D. 1821. With the udāhṛti of Viśvanātha. Incomplete.

Benares (1963) 37224. 112ff. Copied in Saṃ. 1879, Śaka 1744 = A.D. 1822. With the udāhṛti of Viśvanātha.

GJRI 948/60. 27ff. Copied in Saṃ. 1879 = A.D. 1822. Incomplete (varṣatantra).

RORI Cat. I 30. 51ff. Copied in Saṃ. 1879 = A.D. 1822. With the ṭīkā of Viśvanātha. Incomplete (saṃjñātantra).

LDI 7199 (68). Copied by Ṛṣi Bihārī, the pupil of Vimalacanda Svāmin, at Amṛtasaranagara in Saṃ. 1880 = A.D. 1823 during the reign of Raṇajīta Siṃha (1799/1839).

RORI Cat. II 9392. Ff. 48–98. Copied by Bagasūrāma Tivāḍī at Kāśī in Saṃ. 1880 = A.D. 1823. With the *Rasālā* of Govinda.

CP, Kielhorn XXIII 79. 20ff. Copied in Saṃ. 1881 = A.D. 1824. Incomplete (*Praśnakaumudī*). Property of Lakṣmaṇa Śāstrī of Sāgar.

Benares (1963) 35195. Ff. 1–2 and 4–13. Copied in Saṃ. 1882 = A.D. 1825. Incomplete.

Oxford 1551 (Sansk. c. 104) = Hultzsch 292. 26ff. Copied in Saṃ. 1882 = A.D. 1825. Incomplete (*Jyautiṣakaumudī*).

Poleman 5008 (Columbia, Smith Indic 160, pt. 2). 15ff. Copied in Saṃ. 1884 = A.D. 1827. Incomplete (saṃjñātantra).

Poleman 5002 (Columbia, Smith Indic 160, pt. 3).

13ff. Copied in Saṃ. 1885, Śaka 1750 = A.D. 1828. Incomplete (praśnatantra).

RORI Cat. II 5768. 28ff. Copied by Vīrabala at Amṛtasara in Saṃ. 1888 = A.D. 1831. Incomplete (praśnatantra).

Florence 332. 18ff. Copied in Saṃ. 1889 = A.D. 1832. Incomplete (saṃjñātantra).

GVS 2827 (2677). 43ff. Copied on Thursday 6 śuklapakṣa of Pauṣa in Saṃ. 1899 (read 1889), Śaka 1755 = 27 December 1832.

LDI 6898 (2704). 18ff. Copied by Rūpacandra at Nāgapura in Saṃ. 1889 = A.D. 1832.

RORI Cat. II 7945. 34ff. Copied at Ahipura in Saṃ. 1889 = A.D. 1832. With the *Rasālā* of Govinda. Incomplete (bhāvādhyāya).

Benares (1963) 37180. 25ff. Copied in Saṃ. 1890 = A.D. 1833. Incomplete (varṣatantra).

Mithila 129 F. 26ff. Maithilī. Copied in Śaka 1755 = A.D. 1833. Property of Paṇḍita Gokulanātha Jhā of Nanaur, Tamuria, Darbhanga.

Oxford 1546 (Sansk. e 82) = Hultzsch 328. 41ff. Copied in Saṃ. 1890 = A.D. 1833. Incomplete (varṣatantra).

RORI Cat. II 8683. 20ff. Copied in Saṃ. 1890 = A.D. 1833. Incomplete (saṃjñātantra).

Benares (1963) 35351. 35ff. Copied in Saṃ. 1891 = A.D. 1834.

Benares (1963) 35694. 35ff. Copied in Saṃ. 1891 = A.D. 1834. Incomplete (varṣatantra).

Benares (1963) 37158. 16ff. Copied in Saṃ. 1891 = A.D. 1834. Incomplete (saṃjñātantra).

PUL II 3497. 42ff. Copied in Saṃ. 1891 = A.D. 1834.

RORI Cat. III 10548. 75ff. Copied by Bihārī Lāla in Saṃ. 1891 = A.D. 1834. With the ṭīkā of Viśvanātha. Incomplete (saṃjñātantra).

RORI Cat. II 8200. 16ff. Copied in Saṃ. 1892 = A.D. 1835. Incomplete (*Jyotiṣakaumudī*).

AS Bengal 7089 (G 3466). 47ff. Copied by Mādhavacandra Śiromaṇi, the son of Darpanārāyaṇa Śarman, the son of Kṛṣṇa, at Kuṭanīgrāma on 17 Caitra in Śaka 1758, BS 1243 = *ca.* 2 April 1836. Incomplete (saṃjñātantra and varṣatantra).

Kerala 6015 (5407). 525 granthas. Copied in Saṃ. 1894, Śaka 1759 = A.D. 1837. Incomplete (*Jyotiṣakaumudī*).

PUL II 3496. 40ff. Copied in Saṃ. 1895 = A.D. 1838.

VVRI 4479. 19ff. Copied in Saṃ. 1895 = A.D. 1838.

BORI 1011 of 1886/92. 27ff. Copied in Saṃ. 1896 = A.D. 1839. With the ṭīkā of Mādhava. Incomplete (samāviveka).

Leipzig 1127. 13ff. Copied by Rāmacandra in A.D. 1839. Incomplete (varṣatantra 1–4).

Benares (1963) 36263. 24ff. Copied in Saṃ. 1898, Śaka 1763 = A.D. 1841. Incomplete (saṃjñātantra).

Benares (1963) 36264. Ff. 1–38 and 40–45. Copied in Saṃ. 1898 = A.D. 1841. Incomplete (varṣatantra).

Poleman 5001 (Harvard 316). 18ff. Copied in Saṃ. 1898 = A.D. 1841. Incomplete (ṣoḍaśayoga).

VVRI 1172. 36ff. Copied in Saṃ. 1898 = A.D. 1841.

Baroda 9472. 101ff. Copied in Saṃ. 1901 = A.D. 1844. With the udāhṛti of Viśvanātha.

Benares (1963) 34501. 30ff. Copied in Saṃ. 1903 = A.D. 1846. Incomplete (varṣatantra).

RORI Cat. II 5265. 58ff. (ff. 1–4 missing). Copied by Kiśorīlāla in Saṃ. 1903 = A.D. 1846. Incomplete (*Jātakapaddhati*).

RORI (Jaipur) II 25. 46ff. Copied in Saṃ. 1904 = A.D. 1847.

RORI Cat. II 9157. 17ff. Copied by Jayadeva at Phāgī, Jayapura, in Saṃ. 1905 = A.D. 1848.

Benares (1963) 36335. Ff. 1–19 and 21–41. Copied in Saṃ. 1906 = A.D. 1849. Incomplete.

RORI Cat. I 2936. 42ff. Copied in Saṃ. 1908 = A.D. 1851.

BORI 417 of 1895/98. 23ff. Copied in Saṃ. 1909 = A.D. 1852. Incomplete (*Jyotiṣkaumudī*).

Benares (1963) 35340. 17ff. Copied in Saṃ. 1909 and 1910 = A.D. 1852 and 1853. Incomplete (saṃjñātantra and varṣatantra). This is probably Benares (1897–1901) 220. 36ff. Copied in Saṃ. 1910 = A.D. 1853.

Benares (1963) 36336. Ff. 1–17, 20–32, 41, 43–45, and 47–61. Copied in Saṃ. 1910 = A.D. 1853. With the ṭīkā of Viśvanātha. Incomplete.

RORI Cat. II 5185. 54ff. Copied in Saṃ. 1910 = A.D. 1853. With the ṭīkā of Viśvanātha. Incomplete (saṃjñātantra).

Baroda 3142. 44ff. Copied in Saṃ. 1911 = A.D. 1854. Incomplete (saṃjñātantra).

RORI Cat. III 17046. 74ff. Copied by Rāmacandra Jośī, the son of Viradhīcanda, at Stanamaṇi in Saṃ. 1913 = A.D. 1856. With the ṭīkā of Viśvanātha.

Leipzig 1129. 5ff. Copied in A.D. 1858. Incomplete (ṣoḍaśayoga).

Leipzig 1133. 27ff. Copied by Mūlaśaṅkara, the son of Abhayaśaṅkara, in A.D. 1858. Incomplete (praśnatantra).

Mithila 129 J. 64ff. Maithilī. Copied in Śaka 1780, Sāl. San. 1206 = A.D. 1858. Property of Śaśinātha Miśra of Tarauni, Sakri, Darbhanga.

RORI Cat. II 9138. 36ff. (f. 1 missing). Copied in Saṃ. 1915 = A.D. 1858.

Baroda 7684. 41ff. Copied in Saṃ. 1916 = A.D. 1859.

RORI Cat. II 8201. 37ff. Copied by Baladeva, a resident of Nandagrāma, at Karaulī in Saṃ. 1916 = A.D. 1859.

Benares (1963) 35658 = Benares (1905) 1503. 26ff. Copied in Saṃ. 1917 = A.D. 1860. Incomplete (*Praśnakaumudī*). No author mentioned.

Goṇḍal 440. 5ff. Copied by Jyeṭhṭhārāma Raghunātha Rāvala at Ṭāṅkārā on 5 kṛṣṇapakṣa of Vaiśākha in Saṃ. 1918 = *ca.* 23 May 1861. Incomplete (sahama from the *Tājakabhūṣaṇa*).

LDI 7028 (1181). 68ff. Copied in Saṃ. 1918 = A.D. 1861. With the vṛtti of Viśvanātha. Incomplete (saṃjñātantra).
RORI Cat. II 8645. 70ff. (f. 48 missing). Copied by Dvārikānātha in Saṃ. 1919 = A.D. 1862. With the ṭīkā of Viśvanātha.
RORI Cat. III 10914. 138ff. Copied by Muralīdāsa Pujāri in Saṃ. 1921 = A.D. 1864.
SOI 3378 = SOI Cat. II: 1014–3378. 27ff. Copied in Saṃ. 1921 = A.D. 1864.
Nagpur 807 (2023). No ff. given. Copied in Śaka (read Saṃ.) 1922 = A.D. 1865. From Nagpur.
LDI (LDC) 3049. 52ff. Copied in Saṃ. 1926 = A.D. 1869. With a vivaraṇa. Incomplete (varṣatantra).
Mithila 129 E. 43ff. Maithilī. Copied in Śaka 1794 = A.D. 1872. Property of Paṇḍita Rudramaṇi Jhā of Mahinathapur, Deodha, Darbhanga.
Calcutta Sanskrit College 189. 19ff. Copied in Saṃ. 1953 = A.D. 1896. Incomplete (varṣatantra).
ABSP 69. 42ff. Incomplete.
ABSP 136. 17ff. Incomplete (saṃjñāviveka).
ABSP 1175. Ff. 1–19. Incomplete.
ABSP 1233. 1f. Incomplete.
Alwar 1790. Incomplete (*Jyotiṣkaumudī*). 2 copies.
Alwar 1801. With the *Rasālā* of Govinda. Incomplete (saṃjñātantra). 2 copies.
Alwar 1802. With the ṭīkā of Mādhava.
Alwar 1988. With the ṭīkā of Viśvanātha. Incomplete (saṃjñātantra).
Ānandāśrama 1872. With a ṭīkā. Incomplete (varṣatantra).
Ānandāśrama 1873. With a ṭīkā. Incomplete (saṃjñātantra).
Ānandāśrama 1876. Incomplete (saṃjñātantra and varṣatantra).
Ānandāśrama 1984. Incomplete (saṃjñātantra).
Ānandāśrama 1985. Incomplete (varṣatantra).
Ānandāśrama 2748. With a ṭīkā. Incomplete (saṃjñātantra and varṣatantra).
Ānandāśrama 3364. Incomplete (varṣatantra).
Ānandāśrama 4251. Incomplete (varṣatantra).
Ānandāśrama 4801.
Ānandāśrama 5629. Incomplete (varṣatantra).
Ānandāśrama 5630. Incomplete (saṃjñātantra).
Ānandāśrama 6664.
Ānandāśrama 7640.
Ānandāśrama 7784. Incomplete (saṃjñātantra).
Anup 4623. 15ff. Incomplete (*Jātakapaddhati*).
Anup 4710. 25ff. Incomplete.
Anup 4713. 11ff. Incomplete.
Anup 4714. 4ff. Incomplete. Formerly property of Maṇirāma Dīkṣita (*fl. ca.* 1675/1700).
Anup 4715. 3ff. Incomplete (sahamaphala). Formerly property of Anūpasiṃha (*fl.* 1674/1698).
Anup 4864. 28ff. Incomplete (*Praśnakaumudī*). Formerly property of the Jyotiṣarāja (b. 1613).
Anup 4865. 22ff. Incomplete (*Praśnakaumudī*). Formerly property of the Jyotiṣarāja (b. 1613).
Anup 4875 = Bikaner 707. 24ff. Incomplete (praśnatantra). Formerly property of Maṇirāma Dīkṣita (*fl. ca.* 1675/1700).
AS Bengal 7091 (G 5504). 130ff. With the *Śrīphalavardhinī* of Harṣadhara. Incomplete (saṃjñātantra).
AS Bengal 7092 (G. 4307). 66ff. With the ṭīkā of Viśvanātha on the saṃjñātantra and the ṭīkā of Mādhava on the varṣatantra. Incomplete (saṃjñātantra and varṣatantra).
AS Bengal 7093 (G 5564). 56ff. and 46ff. With the ṭīkā of Viśvanātha. Incomplete (saṃjñātantra and varṣatantra).
AS Bengal 7106 (G 3442) III. No ff. given.
AS Bengal 7312 (G 7927). 33ff. With the ṭīkā of Viśvanātha. Incomplete (ṣoḍaśayoga).
AS Bombay 261. 26ff. Incomplete (varṣatantra). From Bhāu Dājī.
AS Bombay 262. 26ff. Incomplete (saṃjñātantra). From Bhāu Dājī.
Baroda 3175. 31ff. With the udāhṛti of Viśvanātha.
Baroda 7698. 45ff. Incomplete (varṣatantra).
Baroda 13997. 76ff. With the ṭīkā of Viśvanātha. Incomplete (saṃjñātantra).
Benares (1963) 34503. 7ff. Incomplete.
Benares (1963) 34504. 10ff. Incomplete.
Benares (1963) 34505. 38ff. Incomplete.
Benares (1963) 34506 = Benares (1878) 43 = Benares (1869) X 1. 12ff. Incomplete (saṃjñātantra in Benares (1878) and (1869)).
Benares (1963) 34507. 2ff. Incomplete.
Benares (1963) 34508 = Benares (1878) 45. 18ff. This is probably Benares (1869) X 2. 17ff. Incomplete (varṣatantra in Benares (1878) and 1869)).
Benares (1963) 34669. 16ff. Incomplete.
Benares (1963) 34734. Ff. 1–77, 79–85, and 87–97. With the ṭīkā of Viśvanātha. Incomplete.
Benares (1963) 34842. Ff. 2–4. Incomplete.
Benares (1963) 34843 = Benares (1869) XXXIX 4. 17ff. With the *Rasālā* of Govinda. Incomplete (varṣamāsadinapraveśavicāra).
Benares (1963) 34844. Ff. 1–4 and 7–57. With the *Rasālā* of Govinda. Incomplete (saṃjñāviveka).
Benares (1963) 35099. Ff. 1, 7–8, 10–18, 20–22, and 24. Incomplete (varṣatantra).
Benares (1963) 35102. 19ff. Incomplete.
Benares (1963) 35108. Ff. 2–20. Incomplete.
Benares (1963) 35137. Ff. 3–6, 8–11, and 13–18. Incomplete.
Benares (1963) 35138. Ff. 8–21. Incomplete.
Benares (1963) 35139. Ff. 1–9, 26–27, 29–33. Incomplete.
Benares (1963) 35194. 27ff. Incomplete.
Benares (1963) 35196. Ff. 1–38, 40–56, and 56b–61, and 1f. Incomplete.
Benares (1963) 35404. Ff. 1–44, 1–35, and 1–38. With the udāhṛti of Viśvanātha. Incomplete.
Benares (1963) 35443. Ff. 1–5 and 7–9. Incomplete (sahamavicāra).

Benares (1963) 35572. 14ff. Incomplete (saṃjñāviveka).
Benares (1963) 35832. 19ff. With the ṭīkā of Mādhava. Incomplete (varṣatantra).
Benares (1963) 36099. Ff. 2–6. Incomplete (*Praśnakaumudī*). No author mentioned.
Benares (1963) 36171. Ff. 17 and 24–27. Incomplete (*Jyautiṣakaumudī*).
Benares (1963) 36173. 76ff. Incomplete.
Benares (1963) 36197. 5ff. Incomplete.
Benares (1963) 36265. 45ff. With a ṭīkā. Incomplete.
Benares (1963) 36285. 5ff. Incomplete.
Benares (1963) 36286. Ff. 24–35 and 37. Incomplete.
Benares (1963) 36287. Ff. 1–6 and 8–22. Incomplete.
Benares (1963) 36330 = Benares (1878) 44 = Benares (1869) X 4. 13ff. Incomplete (saṃjñātantra). In Benares (1869) said to have been copied in Saṃ. 1800 = A.D. 1743.
Benares (1963) 36333. Ff. 2–18. Incomplete.
Benares (1963) 36334. Ff. 5–34. Incomplete.
Benares (1963) 36337. Ff. 1–44 and 1f. With the ṭīkā of Viśvanātha. Incomplete.
Benares (1963) 36339. 18ff. Incomplete.
Benares (1963) 36340. 23ff.
Benares (1963) 36341 = Benares (1878) 43 = Benares (1869) X 3. Ff. 1–8 and 10–15. Incomplete. In Benares (1869) said to have been copied in Saṃ. 1725 = A.D. 1668.
Benares (1963) 36342. 12ff. Incomplete.
Benares (1963) 36343. Ff. 13–14 and 16–32. Incomplete.
Benares (1963) 36344. 9ff. Incomplete.
Benares (1963) 36345. 12ff. Incomplete.
Benares (1963) 36346. Ff. 4–17. Incomplete (saṃjñātantra).
Benares (1963) 36348. Ff. 1–2, 4–21, and 23. Incomplete.
Benares (1963) 36349. Ff. 2–26. With the *Rasālā* of Govinda. Incomplete.
Benares (1963) 36350. Ff. 3–12. Incomplete (varṣatantra).
Benares (1963) 36352. 15ff. Incomplete.
Benares (1963) 36395. 6ff. Bengālī. Incomplete (*Praśnakaumudī*). Ascribed to Divākara.
Benares (1963) 36599. Ff. 2–6 and 23–26. Incomplete.
Benares (1963) 36600. Ff. 3–12. Incomplete.
Benares (1963) 36601. Ff. 1 and 11–32. Incomplete.
Benares (1963) 36603. 17ff. Incomplete.
Benares (1963) 36604. Ff. 2–10, 13–23, and 38–42. Incomplete.
Benares (1963) 36605. 20ff. Incomplete.
Benares (1963) 36606. 18ff. Incomplete.
Benares (1963) 36766. 17ff. Incomplete (*Jyautiṣakaumudī*).
Benares (1963) 36767. 23ff. Incomplete (*Jyautiṣakaumudī*).
Benares (1963) 36809. Ff. 22–23. Incomplete (varṣatantra).
Benares (1963) 36820. 20ff. Incomplete.
Benares (1963) 36873. 4ff. Incomplete.
Benares (1963) 36889. Ff. 21–22. Incomplete.
Benares (1963) 37151. 18ff. Incomplete (*Jyautiṣakaumudī*).
Benares (1963) 37157. 33ff. Incomplete.
Benares (1963) 37181. 17ff. Incomplete.
Benares (1963) 37236. 17ff. With the udāhṛti of Viśvanātha. Incomplete.
Benares (1963) 37237. Ff. 1–11 and 35–43. With the *Rasālā* of Govinda. Incomplete.
Berlin 876 (Chambers 688a). 18ff. Incomplete (saṃjñātantra and varṣatantra).
Bharatpur S 11. Incomplete (varṣatantra).
BM 494 (Add. 14,359a). 23ff. With the ṭīkā of Viśvanātha. Incomplete (saṃjñātantra). From Major T. B. Jervis.
BM Or. 6825.
Bombay U 415. 20ff. Incomplete (saṃjñātantra).
Bombay U 416. 4ff. Incomplete (saṃjñātantra).
Bombay U 417. 80ff. With the udāhṛti of Viśvanātha. Incomplete (saṃjñātantra).
Bombay U 418. 57ff. With the udāhṛti of Viśvanātha. Incomplete (saṃjñātantra, ending in sahamādhyāya).
Bombay U Desai 1371. Ff. 1–12 and 12b–23. Incomplete (saṃjñātantra).
Bombay U Desai 1463. 30ff. Incomplete (*Jyotiṣakaumudī*).
BORI 414 of 1884/86. 25ff. Incomplete (varṣatantra).
BORI 874 of 1884/87. 50ff. (41ff. missing). With the *Rasālā* of Govinda. Incomplete (saṃjñāviveka). From Gujarāt.
BORI 904 of 1884/87. 15ff. (f. 3 missing). Incomplete (varṣatantra). From Mahārāṣṭra.
BORI 821 of 1887/91. 31ff. Incomplete (saṃjñātantra). From Gujarāt.
BORI 529 of 1899/1915. 41ff.
BORI 544 of 1899/1915. 44ff.
Calcutta Sanskrit College 18. 36ff. Incomplete (varṣatantra).
Calcutta Sanskrit College 50. 14ff. Incomplete (saṃjñātantra).
Calcutta Sanskrit College 52. 12ff. Incomplete (saṃjñātantra).
Cambridge University 153.
CP, Hiralal 1993. Property of Nāgnāth Vināyak of Bāsīm, Akolā.
CP, Hiralal 1994. Property of Bhagvāndās of Bārhā, Narsinghpur.
CP, Hiralal 1995. Property of the Bhonsalā Rājās of Nāgpur.
CP, Hiralal 1996. Property of Śāligrām of Hoshangābād.
CP, Hiralal 1997. Property of Rāmeśvar of Hoshangābād.
CP, Hiralal 1998. Property of Govind Joshī of Jubbulpore.

CP, Hiralal 1999. Property of Govindbhaṭ of Jubbulpore.
CP, Hiralal 2000. Property of Vāsudevrāv Golvalkar of Maṇḍlā.
CP, Hiralal 2671. Property of Govind Śāstrī of Mañgalā, Bilāspur.
CP, Hiralal 2672. Property of Sādhurām Brāhmaṇ of Salemanābād, Jubbulpore.
CP, Hiralal 2673. Property of Govindprasād Śāstrī of Jubbulpore.
CP, Hiralal 2674. Property of Śivrām of Hoshangābād.
CP, Hiralal 2675. Property of Bhagvāndās of Māthon, Saugor.
CP, Hiralal 2676. Property of Kanhaiyālāl Guru of Saugor.
CP, Hiralal 2677. Property of Rāmkrishṇa Guṇvant of Mangrulpir, Akolā.
CP, Hiralal 2678. Property of Gopāl Jaikrishṇa of Kuṭāsā, Akolā.
CP, Hiralal 2679. Property of Tukārām Pāṭhak of Yeodā, Amraotī.
CP, Hiralal 2680. Property of Lakshmaṇbhaṭṭ of Brahmapurī, Chāndā.
CP, Hiralal 2681. Property of Devnāth of Ḍoñgargaon, Bhaṇḍārā.
CP, Hiralal 2682. Property of Gaṇeśabhaṭṭ Dakshiṇadās of Haṭṭā, Damoh.
CP, Hiralal 2683. Property of Mādhavrāv of Damoh.
CP, Hiralal 3159. Incomplete (*Praśnakaumudī*). Property of the Chaube family of Gaṛhā, Jubbulpore.
CP, Hiralal 5018. With the ṭīkā of Viśvanātha. Incomplete (varṣatantra). Property of Govindbhaṭṭ of Jubbulpore.
CP, Hiralal 6155. Incomplete (saṃjñāviveka). Property of Śrīnivāsrāv of Ratanpur, Bilāspur.
CP, Kielhorn XXIII 69. 114ff. With the ṭīkā of Viśvanātha. Property of Javāhara Śāstrī of Chāndā.
DC 3300. 37ff. Incomplete (varṣatantra).
DC 3332. 46ff. Incomplete (varṣatantra).
DC (Gorhe) App. 191. Property of Śañkara Bālakṛṣṇa Lumpāṭhakī of Puṇatāmbe, Ahmadnagar.
Dharwar 702 (692). 81ff. With a *Prakāśikā*, presumably of Viśvanātha. Incomplete (saṃjñātantra).
Florence 333. 21ff. Incomplete (saṃjñātantra).
Florence 334. 50ff. With the ṭīkā of Viśvanātha. Incomplete (saṃjñātantra).
GJRI 949/61. Ff. 1–5 and 7–9. Incomplete (saṃjñātantra).
GJRI 950/62. 11ff. With the ṭīkā of Viśvanātha. Incomplete.
GJRI 965/77. 78ff. Maithilī.
GJRI 966/78. 4ff. Maithilī. Incomplete.
GJRI 1086/198. Ff. 14–37. Incomplete (varṣatantra).
GJRI 1087/199. Ff. 2–18. Incomplete (varṣatantra).
GJRI 2952/308. Ff. 21–39. Incomplete.
GJRI 3152/364. 18ff. Maithilī.
GJRI 3153/365. 6ff. Maithilī. Incomplete.
GJRI 3154/366. 62ff. Maithilī.
GJRI 3155/367. 54ff. Maithilī. Incomplete.
GJRI 3156/368. 22ff. Incomplete.
GJRI 3157/369. 74ff. With the ṭīkā of Viśvanātha. Incomplete.
GOML Madras D 14047. 114pp. Telugu. With a vyākhyāna. Incomplete.
GOML Madras D 14074. 231pp. Telugu and Grantha. With the vyākhyāna of Viśvanātha. Incomplete (saṃjñātantra).
GVS 2825 (1567). 33ff. Incomplete.
GVS —— (3402). Ff. 1 and 3–9. Incomplete.
GVS —— (3814). Ff. 15–35 and 29ff. Incomplete (*Jātakapaddhati*).
GVS —— (3838 C). No ff. given.
GVS —— (3861) Ff. 26–31 and 37. Incomplete.
GVS —— (4534). Ff. 5–38. Incomplete.
IO 3049 (1519d). 32ff. Bengālī. Incomplete (varṣatantra). From H. T. Colebrooke.
IO 3051 (2491). 97ff. With the udāhṛti of Viśvanātha. Incomplete (saṃjñātantra). From Gaikawar.
IO 6347 (Mackenzie II. 47a). 49ff. From Colin Mackenzie.
Jaipur (II). Incomplete (*Jyautiṣakaumudī*).
Jaipur (II). 22ff.
Jaipur (II). 36ff. Incomplete (varṣatantra).
Jaipur (II). 35ff. Incomplete (varṣatantra).
Jaipur (II). 15ff. Incomplete (saṃjñātantra).
Jaipur (II). 18ff. Incomplete (varṣatantra).
Jaipur (II). Ff. 4–18. Incomplete (saṃjñātantra).
Jaipur (II). 29ff. Incomplete (varṣatantra).
Jaipur (II). 8ff. Incomplete (saṃjñātantra).
Jammu and Kashmir 2781. 59ff. With the udāhṛti of Viśvanātha. Incomplete (saṃjñāviveka).
Jammu and Kashmir 2967. 120ff. With the *Śrīphalavardhinī* of Harṣadhara. Incomplete (ends with prakaraṇa 2).
Jammu and Kashmir 2968. 60ff. With the *Śrīphalavardhinī* of Harṣadhara. Incomplete (varṣaphala).
Jammu and Kashmir 4989. 21ff. With the *Śiśubodhinī* of Mādhava. Incomplete (sahamaprakaraṇa).
Kavīndrācārya 887. With an udāharaṇa. Incomplete (jātaka).
Kathmandu (1960) 133 (III 338). 25ff. Nevārī. Incomplete (*Jyautiṣakaumudī*).
Kathmandu (1960) 165 (III 432). 56ff. With the ṭīkā of Mādhava. Incomplete (saṃjñātantra).
Kathmandu (1960) 397 (I 1414). 69ff. Nevārī. Incomplete (varṣatantra).
Kathmandu (1960) 452 (III 432). 46ff. With the *Rasālā* of Govinda. Incomplete (samāviveka).
Kerala 6016 (10184). 500 granthas. Incomplete (*Jyotiṣakaumudī*).
Kerala 6715 (1707). 1500 granthas. With the *Rasālā* of Govinda. Incomplete.
Kotah 242. 29pp.
Kurukṣetra 393 (50129).

LDI (LDC) 1185. 31ff. Incomplete (saṃjñātantra).
LDI (LDC) 2521. 56ff. Incomplete (varṣatantra).
LDI (LDC) 2590. 158ff. Incomplete (*Jyotiṣakaumudī*).
LDI (LDC) 2718. 11ff. Incomplete (varṣatantra).
LDI (LDC) 3124. No ff. given. Incomplete (varṣatantra).
LDI (LDC) 3329/119. Ff. 244–265. Incomplete (*Paddhati*).
LDI (LDC) 4269. 34ff. Incomplete (varṣatantra).
Leipzig 1126. 53ff. Copied by Nandikiśora. Incomplete (saṃjñātantra and varṣatantra).
Leipzig 1128. 12ff. Incomplete (saṃjñātantra).
Lucknow 520. N 62 N (45706).
Mithila 129 A. 14ff. Maithilī. Incomplete. Property of Paṇḍita Raghunātha Jhā of Sanakorthu, Manigachi, Darbhanga.
Mithila 129 B. 31ff. Maithilī. Incomplete. Property of Paṇḍita Babujan Jhā of Sasipur, Jogiara, Darbhanga.
Mithila 129 D. 31ff. Maithilī. Property of Paṇḍita Mahīdhara Miśra of Lalabag, Darbhanga.
Mithila 129 H. 66ff. Maithilī. Incomplete. Property of Paṇḍita Sītārāma Pāṭhaka of Karnpur, Sukpur, Bhagalpur.
Mithila 384. 36ff. Maithilī. Incomplete (saṃjñātantra). Property of Babu Cetamaṇi Siṃha of Sukpur, Bhagalpur.
Mithila 384 A. 28ff. Maithilī. Incomplete (saṃjñāviveka). Property of Babu Satyanārāyaṇa Miśra of Balava, Nowhatta, Bhagalpur.
Mithila 384 B. 13ff. Maithilī. Incomplete (saṃjñāviveka). Property of Paṇḍita Jayānanda Miśra of Parsarma, Bhagalpur.
Mysore (1922) 989. 22ff.
Mysore (1911 + 1922) C 583. 143ff. With the vyākhyā of Mādhava. Incomplete (saṃjñātantra and varṣatantra).
Mysore and Coorg 298. No granthas given. Property of Mahādeva Joyisa of Sringeri.
Nagpur 1014 (1433). 23ff. From Nasik.
Nagpur 1015 (1769). 73ff. From Nagpur.
Nagpur 1016 (2369). 80ff. From Nagpur.
N-W P I (1874) 97. 50ff. Incomplete (*Praśnakaumudī*). No author mentioned. Property of Rāmakṛṣṇa of Benares.
N-W P II (1877) B 28. 11ff. Incomplete (varṣatantra). Property of Khuśālpurī of Benares.
N-W P II (1877) B 29. 47ff. Incomplete (praśnatantra). No author mentioned. Property of Khuśālpurī of Benares.
Oppert I 8042. Property of Paravastu Veṅkaṭaraṅgācāryār of Viśākhapaṭṭana, Vizagapatam.
Oppert II 1979. 14pp. Telugu. Property of Veṅkaṭeśvarajosya of Siddhavaṭa, Kaḍapa.
Oppert II 3181. Property of Taḍakamalla Veṅkaṭakṛṣṇarāyar of Tiruvallikeṇī, Madras.
Oppert II 5218. Property of Piccudīkṣitar of Akhilāṇḍapuram, Tanjore.
Osmania University B. 46/2. 113ff. With the *Śiśubodhinī* of Mahādeva (i.e., Mādhava).
Oudh XX (1888) VIII 23. 114pp. Property of Paṇḍita Pratāpa Nārāyaṇa of Allahabad Zila.
Oudh XX (1888) VIII 24 = VIII 25. 16pp. Property of Paṇḍita Pratāpa Nārāyaṇa of Allahabad Zila.
Oudh XX (1888) VIII 28. 30pp. Property of Paṇḍita Pratāpa Nārāyaṇa of Allahabad Zila.
Oudh XX (1888) VIII 29. 82pp. Property of Paṇḍita Pratāpa Nārāyaṇa of Allahabad Zila.
Oudh XX (1888) VIII 30. 50pp. Property of Paṇḍita Pratāpa Nārāyaṇa of Allahabad Zila.
Oudh XX (1888) VIII 31. 34pp. Property of Paṇḍita Pratāpa Nārāyaṇa of Allahabad Zila.
Oudh XX (1888) VIII 32. 180pp. Property of Paṇḍita Pratāpa Nārāyaṇa of Allahabad Zila.
Oudh XXI (1889) VIII 16. 449pp. Property of Raghuvara Prasāda of Gonda Zila.
Oxford 1562 (Sansk. c. 116) = Hultzsch 327. 43ff. Incomplete (varṣatantra).
Oxford CS d. 778 (vii). 9ff. Incomplete (varṣaphala).
Oxford CS d. 808 (v). 25ff.
PL, Buhler IV E 437. 15ff. Incomplete (varṣatantra). Property of Caturbhujabhaṭṭa of Navānagara.
Poleman 5000a (U Penn 1808). 42ff.
Poleman 5000b (U Penn 1867). 50ff.
Poleman 5000c (U Penn 1916). 5ff. Incomplete.
Poleman 5006 (U Penn 651). Ff. 5–18. Incomplete (varṣaphala).
Poleman 5010 (U Penn 1825). 16ff. Incomplete (saṃjñātantra).
Poleman 5011 (Harvard 1107). 24ff. With the udāharaṇa of Viśvanātha. Incomplete (saṃjñātantra).
Probstain 50. 58ff. With the ṭīkā of Viśvanātha.
PrSB 484 (or. fol. 3356; now at Marburg). 20ff. Incomplete (saṃjñātantra and varṣatantra).
PUL II 3256. 2ff. Incomplete (ariṣṭādhyāya).
PUL II 3498. 37ff.
PUL II 3499. 51ff.
PUL II 3500. 27ff. (f. 4 missing).
PUL II 3501. 9ff. Incomplete (saṃjñātantra).
PUL II 3503. Ff. 6–22. Incomplete (saṃjñātantra).
PUL II 3504. 28ff. Incomplete (varṣatantra).
PUL II 3525. 60ff. Grantha (*Tājikaratnākara*).
PUL II 3997. 2ff. Incomplete (ṣoḍaśayoga).
Rajputana, p. 46. Incomplete (varṣatantra). From Bikaner.
Rajputana, p. 47. Incomplete (saṃjñātantra). From Bikaner.
RAS (Tod) 23. 16ff. Incomplete (saṃjñātantra).
RJ 1688 (vol. 2, p. 273). 35ff. Incomplete (varṣatantra). Property of Baḍā Terahapanthiyoṃ of Jayapura.
RJ 452 (vol. 3, p. 245). 56ff. Property of Ṭholiyoṃ of Jayapura.
RJ 3017 (vol. 4, p. 285). 14ff. Incomplete.

RORI Cat. I 3717. 48ff. Incomplete (varṣatantra).
RORI Cat. II 4864. 44ff. Incomplete (varṣatantra).
RORI Cat. II 4996. 21ff. Incomplete (*Jātakapaddhati*).
RORI Cat. II 5350. 23ff. With the ṭīkā of Viśvanātha. Incomplete.
RORI Cat. II 5530. 89ff. (f. 4 missing). With a ṭīkā. Incomplete (*Jātakapaddhati*).
RORI Cat. II 5582. 114ff. With the *Rasālā* of Govinda. Incomplete (varṣatantra).
RORI Cat. II 5583. 123ff. With the *Rasālā* of Govinda. Incomplete (saṃjñātantra).
RORI Cat. II 6049. 36ff. With the *Rasālā* of Govinda. Incomplete (saṃjñātantra).
RORI Cat. II 6660. 20ff. Incomplete (saṃjñātantra).
RORI Cat. II 8216. Ff. 6–21. Incomplete (*Jyotiṣakaumudī*).
RORI Cat. II 9391. 31 (read 81?)ff. (ff. 23–50, 62, 67, and 69 missing). With the *Rasālā* of Govinda. Incomplete.
RORI Cat. II 9393. 38ff. With the *Rasālā* of Govinda. Incomplete (saṃjñātantra).
RORI Cat. III 10167. 25ff. (f. 24 missing). With the ṭīkā of Viśvanātha. Incomplete (saṃjñātantra).
RORI Cat. III 10252. 10ff. Incomplete (saṃjñātantra).
RORI Cat. III 10673. 35ff. With the ṭīkā of Viśvanātha. Incomplete (varṣatantra).
RORI Cat. III 10674. 53ff. With the ṭīkā of Viśvanātha. Incomplete (saṃjñātantra).
RORI Cat. III 11146. 107ff. (ff. 1–20, 30–39, 75, 82–83, and 90 missing). With the ṭīkā of Mādhava. Incomplete (uttarārdha).
RORI Cat. III 11147. 123ff. (ff. 1, 4, 6–7, 9–10, 12, 17, 42, 50, 56, 58, 70, 92–93, 95–104, and 113–114 missing). With the ṭīkā of Mādhava. Incomplete (pūrvārdha).
RORI Cat. III 12433. 14ff. With a ṭīkā. Incomplete (ṣoḍaśayogavicāra).
RORI Cat. III 12632. 67ff. (ff. 1–5 and 38–43 missing). Incomplete (saṃjñātantra and varṣatantra).
RORI Cat. III 13536. 7ff. Incomplete.
RORI Cat. III 13934. 51ff. With the ṭīkā of Viśvanātha. Incomplete (saṃjñātantra).
RORI Cat. III 15364. 40ff. (ff. 13 and 20 missing). Incomplete (varṣatantra).
RORI Cat. III 15377. 76ff. With the ṭīkā of Viśvanātha.
RORI Cat. III 15381. 78ff. (ff. 1–5, 11, and 64–73 missing). With the ṭīkā of Viśvanātha. Incomplete (varṣatantra).
RORI Cat. III 15421. 57ff.
RORI Cat. III 17085. 29ff. With the ṭīkā of Viśvanātha. Incomplete (saṃjñātantra).
RORI Cat. III 17088. 54ff. With the ṭīkā of Viśvanātha. Incomplete (saṃjñātantra).
RORI Cat. III 17095. 23ff. Incomplete (saṃjñātantra and varṣatantra).
RORI Cat. III 18058. 3ff. Incomplete (bhāvagrahaphalāni).
SOI 2593 = SOI Cat. II: 1029–2593. 46ff. With the ṭīkā of Viśvanātha. Incomplete (varṣatantra).
SOI 2594 = SOI Cat. II: 1031–2594. 23ff. With the *Rasālā* of Govinda.
SOI 2907.
SOI 3936 = SOI Cat. II: 1015–3936. 38ff.
SOI 4406. With a ṭīkā.
SOI 6502.
SOI 7225. With an udāharaṇa.
SOI 8390. With a ṭīkā.
SOI 9560. Incomplete (saṃjñātantra).
SOI 9562. Incomplete (praśnatantra).
SOI 10834.
SOI 11089.
Tanjore D 11431 = Tanjore BL 4211. 10ff. Incomplete (varṣatantra).
Tanjore D 11432 = Tanjore 13905. No ff. given. Incomplete.
VVRI 1567. 19ff. With the ṭīkā of Viśvanātha. Incomplete.
VVRI 2379. 17ff. Incomplete.
VVRI 2393. 22ff. With the ṭīkā of Viśvanātha. Incomplete.
VVRI 2470. 10ff. Incomplete.
VVRI 2482. 17ff. Incomplete.
VVRI 2487. 10ff. With the ṭippaṇī of Harṣadhara. Incomplete.
VVRI 5413. 16ff. Incomplete (*Jyotiṣakaumudī*).
VVRI 2549. 142ff. With the ṭippaṇī of Harṣadhara. Incomplete.
VVRI 6923. 121ff. Incomplete.
WHMRL G. 111. r. Incomplete (varṣaphala).
WHMRL H. 2. i.
WHMRL K. 5. e.

The *Tājikanīlakaṇṭhī* has often been published:

the saṃjñātantra and varṣatantra at [NP] in Saṃ. 1908 = A.D. 1851 (IO 9.B.21);
the saṃjñātantra with the ṭīkā of Viśvanātha at Mumbā in Śaka 1783 = A.D. 1861 (BM);
with the ṭīkā of Viśvanātha and the praśnatantra at Vārāṇasī in Saṃ. 1922 = A.D. 1865 (BM);
the saṃjñātantra and the varṣatantra with the *Pṛcchoddeśagaṇakabhūṣaṇa* of Samarasiṃha at Meraṭha in Saṃ. 1923 = A.D. 1866 (BM);
all 3 tantras at Delhi in Saṃ. 1925 = A.D. 1868 (IO 328);
with the ṭīkā of Viśvanātha at Delhi in 1871 (BM and IO 610);
all 3 tantras at Meerut in Saṃ. 1932 = A.D. 1875 (IO 328);
with the *Rasālā* of Govinda at Delhi in Saṃ. 1934 = A.D. 1877 (IO 465);
with the *Rasālā* of Govinda at Benares in Saṃ. 1936 = A.D. 1879 (BM and IO 1.C.12);

with the ṭīkā of Viśvanātha at Bombay in 1879 (BM and IO 13.E.2);
at Bombay in 1893 (NL Calcutta 180.Ka.89.1);
with the Hindī translation of Śaktidhara Śukula at Lucknow in 1894 (NL Calcutta 180.Kb.89.7);
with the ṭīkā of Viśvanātha and the praśnatantra at Mumbaī in Saṃ. 1957, Śaka 1822 = A.D. 1900 (copy at the Warburg Institute, London);
with the ṭīkā of Viśvanātha, the praśnatantra, and the ṭīkā, *Gaṇitaviṣayopapatti*, of Sītārāma Jhā, edited by Anūpa Miśra, Kāśī Saṃ. 1978 = A.D. 1921 (BM 14055.d.34; IO San. D. 559; and IO San. D. 594); reprinted at Benares in 1930 (IO San. D. 1124); this is probably *HNM* 9, Benares 1930 (NL Calcutta 180.Kc.93.1);
with the ṭīkā of Viśvanātha at Bombay in Saṃ. 1980 = A.D. 1923 (IO San. D. 728);
with a Hindī translation, edited by Sītārāma Śarman, Benares 1936 (NL Calcutta 180.Kc.93.15);
with the Saṃskṛta, *Jaladagarjanā*, and Hindī, *Candrikā*, ṭīkās of Gaṅgādhara Miśra, edited by Kapileśvara Caudhuri Śāstrin, *HSS* 143, Benares 1941 (NL Calcutta 180.Kc.94.2); reprinted at Banārasa in 1950.

The praśnatantra was translated into English as *Hindu Horary Astrology* by G. Sri Rama Murthi, Tekkali 1960, and was edited with an English translation by B. V. Raman, Bangalore 1970.

Verses 52–55 of the varṣatantra (verses 52–54 also occur at the end of the saṃjñātantra) are:

āsīd asīmaguṇamaṇḍitapaṇḍitāgryo
vyākhyad bhujaṅgapagavīḥ śrutivit suvṛttaḥ/
sāhityarītinipuṇo gaṇitāgamajñaś
cintāmaṇir vipulagargakulāvataṃsaḥ//
tadātmajo ᵓnantaguṇo ᵓsty ananto
yo ᵓdhok saduktiṃ kila kāmadhenum/
sattuṣṭaye jātakapaddhatiṃ ca
nyarūpayad duṣṭamataṃ nirasya//
padmāmbayāsāvi tato vipaścic
chrīnīlakaṇṭhaḥ śrutiśāstraniṣṭhaḥ/
vidvacchivaprītikaraṃ vyadhāsīt
samāvivekaṃ mṛgayāvataṃsam//
śāke nandābhrabāṇendumita āśvinamāsake/
śukle ᵓṣṭamyāṃ samātantraṃ nīlakaṇṭhabudho
ᵓkarot//

Nīlakaṇṭha also wrote a vivṛti, *Bhāvaprakāśa*, on the *Makaranda* of Makaranda (*fl.* 1478). Manuscripts:

Anup 4955. 11ff.
Anup 4956. 3ff.
LDI (LDC) 3026. 2ff.
RORI Cat. II 7519. 5ff.

NĪLAKAṆṬHA BHAṬṬA (*fl.* 1650)

The son of Śaṅkara Bhaṭṭa, the son of Nārāyaṇa Bhaṭṭa (b. 1513), the son of Rāmeśvara Bhaṭṭa of Pratiṣṭhāna, Nīlakaṇṭha wrote an enormous *Bhagavantabhāskara* in 12 mayūkhas for Bhagavanta, a Bundella rājā of the Seṅgaras ruling at Bhareha near the confluence of the Yamunā and the Cambala rivers; see P. V. Kane [1930/62] vol. 1, pp. 438–440. He completed this on 12 śuklapakṣa of Caitra in Saṃ. 1706 = *ca.* 2 April 1650 according to the following verse:

śrīnīlakaṇṭharacitaḥ smṛtibhāskarākhyo
granthaḥ papāra rasakharṣikusaṃmite ᵓbde/
caitre site ravitithau ravipādapadma-
padmīkṛto vikasatāṃ janatopakṛtyai//

The third section is the *Samayamayūkha* = *Kālamayūkha*. Manuscripts:

Calcutta Sanskrit College (Smṛti) 223. 132ff. Copied in Saṃ. 1772 = A.D. 1715. This is CP, Kielhorn XIX 296. 132ff. Copied in Saṃ. 1772 = A.D. 1715. Property of Jānojī Mahārāja of Nagpur.
AS Bombay 743. 106ff. Copied by Nārāyaṇa Daivajña in Śaka 1663 = A.D. 1741.
Benares (1956) 13994. 65ff. Copied in Saṃ. 1800 = A.D. 1743.
BORI 123 of 1882/83. 153ff. Copied in Saṃ. 1809 = A.D. 1752. From Gujarāt.
Berlin 1171 (Chambers 254b). 108ff. Copied in Saṃ. 1835 = A.D. 1778.
AS Bombay 742. 104ff. Copied by Vyaṅkaṭeśa at Baroda in Śaka 1704 = A.D. 1782. From Bhāu Dājī.
Baroda 140. 114ff. Copied in Saṃ. 1842 = A.D. 1785.
Baroda 4080. 103ff. and 3ff. Copied in Saṃ 1941 = A.D. 1884.
Adyar Index 1247 = Adyar Cat. 34 M 30. 332ff.
Alwar 1526.
Anup 2519. 86ff.
Anup 2520. Ff. 1–69 and 1–9.
AS Bengal 2046 (G 5725). 106ff. Copied at Velāpuragrāma on Sunday 1 kṛṣṇapakṣa of Pauṣa in the Sarvandhārīsaṃvatsara, Śaka 16—.
AS Bombay 741. 114ff. From Bhāu Dājī.
Baroda 219. 80ff.
Baroda 8764. 90ff.
Benares (1956) 11899. 75ff. Incomplete. No author mentioned.
Benares (1956) 12216. Ff. 1–59 and 59b.
Benares (1956) 12281. Ff. 2–70. Incomplete.
Benares (1956) 12805. Ff. 2–124. Incomplete. No author mentioned.
Benares (1956) 12910. 80ff. Incomplete.
Benares (1956) 13667. Ff. 1–136 and 136b–142. Incomplete.
Benares (1956) 13949. 91ff.
Berlin 1172 (Chambers 792s, 4). 1f. Incomplete.
Bikaner 970. No ff. given.
BM 180 (Or. 3341). Ff. 1–75, 75b, and 76–94. With the *Ācāramayūkha*. From Dr. C. D. Ginsburg.
Bombay U Desai 258. 142ff.

BORI 372 of 1875/76. 132ff. From Dilhī.
BORI 61 of A 1879/80. 90ff.
BORI 300 of 1884/87. 127ff. From Mahārāṣṭra.
BORI 132 of Vishrambag I. 109ff.
Calcutta Sanskrit College (Smṛti) 222. 172ff.
Calcutta University 296. 42ff.
CP, Hiralal 835. Property of Dinkarbhaṭ of Multāi, Betūl.
CP, Hiralal 6287, 6288, and 6289. Property of the Bhonsalā Rājās of Nāgpur.
CP, Hiralal 6290. Property of Nārāyaṇ Veñkaṭeś of Bāsim, Akolā.
CP, Hiralal 6291. Property of Śrīkrishṇa Monahar of Bāsim, Akolā.
CP, Hiralal 6292. Property of Dīnānāth of Singharī, Bilāspur.
DC (Gorhe) App. 412. Property of Gañgādhara Rāmakṛṣṇa Dharmādhikārī of Puṇatāmbe, Ahmadnagar.
GOML Madras D 3107. 147pp. Grantha. Incomplete.
GOML Madras D 3108. 166pp. Grantha. Incomplete.
IO 1441 (1132c). 115ff. From H. T. Colebrooke.
IO 5487 A (Burnell 238) III. 80ff. From A. C. Burnell.
IO 5489 (Bühler 318). 117ff. From G. Bühler.
Jammu and Kashmir 4009. 112ff.
Jammu and Kashmir 4731. 156ff.
Kurukṣetra 1212 (50073).
Mithila I 416. 85ff. Property of Paṇḍita Sureśa Miśra of Saurāth, Madhubani, Darbhanga.
Mysore and Coorg 2030. 10,000 granthas. Telugu. Ascribed to Bhāskarabhaṭṭa. Property of Sāmba Śāstrī of Koratagiri.
Nagpur, Deo Collection 88. See NCC, vol. 4, p. 33.
Oppert I 793. 1000pp. Grantha. Property of Nivṛtti Subrahmaṇyaśāstrī of Kāñcīpuram, Chingleput.
Oppert II 6650. Property of P. Rañgācāryar of Kumbhaghoṇam, Tanjore.
Oppert II 6747. Property of the Śañkarācārya Maṭha at Kumbhaghoṇam, Tanjore.
Oudh III (1873) IX 12. 370pp. Property of Paṇḍit Beṇimādhava of Oonao Zillah.
Oudh XV (1882) IX 4. 186pp. Property of Prayāgaprasāda of Rae Bareli Zila.
Radh. 20. See NCC.
RORI Cat. I 228. 94ff.
Tanjore D 18255 = Tanjore BL 445. 162ff.
Tanjore D 18256 = Tanjore BL 446. 126ff.
Tanjore D 18257 = Tanjore BL 447. 113ff.
Tanjore D 18258 = Tanjore BL 448. 96ff.
Tanjore D 18259 = Tanjore BL 449. 105ff.
Tanjore D 18260 = Tanjore BL 450. 93ff.
Tanjore D 18261 = Tanjore BL 451. 116ff.
Tanjore D 18262 = Tanjore 16329. 95ff.
Tanjore D 18263 = Tanjore TS 523. 87ff.
Tuljashankar 251.
Ujjain (List) 225 = Ujjain Cat. II, p. 20. See NCC.
VVRI 3892. 39ff. Telugu. Incomplete.

The *Kālamayūkha* was published at Benares in 1880 (IO 434 & 372), and was edited by J. R. Gharpure, Bombay 1927 (BM 14038.e.17, vol.17, and IO 22.K.24/3).

Verse 2 is:

śrutīḥ smṛtīr vīkṣya purāṇajātaṃ
tattannibandhān api sannibandhān/
śrīśañkarasyātmajanīlakaṇṭhas
tithyādikṛtyaṃ vivṛṇoti sarvam//

The colophon begins: iti śrīseñgaravaṃśāvataṃsamahārājādhirājaśrībhagavantadevādhiṣṭhaśrījagadgurubhaṭṭanārāyaṇasūrisūnupaṇḍitaśiroratnamīmāṃsakaśañkarabhaṭṭātmajabhaṭṭanīlakaṇṭhena.

The last mayūkha is the *Śāntimayūkha*. Manuscripts:

Baroda 8731c. 104ff. Copied in Śaka 1611 = A.D. 1689.
RORI Cat. II 4948. 158ff. Copied in Saṃ. 1779 = A.D. 1722.
Jammu and Kashmir 4727. 29ff. Copied in Saṃ. 1817 = A.D. 1760. Incomplete.
Bombay U 1124. 67ff. Copied on Sunday 11 kṛṣṇapakṣa of Kārttika in Śaka 1683 = 22 November 1761. Incomplete.
IO 1462 (2553). 89ff. Copied in A.D. 1810. From Colin Mackenzie.
RORI Cat. II 6075. 87ff. (ff. 46–50 and 81 missing). Copied by Sadāsukha in Saṃ. 1877 = A.D. 1820.
ABSP 1097. 123ff. Copied in Saṃ. 1890 = A.D. 1833.
RORI Cat. I 1265. 154ff. Copied by Puruṣottama, the son of Ḍośā, in Saṃ. 1900 = A.D. 1843.
Nagpur 2183 (1344). 98ff. Copied in Saṃ. 1925 = A.D. 1868. From Nasik.
Anup 2220. 87ff. (ff. 60–68 (read 63) and 69–71 missing).
Anup 2221. 90ff.
Anup 2529. Ff. 1–63, 64*–71*, and 64–82.
AS Bengal 2059 (G 9328). 68ff. Incomplete.
AS Bengal 2060 (G 868) = Mitra, Not. 1788. 136ff.
AS Bombay 727 145ff. From Bhāu Dājī.
AS Bombay 728. 54ff. Incomplete. From Bhāu Dājī.
Baroda 8594. 75ff. Incomplete.
Berlin 1243 (Chambers 464). 104ff. (ff. 4, 7, and 76 missing).
Bikaner 980. 92ff.
Bombay U 1123. 159ff.
Bombay U 1125. 40ff. Incomplete.
BORI 63 of A 1879/80. 116ff.
BORI 143 of 1892/95. 131ff.
BORI 119 of 1895/98. 119ff.
Calcutta Sanskrit College (Smṛti) 365. 351ff.
CP, Hiralal 5589. Property of Bājirāv Śāstri of Murtizāpur, Akolā.
CP, Hiralal 5590. Property of Gaṇesdatt Pāṭhak of Maṇḍlā.

CP, Hiralal 5591. Property of Nārāyaṇ Veṅkaṭeś of Bāsim, Akolā.
CP, Hiralal 5592. Property of Śyāmrāj Rāmkrishṇa of Pāthroṭ, Amraoti.
CP, Hiralal 5593. Property of Baḍhā Dharmādhikārī of Daryāpur, Amraoti.
CP, Hiralal 5594. Property of Govindbhaṭṭ of Jubbulpore.
CP, Hiralal 5595. Property of Rāmrāj Vaidya of Pāthroṭ, Amraoti.
CP, Hiralal 5596. Property of Mādhav Nārāyaṇ Bhope of Warorā, Chāndā.
CP, Hiralal 5597. Property of Jagmatibāi of Uḍatum, Bilāspur.
CP, Hiralal 5598, 5599, 5600. Property of the Bhonsalā Rājās of Nāgpur.
Darbhanga 116 (Dh. 100) 70ff. Incomplete.
Darbhanga 130 (S 5). 94ff. (ff. 46–47 missing).
IO 1463 (167). 109ff. From H. T. Colebrooke.
IO 5487 C (Burnell 240) XII. 104ff. From A. C. Burnell.
Jammu and Kashmir 4719. 73ff.
Oudh VI (1875) IX 4. 184pp. Property of Paṇḍita Rāmacharaṇa of Bārābānki Zillah.
Oudh XV (1882) IX 7. 366pp. Property of Prayāgaprasāda of Rae Bareli Zila.
RORI Cat. I 227. 106ff. (ff. 46–49 missing).
RORI Cat. III 15195. 48ff.

The *Śāntimayūkha* was published at Kāśī in Saṃ. 1924 = A.D. 1867 (BM); at Benares in 1879 (IO 13.K.7); and edited by J. R. Gharpure, Bombay 1924 (BM 14038.e.17, vol. 25, and IO 22.K.24/12).

The last two sections of the *Ācāramayūkha* form the *Svapnādhyāya*, published at Benares in 1889 (IO 373) and at Murādābāda in 1899 (BM 14053. b.17.(7)).

He also wrote a separate work on śānti, the *Śāntikaustubha*. Manuscript:

Bombay U 1162. 103ff. Copied by Vāsudeva Bhaṭṭa, the son of Kāsibhaṭṭa Gavānda of Junnarapattana near Śivagiri, on Wednesday 10 kṛṣṇapakṣa of Āśvina in Śaka 1698 = 6 November 1776.

NĪLAKAṆṬHA (*fl.* 1663)

Author of a *Jyotiṣasaṅgraha* in Saṃ. 1720 = A.D. 1663. Manuscript:

RORI Cat. I 588. 53ff.

NĪLAKAṆṬHA CATURDHARA (*fl. ca.* 1675/1700)

The son of Govinda Sūri, Nīlakaṇṭha, who composed a ṭīkā on the *Gaṇeśagītā* in 1694, wrote also a *Saurapaurāṇikamatasamarthana* in 18 verses. Manuscripts:

Benares (1963) 37122. 16ff. Copied in Saṃ. 1736 = A.D. 1679.
Benares (1963) 35088. 22ff. Copied in Saṃ. 1844 = A.D. 1787. (*Paurāṇikajyautiṣa*).
AS Bengal 3094 (G 10611). 4ff. Copied by Bhāībhaṭṭa at the Bhairavasannidhi in Kāśī. With a ṭīkā.
AS Bombay 298. 7ff. From Bhāu Dājī.
IO 2885 (1051a). 18ff. With a ṭīkā. From H. T. Colebrooke.

The colophon begins: iti śrīmatpadavākyapramāṇamaryādādhurandharacaturdharavaṃśāvataṃsagovindasūrisūnoḥ nīlakaṇṭhasya.

NĪLAKAṆṬHA (*fl.* 1680)

Author of an abridgment of the *Muhūrtamārtaṇḍa* of Nārāyaṇa (*fl.* 1571/1572) in Saṃ. 1737 = A.D. 1680. Manuscript:

RORI Cat. III 15829(10). 31ff. Copied in Saṃ. 1809 = A.D. 1752.

NĪLAKAṆṬHA REGMĪ (*fl.* 1754)

The son of Jayaśarman Sūri, Nīlakaṇṭha wrote a ṭīkā, *Subodhinī*, on the *Upadeśasūtra* (apparently only on adhyāyas I–II and the beginning of III) of Jaimini in Śaka 1676 = A.D. 1754 for Raṇajit, the Mahārāja of Bhatgaon in Nepāla from 1722 to 1769. Manuscripts:

BORI 883 of 1884/87. 32ff. Copied in Saṃ. 1847 = A.D. 1790. (adhyāyas I–II). From Gujarāt.
PUL II 3446. 43ff. Copied in Saṃ. 1855 = A.D. 1798.
Benares (1963) 34410. 63ff. Copied in Saṃ. 1880 = A.D. 1823. Incomplete (to II 4).
Mithila 419. 41ff. Maithilī. Copied on Tuesday pūrṇimā of śuklapakṣa of Jyeṣṭha in Śaka 1747 = 31 May 1825 (adhyāyas I–II). Property of Paṇḍita Dharmadatta Miśra of Babhanagama, Supaul, Bhagalpur.
Jammu and Kashmir 882. 46ff. Copied in Saṃ. 1885 = A.D. 1828. (adhyāyas I–II).
Mithila 112. 49ff. Maithilī. Copied by Pakṣadhara at Kāśī on Sunday 9 kṛṣṇapakṣa of Śrāvaṇa in Saka 1760 = 12 August 1838 (adhyāyas I–II). Property of Paṇḍita Mahīdhara Miśra of Lalabag, Darbhanga.
Mithila 112 B. 38ff. Maithilī. Copied in Śaka 1760 = A.D. 1838. Property of Paṇḍita Anantalāla Jhā of Nanaur, Tamuria, Darbhanga.
Benares (1963) 36154. 52ff. Copied in Saṃ. 1906 = A.D. 1849. (adhyāyas I–II).
Benares (1963) 35186. 54ff. Copied in Saṃ. 1907 = A.D. 1850. (adhyāyas I–II).
CP, Kielhorn XXIII 179. 40ff. Copied in Saṃ. 1909 = A.D. 1852. Property of Lakṣmaṇa Śāstrī of Sāgar.
RORI Cat. II 6290. 27ff. Copied by Bālamukunda Gosvāmin in Saṃ. 1911 = A.D. 1854.

Oudh XIII (1881) VIII 9. 58pp. Copied in A.D. 1855. Property of Dīna Dayāla of Rae Bareli Zila.
PUL II 3447. 33ff. (ff. 1–6, 10, and 22 missing). Copied in Saṃ. 1913 = A.D. 1856.
PL, Buhler IV E 127. 54ff. Copied in Saṃ. 1923 = A.D. 1866. Property of Maṅgala Śaṅkara of Ahmadābād.
RORI Cat. II 5533. 42ff. Copied by Lalitādāsa Vyāsa at Vṛndāvana in Saṃ. 1924 = A.D. 1867.
DC 7932. 66ff. Copied in Śaka 1794 = A.D. 1872.
VVRI 4477. 39ff. Copied in Saṃ. 1950 = A.D. 1893.
AS Bengal 6955 (G 10462). 83ff. Incomplete (ends in III).
Baroda 3136. 27ff. (adhyāyas I–II).
Benares (1963) 34376 = Benares (1909) 1823. 33ff. (adhyāyas I–II).
Benares (1963) 35205. 28ff. Incomplete.
Benares (1963) 35285 = Benares (1897–1901) 553. 41ff.
Benares (1963) 35467. 16ff. Incomplete.
Benares (1963) 36155. Ff. 1 and 3–58. Incomplete (adhyāyas I–II).
Benares (1963) 36156. 32ff. (adhyāya III).
Benares (1963) 36157. 28ff. (adhyāya I).
BORI 909 of 1886/92. 39ff.
BORI 910 of 1886/92. 35ff.
GJRI 2984/317. 22ff. Maithilī. Incomplete (to II 3).
Jammu and Kashmir 2874. 50ff.
Jammu and Kashmir 2875. 28ff. Incomplete (to III 1).
Kathmandu (1960) 127 (I 164). 75ff. Nevārī. Incomplete (to II 4).
Kathmandu (1960) 128 (III 109). 45ff. Incomplete (to II 4).
Kerala 5918 (1695). 1100 granthas. Incomplete.
Kurukṣetra 345 (19720).
Mithila 112 A. 9ff. Maithilī. Incomplete. Property of Paṇḍita Mahīdhara Miśra of Lalabag, Darbhanga.
Mithila 112 C. 6ff. Maithilī. Incomplete. Property of Babu Candradeva Jhā of Mahinathapura, Jhanjharpur Bazar, Darbhanga.
Mithila 112 D. 10ff. Maithilī. Incomplete. Property of Babu Puṇyānanda Jhā of Chanaur, Manigachi, Darbhanga.
N-W P II (1878) B 14. 109ff. Property of Mākhanji of Mathurā.
Oudh IV (1874) VIII 5. 5000 ślokas. (*Nīlakaṇṭhīṭīkā Subodhinī*). Property of Nandarāma of Kheri Zila.
Oudh XXII (1890) VIII 15. 272pp. Property of Kedāranātha of Āgrā Zila.
RORI Cat. II 8019. 33ff. Incomplete.
RORI Cat. III 10987. 54ff. Incomplete (to II 4).
SOI 2297 = SOI Cat. II: 1458–2297. 46ff.
SOI 9494.
SOI 10029.
VVRI 4008. 5ff. Incomplete.
WHMRL X. 84.

The *Subodhinī* has been published at Kāśī in Saṃ. 1931 = A.D. 1874 (BM) and at Kāśī in Saṃ. 1934 = A.D. 1877 (BM); edited by Rasikamohana Caṭṭopādhyāya, Kalikātā Saṃ. 1941 = A.D. 1884 (BM 14053.dd.6); at Mumbaī in 1888 (BM); at Allahabad in 1888 (IO 3.B.6) (adhyāyas I–II); and at Calcutta in Śaka 1848 = A.D. 1926 (IO San.B.990(d)).

Verse 2 is:

vivasvadvaṃśāgryāvanipativaraśrīraṇajito
dayāmbhodher vācā tava karuṇayā jaiminimuneḥ/
nirālambe śāstre janani girije yāsyati na kiṃ
suṭīkāṃ me kāmo hṛdi samabhipūrtiṃ racayitum//

The last verse is:

śrīśāke rasasaptabhūpatimite nepālakhaṇḍe vare
śrīśrīmadraṇajinnṛpālakavare rājyaṃ prakurvaty asau/
regmī śrījayaśarmasūritanujaḥ śrīnīlakaṇṭho dvijaḥ
śāstre jaimininākṛte suvivṛtiṃ bhūpājñayā vyākarot//

Nīlakaṇṭha also wrote a *Grahaṇādhikāra* on solar and lunar eclipses with a ṭīkā for Raṇajit. Manuscripts:

Kathmandu (1960) 74 (I 1208). 13ff. Nevārī. Incomplete.
Kathmandu (1960) 93 (I 1211). 9ff. Nevārī (*Candragrahaṇādhikāraṭīkā*).
Kathmandu (1960) 497 (I 1208). 3ff. Nevārī (*Sūryagrahaṇādhikārodāharaṇa*).

Verse 2 at the end of the ṭīkā is:

sūryavaṃśatilakasya bhūpater
ājñayā raṇajito mahāmateḥ/
nīlakaṇṭhavibudhena tanyate
ṭīkikā ravihimāṃśuparvaṇaḥ//

The last verse of the ṭīkā is:

śrīdevīcaraṇāravindamadhupasya dviḍgaṇadhvaṃsinaḥ
śrīmadbhāskaravaṃśadīparaṇajidbhūmīpates tuṣṭaye/
triskandhādhyayanī paropakṛtaye śrīnīlakaṇṭho dvijaś
candrārkagrahaṇopayogigaṇite ṭīkāṃ svakīye ᵓkarot//

NĪLAKAṆṬHA VINĀYAKA CHATRE (*fl.* 1885/1886)

Author, with Pāṇḍuraṅga Ābā Moghe Vasaīkar, of pañcāṅgas for Śaka 1807–1808 = A.D. 1885–1886, published at Ratnāgiri in 1885 and 1886 (BM 14096.a.5).

NĪLAKAṆṬHA ŚARMAN (*fl.* 1900/1902)

The son of Viśveśvaranātha of Aminabad, Lucknow, Nīlakaṇṭha wrote a Hindī vyākhyā on the *Muhūrta-*

cintāmaṇi of Rāma (*fl.* 1600), published at Lakṣaṇapura in Saṃ. 1957 = A.D. 1900 (BM 14053.ccc. 21(1) and 14053.ccc.44); 2nd ed., Lucknow Saṃ. 1965 = A.D. 1908 (IO 20.H.2); 3rd ed., [Lucknow] Saṃ. 1972 = A.D. 1915 (IO San.F.58(b)). He also wrote a *Bṛhajjyotissāra*, published with a ṭīkā and a Hindī translation at Lucknow in 1902 (BM 14053. dd.4.(1)).

PUNNAŚŚERI NAMPI NĪLAKAṆṬHA ŚARMAN (*fl.* 1926)

Teacher at the Sārasvatodyotinī Saṃskṛta Mahāpāṭhaśālā at Paṭṭampi, Nīlakaṇṭha wrote a ṭippaṇi on the *Praśnamārga* of Nātha (?) (*fl.* 1650), I–XVI, published at Pālakkāṭ-Kalpātti in 1926.

NĪLAKAMALA VIDYĀNIDHI BHAṬṬĀCĀRYA (*fl.* 1892/1901)

Author of the following works:

1. A Bengālī translation of the *Bṛhajjātakacandrikā* of Rāmaśaṅkara Deva, published at Calcutta in 1892 (BM 14053.c.63).

2. A *Jyotiṣatattvavāridhi*, edited with a ṭīkā and Bengālī translation by Akṣayakumāra Siddhāntaratna, Calcutta 1894 (BM 14053.ccc.1).

3. A Bengālī translation of the *Śuddhidīpikā* of Śrīnivāsa, edited by Gurunātha Vidyānidhi Kāvyatīrtha at Calcutta in 1901 (BM 14053.ccc.23); 2nd ed., Calcutta 1927 (IO San. B. 1002(b)).

NĪLAGOVINDA

Author of a ṭīkā, *Śiśubodhinī*, on the saṃjñātantra of the *Tājikanīlakaṇṭhī* (1587) of Nīlakaṇṭha (*fl.* 1569/1587). Manuscript:

Mithila 367. 25ff. Maithilī. Property of Paṇḍita Mahīdhara Miśra of Lalbag, Darbhanga.

NĪLĀMBARA

Author of a *Kālakaumudī;* see B. Shastri [A3. 1969] 255. Manuscript:

Mitra, Not. 2905. 72ff. Maithilī. Property of Paṇḍita Halī Jhā of Harinagara, Madhuvanī, Darbhāṅgā.

The colophon begins: iti śrīnīlāmbarācāryavinirmitā.

NĪLĀMBARA JHĀ (b. 18 July 1823)

The son of Śambhunātha Jhā, a Maithilī Brāhmaṇa residing in Pāṭaliputra, Nīlāmbara was born on 11 śuklapakṣa of Āṣāḍha in Śaka 1745 = 18 July 1823 as the younger brother of Jīvanātha Jhā (*fl. ca.* 1846/1900) and studied jyotiḥśāstra under Lajjāśaṅkara (b. 1804). Nīlāmbara was astrologer at the court of Śivadāna, the Mahārāja of Alavara, for whom he translated into Sanskrit and enlarged an English textbook on plane and spherical trigonometry brought to Alwar by the political agent, Captain Thomas Cadell. He died at the Maṇikarṇikāghaṭṭa in Kāśī on 11 śuklapakṣa of Vaiśākha in Śaka 1805 = 16 May 1883. See S. Dvivedin [1892] 129–131 and S. B. Dikshit [1896] 301. His works include:

1. The text on plane and spherical trigonometry mentioned above, the *Golaprakāśa*, in which there are 5 adhyāyas:

1. jyotpatti.
2. trikoṇamiti.
3. golīyarekhāgaṇita.
4. cāpīyatrikoṇagaṇita.
5. praśnāḥ.

The *Golaprakāśa* was edited by Bāpū Deva Śāstrin, Benares 1872; the *Golīyarekhāgaṇita* from it was published with the ṭīkā, *Vikāśikā*, of Rājavaṃśī Jhā, Benares 1925 (IO San. D. 950(n)), reprinted Benares 1925 (IO San. D. 1063(b)); and with the upapattika, *Ruṣā*, of Mīṭhālāla Śāstrin as *MM* 245, Kāśī 1954; the *Cāpīyatrikoṇagaṇita* from it was edited by Muralīdhara Ṭhakkura, Benares 1924 (BM 14055.d. 33.(3)), and published with the ṭīkā, *Vividhavāsanā*, of Acyutānanda Jhā as *KSS* 139, Banāras 1944. Verses 1–4 at the end are:

gaṅgāgaṇḍakisaṅgame hariharakṣetraṃ kṣitau viśrutaṃ
tasmāt krośayugāntare suranadītīre pare dakṣiṇe/
khyāte pāṭaliputrasaṃjñanagare vidvajjanair maṇḍite
jyotirdarśanasarvadarśanayaśā mānyo vadānyo vidām//
śrīśambhunāthaḥ sukṛtaikagāthaḥ
kṛpaikapāthaḥ patir īśanāthaḥ/
abhūd dvijendraḥ sucakoracandraḥ
kṣitāv upendrasmaraṇe vitandraḥ//
tadaṅgajaḥ sarvavidā samānas
triskandhapāraṅgamatābhimānaḥ/
cakāsti śiṣyādivirājamānaḥ
śrījīvanāthaḥ kavilabdhamānaḥ//
nīlāmbarākhyo ᵓkṛta tatkaniṣṭhas
tallabdhabodhaḥ paramaikaniṣṭhaḥ/
golaprakāśābhidham āśubodhaṃ
nirastamandehamanonurodham//

2. A *Kṣetraparibhāṣā* in 33 verses, a Sanskrit rendering of a Hindī *Rekhāgaṇita*. Manuscript:

Mithila 37. 2ff. Maithilī. Property of Paṇḍita Muktinātha Jhā of Baruāri, Parsarmā, Bhagalpur.

3. A *Janmapatrodāharaṇa*, which discusses a series of horoscopes according to the rules of the *Laghu-*

jātaka of Varāhamihira (*fl. ca.* 550); the first is dated Friday 14 kṛṣṇapakṣa of Māgha in Śaka 1698 = 7 March 1777. Manuscript:

Mithila 72. 9ff. Maithilī. The date of copying according to Mithila — Thursday 13 kṛṣṇapakṣa of Vaiśākha in Śaka 1744 = 16 May 1822 — is clearly impossible, and may be rather the date of the last horoscope. Property of Paṇḍita Muktinātha Jhā of Baruari, Parsarma, Bhagalpur.

4. A vyākhyā on the *Jyotpatti* from the *Siddhāntaśiromaṇi* of Bhāskara (b. 1114). Manuscript:

Mithila 124. 21ff. Maithilī. Property of Paṇḍita Gaṅgādhara Jhā of Jonki, Deodha, Darbhanga.

The last verse is:

vidvaryaśrījīvanāthānujanmā
golajñānakṣamanīlāmbarākhyaḥ/
praśnādhyāye bhāskarīye suramyāṃ
jyotpattyākhye cāpi ṭīkām akārṣīt//

5. A vyākhyā on the *Praśnādhikāra* from the *Siddhāntatattvaviveka* of Kamalākara (*fl.* 1658). Manuscripts:

Mithila 128. 25ff. Maithilī. Property of Paṇḍita Jayakṛṣṇa Jhā of Champa, Benipati, Darbhanga.
Mithila 202. 18ff. Maithilī. Incomplete. Property of Paṇḍita Sītārāma Pāṭhaka of Karnpur, Sukpur, Bhagalpur.

Verse 1 is:

śrīśaṅkaraṃ naumi karomi ṭīkāṃ
siddhāntasambhrāntanirastaśaṅkām/
praśnādhikāre kamalākarīye
nīlāmbaro ᵓhaṃ sumanīṣituṣṭyai//

6. A vyākhyā on the *Dṛkkarma* from the *Siddhāntaśiromaṇi* of Bhāskara (b. 1114). Manuscripts:

Mithila 142. 24ff. Maithilī. Copied by Phekanaśarma on Tuesday 3 kṛṣṇapakṣa of Pauṣa in Śaka 1794, Sāl. San. 1280 = 14 January 1873. Property of Paṇḍita Sītārāma Pāṭhaka of Karnpur, Sukpur, Bhagalpur.
Mithila 142 A. 22ff. Maithilī. Incomplete. Property of Paṇḍita Jayakṛṣṇa Jhā of Champa, Benipati, Darbhanga.

Verses 1–3 at the end are:

śrīśambhunāthaḥ sukṛtaikanāthaḥ
kṛpaikanāthaḥ patir īśanāthaḥ/
abhūd dvijendraḥ sucakoracandraḥ
kṣitāv upendraḥ smaraṇe vitandraḥ//
naipuṇyapuṇyā dyutir asya sūnuḥ
saujanyajanyaprabhayā sametaḥ/
lāvaṇyavanyaḥ kṛtivṛndamadhye
mūrdhanyadhanyaḥ sa virājate yaḥ//
śrījīvanāthaḥ prathitaḥ pṛthivyāṃ
tasyānujanmā tadavāptavidyaḥ/
nīlāmbaro bhāskaragītanānā-
dṛkkarmasandhānasavāsanāṃ hi//

7. A vāsanā on the *Makaranda* of Makaranda (*fl.* 1478), using as epoch Saka 1512, Kali 4691 = A.D. 1590. Manuscripts:

Mithila 163. 6ff. Maithilī. (*Pañcāṅgavāsanā*). Property of Paṇḍita Gaṅgādhara Jhā of Jonkī, Deodha, Darbhanga.
Mithila 248. 5ff. Maithilī. Property of Paṇḍita Sītārāma Pāṭhaka of Karnpur, Sukpur, Bhagalpur.
Mithila 248 A. 8ff. Maithilī. Property of Paṇḍita Adhikalāla Miśra of Balava, Nawhatta, Bhagalpur.

8. A vyākhyā on the *Praśnottara* from the *Siddhāntaśiromaṇi* of Bhāskara (b. 1114). Manuscripts:

Mithila 205. 37ff. Maithilī. Property of Paṇḍita Sītārāma Pāṭhaka of Karnpur, Sukpur, Bhagalpur.
Mithila 258. 43ff. Maithilī. Copied by Nandalāla Śarman at Kāśī. Property of Paṇḍita Jayakṛṣṇa Jhā of Champa, Benipatti, Darbhanga.

Verse 2 is:

praśnottarārthavijñāni kliśyeran mandabuddhayaḥ/
nīlāmbaraḥ subodhaṃ taṃ tair ahaṃ kartum udyataḥ//

9. A vyākhyā on the *Valana* from the *Siddhāntaśiromaṇi* of Bhāskara (b. 1114). Manuscripts:

Mithila 207 D. 55ff. Maithilī. Copied by Nandalāla Śarman at Kāśī in Sāl. San. 1279 = A.D. 1871. Property of Paṇḍita Jayakṛṣṇa Jhā of Champa, Benipatti, Darbhanga.
Benares (1963) 34467. 93ff. Copied in Saṃ. 1932 = A.D. 1875.
Mithila 207. 10ff. Maithilī. Incomplete. Property of Paṇḍita Vāsudeva Jhā of Sukpur, Bhagalpur.
Mithila 207 A. 8ff. Maithilī. Incomplete. Property of Paṇḍita Śrīnandana Miśra of Kanhauli, Sakri, Darbhanga.
Mithila 207 B. 11ff. Maithilī. Incomplete. Property of Babu Candradeva Jhā of Mahinathapur, Jhanjharpur Bazar, Darbhanga.
Mithila 207 C. 57ff. Maithilī. Property of Paṇḍita Gaṅgādhara Jhā of Jonki, Deodha, Darbhanga.
Mithila 333. 51ff. Maithilī. Property of Paṇḍita Sītārāma Pāṭhaka of Karnpur, Sukpur, Bhagalpur.

Verse 1 is:

umeśaṃ rameśaṃ gaṇeśaṃ dineśaṃ
praṇamyātiramyā sugamyā ca ṭīkā/
mayā tanyate vālane bhāskarīye
budhānāṃ vinodāya nīlāmbareṇa//

10. An upapatti or udāharaṇa on the *Līlāvatī* of Bhāskara (b. 1114). Manuscripts:

Mithila 327. 13ff. Maithilī. Incomplete. Property of Paṇḍita Jayakṛṣṇa Jhā of Champa, Benipati, Darbhanga.
Mithila 327 A. 20ff. Maithilī. Incomplete. Property of Paṇḍita Muktinātha Jhā of Baruari, Parsarma, Bhagalpur.
Mithilā. See NCC, vol. 1, rev. ed., p. 60.

Verse 1 is:

śrīśaṅkaraṃ namya vadāmi yuktiṃ
chandaḥprabhedakriyayā vicitrām/
athāṅkapāśe gaṇite subodhāṃ
nīlāmbaro ᵓhaṃ budharañjanārtham//

ĀKUMALLA NṚSIṂHA

The author of an *Arthaprakāśikā* on the *Upadeśasūtra* of Jaimini; *cf.* the vyākhyā on the *Upadeśasūtra* by Nṛsiṃha Kheḍakara. There are 8 paṭalas:

1. not given.
2. kārakāṃśaphala.
3. padādhikāra.
4. upapadādhikāra.
5. āyurdāya.
6. āyurdāyadaśā.
7. pitrādyāyurdāya.
8. āyurdāyadaśāphalagocaraphalādi.

Manuscripts:

Baroda 13453(a). Ff. 1–17. Nandināgarī.
Baroda 13502. 20ff. Telugu.
GOML Madras D 13740. Ff. 1–51 (?). Telugu.
GOML Madras D 13741. Ff. 23–28. Grantha. Incomplete (paṭalas 7–8).
Mysore (1922) B 144. 5 and 48ff.
Mysore (1911 + 1922) B 592. 46ff.

Verse 2 is:

jaiminiṃ munim ānamya tatsūtrārthaprakāśikā/
ślokair anuṣṭubhair eṣā nṛsiṃhena viracyate//

The colophon begins: iti śrīmadākumallanṛsiṃhasūriviracitāyāṃ.

NṚSIṂHA

Alleged author of a *Kālacakra; cf.* the *Tithicakra* of Narasiṃha (*fl.* between 1807 and 1866?). Manuscript:

Oppert II 7276. Property of Subrahmaṇyaśāstrī of Nalluceri, Tanjore.

NṚSIṂHA

The son of Varadārya of the Bhāradvājagotra and a resident of Poḷipākkam = Prauḍharāyapura, Nṛsiṃha wrote a *Kālaprakāśikā* in more than 40 adhyāyas. Manuscripts:

Kerala 3428 (3172 B). 2500 granthas. Grantha. Copied in ME 1053 = *ca.* A.D. 1877.
Adyar List. 10 copies = Adyar Index 1244 =
 Adyar Cat. 22 I 39. 160ff. Grantha. Incomplete (ends at amāvāsyanirṇaya).
 Adyar Cat. 22 I 40. 218ff. Grantha. Incomplete (adhyāyas 1–30).
 Adyar Cat. 22 I 41. 68ff. Grantha and Telugu. Incomplete (adhyāyas 1–6 and 14–20).
 Adyar Cat. 22 I 42. 8ff. Grantha. Incomplete (adhyāya 34).
 Adyar Cat. 22 I 43. 66ff. Grantha. Incomplete (adhyāyas 26–31).
 Adyar Cat. 22 I 44. 28ff. Grantha. Incomplete (adhyāya 5).
 Adyar Cat. 25 E 31. 118ff. Grantha. Incomplete (adhyāyas 1–35).
 Adyar Cat. 26 B 22. 238ff. Grantha. Incomplete (ends at ṛṇamokṣa).
 Adyar Cat. 26 C 22. 140ff. Grantha. Incomplete (ends at adhyāya 45).
 Adyar Cat. 29 I 43. 118ff. Grantha. Incomplete (adhyāyas 2–7).
 Adyar Cat. 33 E 10. 20ff. Grantha. Incomplete (ends at adhyāya 4).
 Adyar Cat. 34 A 13. 188ff. Grantha. Incomplete (ends at ṛṇamokṣa).
 Adyar Cat. 34 I 23. 10ff. Grantha.
 Adyar Cat. 40 F 8. 116ff. Grantha.
Baroda 6207. 56ff. (f. 55 missing). Grantha. Incomplete (ends in adhyāya 31).
Baroda 6348. 75ff. Grantha. Incomplete.
Baroda 6845(a). 76ff. Grantha. Incomplete (adhyāyas 1–41).
Baroda 7955(c). Ff. 42–64. Grantha. Incomplete. No author mentioned.
Baroda 7955(d). Ff. 65–70. Grantha. Incomplete (adhyāya 26).
Baroda 9843(d). 9ff. Grantha. Incomplete.
Baroda 10136(b). 16ff. Grantha. Incomplete (7 adhyāyas).
Cocanada, Telugu Academy 701. See NCC, vol. 4, p. 31.
GOML Madras D 13519. 267pp. Grantha. Incomplete (40 adhyāyas).
GOML Madras D 13520. 132pp. Grantha.
GOML Madras D 13521. 48pp. Telugu. Incomplete (ends in adhyāya 17).
GOML Madras D 13522. Ff. 39–84. Grantha. Incomplete (ends in adhyāya 22).
GOML Madras D 13523. Ff. 1–28. Grantha. Incomplete (adhyāyas 3–23).
GOML Madras D 13524. 116pp. Grantha. Incomplete (ends in adhyāya 29).

GOML Madras D 13525. 60pp. Telugu. Incomplete (adhyāyas 1–15).
GOML Madras D 13526. Ff. 20–41. Grantha. Incomplete (adhyāyas 11–32).
GOML Madras D 13527. Ff. 2–98. Grantha. Incomplete (ends in adhyāya 32).
GOML Madras D 13528. Ff. 22–65. Grantha. Incomplete (adhyāyas 13–22).
GOML Madras D 13529. Ff. 35–44. Grantha and Tāmil. Incomplete (adhyāyas 3–5).
GOML Madras D 13530. Ff. 14–23. Grantha. Incomplete.
GOML Madras D 14024. Ff. 46–53. Telugu. Incomplete (adhyāya 5).
GOML Madras D 14025. Ff. 62–68. Telugu. Incomplete (adhyāya 5).
GOML Madras R 740. 50ff. Grantha. Incomplete (ends in adhyāya 28). Presented in A.D. 1912/13 through Śeṣaśāyi Ayyaṅgār of Kottaiyur.
GOML Madras R 4086. 83ff. Grantha. Presented in A.D. 1921/22 by N. C. Narasiṃhācāriyar of Karur, Trichinopoly.
GOML Madras R 4092(a). Ff. 3–162. Grantha. Presented in A.D. 1921/22 by N. C. Narasiṃhācāriyar of Karur, Trichinopoly.
GOML Madras R 4594. 90ff. Grantha. Presented in A.D. 1924/25 by the Trustees of the Śrī Yogi Pārthasārathi Ayyaṅgar's Charities of Triplicane, Madras.
GOML Madras R 6024. 122ff. Grantha. Incomplete (ends in adhyāya 36). Purchased in A.D. 1937/38 from E. Śrīnivāsācāryar of Srīperumbūdūr, Chingleput.
GOML Madras R 6771. 121ff. Grantha. Incomplete (ends in adhyāya 38). Purchased in A.D. 1938/39 from Cakravarti Jogannathācarair of Kumbakonam.
GOML Madras R 7447. 87ff. Grantha. With a vyākhyā. Incomplete. Purchased in A.D. 1939/40 from T. S. Krishna Aiyer of Triplicane, Madras.
GOML Madras R 7468. Ff. 18–34. Grantha and Tāmil. Incomplete (adhyāyas 13–21). Purchased in A.D. 1939/40 from T. S. Krishna Aiyer of Triplicane, Madras.
Hultzsch 2. 1140. 64ff. Grantha. Incomplete. Property of the Temple Library at Tiruviḍaimarudūr.
IO 5604 (Mackenzie III. 77b). Ff. 1–66 and 1–86. Telugu. From Colin Mackenzie.
IO 5605 (Mackenzie VIII. 53a). 73ff. Grantha. From Colin Mackenzie.
IO 6332 (3660b). 2ff. Grantha. Incomplete (adhyāya 7). Acquired 5 December 1921.
Kerala 3426 (60). 2500 granthas. Grantha.
Kerala 3427 (2422). 2500 granthas. Grantha.
Kerala 3429 (5435). 2500 granthas. Grantha.
Kerala 3430 (3577 A). 2500 granthas. Grantha.
Kerala 3431 (3583). 2500 granthas. Grantha.
Kerala 3432 (1322 B). 400 granthas. Grantha. Incomplete.
Kerala 3433 (1369). 1800 granthas. Grantha. Incomplete.
Kerala 3434 (1382). 850 granthas. Grantha. Incomplete.
Kerala 3435 (2354 C). 275 granthas. Grantha. Incomplete.
Kerala 3436 (2379 A). 675 granthas. Grantha. Incomplete.
Kerala 3437 (2862 A). 220 granthas. Grantha. Incomplete.
Kerala 3438 (2931). 1950 granthas. Grantha. Incomplete.
Kerala 3439 (2948 A). 1200 granthas. Grantha. Incomplete.
Kerala 3440 (3041). 2200 granthas. Grantha. Incomplete.
Kerala 3441 (3586). 2000 granthas. Grantha. Incomplete.
Kerala 3442 (4032 B). 700 granthas. Grantha. Incomplete.
Kerala 3443 (5963). 1500 granthas. Grantha. Incomplete.
Kerala 3444 (8489). 1400 granthas. Grantha. Incomplete.
Kerala 3445 (13755). 2325 granthas. Grantha. Incomplete.
Kerala 3446 (13805). 300 granthas. Grantha. Incomplete.
Kerala 3447 (C. 2520 E) = Kerala C 682 E. 13ff. Grantha. Incomplete (adhyāya 5). Formerly property of Brahma Śrī Kāśi Vādhyār of Mahādānapuram.
Madras Univ. R.A.S. 77(a). See NCC.
Madras Univ. R.K.S. 97(b). See NCC.
Mysore 453 (490).
Mysore 454 (475).
Mysore (1922) 824. 114ff.
Mysore (1922) 1624. 77ff.
Mysore (1922) 3488. 37ff.
Mysore (1922) 3900. 100ff.
Mysore (1922) 4096. 91ff.
Mysore (1922) 4317. Ff. 4–53.
Mysore and Coorg 267. 2000 granthas. Property of Mahādeva Joyisa of Sringeri.
Mysore and Coorg 268. 2000 granthas. Property of Nārāyaṇa Dīkṣita of Bommarasaiyana Agrahara.
Mysore and Coorg 269. Incomplete (39 adhyāyas). Property of the Śṛṅgeri Maṭha at Sringeri. This is Śṛṅgerī Mutt 206 (2); see NCC. See also Oppert II 4519.
Oppert I 38. 150pp. Grantha. Property of Narasiṃhācāryār of Ammaṇapākam, Chingleput.
Oppert I 151. 300pp. Grantha. Property of Varadācāryār of Ammaṇapākam, Chingleput.
Oppert I 882. 164pp. Grantha. Property of Veṅkaṭavarada Tātācāryār of Kāñcīpuram, Chingleput.

Oppert I 1213. Property of Vaṅkīpuram Śrīnivāsācāryār of Tiruvallūr, Chingleput.
Oppert I 1677. Grantha. Property of Kṛṣṇabhaṭṭa Śrauti of Bhavāni, Coimbatore.
Oppert I 2296. 90pp. Grantha. Property of the Śrī Sarasvatī Bhaṇḍāram Committee of Tiruvallikkeṇi, Madras.
Oppert I 3554. 40pp. Grantha. Property of Narasiṃhapuram Rāghavācāryār of Kumbhaghoṇam, Tanjore.
Oppert I 4521. 225pp. Grantha. Property of Kṛṣṇa Jyośyar of Pudukoṭa, Tanjore.
Oppert I 5009. Property of Āttān Alakappaṅgār of Ālvār Tirunahari, Tinnevelly.
Oppert I 7895. Property of Paravastu Veṅkaṭaraṅgācāryār of Viśākhapaṭṭana, Vizagapatam.
Oppert II 2324. 200pp. Grantha. Property of Anantakṛṣṇaśrauti of Kaṇiyūr, Uḍumalapeṭa, Koimbatore.
Oppert II 2426. 200pp. Grantha. Property of K. Aṇṇāsvāmisāstrī of Kojumam, Uḍumalapeṭa, Koimbatore.
Oppert II 2594. 80pp. Grantha. Property of Rāmappayya of Kumāraliṅgam, Uḍumalapeṭa, Koimbatore.
Oppert II 2630. 100pp. Grantha. Property of Maṇḍalam Rāmasvāmiśāstrī of Kumāraliṅgam, Uḍumalapeṭa, Koimbatore.
Oppert II 2650. 115pp. Grantha. Property of Śaṅkaraśāstrī of Kumāraliṅgam, Uḍumalapeṭa, Koimbatore.
Oppert II 3473. Property of Gopālatātācāryār of Vembūr, Madura.
Oppert II 4519. Property of the Śaṅkarācāryasvāmimaṭha at Śṛṅgeri, Cikkamogulūr, Mysore. See Mysore and Coorg 269.
Oppert II 6025. Property of Gurusvāmi Śāstrī of Kumbhaghoṇam, Tanjore.
Oppert II 7277. Property of Subrahmaṇyaśāstrī of Nalluceri, Tanjore.
Oppert II 7311. Property of Vaidyanāthaśāstrī of Nalluceri, Tanjore.
Oppert II 7521. 71pp. Grantha. 2 copies. Property of the Mahārāja of Pudukoṭa, Tanjore.
Oppert II 8118. Property of Sāmiśāstrī of Sūryanārkovil, Kumbhaghoṇam, Tanjore.
Oppert II 8452. 89pp. Grantha. Property of Gopālaviśvanātha Śāstriyār of Taṇḍānkorai, Tanjore.
Oppert II 9710. Property of Nārāyaṇopādhyāya of Vedāraṇyam, Tanjore.
Oppert II 10118. Property of Rāmasvāmidīkṣitar of Pinnaivāśal, Trichinopoly.
Paliyam 116, 118, and 621. See NCC.
PUL II 3297. 42ff. Grantha. Incomplete (adhyāyas 1–32 and 40).
PUL II 3298. 28ff. Grantha. Incomplete (24 adhyāyas).
Tanjore D 11338 = Tanjore BL 11022. Ff. 4–115. Grantha. Incomplete.
Tanjore D 11339 = Tanjore BL 11025. 85ff. Telugu. Incomplete.
Tanjore D 11340 = Tanjore BL 11024. 78ff. Grantha.
Tanjore D 11341 = Tanjore BL 11023a. 99ff. Grantha. Incomplete.
Tanjore D 11342 = Tanjore 15651a. No ff. given. Grantha. Incomplete.
Tanjore D 11343 = Tanjore BL 11048. 80ff. Grantha. Incomplete.
Tanjore D 11344 = Tanjore BL 11027. 20ff. Grantha. Incomplete.
Tanjore D 11345 = Tanjore BL 4309. 8ff. Incomplete.
Tanjore D 11346 = Tanjore BL 11026a. 126ff. (ff. 77–78 missing). Grantha. Incomplete.
Tanjore D 11347 = Tanjore BL 10984d. 25ff. Grantha. Incomplete.
Tanjore D 11683 = Tanjore 15700. 2ff. Grantha and Tāmil. (*Kālapradīpikā*). No author mentioned, but see NCC.
VVRI 6187. 86ff. Grantha.
VVRI 6292. 20ff. Grantha. No author mentioned.

The *Kālaprakāśikā* was edited by T. Rāmasvāmi Śāstrī, Madras 1915 (BM 14055.d.26 and IO 25.C.7); published at Srirangam in 1917 (NCC, vol. 4, p. 32); edited with an English translation by N. P. Subrahmanyam, Tanjore 1917 (NCC); and published at Madras in 1923 (NCC). Verses 3–8 are:

vādhūlavaradācāryapādapaṅkajam āśraye/
yadāśritānāṃ bhuktiś ca muktiś ca sulabhā bhavet//
vandāmahe nṛsiṃhāryaṃ vādhūlānvayanāyakam/
kāntopayantṛyogīndrakaruṇāpātratāṃ gatam//
padavākyapramāṇajñaṃ vande samarapuṅgavam/
kṛtārthāḥ prāṇinaḥsarve yasyāvataraṇād bhuvi//
poḷipākka iti khyāte prauḍharāyapure vasan/
ābhijātyena vṛttena vidyayā cātiśāyinaḥ//
putro ᵓhaṃ varadāryasya bhāradvājakulodbhavaḥ/
śrīnṛsiṃha iti khyāto jyotiḥśāstrābdhitārakaḥ//
pravīṇāni vidhānāni samyag vīkṣya samantataḥ/
saṅkṣipya teṣu sāro ᵓyaṃ vakṣye kālaprakāśikām//

NṚSIṂHA

Author of a *Grahaṇadvayasādhana*. Manuscript:

IM Calcutta 6925 B. Incomplete. See NCC, vol. 6, p. 248.

NṚSIṂHA

Author of a *Grahasamullāsa*. Manuscript:

Mysore (1922) 1798. 6ff.

NṚSIṂHA

Author of a vāsanā on the *Grahasādhanopapatti* of Dyumaṇi; this may be part of the *Saurabhāṣya* of

Nṛsiṃha (b. 1586). Manuscript:

RORI Cat. III 12636. 56ff. Incomplete.

NṚSIṂHA

The son of Nāganātha of the Maudgalyagotra, Nṛsiṃha wrote a *Jātakamañjarī*. Manuscripts:

Oudh (1876–1878) VIII 3. 122pp. Copied in A.D. 1830. Ascribed to Śivasahāya. Property of Mannālāla of Tirwā, Lucknow Zila.
AS Bengal 7038 (G 1832) = Mitra, Not. 2455. 44ff.
GOML Madras D 13693. Ff. 12–18. Telugu. Incomplete (adhyāya 5).
N-W P IX (1885) A 30. 10ff. No author mentioned. Property of Pandit Śyāmā Caraṇa of Benares.
Oudh XVIII (1885) VIII 4. 55pp. Property of Kuñjabihārī Lāla of Sultanpur Zila.

Verse 6 is:

śrīmanmaudgalyagotro budhalasanaśaśī vidvanendrasya śiṣyo
daivajñānāṃ variṣṭhaḥ sakalaguṇanidher nāganāthasya sūnuḥ/
sūryāl labdhā varāṇi tribhuvanahitakṛt sūryasiddhāntasāraṃ
saṅgṛhyāsau nṛsiṃhaḥ saraṇimatitarāṃ jātakākhyaṃ pracakhyau//

NṚSIṂHA = NṚHARI

Author of a vast compilation in 93 adhyāyas, the *Jātakasāradīpa* = *Jātakasāra*. Manuscripts:

Kerala 5815 (11787). 2800 granthas. Copied in Saṃ. 1694 = A.D. 1637. Incomplete.
Anup 4646. 125ff. Copied by Lakṣmaṇa, the son of Kamalākara Bhaṭṭa, the son of Padmākara in Saṃ. 1708 = A.D. 1651.
Paris, BN 970 I (Sans. Dév. 445). Copied in A.D. 1696.
Bombay U Desai 1361. 204ff. Copied in Saṃ. 1786 = A.D. 1729.
BORI 516 of 1895/1902. 143ff. (ff. 2 and 4 missing). Copied in Saṃ. 1880 = A.D. 1823.
Alwar 1768.
Anup 4643. 15ff. Incomplete (romakācāryamatatājika).
Anup 4644. 222ff. Copied by Udho Kāyastha at Ādamapura. Formerly property of Maṇirāma Dīkṣita (*fl. ca.* 1650/1700).
Anup 4645. 145ff.
Anup 4647. 192ff. Incomplete.
Anup 4648. 10ff. Incomplete.
Anup 4649. 9ff. Incomplete.
Baroda 9282. 45ff. Incomplete.
Benares (1963) 34611. Ff. 1–151 and 153–172. Incomplete. This is probably Benares (1903) 1064. 137ff.
Benares (1963) 36379. Ff. 2–16, 23–28, and 33–118. Incomplete.
Benares (1963) 36810. 68ff. Incomplete.
BORI 471 of 1892/95. 97ff. Incomplete.
GVS —— (3852). 61ff.
Oppert I 5980. Property of the Mahārāja of Travancore.
PL, Buhler IV E 119. No ff. given. Incomplete. Property of Śrīdhara Bhaṭṭa of Śondurṇī. Buhler notes 2 other copies.
SOI 9522.
Tanjore D 11397 = Tanjore BL 4218. 164ff.

The *Jātakasāradīpa* was edited from Tanjore D 11397 by Lakṣmīnārāyaṇa Upādhyāya with his own vivṛti, *Durghaṭārtha*, as *TSMS* 45 = *Madras GOS* 64, Tanjore 1951. Verse 4 is:

nijatātapadāravindayugmaṃ
svamanonīrajapañjare nidhāya/
likhanakramasaṃyutaṃ pravakṣye
nṛharir jātakasāradīpam asmāt//

The colophon begins: iti sakalāgamācāryaśrīnṛsiṃhadaivajñakṛtaḥ.

NṚSIṂHA

Author of a *Tithipradīpikā*. Manuscripts:

GOML Madras R 5643. 95ff. Copied in A.D. 1932/33 from GOML Madras D 3122.
GOML Madras D 3122. 174pp. Telugu.

NṚSIṂHA

Author of an enormous *Nibandhaśiromaṇi* which contains much astrological information. Manuscripts:

Baroda 4012. 650ff.
Baroda 9212. 17ff. and 658ff.
DC 367. 21ff. Incomplete (saṃvatsaraphala). From Dīkṣit (A) Collection.

NṚSIṂHA

Author of a *Nūtanatithicakra* = *Pañcāṅgasādhana*. Manuscript:

Anup 4817. 3ff.

NṚSIṂHA

Assumed author of a *Nṛsiṃhakaraṇa*. Manuscript:

Oppert I 8045. No author mentioned. Property of Paravastu Veṅkaṭaraṅgācāryār of Viśākhapaṭṭana, Vizagapatam.

NṚSIṂHA

Author of a *Nṛsiṃhagaṇita*. Manuscript:

Oppert I 6933. Property of Puligaḍḍa Aruṇācalaśāstrī of Kottapeṭa, Vijayanagaram, Vizagapatam.

NṚSIṂHA

A resident of Gurjaramaṇḍala, Nṛsiṃha wrote a *Phalakalpalatā;* this may be the *Varṣaphala* of Nṛsiṃha (b. 1548). Manuscripts:

Baroda 2478. 12ff. Copied in Saṃ. 1706 = A.D. 1649.
Baroda 11841. 8ff. (f. 1 missing). Incomplete.
Benares (1963) 35538. 5ff. Incomplete. No author mentioned.
Benares (1963) 35809 = Benares (1913–1914) 2344. 10ff. No author mentioned.
PL, Buhler IV E 273. 12ff. No author mentioned. Property of Maṇiśaṅkara Jośī of Aṅkaleśvara.
Rajputana, p. 30. From Jaisalmer.

NṚSIṂHA

Author of a *Brahmatulyādipātasādhanavāsanā.* Manuscript:

Baroda 3217. 9ff.

NṚSIṂHA

The son of Śiṅganārya, the son of Tripurāntaka Somāsī, the purohita of Narasiṃha, a mahārāja, and a resident of Taṭidala agrahāra on the south bank of the Kuśasthalī, 3 yojanas south of Haripura, Nṛsiṃha wrote a *Veṅkaṭādrināthīya* = *Grahatantra,* following the *Sūryasiddhānta.* Manuscripts:

Adyar List. 2 copies = Adyar Index 5879 =
Adyar Cat. 20 G 55. 10ff. Grantha. Incomplete (ends in sphuṭādhikāra).
Adyar Cat. 20 G. 62. 44ff. Grantha.
Mysore (1911 + 1922) 2559. Ff. 9–26.
Tanjore D 11614 = Tanjore BL 4270. 30ff. Incomplete.
Tanjore D 11615 = Tanjore BL 11007. 63ff. Telugu. Incomplete.

Verses 2–8b are:

prāktriṃśadyojanair bhāti bhūmadhyād veṅkaṭo giriḥ/
tatra sthite haripure viṣuvadvāguṇāṅgulā//
tasya dakṣiṇadigbhāge triyojanamite nadī/
kuśasthalīti vikhyātā taddakṣiṇataṭe sthite//
agrahāre taṭidale vidvajjanayute vasan/
narasiṃhamahārājapurohitaśatāguṇaiḥ (?)//
tripurāntakasomāsī triskandhajñānavān sudhīḥ/
tasyātmajo ᵓbhūd vedajñaḥ padavākyapramāṇavit//
śiṅganārya iti khyātaḥ śrīnṛsiṃhaprasādajaḥ/
nṛsiṃhavarajātatvāt siṃhād utpannasiṃhavat//
prativādimahādantisiṃho ᵓbhūd bhūmigolavit/
tatsutaḥ śrīnṛsiṃho ᵓhaṃ sūryasiddhāntasammatam//
grahatantraṃ veṅkaṭādrināthīyākhyaṃ samārabhe/

NṚSIṂHA

Author of a *Siṃhalatājikoktāḥ ṣoḍaśayogāḥ,* on which there is a ṭīkā, *Praśnasāra,* by Vyaṅkaṭeśa. Manuscript:

Benares (1963) 34887. 5ff. Copied in Saṃ. 1949 = A.D. 1892. With the ṭīkā of Vyaṅkaṭeśa.

This was edited by Mīṭhālāla Ojhā, Vārāṇasī.

GĀLI NṚSIṂHA KAVI

Author of an *Ahobalapaṇḍitīya.* Manuscripts:

Cocanada, Telugu Academy 871 and 1857/3. See NCC, vol. 1, rev. ed., p. 489.

NṚSIṂHA KHEḌAKARA

A Brāhmaṇa of the Bhāradvājagotra, Nṛsiṃha wrote a vyākhyā or vṛtti on the *Upadeśasūtra* of Jaimini; *cf.* the *Arthaprakāśikā* of Ākumalla Nṛsiṃha. Manuscripts:

Baroda 1205. 68ff.
Mysore 452 (473). Ascribed to Nṛsiṃhadeva.
Mysore (1911 + 1922) B 593. 44ff. and 18ff.
Mysore (1955) 5222. 34ff. Telugu. Incomplete (adhyāya I). Ascribed to Kheṭa Oṃkāra Nṛsiṃhadeva.
Mysore and Coorg 284. 1000 sūtras. (*Jaiminisūtra* of Narasiṃhācārya). Property of Mahādeva Joyisa of Sringeri.

NṚSIṂHA BHAṬṬA

Author of a *Vidhānamālā.* Manuscripts:

Baroda 10449. 172ff. (ff. 1–28 missing). Copied in Saṃ. 1622 = A.D. 1565.
Anup 2573. 141ff. Copied in Śaka 1510 = A.D. 1588. Property of Dinakarabhaṭṭa.
Anup 4453. 2ff. Copied in Saṃ. 1720, Śaka 1585 = A.D. 1663. Incomplete (saptarṣitaraṅga). No author mentioned.
Baroda 2015. 145ff. Copied in Saṃ. 1850 = A.D. 1793.
Baroda 1484. 4ff. and 137ff. Copied in Śaka 1748 = A.D. 1826.
Jammu and Kashmir 4108. 151ff. Copied in Saṃ. 1941 = A.D. 1884.
Anup 2574. 241ff. = Bikaner 1058 A. 247ff.
Anup 2575. 157ff. = Bikaner 1058 B. 150ff.
Anup 2576. 119ff. Incomplete.
Anup 2577. 70ff. Incomplete.
Anup 2578. 77ff. Incomplete.
Anup 2579. 80ff. Incomplete.
Anup 2580. 101ff. Incomplete.
Anup 2581. 23ff. Incomplete (candrapūjā to svastivācana).
Anup 2582. 36ff. Incomplete (to yajñavisarjana).
Baroda 4061. 243ff.
Baroda 9601. 85ff. Incomplete.
Baroda 10583. 43ff. Incomplete.

This was edited by Śaṅkara Śāstrin Mārulkar as *ASS* 86, Poona 1920. The colophon begins: iti śrīnṛsiṃhabhaṭṭaviracitāyāṃ.

NṚSIṂHA SŪRI

The son of Nīlakaṇṭha of the Vatsagotra, Nṛsiṃha wrote a *Jātakayogāvalī* = *Triṃśadyogāvalī*. Manuscripts:

Adyar List. Telugu. (*Yogāvalī*). Not found in Adyar Index or Adyar Cat.
Ānandāśrama 6396.
Ānandāśrama 7735.
Dharwar 703 (693). 17ff.
GOML Madras D 13697. Ff. 1–14. Telugu.
GOML Madras D 13698. Ff. 25–35. Karṇāṭakī.
GOML Madras D 13699. Ff. 25–31. Telugu. Incomplete.
GOML Madras D 13766. Ff. 78–88. Telugu. Incomplete.
Kerala 5809 (5739 C). 175 granthas. Grantha.
Mysore (1922) 299. 9ff.
Mysore (1922) 370. Ff. 110–117.
Mysore (1922) 1804. Ff. 11–57.
Mysore (1922) 1813. 12ff.
Mysore (1911 + 1922) 2053. 142ff.
Mysore (1911 + 1922) 2589. Ff. 15–29.
Mysore (1922) 4441. 12ff.
Mysore (1922) 4751. 88ff.
Oppert I 361. 16pp. Grantha. No author mentioned. Property of Koṇḍaṅgi Anantācāryār of Kāñcīpuram, Chingleput.
Oppert II 3159. No author mentioned. Property of Taḍakamalla Veṅkaṭakṛṣṇarāyar of Tiruvallikeni, Madras.

Verses 1–3 are:

śrīmannṛsiṃhena sarojajātā
seyaṃ sarojālayalālanīyā/
itīva nityaṃ hṛdayāravinde
niveśitaṃ māṃ kamalāya x vyāt//
śrīvatsagotrodbhavagranthakartā
nṛsiṃhanāmākhilaśāstravettā/
siddhāntaśāstre bahudhā samartho
yogāvalījātakabhāvahartā//
sarvārthacintāmaṇisūtrabhāvau
lakṣmīpatir jātakaśastyabhāvau/
ityādigranthānvayasaṃmatena
karomi yogāvalim ādareṇa//

The colophon begins: nīlakaṇṭhātmajanṛsiṃhasūriviracite.

NṚSIṂHA SŪRI

Author of a *Jyotiṣārthasaṅgraha*. Manuscript:

Tanjore D 11422 = Tanjore BL 11053k. Ff. 40–47. Grantha. Incomplete.

CHALĀRI NṚSIṂHA (*fl.* 1198)

The pupil of Madhva, Nṛsiṃha wrote in Śaka 1120 = A.D. 1198 a *Smṛtyarthasaṅgraha* = *Smṛtyarthasāgara*, of which one section is the *Kālataraṅga;* see Chalāri. Manuscripts:

Baroda 5852. 35ff. Copied in Saṃ. 1811 = A.D. 1754.
Anup 2656. 28ff.
CP, Kielhorn XIX 38. 62ff. Property of Lakṣmaṇa Śāstrī of Sāgar.
Oxford 669 (Wilson 204a). Ff. 1–28.
PUL I 144. 93ff.

The *Smṛtyarthasaṅgraha* was published with the ṭippanī of Uddhava Bālācārya Aināpure at Bombay in 1885 (IO 2.E.23).

Verses 1–5b are:

śrīnṛsiṃhaṃ madhvaguruṃ śrīvyāsaṃ naumi sadgurum/
satsukhasvātmarūpaṃ taṃ śubhakarmaphalapradam//
kalau pravṛttaṃ bauddhādimataṃ rāmānujaṃ tathā/
śake hy ekonapañcāśadadhikābdasahasrake//
nirākartuṃ mukhyavāyuṃ sanmatakhyāpanāya ca/
ekādaśaśate śake viṃśatyabdayute gate//
avatīrṇaṃ madhvaguruṃ sadā vande mahāguṇam/
guṇāḍhyān bhagavadbhaktān jayatīrthādikān gurūn//
saṃnatya kurmas tattuṣṭyai spaṣṭaṃ smṛtyarthasāgaram/

The colophon begins: chalārinṛsiṃhācāryakṛtasmṛtyarthasāgare.

NṚSIṂHA = NARASIṂHA (*fl.* between *ca.* 1360 and 1435)

A native of Karṇāṭaka and a Brāhmaṇa of the Kauṇḍinyagotra, Nṛsiṃha wrote a *Prayogapārijāta* in 5 kāṇḍas; a part of one of these, the *Ṣoḍaśakarma*, is the *Grahayajñaprayoga*. Manuscripts:

Anup 2492. 4ff. (*Pariśiṣṭoktagrahaprayoga*).
Calcutta Sanskrit College (Smṛti) 326. 8ff.
IO 1396 (1795). 258ff. (ff. 80–115 and 157–158 missing). (*Pākayajñakāṇḍa* and *Ṣoḍaśakarmakāṇḍa*). From Dr. J. Taylor.
IO 1397 (776b). 39ff. From H. T. Colebrooke.
IO 5467 (Mackenzie II. 27). 27ff. Nandināgarī. From Colin Mackenzie.
IO 5468 (Mackenzie II. 62a). 31ff. Nandināgarī. From Colin Mackenzie.

The *Prayogapārijāta* was partially edited by C. Rāma Śāstrī, Dvivedi Subrahmaṇya Avadhānī, and C. Narahari Jyotirvid, Mysore 1908–1911 (BM 14028. bbb.17); and by Vāsudeva Śarman, Bombay 1916 (IO 25.B.3).

NṚSIṂHA = MĀMIḌI ŚIṄGAYA (*fl. ca.* 1400)

The son of Peddanārya, the son of Māmiḍi Mantrin, the son of Ciṭṭaya, and a descendent of Pedda Tukkaya Mantrin of the Bharadvājagotra, Nṛsiṃha, a mantrin of Pedda Komaṭi Vemendra, the Reḍḍi who ruled Koṇḍavīḍu from 1398 to 1415, wrote a ṭīkā, *Gūḍhārthadīpikā*, on the *Somasiddhānta*. Manuscript:

GOML Madras R 1715. 56ff. Telugu. Copied in A.D. 1915/16 from a manuscript copied by Veṅkaṭācārya and belonging to Bhamiḍipāṭi Acyutarāmasomayājulugāru of Irusumanda, Godāvari.

Verses 4–8 are:

asti śrīmadbharadvājagotre śrotriyabhūṣaṇam/
peddatukkayasanmantrivaṃśāmbodhisudhākaraḥ//
naptā śrīcaṭṭayākhyasya pautro māmiḍimantriṇaḥ/
tanujaḥ peddanāryasya śrīnṛsiṃho mahāyaśāḥ//
mantriṇo yasya mantreṇa narendrā vairiṇo bhuvi/
citraṃ giribilānteṣu bhajante vanavāsitām//
peddakomaṭivemendramantriṇā tena dhīmatā/
jyotirvidyānirāghāṭasarasvatpāradṛśvanā//
śiṅgayāmātyaratnena bhuvi lokahitaiṣiṇā/
kriyate somasiddhāntavyākhyā gūḍhārthadīpikā//

The colophon begins: iti sakalasiddhāntamatānusāreṇa māmiḍiśiṅgayāryeṇa (also: māmiḍiśiṅgaṇācāryeṇa) viracitāyāṃ.

NṚSIṂHA (*fl.* 1409)

The son of Rāmacandra (*fl. ca.* 1400), the son of Kṛṣṇa, Nṛsiṃha wrote a vivaraṇa on his father's *Kālanirṇayadīpikā*, apparently in Śaka 1331 = A.D. 1409. Manuscripts:

IO 1662 (2644). 145ff. Copied on Sunday 11 kṛṣṇapakṣa of Māgha in Saṃ. 1604 = 4 February 1548. From Gaikawar.

Anup 1680. 223ff. Copied by Tapasyārya Nṛsiṃha in Saṃ. 1609 = A.D. 1552. Formerly property of Śrīvallabha, Bālakṛṣṇa Dīkṣita, and Anūpasiṃha (*fl.* 1674/1698).

Baroda 5880. 112ff. Copied in Saṃ. 1611 = A.D. 1554.

BORI 91 of 1882/83. Ff. 1–58 and 61–172. Copied in Saṃ. 1621 = A.D. 1564. From Gujarāt.

Oudh V (1875) IX 4. 298pp. Copied in A.D. 1573. Property of Rājā Kāśīnātha of Faizabad Zillah.

BORI 222 of 1879/80. 191ff. Copied in Saṃ. 1641 = A.D. 1584.

Bombay, Bhandarkar 12. 53ff. Copied on Saturday 1 kṛṣṇapakṣa of Āśvina in Saṃ. 1647 = 3 October 1590 Julian.

Jammu and Kashmir 2457. 153ff. Copied in Saṃ. 1648 = A.D. 1591.

BORI 92 of 1882/83. 111ff. Copied in Saṃ. 1651 = A.D. 1594. Incomplete.

Anup 1681. 117ff. Copied by Vasanta Kāyastha of Kāśī in Saṃ. 1652 = A.D. 1595. This is probably Bikaner 859 A. 124ff.

Oxford 1496 (Sansk. d. 137) = Hultzsch 198. 98ff. Copied by Viṭṭhala Brāhmaṇa of Kāśī on Sunday 5 (?) (read 3) śuklapakṣa of Caitra in Saṃ. 1652 = 2 March 1595 Julian. This is CP, Kielhorn XIX 43. 98ff. Copied in Saṃ. 1652 = A.D. 1595. Property of Nānā Śāstrī of Sāgar.

Leningrad (1918) 51. 101ff. Copied by Viṣṇujīka, the son of Śrīkaṇṭha Bhaṭṭa of the Gauḍajāti, a resident of Stambhatīrtha, at Brahmapurī on Thursday 1 kṛṣṇapakṣa of Āśvina in Saṃ. 1655 = 5 October 1598 Julian.

Baroda 592. Ff. 29–158. Copied in Saṃ. 1672 = A.D. 1615. Incomplete.

Baroda 9706. 131ff. Copied in Śaka 1538 = A.D. 1616.

Baroda 12240. Ff. 12–150. Copied in Saṃ. 1683 = A.D. 1626.

Benares (1956) 13690. Ff. 1–5 and 7–100. Copied in Saṃ. 1684 = A.D. 1627. Incomplete.

Udaipur 136. 136ff. Copied in Saṃ. 1707 = A.D. 1650. See G. N. Sharma [1965] 67.

AS Bengal 2660 (G 5752). 223ff. Copied on Thursday 14 śuklapakṣa of Āṣāḍha in Saṃ. 1716 = 23 June 1659 Julian.

Calcutta Sanskrit College (Smṛti) 53. 133ff. Copied by Rāmeśvara, the son of Haṃsarāma, the son of Urvīdhara Miśra, on Thursday 10 (read 15) kṛṣṇapakṣa of Mārgaśīrṣa in Saṃ. 1727 = 1 December 1670 Julian.

Baroda 2239. 96ff. Copied in Saṃ. 1730 = A.D. 1673.

Baroda 1459. 86ff. Copied in Śaka 1600 = A.D. 1678.

IO 1663 (181a). 134ff. Copied in A.D. 1678. From H. T. Colebrooke.

Baroda 9031(b). 138ff. Copied in Saṃ. 1740 = A.D. 1683.

Baroda 10559. 121ff. Copied in Saṃ. 1767 = A.D. 1710.

Bombay U 1022. 119ff. Copied on Wednesday 3 śuklapakṣa of Māgha in Śaka 1700 = 20 January 1779. Formerly property of Bālakṛṣṇa Āraṃvakara.

RORI Cat. II 10021. 68ff. (f. 1 missing). Copied by Bihārī Lāla in Saṃ. 1838 = A.D. 1781 from a manuscript belonging to Mayārāma.

Baroda 3872. 153ff. Copied in Saṃ. 1839 = A.D. 1782.

BORI 161 of 1886/92. Ff. 1–34, 37–49, and 51–63. Copied in Saṃ. 1855 = A.D. 1798.

IO 1661 (1323). 82ff. Copied in A.D. 1805. From H. T. Colebrooke.

Benares (1956) 13346. 151ff. Copied in Saṃ. 1898 = A.D. 1841.

Adyar Index 1242 = Adyar Cat. 38 E 27. 212ff.

Alwar 1289.

Anup 1682 = Bikaner 859 B. Ff. 1–11 and 11b–114.

Baroda 762. 148ff. Incomplete.

Baroda 8734. 138ff.

Baroda 10410. 137ff.

Benares (1956) 12065. Ff. 4–5, 7–22, and 22b–45. Incomplete. No author mentioned.
Benares (1956) 12126. Ff. 1–47, 49–114, 114b, and 120. Incomplete. No author mentioned.
Benares (1956) 13226. Ff. 2–70, 73–119, and 121–250, and 1f. Incomplete.
Benares (1956) 13977. Ff. 1 and 1b–57.
Bhor 46.
Bhor 47.
BORI 99 of 1871/72. 170ff.
BORI 327 of 1880/81. 119ff.
BORI 252 of A 1881/82. 92ff.
BORI 524 of 1883/84. 112ff. Incomplete. From Mahārāṣṭra.
BORI 290 of 1884/87. 157ff. From Mahārāṣṭra.
BORI 66 of 1895/98. 92ff.
BORI 139 of Vishrambag I. 133ff.
Calcutta Sanskrit College (Smṛti) 51. 34ff. Incomplete.
Calcutta Sanskrit College (Smṛti) 52. 153ff. Incomplete.
DC 4210. Ff. 6–14, 16–18, 20–28, and 30. Incomplete. From Dīkṣit (A) Collection.
Florence 119. 51ff. Incomplete.
IL Calcutta 242. See NCC, vol 4, p. 29.
IO 1660 (2513). 157ff. From Gaikawar.
Madras Univ. R.A.S. 187. See NCC.
Mitra, Not. 140. 153ff. (ff. 31–32 and 83–84 missing). Property of the Asiatic Society of Bengal, but not in AS Bengal.
Mitra, Not. 2282. 144ff. Property of Mahārāja Rājendrakiśora Siṃha, Bahādur, of Bettiyā.
Mysore (1922) pp. 102–103 (4 manuscripts of which 2 are incomplete). See NCC.
Nagpur, Deo Coll. 102. See NCC.
N-W P V (1880) Dharmaśāstra II 10. 57ff. Property of Ḍhuṇḍhirāja Śāstrī of Benares.
PL, Buhler III E 62. 98ff. Property of Bālambhaṭṭa of Surat. Buhler notes 3 other copies.
PL, Buhler III E 63. 95ff. Property of Maṅgalaśaṅkara of Ahmadābād.
Rajapur, Saṃskṛta Pāṭhaśālā 244 and 564. See NCC.
Rajputana, p. 7. From Ujjain.
RORI Cat. II 9963. 12ff. Incomplete.
Tanjore D 18571 = Tanjore BL 214. 216ff.
Tanjore D 18572 = Tanjore BL 125. 109ff.
VVRI 6728. 10ff. Incomplete (parvanirṇaya). No author mentioned.
WHMRL M. 14. b.

Verses 1–4 are:

śrīviṭṭhalaṃ śrutigiraḥ prathitaprabhāvaṃ
bhāvārdramānasasarovararājahaṃsam/
śrīrāmacandragurum ekam anekaśāstra-
pārīṇakovidadhurīṇam ahaṃ namāmi//
yena vyākaraṇārṇavaikataraṇiḥ sa prakriyākaumudī
vedāntānumatā ca vaiṣṇavamahāsiddhāntasandīpikā/
kālajñānavidhau vyādhāyi vibudhānandi prabandhatrayaṃ
kṛṣṇācāryasutaḥ sa naḥ sukhayatu śrīrāmacandro guruḥ//
śrīrāmacandrakṛtinā nidhinā kalānām
ānītam etad adhunā vasudhāsudhāyāḥ/
śrīkālanirṇayakṛduktisudhābdhisāram
āpīyam ākarṇapaṭakaiḥ kalayantu kālam//
tatsūnunā samayanirṇayadīpikāyāṃ
gāmbhīryagarbhapadapadmaguṇānvitāyām/
nirṇīyate vivaraṇaṃ kaṇaśaḥ pravīṇair
ā cūḍamūlam avalokya vivecanīyam//

The last verse is:

tajjyeṣṭhabhrātṛputraḥ parikalitakalaḥ śrīguro rāmacandrāt
kāvyānāṃ yena ṭīkā vyaraci sa karuṇāmbhonidhir jñānasindhuḥ/
śrīkṛṣṇācāryasaṃjño gurur ayam avatān māṃ kṛpāṃ prāpya yasya
śrīrāmācāryasūnur vivaraṇam akarod dīpikāyāṃ nṛsiṃhaḥ//

The colophon begins: iti śrīmatsarvaśāstrajñasakalamahāgamācāryaparamahaṃsaparivrājakācāryaśrīgopālagurupūjyapādapriyaśiṣyaśrīrāmacandrācāryasutaśrīnṛsiṃhācāryakṛta.

The verse giving the date of composition, Śaka 1331 = A.D. 1409, is found in some manuscripts:

śāke śaśāṅkānilaviśvasaṃmite
virodhivarṣe sitapakṣake ca/
some nabhasy āryanṛsiṃhanāmabhiḥ
saddīpikāyā vivṛtiḥ samāptā//

Nṛsiṃha is also the author of a vyākhyā on Rāmacandra's *Tithinirṇayasaṅgraha*. Manuscripts:

Baroda 10552(b). 21ff. Copied in Saṃ. 1657 = A.D. 1600.
Baroda 1524. 21ff. Copied in Saṃ. 1683 = A.D. 1626.
BORI 192 of 1886/92. 39ff. Copied in Saṃ. 1684 = A.D. 1627.

NṚSIṂHA (b. 1548)

The son of Rāma (*fl. ca.* 1525/1550), the son of Keśava (*fl.* 1496/1507) of the Kauśikagotra, Nṛsiṃha was born at Nandigrāma in Śaka 1470 = A.D. 1548 and studied jyotiḥśāstra under his uncle, Gaṇeśa (b. 1507). See S. B. Dikshit [1896] 317. He wrote the following works on jyotiḥśāstra.

1. The *Grahakaumudī*, in which are given 2 epochs: 31 March 1588 and 31 March 1603. See D. Pingree [A2. 1970b] 101 and SATE 118–123. Manuscripts:

IO 2945 (2083d). 3ff. From Gaikawar. See SATE 27–28.
IO 2946 (2083e). 71ff. From Gaikawar. See SATE 27–28.

Verses IV 11–12 are:

sahyādrer adharāparāntaviṣaye kṣarāmbudheḥ
prāktaṭe
grāme nandipadādime sukadalīsaśīrṣapūgānvite/
āsīt kauśikavaṃśabhūṣaṇamaṇiḥ śrīkeśavo daivavin
nānāśāstrakalākalāpacaturaḥ saujanyaratnākaraḥ//
tatputro vividhāgamārthakuśalo rāmo grahajñāmaṇis
tatputro ᵓjani khādrivāsavamite śāke
nṛsiṃhābhidhaḥ/
sadbuddhiḥ svapitṛvyato guruganeśāt prāpya
bodhāṃśakaṃ
teneyaṃ grahakaumudī viracitā daivajñasantuṣṭaye//

2. A *Kheṭamuktāvalī*. Manuscripts:

Anup 4502. 38ff. Copied by Kṛṣṇaśāmati Mala in Śaka 1587 = A.D. 1665. With sāraṇī.
IM Calcutta 1461. No author mentioned. See NCC, vol. 5, p. 190.
Poleman 4724 (Harvard 54). Ff. 1–4 and 1–9. With a ṭīkā.

3. A *Grahadaśāphala* in 86 verses. Manuscripts:

Bombay U 404. 4ff. Copied by Yajñeśvara Sānye on 2 kṛṣṇapakṣa of Phālguna in Śaka 1724 = *ca.* 9 March 1803.
Bombay U 403. 6ff. Copied by Yajñeśvara Dīkṣita Sānye on 10 kṛṣṇapakṣa of Bhādrapada in Śaka 1732 = *ca.* 22 September 1810.
Adyar Index 1945 = Adyar Cat. 11 B 46. 9ff.

Verse 1 is:

gaṇanāthaṃ praṇamyādau grahān sūryādikān gurūn/
rāmacandrātmajo brūte nṛsiṃho janmajaṃ phalam//

Verse 86 is:

parodadheḥ pūrvagatīrasaṃsthaḥ
śrīnandipuryāṃ śrutimān grahajñaḥ/
rāmātmajaśrīnṛharir dvijāgryas
tenoditaṃ janmaphalaṃ grahāṇām//

4. A *Grahadīpikā*. Manuscript:

Anup 4532 = Bikaner 633. 5ff. Copied by Manohara Vyāsa in Saṃ. 1725 = A.D. 1668.

Verse 1 is:

gaṇapaticaraṇāravindayugme
nijakathaya bhramarāpi neddhi (?)/
dhāsugaṇakavararājo (?)
gaṇakumudagrahadīpikāṃ praṇuve//

The colophon begins: iti sakalāgamācāryavaryaśrīrāmadaivajñātmajaśrīnarasiṃhadaivajñaviracitā.

5. A *Varṣaphaladīpikā; cf.* the *Phalakalpalatā* of Nṛsiṃha. Manuscripts:

Baroda 3289. 5ff. (*Varṣaphala*).
Tanjore D 11593 = Tanjore BL 4210. 3ff.

The last verse is:

rāmātmajo nandipurādhivāsī
śrīmannṛsiṃhaḥ khagavipravaryaḥ/
tenoditaṃ varṣaphalaṃ sabhāyāṃ
vaktuṃ sphuṭaṃ daivavidāṃ mukhena//

6. A ṭīkā, *Harṣakaumudī*, on the *Grahalāghava* (1520) of his uncle Gaṇeśa (b. 1507). Manuscripts:

VVRI 2654. 88ff. Copied by Dharmacandra for Gaṇḍā Miśra on Monday 10 kṛṣṇapakṣa of Phālguna in Saṃ. 1816 = 10 March 1760.
ABSP 1089. 33ff.
Ānandāśrama 7804.
Anup 4542. 41ff. Incomplete.
BORI 509 of 1895/1902. 106ff.
DC (Gorhe) App. 112. Property of Gaṅgādhara Rāmakṛṣṇa Dharmādhikārī of Puṇatāmbe, Ahmadnagar.
LDI (VDS) 1294 (9856) = LDI (DSC) 9856. 18ff.

The last 2 verses are:

āsīt kauśikavaṃśabhūṣaṇamaṇiḥ śrīkeśavasyātmajaḥ
kṣīrāmbhonidhipūrvatīrakagatāyāṃ nandipuryāṃ
vasan/
nānāśāstrakalākalāpacaturaḥ śiṣyādigītastutir
jyotirvittilako guṇaikavasatiḥ śrīrāmacandraḥ
sudhīḥ//
tatsūnur gaṇakāgraṇir dvijavaraḥ
śrīmannṛsiṃhābhidhaś
cakre tadvivṛtiṃ sphuṭāṃ suvimalāṃ
bālāvabodhapradām/
yat sphuṭaṃ grahalāghavākhyakaraṇaṃ
śrīmadgaṇeśo gurur
gūḍhārthaṃ parimandabuddhibhir
avijñānārthabodhaṃ yataḥ//

7. A *Hillājadīpikā*. Manuscripts:

Benares (1963) 36930. 10ff. Copied in Śaka 1555 = A.D. 1633.
Tanjore D 11594 = Tanjore BL 4217. 11ff. Copied at Kāśī on Tuesday 7 śuklapakṣa of Pauṣa in Śaka 1563 = 28 December 1641 Julian.
Baroda 3365. 15ff. Copied in Saṃ. 1803 = A.D. 1746.
BORI 891 of 1884/87. 5ff. Copied in Saṃ. 1860 = A.D. 1803. From Gujarāt.
RORI Cat. II 5718. 14ff. Copied in Saṃ. 1865 = A.D. 1808.
PL, Buhler IV E *461. 13ff. Copied in Saṃ. 1871 = A.D. 1814. Property of Harirāmaśāstrī of Aṅkaleśvara.
Benares (1963) 35452 = Benares (1905) 1509. 9ff. Copied in Saṃ. 1889 = A.D. 1832.
Benares (1963) 34473. 8ff. Copied in Saṃ. 1905 = A.D. 1848.
Benares (1963) 34895. 12ff. Copied in Saṃ. 1970, Śaka 1836 = A.D. 1913.

Alwar 2031.
Anup 5381. 11ff.
AS Bengal 7351 (G 10461). 5ff. Incomplete.
AS Bengal 7352 (G 2912) = Mitra, Not. 4095. 11ff.
Benares (1963) 35573. 9ff.
Benares (1963) 36656. 12ff.
CP, Kielhorn XXIII 189. 11ff. Property of Javāhara Śāstrī of Chāndā.
Jammu and Kashmir 4057. 21ff. Copied from Alwar 2031.
Kathmandu (1960) 164 (III 104). 27ff. Copied by Durgādatta.
Mithila 436. 8ff. Maithilī. Property of Paṇḍita Rudramaṇi Jhā of Mahinathapur, Deodha, Darbhanga.
Oudh XX (1888) VIII 75. 20pp. Said to have been copied in A.D. 1498 (read 1849?). Property of Paṇḍita Pratāpa Nārāyaṇa of Allahabad Zila.
Oudh XX (1888) VIII 88 = VIII 162. 22pp. Property of Paṇḍita Pratāpa Nārāyaṇa of Allahabad Zila.
Poleman 5176 (Columbia, Smith Indic 128). 12ff. No author mentioned.
PUL II 4099. 9ff.
RORI Cat. II 4860. 10ff.

Verses 1–2 at the end are:

kṣoṇīmaṇḍalamaṇḍanaṃ dvijakulālaṅkārahārakṣitau
śrīmatkauśikavaṃśabhūṣaṇamaṇiḥ śrīkeśavas tatsutaḥ/
nandigrāmanivāsy anekaguṇavān rāmābhidho daivavin
nānāśāstrakalākalāpacaturaḥ saujanyaratnākaraḥ//
tadātmajaḥ sarvajanābhirāmo
nṛsiṃhanāmā grahavidvariṣṭhaḥ/
pitṛvyataḥ śrīguruto gaṇeśād
gaṇeśarūpāt samavāptabuddhiḥ//

NṚSIṂHA (b. 1586)

The son of Kṛṣṇa (*fl. ca.* 1575/1600), the eldest son of Divākara (who had 4 other sons: Viṣṇu (*fl. ca.* 1575/1600), Mallāri (*fl. ca.* 1600), Keśava, and Viśvanātha; Divākara was a pupil of Gaṇeśa (b. 1507)), the son of Bhaṭṭācārya, the son of Rāma of the Bhāradvājagotra, a resident of Golagrāma, Nṛsiṃha studied under his uncles Viṣṇu and Mallāri at Varaṇāsi. See S. Dvivedin [1892] 82–84 and S. B. Dikshit [1896] 283. He wrote the following works on jyotiḥśāstra.

1. A ṭīkā, *Saurabhāṣya*, on the *Sūryasiddhānta*, composed at the age of 25 in A.D. 1611; see the *Grahasādhanopapattivāsanā* of Nṛsiṃha. An example for Saṃ. 1641 = A.D. 1584 may be due to his father, Kṛṣṇa, to whom the *Saurabhāṣya* is sometimes ascribed. Manuscripts:

BORI 601 of 1895/1902. 160ff. Copied in Śaka 1554 = A.D. 1632.
Anup 5350. 101ff. Copied in Saṃ. 1716 = A.D. 1659.
Tanjore D 11661 = Tanjore BL 4279. 19ff. Copied by Ambāji at Jhijharavāḍagrāma on Monday 14 kṛṣṇapakṣa of Phālguna in Śaka 1616 (?) = 4 March 1695 Julian (?).
Benares (1963) 35779 = Benares (1878) 87 = Benares (1869) XVI 2. 68ff. Copied in Saṃ. 1819 = A.D. 1762. Incomplete (ends with pātādhikāra). Ascribed to Kṛṣṇa.
Mithila 426 = Mitra, Not. 1838. 121ff. Maithilī. Copied by Buddhinātha of the Sukaraṇakula in A.H. 1197 = A.D. 1783. Property of the Raj Library at Darbhanga.
DC 6262. Ff. 47–54 and 80–88. Copied in Saṃ. 1842 = A.D. 1785.
Mithila 421 C. 145ff. Copied in Saṃ. 1847 = A.D. 1790. Property of Paṇḍita Babuājī Miśra of Koilakha, Lohat, Darbhanga.
Paris BN 957 (Sans. Bengali 189) V = Guérin 32. Bengālī. Copied in A.D. 1840. Incomplete. Ascribed to Kṛṣṇa.
Paris BN 998 (Sans. Bengali 186) I = Guérin 15. Bengālī. Copied in A.D. 1840.
VVRI 4695. 53ff. Copied in Saṃ. 1901 = A.D. 1844.
Mithila 421 B. 72ff. Maithilī. Copied in Śaka 1772 = A.D. 1850. Property of Paṇḍita Lakṣmīvallabha Jhā of Bhakharaini, Madhepur, Darbhanga.
Alwar 2020.
Baroda 9306. 96ff.
Benares (1963) 34458. 57ff. Incomplete.
Benares (1963) 35777 = Benares (1878) 123 = Benares (1869) XXVI 3. 67ff.
BORI 602 of 1895/1902. 52ff. (f. 34 missing). Ascribed to Kṛṣṇa.
Cambridge R. 15. 103. 144ff.
Cambridge R. 15. 104. 24ff. Bengālī. Incomplete.
CP, Kielhorn XXIII 181. 136ff. Property of Lakṣmaṇa Śāstri of Sāgar.
IO 2778 (1755). 204ff. From H. T. Colebrooke.
IO 2779 (2264). 88ff. Copied from IO 1755. From Calcutta.
IO 6283 (Mackenzie II. 39b). Ff. 29–177. From Colin Mackenzie.
Mithila 421. 84ff. Maithilī. Property of Paṇḍita Umādatta Miśra of Salampur, Ghataho, Darbhanga.
Mithila 421 A. 44ff. Maithilī. Incomplete. Property of Babu Candradeva Jhā of Mahinathapur, Jhanjharpur Bazar, Darbhanga.
Mysore (1922) 16. 91ff.
Mysore and Coorg 388. Ascribed to Kṛṣṇa. (*Siddhāntavyākhyāna*). Property of Mahādeva Joyisa of Sringeri.
Oppert II 3554. Property of Gomaṭham Guñjā Narasiṃhācāryār of Melkoṭa, Mysore.
Poleman 4931 (Columbia, Smith Indic 210). 22ff.
RORI Cat. II 4652. 61ff.

RORI Cat. III 11328. 106ff. (ff. 1–4 and 6–14 missing). Incomplete.
RORI Cat. III 12620. 143ff. (ff. 1–37, 39, 50–52, and 56 missing). Incomplete.
Tanjore D 11660 = Tanjore BL 4277. 49ff. Incomplete (ends with mānādhikāra).

Verse 7 is:

śrīviṣṇusaṃjñakapitṛvyamukhāravindān
mallārisaṃjñavadanād adhigatya vidyām/
saurāgamābdhitaraṇāya nṛṇāṃ karomi
bhāṣyaplavaṃ bahuvicāraviśeṣaramyam//

Verses 1–5 at the end are:

bhāradvājamaharṣivaṃśajavaras tīre sugodottare
golagrāmasamāhvaye sunagare deśe ca pārthābhidhe/
āsīt tatra gaṇeśasaṃjñakaguror labdhāśubodhāṃśako
bhaṭṭācāryasuto divākara iti khyātaḥ kṣitīśārcitaḥ//
tasyātmajāḥ pañca babhūvur eṣāṃ
jyeṣṭhas tu kṛṣṇo gaṇakāgravandyaḥ/
sūtrātmakaṃ bījam akāri yena
sa viṣṇunāmā gaṇako dvitīyaḥ//
yaṃ brahmaguptagaṇakāryavarāhalalla-
śrīkeśavācāryaguruvaryagaṇeśatulyam/
śrībhāskareṇa ca samaṃ gaṇakā vadanti
mallārisaṃjñakasuto ᵓkhilaśāstrakartā//
putrau tathānye tu divākarasya
mallārisaṃjñānusamudbhavau ca/
śrīkeśavo daivavidāṃ variṣṭhaḥ
śrīviśvanāthas tadanu pradiṣṭhaḥ//
daivajñāryadivākarātmajavaraḥ śrīkṛṣṇanāmā dvijo
yo ᵓbhūt tattanayo nṛsiṃhagaṇakaḥ sadyuktibhāṣyaṃ vyadhāt/
brahmeśānajanārdanaprabhṛtibhiḥ sevyena tigmāṃśunā
siddhāntasya mayāsurāya kathitasyājñānadośāpaham//

2. A ṭīkā, *Vāsanāvārttika*, on the *Siddhāntaśiromaṇi* of Bhāskara (b. 1114), composed in Śaka 1543 = A.D. 1621 at the age of 35. Manuscripts:

Oudh (July–Sept. 1875) VIII 3 = Oudh XI (1878) VIII 6. 312 pp. Copied in A.D. 1659. Property of Rājā Rāmanātha of Faizābād Zila.
IO 2857 and 2858 (1648 and 1706). Ff. 1–133 and ff. 1–43 and 43b–85. Copied in A.D. 1751. From H. T. Colebrooke.
AS Bombay 282. 111ff. Copied by Rāma at Nimba on 14 kṛṣṇapakṣa of Māgha in Śaka 1700 = *ca.* 13 February 1779. Incomplete (gaṇitādhyāya). From Bhāu Dājī.
Mithila 414. 88ff. Copied on 11 śuklapakṣa of Pauṣa in Saṃ. 1920 = *ca.* 19 January 1864. Incomplete (gaṇitādhyāya). Property of Paṇḍita Gaṅgādhara Jhā of Jonki, Deodha, Darbhanga.
Alwar 2010. Incomplete (golādhyāya).
Anup 5325. 93ff. Associated with Gaṇeśa and Ballāla.
Anup 5326. 37ff. Associated with Keśava, Gaṇeśa, and Ballāla. Incomplete.
Anup 5327. 27ff. Associated with Gaṇeśa and Ballāla. Incomplete.
AS Bombay 283. 26ff. Incomplete (gaṇitādhyāya ending in spaṣṭādhikāra).
AS Bombay 286. 74ff. Incomplete (golādhyāya). From Bhāu Dājī.
Benares (1963) 35628 = Benares (1878) 102 = Benares (1869) XXI 3. 75ff. Incomplete (golādhyāya).
Benares (1963) 35761 = Benares (1878) 103 = Benares (1869) XXI 4. 149ff. Incomplete (gaṇitādhyāya).
BORI 879 of 1884/87. 111ff. Incomplete (gaṇitādhyāya). From Gujarāt.
BORI 853 of 1887/91. 91ff. (ff. 1–3 missing). Incomplete (gaṇitādhyāya). From Gujarāt.
BORI 594 of 1895/1902. 51ff. Incomplete.
Calcutta Sanskrit College 166. 107ff.
IO 2859 and 2860 (2289 and 2283). 87 and 58ff. Copied from IO 1648 and 1706. From Calcutta.
IO 2861 (1939). 130ff. (ff. 39–42 and 45–50 missing). Incomplete (gaṇitādhyāya). From Dr. John Taylor.
IO 6294 (Mackenzie II. 44). 112ff. Incomplete (golādhyāya). From Colin Mackenzie.
Jammu and Kashmir 2783. 44ff. Incomplete (golādhyāya).
Jammu and Kashmir 2784. 68ff. Incomplete (gaṇitādhyāya).
Kurukṣetra 727 (19585). Incomplete (bhuvanakośa from golādhyāya).
Mithila 414 A. 67ff. Maithilī. Incomplete (gaṇitādhyāya). Property of Paṇḍita Jayakṛṣṇa Jhā of Champa, Benipatti, Darbhanga.
Mysore (1955) 5154. Ff. 37–109. Telugu. Incomplete (golādhyāya). Ascribed to Kṛṣṇa.
N-W P IX (1885) A 8. 54ff. Incomplete (gaṇitādhyāya). Property of Paṇḍita Vāmanācārya of Benares.
PL, Buhler IV E 524. 158ff. Property of Bālakṛṣṇa Jośī of Ahmadābād.
RORI Cat. II 5629. 21ff. Incomplete (pūrvārdha).
RORI Cat. II 5630. 50ff. Incomplete (uttarārdha).

The *Vāsanāvārttika* on the madhyamādhikāra of the grahagaṇita was edited by M. Jha [1908/16]. Verses 8–10 are:

nijatātasya kṛṣṇasya kṛtvā pādāmbujaṃ hṛdi/
śāstraṃ pitṛvyato ᵓdhītya vakṣye ᵓpūrvāṃ suvāsanām//
siddhāntavāsanābhāṣyam amitārthaṃ mitākṣaram/
vyākhyāyate nṛsiṃhena gaṇakānandahetave//
vidhāya sūryasiddhāntavāsanābhāṣyam uttamam/
vāsanāvārttikaṃ kartum udyato ᵓsmi śiromaṇeḥ//

The last verse in the gaṇitādhyāya is:

śrīmatkoṅkanavāsikeśavasutaprāptāvabodhād budhād
bhaṭṭācāryasutād divākara iti khyātāj janiṃ prāptavān/
yaḥ kṛṣṇas tanayena tasya racite sadvāsanāvārttike
satsiddhāntaśiromaṇer ayam agāt pātādhikāraḥ sphuṭaḥ//

Verses 2–9 at the end of the golādhyāya are:

guṇavedaśarendusaṃmite śakakāle nagare pureśituḥ/
vasatā varaṇāsimadhyage narasiṃhena vinirmitaṃ tv idam//
nije tattvamite varṣe saurabhāṣyaṃ mayā kṛtam/
pañcatriṃśanmite varṣe vāsanāvārttikaṃ kṛtam//
navīnayuktipratipādanena
pūrvoktatantrād api sadviśeṣam/
narapraṇītān nṛharipraṇītaṃ
svīkāryam āryaiḥ svadhiyā vicārya//
godāvarīsaumyataṭopakaṇṭha-
grāme ca golābhidhayā prasiddhe/
vipro mahārāṣṭrasugītakīrtir
babhūva rāmo gaṇakāryavandyaḥ//
agraṇīs taittarīyāṇāṃ bhaṭṭācāryas tu tatsutaḥ/
āsīn mīmāṃsakaśreṣṭhaḥ kumārila ivāparaḥ//
gotre bharadvājamuneḥ pavitre
divākaras tattanayo babhūva/
vedāntaśāstrābhyasanena kāśyāṃ
yaḥ puṇyarāśyāṃ tanum utsasarja//
sāṃvatsarāryasya divākarasya
śrīkṛṣṇadaivajña iti prasiddhaḥ/
babhūva putraḥ sutarāṃ pavitraḥ
sattīrthakartākhilaśāstravettā//
tajjas tu tasyaiva kṛpālavena
tātānujāvāptasamastavidyaḥ/
sadvāsanāvārttikanāmadheyaṃ
granthaṃ nṛsiṃho racayāṃ babhūva//

3. A ṭīkā on the *Tithicintāmaṇi* (1525) of Gaṇeśa (b. 1507). Manuscripts:

Anup 4733 = Bikaner 746. 51ff. Copied in Śaka 15(2?)9 = A.D. 1607(?).
Benares (1963) 35493 = Benares (1909) 1820. 4ff. Ascribed to Kṛṣṇa.

Verses 1–2 are:

śrīmallārikulādhīśapādapadmaṃ bhajāmy aham//
yatpādāmbujadarśanāt paramatidīpasphuradrūpiṇī
svasvājñānaghanāndhakāram anayā cetogṛhaṃ śuddhyate/
śuddhe vātasi cātmacintanam ato muktiḥ kimasthāparas(?)
tasmāt taṃ gurum ātmarūpam aparaṃ nityaṃ namaskurmahe//

The colophon begins: iti śrīkṛṣṇadaivajñasutanṛsiṃhagaṇakaviracitāyāṃ.

NṚSIṂHA (*fl.* 1644)

Author of a ṭīkā on the *Ratnāvalīpaddhati* of Gaṇeśa (*fl. ca.* 1550/1600) in Saṃ. 1701 = A.D. 1644. Manuscript:

Baroda 3397. 15ff.

YANAMANDRA NṚSIṂHA SŪRI (*fl. ca.* 1650)

The son of Ahobala of the Kauśikagotra, Nṛsiṃha wrote a *Daivajñabhūṣaṇa* in 15 prakaraṇas in which he refers to the 60-year cycle beginning in Śaka 1549 = A.D. 1627. Manuscripts:

Adyar List. 2 copies = Adyar Index 2907 =
 Adyar Cat. 21 M 35. 182ff. Telugu. Incomplete (ends in prakaraṇa 8).
 Adyar Cat. 21 M 36. 404ff. Grantha.
GOML Madras D 13432. 664pp. Copied from a manuscript copied by Nuddum Prasannajosyulu on Thursday 15 Bhādrapada of Kālayuktisaṃvatsara = 23 September 1858.
GOML Madras D 13433. Ff. 2–10. Telugu. Incomplete.
Hultzsch 1. 124. 36ff. Telugu. Incomplete. Property of Koṭra Rāmaliṅga Śāstri of Nellore.
Oppert I 801. 300pp. Grantha. No author mentioned. Property of Nivṛtti Subrahmaṇyaśāstrī of Kāñcīpuram, Chingleput.
VVRI 3889. 102ff. Telugu.

Verses 1–2 are:

śrīlakṣmīśaṃ namaskṛtya bhāskaraṃ bhāratīṃ gurum/
ahobalāryaṃ pitaraṃ gaṇeśaṃ ca muhur muhuḥ//
gotre ᵓsmin kauśike jāto yena mantrābdhicandramāḥ/
nṛsiṃhasūriḥ kurute daivajñānāṃ subhūṣaṇam//

The colophon begins: iti śrīśayanamandrāhobaladaivajñātmajaśrīnṛsiṃhasūriviracite.

NṚSIṂHA (b. 9 November 1821)

See Bāpū Deva Śāstrī (b. 9 November 1821).

NṚSIṂHADATTA MIŚRA (*fl.* 1837)

The son of Haradatta, Nṛsiṃhadatta wrote an upapatti on the *Makaranda* of Makaranda (*fl.* 1478) in Śaka 1759 = A.D. 1837. Manuscript:

Mithila 250. 10ff. Maithilī. Copied on Wednesday 8 kṛṣṇapakṣa of Bhādrapada in Śaka 1780, Sāl. San. 1265 = 29 September 1858.

Verse 1 is:

kṛṣṇaṃ natvā nṛsiṃhena makarandasya sādhane/
kandagucchādipatrāṇām upapattir viracyate//

The colophon begins: iti śrīmiśraharadattasutaśrīnṛsiṃhadattaviracita.

He also wrote a *Jātakaratnasaṅgraha* which may be identical with the *Jātakaratna* of Haradatta. Manuscript:

GJRI 3126/338. 5ff. Maithilī.

NṚSIṂHABHAṬṬA

Author of a *Nṛsiṃhabhaṭṭīya*. Manuscript:

Oppert I 6849. Property of Siṃhabhaṭṭa Siddhānti of Naḍupūr, Vizagatam.

NṚSIṂHAVARYA DĪKṢITA

Author of a *Jātakakalānidhi*. There are numerous manuscripts of works of this title in South India; those ascribing it to Nṛsiṃhavarya are:

Adyar List. 3 copies = Adyar Index 2193 =
Adyar Cat. 21 F 23. 14ff. Grantha.
Adyar Cat. 21 F 25. 24ff. Grantha.
Adyar Cat. 21 F 27. 133ff. Telugu. Incomplete (*sic!*).
Kerala 5743 (5746 E). 120 granthas. Grantha.
Oppert II 8216. Property of T. Rāmarow of Tanjore.

The *Jātakakalānidhi* was published with the *Jātakacandrikā* of Veṅkaṭeśvara at Madras in 1863, reprinted Madras 1873 (IO 12. D. 7 and 13. G. 7).

NṚHARI

Author of a *Nibandharatnākara*, of which a part is the *Māsanirṇaya*. Manuscript:

VVRI 1149. 28ff. Incomplete.

NṚHARI SAPTARṢI

Author of a *Vivāhacandrodaya*. Manuscripts:

AS Bombay 330. 8ff. From Bhāu Dājī.
AS Bombay 330 A. 8ff. From Bhāu Dājī.

The colophon begins: iti nṛharisaptarṣiviracito.

NETRĀNANDA

Author of a *Jayayātrā*. Manuscript:

BORI 901 of 1886/92. 15ff.

NEMICANDRA

Author of a *Kṣetragaṇita*. Manuscripts:

Ahmadabad, Dela Upāśraya Bhandar, ground floor 104 (31 and 32). See Velankar, p. 98.
Bhuleśvara, Bombay, Pannalal Jain Sarasvati Bhavan 1501 and 2649. See Velankar and NCC, vol. 5, p. 155.

NEMICANDRA (*fl. ca.* 975)

The pupil of Abhayanandin, Nemicandra wrote a *Trilokasāra* in 1018 Prākṛta verses for Cāmuṇḍarāya, the minister of the Western Gaṅga monarchs Mārasiṃha III (*ca.* 961–974) and Rājamalla IV (974–*ca.* 1004). There are commentaries by his pupil, Mādhavacandra (*fl. ca.* 1000), by Abhayacandra, by Sahasrakīrti, and by Sāgarasena. See B. B. Datta [A3. 1935]. Manuscripts:

RJ 1798 (vol. 2, p. 284). 62ff. Copied in Saṃ. 1529 = A.D. 1472. With the ṭīkā of Sāgarasena. Property of Baḍa Terahapanthiyoṃ of Jayapura.
RJ 1793 (vol. 2, p. 284). 29ff. Copied on 11 kṛṣṇapakṣa of Caitra in Saṃ. 1542 = *ca.* 10 April 1485. Incomplete. Property of Baḍā Terahapanthiyoṃ of Jayapura.
RJ 3368 (vol. 4, p. 320). 66ff. Copied in Saṃ. 1553 = A.D. 1496.
RJ 466 (vol. 2, p. 43). 71ff. Copied on 5 śuklapakṣa of Āṣāḍha in Saṃ. 1617 = *ca.* 27 June 1560. Property of Lūṇakaraṇajī Pāṇḍyā of Jayapura.
RJ 3367 (vol. 4, p. 320). 69ff. Copied by Rāmacandra Kālā at Vasavā on 5 śuklapakṣa of Māgha in Saṃ. 1733 = *ca.* 27 January 1677 during the reign of Mahārāja Rāmasiṃha (who ruled Amber from 1667 to *ca.* 1690).
BORI 268 of A 1883/84. 63ff. Copied in Saṃ. 1795 = A.D. 1738. With the ṭīkā of Sāgarasena.
RJ 375 (vol. 3, p. 375). 88ff. Copied by Narasiṃha Agravāla on 10 śuklapakṣa of Māgha in Saṃ. 179*x*. With the ṭīkā of Sahasrakīrti. Property of Ṭholiyoṃ of Jayapura.
RJ 1796 (vol. 2, p. 284). 26ff. Copied by Narasiṃha Agravāla on 9 kṛṣṇapakṣa of Vaiśākha in Saṃ. 1796 = *ca.* 19 May 1739. Property of Baḍā Terahapanthiyoṃ of Jayapura. This is apparently identical with RJ 373 (vol. 3, p. 234). 26ff. Copied by Narasiṃha Agravāla of 4 kṛṣṇapakṣa of Vaiśākha in Saṃ. 1796 = *ca.* 14 May 1739. Property of Ṭholiyoṃ of Jayapura.
RJ 3362 (vol. 4, p. 320). 81ff. Copied on 11 kṛṣṇapakṣa of Mārgaśīrṣa in Saṃ. 1819 = *ca.* 10 December 1762.
RJ 3364 (vol. 4, p. 320). 92ff. Copied on 5 kṛṣṇapakṣa of Śrāvaṇa in Saṃ. 1829 = *ca.* 18 August 1772.
RJ 3363 (vol. 4, p. 320). 45ff. Copied in 11 śuklapakṣa of Vaiśākha I in Saṃ. 1869 = *ca.* 21 April 1812.
RJ 1792 (vol. 2, p. 283). 214ff. Copied on 7 kṛṣṇapakṣa of Bhādrapada in Saṃ. 1873 = *ca.* 13 September 1816. Incomplete. Property of Baḍā Terahapanthiyoṃ of Jayapura.
RJ 1791 (vol. 2, p. 283). 133ff. Copied on 3 śuklapakṣa of Pauṣa in Saṃ. 1884 = *ca.* 20 December 1827. Property of Baḍā Terahapanthiyoṃ of Jayapura.
RJ 596 (vol. 3, p. 92). 187ff. Copied at Jayapura in Saṃ. 1946 = A.D. 1889. With the ṭīkā of Mādhavacandra. Property of Badhīcandajī of Jayapura.

Arrah, Devanāgarī 19 and 39. See Velankar, p. 162.
Arrah, Kanarese 1004–1007 (1007 with the vṛtti of Mādhavacandra). See Velankar.
AS Bengal Jaina 1512 = Mitra, Not. 2041. 257ff. With the vṛtti of Mādhavacandra.
AS Bombay 1614. 3ff. From Bhāu Dājī.
Bhuleśvara, Bombay, Candraprabha Jain Mandira 41 (with the vṛtti of Mādhavacandra) and 165. See Velankar.
Bhuleśvara, Bombay, Pannalal Jain Sarasvati Bhavan 2, 619 (with the vṛtti of Mādhavacandra), 1306, 1307, and 2023. See Velankar.
BORI 1431 of 1886/92. 241ff. With a ṭīkā.
BORI 1002 of 1887/91. 429ff. With the vṛtti of Mādhavacandra. From Gujarāt.
BORI 1085 of 1895/1902. See Velankar.
CP, Hiralal 7334–7352. Property of the Balātkār Gaṇ Jain Mandir at Kārañjā, Akolā.
CP, Hiralal 7353–7358. Property of the Sen Gaṇ Jain Mandir at Kārañjā, Akolā.
CP, Hiralal 7359. Property of the Jain Mandir at Murwārā, Jubbulpore.
Hebru, South Kanara, Varaṅga Jaina Maṭha 19 and 42. See Velankar.
Humbuccha Katte, Shivamoga, Mysore, Jaina Bhandar 37, 101, and 167 (all with the vṛtti of Mādhavacandra). See Velankar.
Idar, Ahmadabad 22 (12 copies, of which 2 have the vṛtti of Mādhavacandra), 23, and A. 16 (6 copies). See Velankar.
IO 7527 (Burnell 417). 37ff. Karṇāṭakī. From A. C. Burnell.
IO 7528 (Burnell 381). 204ff. Karṇāṭakī. With the vṛtti of Mādhavacandra. From A. C. Burnell.
IO 7529 (1033). Ff. 152–254 and 259. With the vṛtti of Mādhavacandra. Incomplete. From H. T. Colebrooke.
Kolhapur, Bhandar of Lakṣmīsena Bhaṭṭārakaji's Jain Maṭha 12, 13, and 20. See Velankar.
Mudvidri, South Kanara 164. See Velankar.
Mysore, Padmaraj Jain 19 and 134. See Velankar.
Mysore and Coorg 2886. No author given. Property of the Jaina Maṭha at Śravaṇa Beḷgoḷa.
Mysore and Coorg 2887. 2000 granthas. Property of the Jaina Maṭha at Śravaṇa Beḷgoḷa.
Mysore and Coorg 2888. 1000 granthas. Haḷe Kannaḍa. Property of the Jaina Maṭha at Śravaṇa Beḷgoḷa.
Mysore and Coorg 2889. No author mentioned. Haḷe Kannaḍa. Incomplete. Property of Goṅgaḍi Puṭṭappa of Humcha.
Oppert II 319. 200pp. Grantha. Property of Bhadrabāhunainār of Elañkāḍu, Vandavāśi, North Arcot.
RJ 1790 (vol. 2, p. 283). 71ff. Property of Baḍā Terahapanthiyoṃ of Jayapura.
RJ 1794 (vol. 2, p. 284). 64ff. Property of Baḍā Terahapanthiyoṃ of Jayapura.
RJ 1795 (vol. 2, p. 284). 57ff. With the ṭīkā of Sāgarasena. Property of Baḍā Terahapanthiyoṃ of Jayapura.
RJ 1797 (vol. 2, p. 284). 91ff. With the ṭīkā of Sahasrakīrti. Property of Baḍā Terahapanthiyoṃ of Jayapura.
RJ 3365 (vol. 4, p. 320). 72ff.
RJ 3366 (vol. 4, p. 320). 68ff. RJ, vol. 4, p. 321 mentions 5 more copies.
RJ 3386 (vol. 4, p. 322). 63ff.

The *Trilokasāra* was edited by Manohar Lal, Bombay 1918.

NEMICANDRA ŚĀSTRIṆ (*fl.* 1956)

Author of an explanation in Hindī of the *Vratatithinirṇaya* of Siṃhanandin, published in his edition of that work, *JMJSG* 19, Kāśī 1956.